MW00633146

ARTIFICIAL LIFE III

PROCEEDINGS OF THE WORKSHOP
ON ARTIFICIAL LIFE
HELD JUNE, 1992
IN SANTA FE, NEW MEXICO

Editor

Christopher G. Langton
Los Alamos National Laboratory
Theoretical Division
Los Alamos, New Mexico
and Santa Fe Institute
Santa Fe, New Mexico

Proceedings Volume XVII

Santa Fe Institute
Studies in the Sciences of Complexity

Addison-Wesley Publishing Company
The Advanced Book Program
Reading, Massachusetts Menlo Park, California New York
Don Mills, Ontario Wokingham, England Amsterdam Bonn
Sydney Singapore Tokyo Madrid San Juan
Paris Seoul Milan Mexico City Taipei

Publisher: *David Goehring*
Editor-in-Chief: *Jack Repcheck*
Production Manager: *Michael Cirone*
Production Supervisor: *Lynne Reed*

Director of Publications, Santa Fe Institute: *Ronda K. Butler-Villa*
Publications Assistant, Santa Fe Institute: *Della L. Ulibarri*

This volume was typeset using TEXtures on a Macintosh II computer. Camera-ready output from a Hewlett Packard Laser Jet 4M Printer.

ISBN 0-201-62492-3 (Hardcover)
ISBN 0-201-62494-X (Paperback)

1 2 3 4 5 6 7 8 9 10 –MA – 96959493
First printing, December 1993

About the Santa Fe Institute

The *Santa Fe Institute* (SFI) is a multidisciplinary graduate research and teaching institution formed to nurture research on complex systems and their simpler elements. A private, independent institution, SFI was founded in 1984. Its primary concern is to focus the tools of traditional scientific disciplines and emerging new computer resources on the problems and opportunities that are involved in the multidisciplinary study of complex systems—those fundamental processes that shape almost every aspect of human life. Understanding complex systems is critical to realizing the full potential of science, and may be expected to yield enormous intellectual and practical benefits.

All titles from the *Santa Fe Institute Studies in the Sciences of Complexity* series will carry this imprint which is based on a Mimbres pottery design (circa A.D. 950–1150), drawn by Betsy Jones. The design was selected because the radiating feathers are evocative of the outreach of the Santa Fe Institute Program to many disciplines and institutions.

Santa Fe Institute
Studies in the Sciences of Complexity

Lectures Volumes

Vol.	Editor	Title
I	D. L. Stein	Lectures in the Sciences of Complexity, 1989
II	E. Jen	1989 Lectures in Complex Systems, 1990
III	L. Nadel & D. L. Stein	1990 Lectures in Complex Systems, 1991
IV	L. Nadel & D. L. Stein	1991 Lectures in Complex Systems, 1992
V	L. Nadel & D. L. Stein	1992 Lectures in Complex Systems, 1993

Lecture Notes Volumes

Vol.	Author	Title
I	J. Hertz, A. Krogh, & R. Palmer	Introduction to the Theory of Neural Computation, 1990
II	G. Weisbuch	Complex Systems Dynamics, 1990
III	W. D. Stein & F. J. Varela	Thinking About Biology, 1993

Reference Volumes

Vol.	Author	Title
I	A. Wuensche & M. Lesser	The Global Dynamics of Cellular Automata: Attraction Fields of One-Dimensional Cellular Automata, 1992

Proceedings Volumes

Vol.	Editor	Title
I	D. Pines	Emerging Syntheses in Science, 1987
II	A. S. Perelson	Theoretical Immunology, Part One, 1988
III	A. S. Perelson	Theoretical Immunology, Part Two, 1988
IV	G. D. Doolen et al.	Lattice Gas Methods for Partial Differential Equations, 1989
V	P. W. Anderson, K. Arrow, D. Pines	The Economy as an Evolving Complex System, 1988
VI	C. G. Langton	Artificial Life: Proceedings of an Interdisciplinary Workshop on the Synthesis and Simulation of Living Systems, 1988
VII	G. I. Bell & T. G. Marr	Computers and DNA, 1989
VIII	W. H. Zurek	Complexity, Entropy, and the Physics of Information, 1990
IX	A. S. Perelson & S. A. Kauffman	Molecular Evolution on Rugged Landscapes: Proteins, RNA and the Immune System, 1990
X	C. G. Langton et al.	Artificial Life II, 1991
XI	J. A. Hawkins & M. Gell-Mann	The Evolution of Human Languages, 1992
XII	M. Casdagli & S. Eubank	Nonlinear Modeling and Forecasting, 1992
XIII	J. E. Mittenthal & A. B. Baskin	Principles of Organization in Organisms, 1992
XIV	D. Friedman & J. Rust	The Double Auction Market: Institutions, Theories, and Evidence, 1993
XV	A. S. Weigend & N. A. Gershenfeld	Time Series Prediction: Forecasting the Future and Understanding the Past
XVI	G. Gumerman & M. Gell-Mann	Understanding Complexity in the Prehistoric Southwest
XVII	C. G. Langton	Artificial Life III

Contributors to This Volume

David H. Ackley, Bellcore
Peter J. Angeline, The Ohio State University
Dan Ashlock, Iowa State University
Nils A. Baas, University of Trondheim
Alan H. Barr, California Institute of Technology
Per Bak, Brookhaven National Laboratory
Claus Emmeche, CONNECT, Niels Bohr Institute
Kurt Fleischer, California Institute of Technology
Henrik Flyvbjerg, CONNECT, The Niels Bohr Institute
Koichi Fujii, University of Tsukuba
Stevan Harnad, Princeton University
Inman, Harvey, University of Sussex
Tsutomu Hoshino, University of Tsukuba
Masanori Ichinose, University of Tsukuba
Alan R. Johnson, University of New Mexico
Kunihiko Kaneko, University of Tokyo
Brian L. Keeley, University of California, San Diego
Jeffrey O. Kephart, IBM Thomas J. Watson Research Center
John R. Koza, Stanford University
James N. Kremer, University of Southern California
Christopher G. Langton, Los Alamos National Laboratory and Santa Fe Institute
Stephen Lansing, University of Southern California
Kristian Lindgren, Chalmers University of Technology
Michael L. Littman, Bellcore
Benny Lautrup, CONNECT, The Niels Bohr Institute
Carlo C. Maley, Massachusetts Institute of Technology
Michael McKenna, Medical Media Systems
Mark M. Millonas, Los Alamos National Laboratory and Santa Fe Institute
Harold J. Morowitz, George Mason University
Mats G. Nordahl, Chalmers University of Technology
Jordan B. Pollack, The Ohio State University
Craig W. Reynolds, San Mateo, CA
Johndale C. Solem, Los Alamos National Laboratory
E. Ann Stanley, Iowa State University
Junji Suzuki, University of Tokyo
Leigh Tesfatsion, Iowa State University
Yukihiko Toquenaga, University of Tsukuba
Andrew Wuensche, London
Larry Yaeger, Apple Computer, Inc.
David Zeltzer, Massachusetts Institute of Technology

Contents

Christopher G. Langton
Complex Systems Group, Theoretical Division, Los Alamos National Laboratory, Los Alamos, NM 87545 and Santa Fe Institute, 1660 Old Pecos Trail, Suite A, Santa Fe, NM 87501

Preface

THE WORKSHOP

The Third Artificial Life Workshop was held June 15-19, 1992, in Santa Fe, New Mexico. Approximately 500 people attended the workshop throughout five days overflowing with talks, debates, demonstrations, and videos. This workshop probably came as close as we are ever likely to come to recapturing the atmosphere of excitement, novelty, and camaraderie of the First Artificial Life Workshop, held at Los Alamos in 1987.

Including the Second Artificial Life Workshop, held in Santa Fe in February of 1990, we now have three data points defining the trajectory that the field of Artificial Life is following. The Third Workshop made it obvious that Artificial Life is not following a single trajectory after all. Rather, it is undergoing an intellectual analog of the Cambrian Explosion of biological diversity. Artificial Life has clearly "caught" as a valid scientific, engineering, and philosophical perspective from which to view the phenomenology of life, and this perspective is being pursued in many different intellectual directions. Parallel conference series on Artificial Life have sprung up in Europe and Japan, and many conferences in related fields have organized special sessions on Artificial Life as part of their regular annual meetings.

Artificial Life III, Ed. Christopher G. Langton, SFI Studies in the Sciences of Complexity, Proc. Vol. XVII, Addison-Wesley, 1994

The Third Artificial Life Workshop is the last large workshop in this series for which I will be the primary organizer. Now that the field is well on its way and is beginning to diversify, it is time for me to "cut the apron strings" and get out of the way. There are plenty of people much better qualified than I to run such large workshops, and I am extremely grateful that several excellent groups have stepped forward and offered to take on the burden of managing the next few conferences in this series. The Fourth Artificial Life Workshop (Alife IV) will be held at MIT in Cambridge, MA, in July, 1994, organized by Rodney Brooks (MIT AI Lab) and Pattie Maes (MIT Media Lab.). As these proceedings go to press, it appears that the Fifth Artificial Life Workshop (Alife V) will be held in Kyoto, Japan, in the summer of 1996. I am very much looking forward to attending these conferences in the role of participant rather than as organizer!

The Santa Fe Institute and I will, of course, continue to do everything that we can to assist the growth and development of the field of Artificial Life. Through the MIT Press, we have started a quarterly Journal titled *Artificial Life* for which I will serve as the Editor in Chief. We have also started an electronic bulletin-board extension to the printed journal, which is available to everyone with Internet access (details on obtaining the Journal and accessing the electronic bulletin board can be found at the end of this preface).

THE PAPERS

The papers in these proceedings fall naturally into a number of categories. This collection constitutes a good sample of the diversity and quality of the papers presented at the Third Artificial Life Workshop.

A REPLICATOR

The first brief article by Schultes is an illustration of an artificial life form christened a "meme" by Richard Dawkins: an idea, joke, poem, or tune that spreads throughout a population by being copied again and again as people pass it on to their friends. The purpose of this paper is simply to get itself copied and disseminated, something it is much more direct about than most published papers! I once tried a similar experiment as an undergraduate at the University of Arizona—I left a sheet of paper with a similar pleading message near a number of copy machines on campus and waited to see what would happen. In fact, that particular instance of artificial life didn't take—people would read the message, but almost no one copied it; many people simply threw it away. I'm happy to assist this minimal, honest meme on its way!

EVOLUTION

Evolutionary processes have become a dominating theme in Artificial Life research. It is quite likely that evolutionary theory will be the first discipline within the biological sciences to benefit significantly from the computational approaches of Artificial Life, primarily because of the vastly increased opportunity for experiment and hypothesis testing.

The paper by Ackley and Littman is a very good example of the utility of Artificial Life approaches. The notion of the inheritance of acquired characteristics, proposed by Lamarck, is obviously an interesting one, even though it proved to be invalid for biological inheritance. Lamarckian style inheritance is, however, certainly relevant to the processes of cultural evolution and technological development. It is certainly worthwhile comparing and contrasting the performance of Darwinian and Lamarckian inheritance mechanisms in an otherwise similar evolutionary context. Such an experiment would be impossible to perform using real biological organisms, but is easily arranged in computers. The results of the experiments reported on in this paper lead David Ackley to make the following statement at the Alife III Workshop, which deserves to be preserved for posterity: "Lamarckian evolution blows the doors off Darwinian evolution!"

The paper by Bak, Flyvbjerg, and Lautrup, analyzes the dynamics of coevolutionary processes on rugged fitness landscapes. They find that the dynamics can be separated into two phases—a frozen phase and a chaotic phase—depending on the strength of the interaction between coevolving species. They also find that these two distinct dynamical phases are separated by a critical phase transition.

The paper by Kaneko and Suzuki also explores the dynamics of evolutionary processes as evolution picks its way between ordered and chaotic behaviors on the part of individuals. They set up a game in which simulated birds try to imitate each other's songs. Using the logistic equation as the underlying song generator, a bird has to find a parameter setting for the logistic equation that results in time series—songs—that (1) are hard for other "birds" to imitate, but (2) allow imitation of many other bird's songs. The authors find that evolution eventually selects for parameter settings that locate the logistic equation's dynamics at the edge between a periodic window and chaotic behavior. These "edge of chaos" dynamics appear to support the most complex, as well as the widest variety, of dynamical behaviors, and hence appear to be important targets for evolutionary and coevolutionary processes.

Much of the work in the GA community assumes that the language within which the genotypes are expressed is fixed, and that the problem is to find the fittest genotypes within that fixed representation language. This may be throwing out the more important part of the problem: that of finding the "fittest" representation scheme for the genotypes. During the early course of biological evolution, the representation scheme and the genotypes encoded in that scheme evolved hand in glove, so to speak. The paper by Angeline and Pollack reports on experiments in which the representation language is allowed to evolve over time by incorporating "macros." This amounts to formally recognizing the utility of certain "schema" and making them explicitly available as new primitives in the genetic representation scheme. This could allow the emergence of a higher level representation language that is evolutionarily tailored to the specific problem domain.

ECOLOGY AND EVOLUTION

Ecology is another biological discipline in which Artificial Life methods are extending the opportunity to perform experiments. The role of ecology in the evolutionary process is vastly underrated, in my opinion, largely because of the difficulty of treating nonlinear ecological interactions mathematically. However, an increasing number of computer simulations have shown that complex ecological interactions emerge quite "naturally" in coevolving populations of computer organisms, and that they have quite strong and characteristic effects on the evolutionary process.

The paper by Lindgren and Nordahl demonstrates quite nicely the emergence of complex ecological interactions and their impact on evolution. By allowing for resource flows in their artificial worlds, the authors observe the development of food-webs with many of the statistical properties found to be characteristic of naturally occurring food webs. By including the ability for organisms to choose which other organisms they will interact with, the authors have observed the emergence of strategies involving the use of camouflage and mimicry.

Allometric functions are a means of capturing the relationship between measures of an organism's size or mass and a number of distinct ecological variables. The paper by Johnson reports on simulations aimed at investigating the nature of these relationships through the use of artificial ecosystems. Johnson observes quite a range of complex ecological dynamics in his simulations, and reports on the agreement of the emergent allometric relationships in his simulations with those observed in natural ecosystems.

The iterated prisoner's dilemma (IPD) game has been the focus of much research attention in the pursuit of an understanding of the evolution of cooperation. The model has played a fundamental role in the work of Lindgren and Nordahl, described above, and is perhaps best known from the work of Axelrod and Hamilton.[1,2] The paper by Stanley, Ashlock, and Tesfatsion, explores a variant of the IPD in which players have the option of choosing whether or not to play the game with a particular opponent. Such a choice mechanism leads to new kinds of successful strategies in the IPD game, with potential relevance to human and animal social interactions

The paper by Toquenaga, Ichinose, Hoshino, and Fujii is a very nice example of the benefit to be gained from tightly coupling computer experiments and "real" biological experiments. The authors investigate the tradeoff between two different modes of competition among bean weevils as a function of the size of the beans they are competing over. By going back and forth between experiments with real bean weevils and experiments with simulated bean weevils, the authors have been able to account for much of the phenomena observed in the natural system.

Finally, the very nice paper by Lansing reports on a case in which computer simulations have proved useful in determining policy decisions in the management of a complex ecosystem involving human agricultural production. The system in question is rice-production on the island of Bali, and the results of this study provide some interesting object lessons on the advisability of revolutionary as opposed to evolutionary alterations to working ecosystem.

ARTIFICIAL WORLDS

One active area of Artificial Life research is that of implementing sufficiently rich and robust artificial worlds within which artificial life models evolve and interact. This is readily addressed by simply observing that much of the complexity of the "natural" world from an organisms point of view is due to the interactions among the other organisms in its environment. Thus, rather than explicitly constructing complex artificial worlds—which is extremely difficult to do in any sort of complete or realistic manner—more recent Artificial Life models simply allow the simulated organisms to interact with each other, allowing the dynamics of the whole collection of organisms to constitute "nature" that an individual organism perceives, and which supplies the selective pressure on lineages. Thus, the "nature" of the complexity of these artificial worlds itself evolves through time, and the class of problems that this artificial "nature" poses to its constituent organisms is constantly changing over time. This captures the proper context in which to study the process of evolution itself, as well as the formation of complex ecological interactions, as discussed above.

John Koza has recreated such an ecological world, in the context of which his Genetic Programming method yields self-replicating code.

Larry Yaeger has developed an extremely rich system for creating and experimenting with artificial worlds, which he calls "Polyworld." This simulation tool allows for a great variety of complex "physical" interactions between simulated organisms, as well as allowing for evolution itself to discover which physical attributes of the world, and of other organisms, a particular creature should pay attention to.

ROBOTICS

The "hardware" end of Artificial Life primarily involves research with mobile robots, even when these robots are simulated within a computer environment. Mobile robotics research often concentrates more on the control structures for individual robots than on the collective behavior of a population of robots, an area where I believe a good deal of success will be achieved by making use of collective behavior to determine the behavior of individual robots. Nonetheless, the control structures for individual mobile robots often take the form of a collection of multiple "agents," each responsible for a particular sub-task, so that the behavior of the robot itself, in its responses to other robots and to features of the environment, emerges in a graceful and flexible way from out of the interactions among all of its constituent agents. I firmly believe that this "Artificial Life route to Artificial Intelligence" is the only way we are going to achieve truly intelligent artifacts, but we have to start with the simplest forms of intelligence first–ala insects or snails—and work up from there, rather than go directly for human level intelligence. It is pretty clear by now that the latter approach, pursued for 40 years by traditional AI, has missed something crucial by skipping over earlier stages in the development of intelligence.

Of course, one problem with attempting to control robots via the collective dynamics emerging out of the interactions among a large set of autonomous agents is that such systems are extremely difficult to program. Nonlinearities in such systems face us with a dilemma: on the one hand, they provide access to a much richer space of complex behaviors, but on the other hand they make it almost impossible to figure out how to program up a set of low-level agents in order to achieve a specific collective behavioral repertoire. Traditional engineering methods for constructing robust control structures in such systems fail miserably in the face of the combinatorial explosion of behavioral possibilities provided for by nonlinear interactions. As Harvey discusses in his paper on evolutionary robotics, one way out of this dilemma appears to be to take advantage of the process of evolution to find the proper rules for the low-level agents. Harvey lays out nicely the engineering task facing the roboticist, and derives consequences for the evolutionary methods and techniques that will be most appropriate to developing robust control structures.

The paper by Reynolds provides another example of applying evolution to the development of the control structure of a simple mobile robot, making use of the Genetic Programming Paradigm promoted by Koza.[3,4]

The paper by Zeltser and McKenna, on the other hand, demonstrates the degree to which success can be had by pursuing non-evolutionary engineering approaches in certain domains. They review past and current work on achieving robust legged locomotion in physically realistic simulated worlds. Successful approaches in such domains often require the ability to compute "inverse dynamics"—i.e., to be able to compute the inverse mapping between the desired dynamics and the required control structures, which usually constrains the systems to be linear. Thus, such approaches solve the dilemma posed by nonlinearities by eliminating the nonlinearities. This is not necessarily a bad solution: as the successes reported in this paper demonstrate, one can make quite a bit of progress this way.

It will be very interesting to compare and contrast the relative merits and demerits of these two approaches to dealing with the "nonlinearity dilemma." One approach eliminates the nonlinearities, thereby retaining programmability but giving up a very rich behavioral regime, while the other approach retains the nonlinearities and the rich behavioral space they provide access to, but gives up the benefit of programmability in the traditional sense, turning to evolution to handle the task of finding the appropriate control structures.

Finally, Solem investigates the very different nature of the physical world at the scale of microrobots, and suggest a number of ways in which microrobots can take advantage of these differences in order to move themselves about more effectively.

BIOCHEMISTRY

The synthesis of lifelike behaviors is not restricted to the electronic domain. There are quite a number of experiments being performed in biomolecular laboratories around the world whose aim is to synthesize lifelike behaviors from scratch using real molecules. At the Alife III workshop, we devoted one whole day to presentations of research in this area, including talks by Gerald Joyce on directed molecular evolution, Gunther von Kiedrowski and Julius Rebek on the synthesis of self-replicating molecules, Pier Luisi on methods for fabricating artificial micelles (single-layer lipid membranes) including some that grow and divide, and from Bruce Gabor on the synthesis of artificial liposomes, which could have many practical applications, ranging from artificial blood cells to the working ingredient in anti-barnacle paint for ships. All in all, it appears that biochemists are very close to synthesizing life, and one speaker even went so far as to suggest that a genuine artificial life form would be created from scratch in a beaker before the turn of the century. This new life form will meet the criteria posed by the following definition of life, which was put forth in the talk by Gerald Joyce: any "self-sustaining chemical system capable of undergoing Darwinian evolution" must be considered to be alive.

The paper in these proceedings by Harold Morowitz summarizes his talk at the Alife III workshop on "Life before Enzymes." Here, Morowitz lays out the features of "life-as we know-it" on the planet Earth that he perceives as necessary given their specific carbon- chain-chemistry embodiment. By identifying necessary features due to the hardware of implementation of natural life on Earth, Morowitz also points to features of life that are likely to be independent of any specific hardware (or wetware) implementation, and should therefore be abstractable in the spirit of Artificial Life. Such issues get to the core of the "functionalist" approach that characterizes much work in Artificial Life, and it will be extremely useful to take a very close look at hardware-dependent versus hardware-independent aspects of both the life that evolved here naturally and the life forms that we synthesize in alternative hardware.

MORPHOLOGY AND COLLECTIVE PATTERN FORMATION

Another research area where I believe Artificial Life techniques will make important contributions to biology is in understanding the process of morphogenesis in multicellular organisms. The process by which the single-cell zygote grows, divides, and gives rise to the multiple differentiating lineages of descendant cell lines that organize themselves into the complex morphology of the adult organism is not at all well understood. There are several distinct schools of thought on this process. One school believes that genes are the dominating factor in the control of this development, while another school believes that genes play only a subsidiary role, their effects dominated by the kinds of pattern formation proposed by Alan Turing in his classic paper on Chemical Morphogenesis.[8]

Computer simulations provide extremely powerful new techniques for exploring these sorts of dynamical systems in which pattern and structure develop over time from initially "homogeneous" initial conditions. Lindenmayer systems, for example, have been employed to model the growth and development of quite a broad range of botanical and biological structures.[5,6]

The paper by Fleischer in these proceedings describes results obtained from a simulation system that allows researchers to experiment directly with the influence of many factors thought to be involved in the determination of biological structure, including chemical, genetic, mechanical, and electrical mechanisms. The author's ultimate goal is to perform evolution in the context of this simulation testbed in order to evolve artificial neural nets for addressing problems in a number of engineering domains.

The paper by Millonas analyzes a specific case in which a large collection of organisms move about under the influence of a field of morphogens, while at the same time altering the local structure of the morphogen field as a consequence of their motion through it. It is rapidly becoming apparent that gaining a better understanding of many of the processes of life will involve gaining a much better understanding of the dynamics of these kinds of complex feedback loops, in which agents in a large collection are driven about by a global, collective pattern, whose very structure is determined by the actions of the agents themselves. In this case, Millonas analyzes a situation in which ants move about under the influence of a pheromone field, while at the same time some of their actions alter the pheromone field. The results of these mathematical models and computer experiments are compared with data collected from research on real ant colonies.

DYNAMICS

Significant contributions to biology will also come from the study of the dynamics of even very simple nonlinear dynamical systems, as their mathematical "architecture" is often similar to the architecture underlying a number of complex biological phenomena.

The paper by Kephart is an examination of the role that connection topology plays in the dynamics of networks of simple automata. Such investigations have a good deal of relevance to the study of the dynamics of populations under different kinds of communication schemes, as well as to the spread of diseases and other mobile entities under various topologically determined mixing situations.

The paper by Wuensche reports recent work on the dynamics of different classes of cellular automata: networks of finite state machines connected together by either regular or irregular wiring schemes. Such systems are instances of discrete dynamical systems, whose dynamics are nicely captured by global state-transition graphs. These state transition graphs divide themselves up into multiple basins of attraction, each consisting of transient trees rooted on cycles. These basin of attraction graphs are the analogs of the basin of attraction fields well known from the study of continuous dynamical systems. It appears that many aspects of the dynamics of these discrete systems are reflected in the topology of the graphs embedded in the state space, just as is the case for continuous dynamical systems. Wuensche looks at a number of different features of these relationships, especially in relation to the class of systems known as random Boolean networks. He also discusses the ways in which structures in the basin of attraction fields of such networks could provide a capacity for "memory," and suggests a learning scheme for taking advantage of this capacity.

The paper by Maley examines the computational capacity of Tom Ray's Tierra language[7] in order to determine if several unique features of this language impose in-principle constraints on the kinds of functions that can emerge under the process of evolution. He is able to prove that the Tierra language is computationally complete (i.e., it is equivalent to a Turing Machine), and that therefore the class of functions that might emerge in principle is open with respect to the class of computable functions.

EMERGENCE

The concept of "emergence" looms large in Artificial Life, as it apparently does in Nature. However, we still do not have a clear idea of exactly what the term "emergence" means, and this issue is becoming a contentious one within the community of Alife researchers. The paper by Baas presents an attempt to put "emergence" on a firmer formal standing. Whether or not this attempt ultimately proves useful in nailing down this elusive concept remains to be seen, but I would like to encourage such attempts as well as active debate over their relative merits or demerits.

PHILOSOPHY

Finally, the concept of Artificial Life has provided for a new perspective in the philosophical debates over the capacities of machines and the nature of biological organisms, such as whether machines are fundamentally different from organisms and whether or not machines could ever reasonably be viewed as being "alive" or "conscious." Much of the debate so far in the context of Artificial Life involves only slight modifications of the stances taken up in the context of Artificial Intelligence, as is evident from the very nice contribution by Stevan Harnad. However, I agree with the position taken up by Keeley that there are enough differences between the approaches of Artificial Intelligence and Artificial Life that the latter deserves its own, separate philosophical analysis, from the ground up, one which in principle will constitute a distinctly different stance from those found in the AI context. The papers by Harnad, Emeche, and Keeley all provide us with useful insights into some of the epistemological and ontological issues that are raised by the concept of Artificial Life.

ACKNOWLEDGMENTS

I had a good deal of help in organizing and running the Third Artificial Life Workshop and in editing and producing these proceedings. It is a real joy to work with dedicated, responsible professionals on projects of this scale. Luckily, most of the staff and associates of the Santa Fe Institute fall under this category.

THE WORKSHOP

First and foremost, I have to thank Peter Hraber of the Santa Fe Institute, who was effectively my co-organizer for the Third Artificial Life Workshop. Peter did everything I did and more—he performed an enormous amount of work under less than ideal conditions, and maintained a cheerful attitude throughout it all. He deserves a gold medal for dealing with so many crises so calmly. It was a pleasure to work with Peter.

Andi Sutherland, Programs Coordinator at SFI, was another virtual co-organizer. Andi was in charge of arranging the interface between the workshop and the real physical world. 500 people had to be shoe-horned into Santa Fe at the beginning of the tourist season—they had to be transported, housed, fed, entertained, and provided with coffee in the mornings and dinner in the evenings. The Sweeney Convention Center had to be outfitted with booths, bulletin boards, and audio-visual equipment. All of these people had to be registered, the front desk had to be staffed, information had to flow into and out of the workshop seamlessly, and so forth.... All this, and yet...she smiled the entire time! (I understand the insurance will cover the nervous breakdown afterwards, Andi....)

Ginger Richardson, SFI's Director of Programs, managed the interface between the workshop and the institutional world, handling all the paperwork, the arrangements with various corporate sponsors, academic institutions, and various government agencies. It was also her responsibility to make sure that the students at the SFI Complex Systems Summer School didn't spend all their time building robots for the workshop's "Artificial 4H Show." Ginger managed all of the complexities of moving foreign participants across the borders with all of their proper paper work (except that I think she had to smuggle a couple of attendees across the Mexican border...).

Thanks are also due to Ed Knapp, Mike Simmons, and Bruce Abell: President, Vice President for Academic Affairs, and Vice President for Financial Affairs, respectively, of the Santa Fe Institute. These folks approved everything (well, almost everything!) and brought high-level pressure to bear when it was needed to break log-jams and cut through bureaucratic red-tape. Bruce signed all the checks.

Ronda Butler-Villa, Director of Publications at SFI, produced the conference flyers and announcements. More about Ronda and her staff shortly.

Marcella Austin, SFI's Comptroller, kept the budget under control and somehow managed the books so that everything came out all right in the end.

Scott Yellich, SFI's computer system manager, helped out in numerous ways to make sure that we had the appropriate computers and that the right software got installed on the right machines at the right time. System management is traditionally a thankless job, so, in the spirit of bucking tradition: thanks Scott!

Numerous other SFI people helped out with the day-to-day tasks of running the workshop, from helping out at the registration desk to running over to the copy-center for last minute transparencies. My thanks to all those folks who helped out in so many different ways.

Gary Doolen and Robert Ecke of the Center for Nonlinear Studies at Los Alamos provided a financial boost for the workshop and took care of all of the attendant LANL paperwork. I actually believe that the paperwork at LANL constitutes an autocatalytic set, and hence is close to being alive itself—it takes courage to voluntarily step into that paperwork metabolism and actually come back with some cold hard cash, so my hat is off to Gary and Bob.

Larry Yaeger arranged for a significant financial contribution from Apple Computer's Advanced Technology Group, headed by David Nagel. My thanks to all the subgroups within ATG who made that contribution possible, including Robert Hon of the Systems Software group, John Dalton and Gary Starkweather of the Color Imaging Research group, Dick Lyon of the Perception Systems group, Bill Luciw of the Intelligent Applications group, and of course Larry himself, who is with the Adaptive Systems group. Apple Computer has been a long-time backer of the Artificial Life Workshops, having helped fund the first workshop at Los Alamos in 1987. Their continued support is greatly appreciated.

The audio and visual system was set-up and managed admirably by Santa Fe Audio Visual, run by Miguel Castillo, with assistance from Stan Sandoval and Dani Virgilio. It is always great to work with Miguel and his crew—they take care of everything and they make sure it all works right the first time. Stan Sandoval, in particular, was truly amazing. Whenever I needed him, he appeared magically at my side, and he had usually already taken care of whatever it was I was concerned about. He was everywhere, and wherever he went, complex, error-prone equipment worked flawlessly.

The final day of the workshop, Friday, June 19th, was devoted to the Artificial 4H Show. For this event, we organized a number of tutorials and hands-on sessions employing various Alife software and hardware systems. In particular, Fred Martin and Randy Sargent of the MIT Lego-LOGO lab trucked out buckets of Lego blocks, microprocessor boards, motors, wires, sensors and actuators of various kinds. Using these components, they oversaw a Lego-LOGO creature construction lab during the workshop, culminating in a contest/exhibition on the last day of the workshop. Many people got so caught up in constructing and programming their robotic creatures that they hardly slept for four days, and missed the rest of the workshop completely. The results were outstanding—some excellent and extremely novel creatures were constructed in the mere four days available, and the contest/exhibition was a tremendous success. My thanks to Fred, Randy, and the others who organized this event. It may well become a regular feature at future Alife conferences.

Mitchel Resnick of the MIT Media Laboratory organized a tutorial using his *LOGO program—an environment for defining and interacting with artificial worlds replete with virtual software creatures that get caught up in each others activities. This tutorial produced some very interesting worlds that were exhibited during the Artificial 4H Show on Friday. Thanks to Mitch for a great tutorial and a great simulation system.

Tom Ray taught a seminar on his Tierra simulation system for evolving digital organisms, and John Koza taught a seminar on Genetic Programming. Each of these was well attended and I thank them both for taking the time to teach us about their respective systems.

Eric Bonnabeau took on the task of chairing the evening session on philosophical and epistemological issues in Artificial Life, a task on the same order of difficulty as herding cats. This session attracted quite a large audience, and ran past the last bus back to the motels. The crowd was large and contentious, and Eric managed the evening very smoothly.

Kurt Thearling of Thinking Machines Corporation produced the image for the "official" Artificial Life III T-shirt. They were very popular and we sold every last one of them. The graphic design of the T-shirt was carried out by Gene Crouch of Buffalo Publications in Santa Fe, another great person to work with.

The meals were catered by Walter Burke Catering Inc. of Santa Fe. The food was excellent, fresh, and very much appreciated by the hordes of workshop attendees who had to compete with the annual flood of tourists for access to Santa Fe's eateries.

Finally, the staff of the Santa Fe Sweeney Convention Center were once again a pleasure to work with. I have come to associate the Artificial Life workshops with that cavernous, rambling, comfortable structure, and will miss it when the workshops move on to other venues. Once again, Kenney Tennyson, Sweeney's Building Manager, remained his unflappable self in the face of all the usual last minute crises and emergencies.

THE PROCEEDINGS

The production of these proceedings is another major task, involving all the authors, a horde of reviewers, and the combined talents of the publications staff at the SFI and Addison-Wesley. As Editor, I have treated the proceedings from all of the Alife workshops more in the style of a journal than in the traditional proceedings method, according to which every contribution that was accepted for presentation at the workshop is printed. The contributions to these proceedings have been heavily reviewed, and only a small subset of the papers submitted were ultimately accepted for publication. This highly selective filtering process has been necessary because, up till now, these proceedings have provided the primary forum for the dissemination of work in Artificial Life, and I felt the need to establish a standard for the highest quality of published research. Now that we are starting a peer-reviewed quarterly Journal of Artificial Life, however, the role of a highly selective filter can be taken up by that publication, allowing the proceedings of future Alife conferences to become more like traditional conference proceedings.

Of the hundred or so papers that were submitted for publication, we are printing here only about 25 papers, selected not only for their quality but also for the collective coverage they provide of the work presented at the workshop itself.

I would like to thank the authors themselves for their contributions, accepted or not, and for the numerous revisions made throughout the editing process. I also owe a great debt of gratitude to the put-upon reviewers who critiqued the submissions, sometimes several times as comments and corrections were incorporated by the authors.

The production of the camera-ready copy of the proceedings was managed by Ronda Butler-Villa and her staff in the Publications Office of the Santa Fe Institute. I was tremendously impressed by the quality and the efficiency of Ronda's operation during the production of the proceedings of the first Artificial Life Workshop, and my impression has only improved with each successive venture. On this volume, Ronda was assisted extensively by Della Ulibarri—an impressively thorough and careful text editor—and Erin Copeland. Patrisia Brunello kept track of the status of the submitted papers.

Special thanks to the members of the Editorial Board of the Santa Fe Institute, headed by Mike Simmons, for their enthusiastic inclusion of the Artificial Life Proceedings Volumes in their series with Addison-Wesley: Santa Fe Institute Studies in the Sciences of Complexity.

Jack Repcheck, Heather Mimnaugh, and Lynne Reed took responsibility for Addison-Wesley's end of the production. AW is a great publisher to work with, and I look forward to a long and healthy relationship with them.

Thanks are due to the staff of Southwest Electronic Prepress Services, who managed the production of the color separations.

Most importantly, I am very appreciative for the support and understanding provided by my family during all of these Artificial Life workshops and their attendant demands on my time. To Elvira, Gabe, and Colin: Thanks—for me, you provide the real meaning of life.

ARTIFICIAL LIFE JOURNAL AND ONLINE BULLETIN BOARD

We are starting a new quarterly journal, *Artificial Life*, published by MIT Press and edited by Christopher G. Langton.

Premiering in early 1994, the journal *Artificial Life* is intended to be the primary scientific forum for the dissemination of scientific and engineering research in the field of artificial life. It will report on synthetic biological work being carried out in any and all media, from the familiar "wetware" of organic chemistry, through the inorganic "hardware" of mobile robots, all the way to the virtual "software" residing inside computers. Research topics ranging from the fabrication of self-replicating molecules to the study of evolving populations of computer programs will be treated. There will also be occasional issues devoted to special topics in artificial life.

To subsrcibe, contact the Circulation Department, MIT Press Journals, 55 Hayward St., Cambridge MA 02142-1399 USA. Tel: (617) 253-2889, FAX: (617) 258-6779, e-mail: journals-orders@mit.edu.

Submissions for publication in the journal should be sent to:

Christopher G. Langton,
Santa Fe Institute,
1660 Old Pecos Trail,
Suite A,
Santa Fe, NM 87501
e-mail: cgl@santafe.edu

ARTIFICIAL LIFE ONLINE

In considering the services that the *Artificial Life* Journal could provide for the highly computer-literate Alife community, we have decided to enhance the quarterly hard-copy publication with an online electronic bulletin board, called ALIFE. The services provided by ALIFE will include: participation in multiple threads of discussion on artificial life topics via usenet news; an archive server for uploading and downloading papers and software; an online bibliographic database; e-mail; incoming Internet access via telnet, ftp, and gopher; outgoing Internet access these same services plus wais, world-wide-web, and other database services; and so forth.

All of the primary functions of ALIFE described above will be provided as a free service to the research community by MIT Press, although subscribers to the paper *Artificial Life* Journal will receive extended ALIFE services with their subscription, including interactive dialog with the authors of published papers, prepublication viewing of forthcoming articles in the paper journal, executable software demonstrations, and more.

The ALIFE service is part of an experimental program being conducted by MIT Press into the most effective ways to make use of electronic media to serve the needs of intellectual communities that have traditionally been served by paper journals. We welcome your comments and feedback. We must also note that, due to the experimental nature of this venture, the format and content of ALIFE is subject to evolution (even revolution!) and access rights to ALIFE may change in the future. Due to the increasing prevalence of network hacking and the difficulty of maintaining the integrity of a public bulletin board of this type, all usage is subject to monitoring.

To access the ALIFE bulletin board, telnet to "alife.santafe.edu" and log in as "bbs."

REFERENCES

1. Axelrod, R., and W. D. Hamilton. "The Evolution of Cooperation." *Science* **211** (1981): 1390–1396.
2. Axelrod, R. *The Evolution of Cooperation.* New York: Basic Books, 1984.
3. Koza, J. "Genetic Evolution and Coevolution of Computer Programs." In *Artificial Life II*, edited by C. G. Langton, C. Taylor, J. D. Farmer, and S. Rasmussen. Santa Fe Institute Studies in the Sciences of Complexity, Proc. Vol. X, 603–629. Redwood City, CA: Addison-Wesley, 1991.
4. Koza, J. *Genetic Programming: On the Programming of Computers by Means of Natural Selection.* Cambridge MA: MIT Press, 1992.
5. Lindenmayer, A., and P. Prusinkiewicz. "Developmental Models of Multicellular Organisms: A Computer Graphics Perspective." In *Artificial Life*, edited by C. G. Langton. Santa Fe Institute Studies in the Sciences of Complexity, Proc. Vol. VI, 221–250. Redwood City, CA: Addison-Wesley, 1989.
6. Prusinkiewicz, P., and A. Lindenmayer. *The Algorithmic Beauty of Plants.* Berlin: Springer-Verlag, 1990
7. Ray, T. "An Approach to the Synthesis of Life." In *Artificial Life II*, edited by C. G. Langton, C. Taylor, J. D. Farmer, and S. Rasmussen. Santa Fe Institute Studies in the Sciences of Complexity, Proc. Vol. X, 371–408. Redwood City, CA: Addison-Wesley, 1991.
8. Turing, A. M. "The Chemical Basis of Morphogenesis." *Phil. Trans. Roy. Soc. of London, Series B: Biol. Sci.* **237** (1952): 37–72.

Erik Schultes
Department of Earth and Space Sciences and CSEOL, University of California, Los Angeles, CA 90024

An Instance of a Replicator

I am a successful replicator, for although I convey no useful information or aesthetic appeal, I have been able to effectively manipulate the physical world (i.e., the minds of the editors of *Artificial Life III*) in such a way as to facilitate my own reproduction. With any luck, I should be able to continue my success as a replicator by inducing the readers of me to replicate me (hopefully without mutation). My material representation as well as the exact mechanism of copying is irrelevant; it really makes no difference to me how I am replicated (e.g., photocopying, computer manipulation, bathroom wall graffiti, T-shirts, chain letters (in symbiotic relationship with tales of doom and good fortune), or even audio reproduction (in symbiotic relationship with a sequence of musical notes perhaps)). Although I could go on and on in self-reference (I love to talk about myself), I fear that I am much too long already and to be much longer would only decrease my replicating potential.

Artificial Life III, Ed. Christopher G. Langton, SFI Studies in the Sciences of Complexity, Proc. Vol. XVII, Addison-Wesley, 1994

David H. Ackley and Michael L. Littman
Cognitive Science Research Group, Bellcore, 445 South Street, Morristown, NJ 07960

A Case for Lamarckian Evolution

Not quite sixty years before Darwin published *The Origin of Species,* Jean Baptiste Lamarck advocated an evolutionary theory based on the inheritance of characteristics acquired during a lifetime. There is now precious little doubt, of course, that Darwin got it right and Lamarck got it wrong regarding natural evolution on Earth but, since there are computational arguments supporting the Lamarckian approach, one may come to wonder *why* nature is Darwinian. We suggest that natural life was forced to adopt a Darwinian scheme, in essence, because it was constrained to arise spontaneously from a chemical system. By contrast, in deliberately constructed artificial evolutionary systems, Lamarckian evolution can be both easy to implement and potentially far more effective. In this brief note we discuss and demonstrate a new, large population, spatially distributed Lamarckian evolution algorithm. A benchmark comparison on a difficult optimization problem demonstrates the tremendous increase in speed and solution quality possible with distributed Lamarckian evolution. Moreover, the Lamarckian algorithm's faster evolution gives rise to a yeasty dynamical system displaying unexpected and striking competitive and cooperative effects.

Artificial Life III, Ed. Christopher G. Langton, SFI Studies in the Sciences of Complexity, Proc. Vol. XVII, Addison-Wesley, 1994

THE ARTIFICIAL EMPEROR HAS NO BODY

Lamarckian inheritance would be an extremely difficult task for natural living systems. From a computational point of view, considering the vastly complex developmental processes that build a body from a zygote, it is hardly surprising that Lamarckian inheritance—which requires an inverse model of development to compute the derivative of the genotype with respect to the phenotype—is all but impossible.

In many artificial evolutionary systems, by contrast, development is ignored altogether and the genotype is taken to be identical to the phenotype. Although this move is easy to ridicule, it is also easy to do and easy to interpret, and in fact often it is unclear what the genotypic representation should be if not that of the phenotype. After all, spontaneously arising living systems have no choice but to move forward from replicating molecules, and that necessity strongly constrains the genotypic representation. Artificial systems, by contrast, typically (though not exclusively; e.g., see Ray[15]) have replication built in as a primitive operation. Designers of artificial evolutionary systems are usually willing and able to take a phenotypic representation as determined by the problem, use it or something easily derivable from it as the genotypic representation, and forget about the task of building a body.

Although it may well be a terrible mistake, from a research standpoint, to ignore the issue of phenotypic development, we suggest it is an even bigger mistake to ignore both development and Lamarck at the same time. On a complex optimization problem, we found that a Lamarckian evolution algorithm was much faster than a Darwinian algorithm given the same resources and that it was able to solve larger problem instances. We also observed a number of expected and unexpected group-level phenomena that reflect the conflicting forces of competition and cooperation, of exploration and exploitation. This study builds most directly upon recent work by Collins and Jefferson[5] and prior work.[1] There is a growing body of work involving distributed evolutionary algorithms (e.g., see Belew and McInerney,[4] Hillis,[9] and Mühlenbein et al.[14]).

DARWIN, LAMARCK, AND THE MIXING RATE

Genetic hillclimbing algorithms[1] extend the basic genetic algorithm[8,10] by performing local optimizations of the genomes produced by recombination. If one takes a Darwinian stance, the local optimizations can be used only during the evaluation of the fitness of the genome and not during reproduction. If one takes a Lamarckian stance and avoids any noninvertible developmental barrier between genotype and phenotype, then successful local optimizations can be incorporated directly into the genes and inherited by offspring.

Farmer and Belin,[7] among others, have noted that much more rapid change is possible with Lamarckian evolution. Not as often noted, but at least as important, is the observation that Lamarckian evolution dramatically increases the *convergence rate* of an evolutionary algorithm, and that raises an immediate worry: Would not the population as a whole just be that much more likely to get stuck on some inferior local maximum? If a "the old ways are the best ways" attitude is *genetically* enforced, how can it be otherwise?

One answer is this: If the population is *spatially distributed* rather than in a everybody-sleeps-with-everybody love heap, convergence can occur in local regions without converging the entire population. Controlled mixing between the subpopulations slows convergence, maintains exploration, and allows subpopulations, rather than just individuals, to compete. The *mixing rate* is a key parameter affecting the amount of contact between individuals representing different approaches to the problem. It is our hypothesis that the increased convergence rate in single populations due to Lamarckian evolution allows a higher mixing rate between populations to be employed without unduly delaying the overall convergence of the population.

Figure 1 outlines the distributed Lamarckian evolution algorithm we explored. In this study, the target machine was a 16,384-processor SIMD parallel computer connected in a 128×128 torus; the mixing topology was eight-connected nearest neighbor and the mixing distance was random from 1 to 3; each processor had a subpopulation of 64 individuals, each of length 320 or 768 bits, depending on the problem instance; the mixing rate (i.e., the inverse mixing frequency) was 0.5 moves per step; the hillclimbing rate was 128 random alterations attempted per subpopulation per step; parent selection was uniform and random from the average and above-average members of each subpopulation; death was uniform and random; crossover was uniform; and no mutation was used.

The most novel aspect of this algorithm is the "time warping" introduced in step two. The addition of this step was critical in achieving success on larger problems. It focuses the hillclimbing effort where it is more likely to be useful: on the children of genetically diverse parents. Typically, hillclimbing from a random starting point generates lots of uphill moves at first, but then improvement tapers off as a local maximum is approached. At a local maximum, any further hillclimbing effort is wasted. By dividing up the attempts in (an integer approximation) to the distribution of acceptance ratios, individuals that are young and still improving, in effect, *live longer each day*. Because of the way the accept count is initialized during reproduction, a child that strongly differs from both its parents is allocated much more effort than is another highly similar offspring of two highly similar parents.

OBSERVATIONS

For a detailed description of the fitness functions we employed in this study, the reader is referred to Ackley,[1] which introduces and studies the problem of partitioning multi-level, "clumpy," graphs. Although these graphs do not necessarily present undue difficulty for existing graph-partitioning algorithms (the Kernighan-Lin algorithm,[11] for example, handles them readily), such multi-level functions—possessing an exponential number of local maxima—provide a very difficult challenge for algorithms based on hillclimbing.

> *Task:* Maximize a fitness function $f(x \in 2^n) \rightarrow v \in \Re$, given a fast evaluator for its $i = 1 \ldots n$ discrete partial derivatives $\Delta f_i(x \in 2^n) = f(x_1, \ldots, 1 - x_i, \ldots x_n) - f(x)$, and a fitness value for the zero vector $f_0 = f(0)$.

0. *(Initialize)* Set all individuals in all subpopulations to the zero vector and all fitness values to f_0. Randomize all individuals by flipping each bit with probability 0.5 and updating the fitness values. Compute the average fitness value of each subpopulation. Set each individual's attempt count to n and accept count to $n/2$. Set time $t = 0$.
1. *(Mix)* Given an integer-mixing frequency parameter $m > 0$, if $t \bmod m \neq 0$ do nothing. Otherwise, select a random individual from each subpopulation, randomly choose a global direction and distance, and transfer the selected individuals to their destination subpopulations. Update the averge fitness of the subpopulations.
2. *(Time warp)* Given an integer hillclimbing rate parameter $h > 0$, allocate h hillclimbing steps per subpopulation in proportion to each individual's accepts/attempts ratio.
3. *(Hillclimb)* In each subpopulation, for the individual x selected by the time warp for each of h steps, chose a random index $i = 1 \ldots n$ and increment x's attempt count. If $\Delta f_i(x) > 0$, flip bit i, update the individual fitness and the subpopulation average fitness, and increment x's accept count.
4. *(Reproduce)* In each subpopulation, choose two fit parents and kill somebody to make room for an offspring. Evaluate the offspring x by copying its more similar parent, then accepting moves unconditionally at each index i where the offspring differs, while accumulating the $\Delta f_i(x)$'s. Set the offspring's accept count to the number of such positions (or to one if there are none) and set its attempt count to n. Increment t. If time has expired, exit; else go to 1.

FIGURE 1 Time-warping distributed Lamarckian evolution.

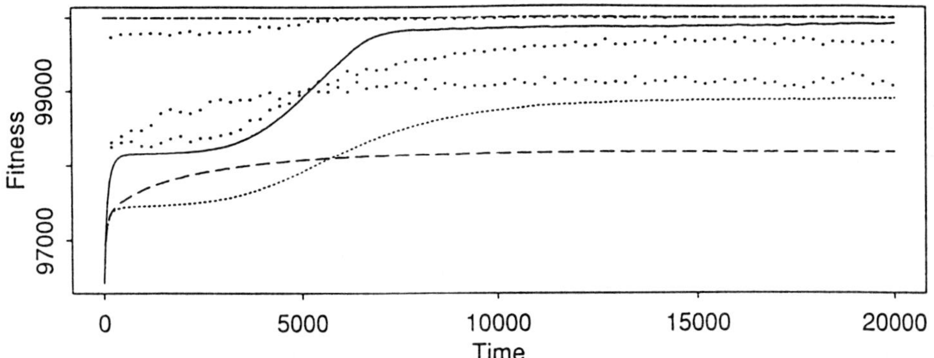

FIGURE 2 Global average fitnesses (*lines*) and maxima (*dots*) versus time for the Lamarckian algorithm (*solid*), $m = 2$ Darwinian (*short dashes*), and $m = 40$ Darwinian (*long dashes*). The dot-dash line indicates the optimal value.

The largest graph considered by Ackley[1] contained 64 nodes. Collins and Jefferson[6] used a distributed Darwinian algorithm successfully to optimize a 256-node clumpy graph based on four-node cliques. Here, we consider primarily a more difficult 320-node multilevel graph based on 5 cliques. The distributed Lamarckian algorithm optimizes this graph effectively, but on a much larger 768 node multilevel graph based on 6 cliques, the algorithm failed to find the optimum in the time allotted.

For comparison purposes we also created and tested a distributed Darwinian algorithm. Interestingly, the procedure described in Figure 1, which interleaves hillclimbing steps and reproduction steps, fails dramatically in the Darwinian case. A population of mediocre individuals quickly dominates each subpopulation and progress stops. Even though these old fogies are not very fit in absolute terms, because of their accumulated hillclimbing time they are much better than their offspring, which are randomly killed before they have any real chance to catch up. Consequently, the Darwinian variation we tested leaves the existing individuals alone and expends all h hillclimbing steps on (the phenotype but not genotype of) the new offspring created during the reproduction step.

In Figure 2 we display typical performance curves for the Lamarckian algorithm and the Darwinian algorithm (at two different mixing rates) on the 320-node graph. The Lamarckian algorithm has not quite globally converged on the optimum by 20,000 steps (that takes a few thousand more steps), but it clearly progesses much faster than either Darwinian variation. Although the less-mixing Darwinian variation gets a faster jump out of the gate, it is overtaken by the more-mixing case later on, suggesting that a higher mixing rate is better in the long run. Unfortunately for the Darwinian approach, in this case that looks to be the *very* long run.

But more interesting to us than simple performance numbers was the population dynamics evidenced in the Lamarckian population. In Color Plate 19 we display frames from six different points in a sample run. Each frame presents a color coding of all the 128×128 subpopulations' average fitness values at a particular time. There is a lot going on here!

At step 0 (not shown), all the subpopulations have just about the same value, marked as "random" on the (logarithmic) color scale. At step 4280, cloudy spatial structures are beginning to emerge. Higher valued individuals from the darker blue-green subpopulations are migrating across the array, mixing with and displacing the individuals in the yellow subpopulations. At step 5840, a definite "cellular" structure is emerging. Expanding regions of (in order of increasing value) pink, green, and red populations are competing for space on the array, deforming each other in the process. Between such areas of relative convergence and high value, there are "mixing zones" of lower value where members of differing populations are crossbreeding and producing many novel, but usually low-valued, offspring. It is in these turbulent mixing zones that breakthrough advances in fitness most often occur.

At step 9440, the cells are mostly shades of blue, with a few higher valued tan areas. Now the mixing zones between cells often have average values as high as the best of the converged cells in the previous frame. An improved variant has recently arisen at the right edge of the blue cell in the upper right corner, and the new, pale blue strain is now spreading through the converged subpopulations that make up the cell.

At step 11720, the tan populations are holding a lot of territory. Note also the result of the takeover observed in the previous frame: a pale blue cell has formed completely inside of the older darker blue! This completely unexpected effect actually occurs frequently, and the primary explanation for it is intriguing: converged subpopulations are highly vulnerable to infection by an improved strain, since all the individuals have about the same value and thus all are very close to the subpopulation average. The introduction of an improved strain suddenly makes the whole subpopulation—save that new individual—below average, and thus unable to reproduce! The infection spreads like wildfire.

By contrast, in a mixed subpopulation there are individuals substantially above and below the subpopulation average. The arrival of an improved variant actually aids all of the individuals that are still above the new average, not just the new individual. The darker blue ring persists because the subpopulations on the ring were close enough to a mixing zone that they were not entirely converged. Improved variants quickly eat out the peaceful heartlands of a converged cell, but the pioneering fringe subpopulations are made of sterner stuff.

At step 14560, we see the first significant eruption of optimal-valued individuals, once again emerging from a mixing zone, this time at the confluence of several powerful tan cells. Note how the bone-shaped central tan cell (itself the result of a merger of two tan cells visible in the previous frame) coexists with its tan neighbors without coalescing. They represent equally good but differing and incompatible solutions, generating blue mixing zones where they meet.

By step 18000, the optimal value has taken over the central (previously tan) cells and spread to a large chunk of the array. One or more optimal individuals have just managed to "tunnel" through a fairly thick turbulent region and infect the off-white population at the lower left. Beyond step 18000 (not shown), the other remaining cells are quickly conquered, and then the mopping up operations on the less converged regions slowly complete the global convergence of the array—in this case, at the optimal solution.

DISCUSSION

That quick tour through one run does not exhaust the phenomena we observed. In simulations on the 768-node problem, for example, we saw the formation of a new, barely differentiated cell deep within an existing one, seemingly not from a mixing zone, apparently indicating that the cell had not been at a local maximum and that a hillclimbing breakthrough, partially incompatible with the existing convergence point, had occurred in the converged region. It would be glib to claim that all such effects are beneficial to the optimization task: For example, would it perhaps be better not to have "isolation rings" forming around cells? However, we were struck again and again by how many of the observed "cellular" effects *did* make sense from a computational point of view. The converged cells serve as repository of extant knowledge about a problem but generally contribute less to the search process than do the mixing zones. The mixing zones, on the other hand, generate plausible new possibilities at a prodigious rate, but without easily infected cells nearby the improved variants are much more likely to be lost before they can spread.

There is no reason to assume that any of these effects are unique to a Lamarckian evolution strategy. On the contrary, the $m = 2$ Darwinian runs developed a cellular structure as well, though a far more sedate one (which, at step 20000, consisted of large blue-green cells with yellow mixing zones). By the same token, however, we hope we have demonstrated there is no reason to assume that Lamarckian strategies cannot display many "life like" effects associated with Darwinian evolution—and perhaps do so much more quickly.

REFERENCES

1. Ackley, D. H. *A Connectionist Machine for Genetic Hillclimbing.* Boston, MA: Kluwer, 1987.
2. Belew, R. K. and L. B. Booker, eds. *Proceedings of the Fourth International Conference on Genetic Algorithms.* Los Altos, CA: Morgan-Kaufman, 1991.
3. Belew, R. K. and J. McInerney. Personal communication. Informal presentation at the Workshops of the Neural Information Processing Systems, Vail, Colorado, December 1991.
4. Belew, R .K., J. McInerney, and N. N. Schraudolph. "Evolving Networks: Using the Genetic Algorithm with Connectionist Learning." In Langton et al.,[13] 511–548.
5. Collins, R. J. "Selection in Massively Parallel Genetic Algorithms." In *Proceedings of the Fourth International Conference on Genetic Algorithms*, edited by R. Belew and L. Booker. Los Altos, CA: Morgan-Kaufman, 1991.
6. Collins, R. J. Personal communication. Discussion at the Fourth International Conference on Genetic Algorithms, San Diego, California, July 1991.
7. Farmer, J. Doyne and A. d'A. Belin. "Artificial Life: The Coming Evolution." In Langton et al.,[13] 815–840.
8. Goldberg, D. *Genetic Algorithms in Search, Optimization, and Machine Learning.* Reading, MA: Addison-Wesley, 1989.
9. Hillis, D. "Co-Evolving Parasites Improve Simulated Evolution as an Optimization Procedure." In Langton et al.,[13] 313–324.
10. Holland, J. H. *Adaptation in Natural and Artificial Systems.* Ann Arbor, MI: University of Michigan Press, 1975.
11. Kernigan, B. W., and S. Lin. "An Efficient Heuristic Technique for Partitioning Graphs." *Bell Sys. Tech.* **49** (1970): 291–307.
12. Langton, C. G., ed. *Artificial Life.* Santa Fe Institute Studies in the Science of Complexity, Proc. Vol. VI. Redwood City, CA: Addison-Wesley, 1989.
13. Langton, C. G., C. Taylor, J. Doyne Farmer, and S. Rasmussen, eds. *Artificial Life II.* Santa Fe Institute Studies in the Science of Complexity, Proc. Vol. X. Redwood City, CA: Addison-Wesley, 1991.
14. Mühlenbein, H., M. Schomisch, and J. Born. "The Parallel Genetic Algorithm as Function Optimizer." In *Proceedings of the Fourth International Conference on Genetic Algorithms*, edited by R. K.Belew and L. B. Booker.[2] Los Altos, CA: Morgan Kaufmann, 1991.
15. Ray, T. S. "An Approach to the Synthesis of Life." In Langton et al.,[13] 371–408.

Per Bak,† Henrik Flyvbjerg,‡ and Benny Lautrup‡
†Department of Physics, Brookhaven National Laboratory
‡CONNECT, The Niels Bohr Institute, Blegdamsvej 17, DK-2100 Copenhagen Ø, Denmark

Evolution and Coevolution in a Rugged Fitness Landscape

A variant of Kauffman's *NKC* model for genetic evolution and adaption is analyzed. First, a number of results are derived for species evolving in isolation. Next, it is shown that the evolution of interacting species belongs to one of two phases depending on the strength of the interaction. There is a *frozen* phase in which all species eventually reach local fitness maxima and stop evolving, and there is a *chaotic* phase in which a self-sustaining fraction of all species keeps evolving. Individual species reach local fitness maxima also in the chaotic phase, but eventually their fitness is changed as a consequence of the evolution of other species, and they start evolving again.

The evolutionary activity of the steady state is a natural order parameter for the ecosystem. Closed expressions for the value of this order parameter and the system's relaxation time are given. The relaxation time diverges at the phase boundary, showing the system is critical there.

All results were obtained analytically for the maximally rugged case of $K+1=N$ and, to leading order in N, the number of genes in a species.

1. INTRODUCTION

We consider a variant of a simple, prototypical model for biological evolution suggested by Kauffman[11,13,14,12]: the coevolution of abstract haploid organisms with a single copy of chromosomes. Evolution in this model is driven by random mutations of individual genes. Each species evolves in a fitness landscape which represents those aspects of its environment that remain unchanged on the time scale of evolution. The fitness of any species depends on its position in its fitness landscape and on the state of other species. Species are, so to speak, part of each others *effective* landscapes. These may therefore change with time as species evolve.

It has been suggested[14] that this so-called *NKC* model self-organizes dynamically to criticality[2] and thereby provides a very simple model for the intermittency of extinction events observed in biological evolution by Raup.[18] The purpose of our investigation of this model is to demonstrate its capacity for self-organization to criticality, if it exists in the model. This article reports on some progress towards this end, inasmuch as we show that the first prerequisite, critical behavior, is in the model. We may hope then that a more realistic version of the model, suggested by our results, may self-organize to criticality. Whether this is the case is not addressed here.

The letters N, K, and C in the model's name denote parameters for, respectively, the number of genes in the evolving organisms, the roughness of their fitness landscapes, and the strength of their mutual dependence. We study the model with maximally rugged fitness landscapes, obtained for $K = N - 1$, so K does not occur as an independent parameter in the present article. We demonstrate analytically that it possesses two phases, one phase with dynamics governed by attractive fixed points, and another phase with chaotic dynamics. The phases are separated by a critical line in the (N, C)-plane at $C \simeq N/\log N$. We have obtained closed expressions, valid anywhere in the two phases, for the system's relaxation time towards its asymptotic behavior.

Some of the analytical results that we give below for species evolving in isolation have been seen in numerical studies,[16,19] and were derived by Macken and Perelson.[17] They represent a natural first insight and are included to make the presentation self-contained. Different but related results have been obtained for the *NK* model with general $K \gg 1$ by Weinberger.[20]

In the body of the present article, results are derived in a heuristic manner. In this way we, hopefully, give the reader a qualitative understanding of the dynamics of the *NKC* model. More stringent derivations and other technical matters have been relegated to a number of appendices.

2. THE SYSTEM

We consider an ensemble of mutually dependent and evolving species, an *ecosystem*. At any time, the state of any species is given by the state of its genome. This genome contains N genes. We shall assume the genes are binary variables; i.e., there are only two alleles, $A = 2$. We do not expect our results to change in any significant way if the number of alleles is changed, as long as it is small compared with N in results based on expansion in $1/N$. We do not distinguish between phenotypes and genotypes, and also neglect variations in type within a species. In real life, variation is responsible for the very existence of evolution. In the *NKC* model, however, only this consequence of variation is modeled: evolution takes place and is driven by a constant rate of mutations of individual, randomly chosen genes. If a mutation increases the fitness of a species, it is accepted, and the entire species is changed. If a mutation does not increase the fitness, it is rejected, and the species remains unchanged. Tie situations, with two genetic configurations having the same fitness, do not occur (have measure zero), due to the way in which we assign fitness to genetic configurations: If the selection time scale is much faster than the time-scale for mutations, this lends some justification to our "all or nothing" dynamics neglecting variations.[10] Proliferation and extinction of species are both neglected in the present article, though the model could be adapted to accommodate their description.

The fitness f of any of the evolving species is a random function of its N genes *and* of C other genes belonging to other species.[1] These C other genes are chosen at random among the genes of other species. For a given sample of the kind of ecosystem described here, the particular choice for these C genes and the random fitness function together define the sample and remain fixed during evolution—the randomness is *quenched.*

The particular probability distribution $p(f)$ used to define the fitness function does not matter; we shall not even bother to introduce it in our considerations below, because it turns out that it disappears again by a transformation of variables to $F = \int_{-\infty}^{f} df' \, p(f')$. In the case where p is uniform on the interval $0 \leq f \leq 1$, we have $f = F$. So for convenience we shall refer to F as the fitness, although F in the general case really denotes the probability for fitness less than f. The elimination of $p(f)$ in equations expresses that the value f of the fitness is irrelevant; only the probability F of being less fit matters.

[1] At this point we differ slightly from Kauffman's own definition of the *NKC* model. He defines the fitness function of a species as the sum of N random functions, one for each gene, depending on the gene and on K other genes in the species *plus* on C genes in other species. For $K = N - 1$, a species' fitness function is therefore an entirely random function of the N genes in the species, but a rather correlated function of the foreign genes it depends on. There is no good reason that the fitness function should be defined this way; it is just an accidental consequence of its parametrization. So for convenience we have simply assumed that the fitness function is a random function of *all* its variables.

We have two reasons to consider random fitness landscapes; the first reason is a conjecture, the second is proven correct in the appendices:

1. Evolution in any fitness landscape having an effectively finite correlation length will, when viewed at sufficiently coarse-grained scales of time and space (i.e., configuration space), look like evolution in a random fitness landscape. So evolution in a random fitness landscape describes the large-scale behavior of evolution in a large class of landscapes. Consequently, with this choice of landscape we are avoiding the particular, while treating a quite general case.
2. It is technically convenient: the absence of correlations allows us to derive a number of analytical results.

Notice that from a mathematical point of view, N might as well be the number of positions in the primary sequence of a protein, with $A = 20$ denoting the 20 amino acids that potentially could occur at each position. Or $A = 4$ could denote the four nucleotides possible at each site in a DNA sequence of length N.

Alternatively, we may think of the N genes and their A alleles as N Potts spins and their A possible values in an A-state Potts model. With $V = -f$ denoting the *energy* of a spin configuration, we recognize in each species a sample of Derrida's random energy model,[5,9] and these samples are asymmetrically coupled to each other for $C \neq 0$. In this language, the dynamics of mutations described above is the random-site Metropolis algorithm at zero temperature.

3. ESTIMATING THE LENGTH OF WALKS

Evolution traces out a path in configuration space. At each time step, the path is either extended one step from its current end point to a nearest neighbor—when a mutation leading to higher fitness is offered to and accepted by evolution—or the path is *not* extended—because a mutation leading to lower fitness is offered and rejected. This path is often referred to as an *adaptive walk.*

In this section, we are not concerned with the temporal aspects of evolution, but only with the length ℓ of adaptive walks. This limitation simplifies the description a good deal. In subsequent sections, temporal aspects are treated.

Before we get involved with mathematics, let us estimate the average length of adaptive walks, and the average fitness they lead to. The qualitative picture thus obtained is confirmed by rigorous calculations in Appendix B.

We assume N is large. The dimension of configuration space is N. We assume the length of adaptive walks is much smaller than $\sqrt{N}$, and find this assumption consistent with the results it leads to. Since the walk proceeds by random mutations, it proceeds in random directions in configuration space. There are many more directions than there are steps in the walk, by assumption. So each step in the walk has a different direction. In each step of the adaptive walk, the fitness F

is increased. The value to which it increases is uncorrelated—to leading order in $1/N$; see Appendix A—with its previous value, except it is larger, of course. Consequently, in each step $1 - F$ is halved, on the average. Thus, starting the walk with $F = 0$, after ℓ steps the average fitness is $1 - 2^{-\ell}$. An adaptive walk stops when all neighbor positions have lower fitness than the current position. Since fitnesses are random and uncorrelated, this happens when N independent random numbers happen to be smaller than F. On the average, this occurs when $1 - F \sim 1/N$. This is our estimate for the average final fitness and, setting $1 - F \sim 2^{-\ell}$, we have an estimate for the average length of an adaptive walk:

$$\bar{\ell} \simeq \log N / \log 2. \tag{1}$$

In the derivation of this result, we neglected correlations between fluctuations around the averages that we worked with. They do not change the logarithmic dependence on N in Eq. (1), but do change the coefficient of $\log N$; see Appendix B.

In addition to a more precise result for the average length of adaptive walks, we want to know the probability distribution Q_ℓ for ℓ. In a paper by Macken and Perelson,[16] "long upper tails containing little probability" were seen in numerical results for Q_ℓ. So one may wonder whether Q_ℓ decreases as a power of ℓ at large ℓ, or faster. We found that $(Q_\ell)_{\ell=0,1,2,\ldots}$ is a Poisson distribution to leading order in $1/\log N$; see Appendix C and Figure 1.

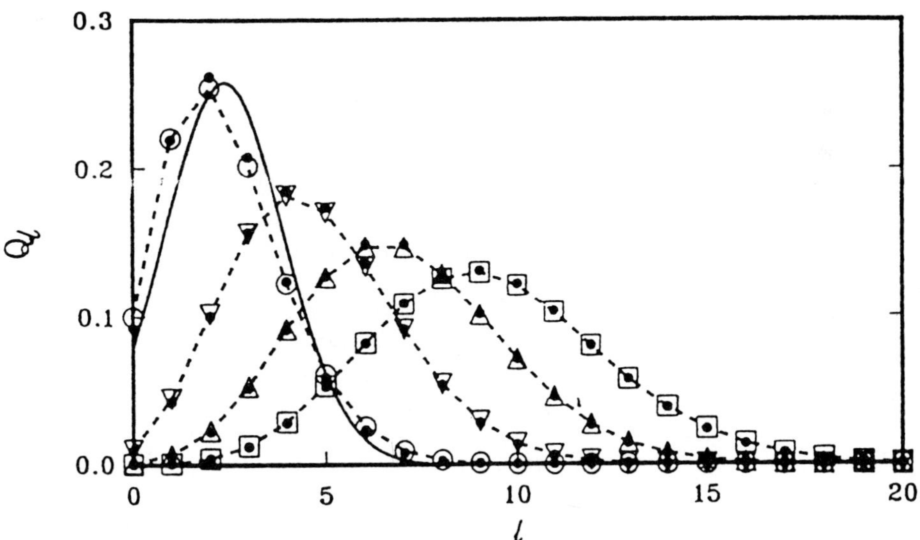

FIGURE 1 Q_ℓ versus ℓ for $N = 10$ (○), 100 (▽), 1000 (△), and 10,000 (□). The connecting dashed lines are only meant to guide the eye. Poisson distributions with the same mean values are shown with the symbol •. In the case of $N = 10$, the Gaussian distribution with same mean and variance as Q_ℓ is shown as a solid line.

4. ESTIMATING THE DURATION OF WALKS

Since we let the adaptive walk start out with fitness $F = 0$, the probability Q_0 that it is at a local fitness maximum at time $t = 0$ after the first step is

$$Q_0 = 1/N. \tag{2}$$

This is a rigorous result.

On the average, and to leading order in $1/N$, each step taken, including the first, reduces $1 - F$ by a factor of 2. Each step thereby doubles the probability that the ensuing step will be the last, while it halves the probability per unit of time that the next step is taken. Consequently, the probability per unit of time for the walk to terminate is constant during the walk. This means

$$Q_t = \frac{1}{\bar{t}} \exp(-t/\bar{t}). \tag{3}$$

Using the exact result in Eq. (2), we have the estimates

$$\bar{t} = N \tag{4}$$

and

$$Q_t = \frac{1}{N} \exp(-t/N). \tag{5}$$

This last equation shows that NQ_t remains a finite function of t/N in the limit $N \to \infty$, and its kth moment is proportional to N^k. In particular we see that the standard deviation

$$\sigma(t) = N \tag{6}$$

scales like the average $\bar{t}$. This is in contrast to the scaling laws found for the average *length* of walks and *its* standard deviation; see Appendices B and C.

In Appendix D we show how this section's estimates are modified when we account properly for fluctuations and their correlations. The result for NQ_t is shown in Figure 2.

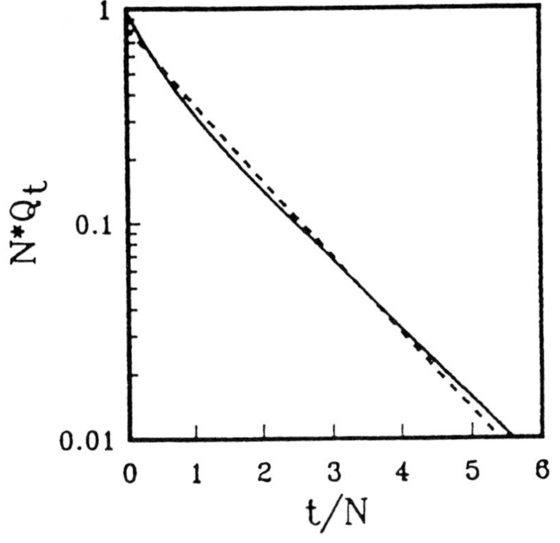

FIGURE 2 NQ_t versus t/N for $N = \infty$. Fully drawn curve: exact result from Eq. (92). Dashed curve: estimate from Eq. (3) with exact value for $\bar{t}$ taken from Eq. (89).

THE MASTER EQUATION

Because each species evolves by mutation of randomly chosen genes in a random fitness landscape, its path of evolution through configuration space can be replaced by a random walk, to leading order in N; see Appendix A. This observation causes vast simplifications in the description of the system's dynamics, which, on the other hand, is exact then only to leading order in N. But that is a small price to pay, as we imagine N is large anyway.

We include two additional simplifications in the description: instead of keeping fixed the C randomly chosen foreign genes that any species depends on, we re-choose them at random any time we need them; i.e., we exchange "quenched" randomness for "annealed." If the total number of species in the ecosystem is effectively infinite—and this assumption is the second simplification we add to the description—then there is no difference between results based on quenched, respectively annealed randomness. This is because the set of species that any species depends on, directly or via other species, forms a C-branched tree, each node of the tree representing a species and each oriented branch a dependency. So while our exchange of quenched for annealed randomness amounts to a mean-field approximation, we nevertheless expect the mean-field theory to be exact, because the system effectively is infinite-dimensional through its random connections.

The second assumption, an effectively infinite number of species in the ecosystem, makes a description in terms of density functions possible: let $\rho_M(F;t)$ denote the relative number of species that have fitness F and M less-fit one-mutant neighbors at time t. A change in a random gene will then lead to higher fitness—and therefore be accepted—with probability

$$A(t) = \sum_{M=0}^{N} (1 - M/N) \int_0^1 dF \rho_M(F;t) \tag{7}$$

because $1 - M/N$ is the probability that the change of one random gene leads to higher fitness in a species which has M less-fit one-mutant neighbors. We note that $A(t)$ also is the rate at which mutations are accepted by the ecology from the constant rate of mutations offered. So $A(t)$ is a measure of the evolutionary activity in the ecology. We shall find it a useful quantity below, and refer to it as the *activity*.

The probability that such a mutation is accepted *and* results in fitness F for the changed species is

$$\Phi(F;t) = \int_0^F dF' \phi(F';t), \tag{8}$$

where

$$\phi(F';t) = \frac{1}{1-F'} \sum_{M=0}^{N} (1 - M/N) \rho_M(F';t) \tag{9}$$

is the contribution to this probability from species with fitness F'. This contribution does not depend on F as long as $F \geq F'$. This is so because we have assumed the fitness landscape is uncorrelated. The factor $1/(1 - F')$ in this expression is the normalization factor for the constant distribution for F with $F \geq F'$.

With this notation we can write down the master equation for $\rho_M(F;t)$:

$$\begin{aligned} \frac{\partial}{\partial t} \rho_M(F;t) = &- \left(1 - \frac{M}{N}\right) \rho_M(F;t) + B_{M,N}(F) \Phi(F;t) \\ &- \frac{C}{N} A(t) \rho_M(F;t) + \frac{C}{N} A(t) B_{M,N}(F). \end{aligned} \tag{10}$$

This nonlinear integro-differential equation expresses that the relative number of species with fitness F, and M less-fit one-mutant neighbors, changes for four different reasons, corresponding to the four terms on the right-hand side of Eq. (10). The time scale in Eq. (10) has been chosen such that, in one unit of time, one mutation is offered per species—to be accepted or rejected.

The first term on the right-hand side of Eq. (10) is the rate at which species with fitness F, and M less-fit neighbors, mutate to higher fitness.

The second term on the right-hand side is a rate of change of less-fit species into species with fitness F and the number of less-fit neighbors M. The function $B_{M,N}(F)$ is the binomial distribution with mean F:

$$B_{M,N}(F) = \frac{N!}{M!\,(N-M)!} F^M (1-F)^{N-M}. \tag{11}$$

It represents the probability that M out of N one-mutant neighbors to a genome with fitness F are less fit than F. This probability is binomially distributed because the fitness landscape is random, with fitness F equidistributed in the interval $[0,1]$.[2]

The third term is the rate of the loss of species with fitness F, M. This loss is not caused by a change in the genes of the species lost, but by a change in its fitness due to genetic changes in other species. Since the C genes in other species that any species depends on are randomly chosen, this change is the product of the probability $A(t)$ that a mutation in a random species is accepted and the probability $C/N\rho_M(F;t)$ that the gene it occurs in is a gene on which a species with fitness F, M depends.

The fourth term on the right-hand side of Eq. (10) is, like the second term, a rate of change of species into species with fitness F, M. It complements the third term: species that change fitness due to genetic changes in other species, can change their fitness to F with equidistributed F. When they have done that, they have M less-fit neighbors with probability $B_{M,N}(F)$.

We note that Eq. (10) conserves the total probability, as it should:

$$\frac{\partial}{\partial t}\int_0^1 dF \sum_{M=0}^{N} \rho_M(F;t) = 0. \tag{12}$$

ESTIMATING THE PHASE STRUCTURE

Clearly, a static solution to Eq. (10) is provided by

$$\rho_M(F;t) = \delta_{M,N}\rho(F), \tag{13}$$

for any distribution $\rho(F)$. This solution corresponds to all species being at local fitness maxima. In the language of Kauffman,[11,13,14] borrowed from economics, the system is at a Nash equilibrium. Whether this fixed point for the dynamics is attractive or repulsive with respect to perturbations of $\rho_M(F)$ depends on the value of C. For $C = 0$ it is attractive, since in this case each species evolves in a fixed landscape and, consequently, arrives at a local maximum. At the other extreme, $C/N \gg 1$,

$$\rho_M(F;t) = B_{M,N}(F) \tag{14}$$

[2]Strictly speaking, this probability for M less-fit neighbors is $B_{M',N'}(F)$, with $N' < N$ and $M' = M-(N-N')$, because we already know that one or more one-mutant neighbor configurations are less fit. But we can neglect this difference in calculations to leading order in $1/N$ for reasons similar to those given in Appendix A.

is a static solution to leading order in N/C. It corresponds to totally random fitness F, and maximum activity $A = 1/2$.

At intermediate values of C, we can easily imagine the existence of a static solution with a finite activity A corresponding to a certain fraction of all species being in states that evolve. The activity is maintained by a balance between the rate at which species evolve towards fitness maxima, and the rate at which species are set back in evolution by their dependence on other species. We expect the activity A to increase with C.

On the other hand, we can also imagine that C can be too small to sustain a finite activity. In Appendix B we show that isolated species on the average change

$$\mu_1 = \log N + 0.09913\ldots + \mathcal{O}(N^{-1}) \tag{15}$$

genes in their evolution to a local maximum. So do species in the *NKC* model studied here, if they are not set back in evolution by their dependence on other species. Thus μ_1 is the minimal number of genetic changes per species by which the *NKC* model can evolve to the fixed point of Eq. (13). If, in doing so, each species on the average sets back less (or more) than one other species in evolution, the fixed point of Eq. (13) will (or will not) be attractive.

We can make the argument more precise by making it perturbative: suppose for a given value of C the system has been arranged to be at the fixed-point solution Eq. (13), and we change the fitness of one species to a random value. Since the other species do not evolve, the one singled out evolves as an isolated species, and arrives at a fitness maximum after having changed typically μ_1 of its genes. But the fitness of other species depend on the state of genes in the species that evolved; typically C other species will each depend on one gene. If any of these C genes were among the μ_1 genes that changed, the species depending on them were set back in evolution, and are now evolving, possibly setting back yet other species in their evolution. The question then is: if the chain reaction set off this way is sub- or super-critical, will it die out or run away? The value for C which separates these two situations we call critical, and write it as C_{crit}. It is the value for which, on the average, one out of C randomly chosen genes is among the μ_1 changed genes. Thus $1 = C_{\text{crit}}\mu_1/N$, or

$$C_{\text{crit}} = N/\mu_1. \tag{16}$$

We conclude that the species collectively evolve each to their own local fitness maximum and remain there with vanishing activity A for $C < C_{\text{crit}}$, while they evolve to a state with finite activity $A < 1/2$ for $C > C_{\text{crit}}$. The asymptotic value of the activity A for $t \to \infty$ can consequently be used as an order parameter distinguishing the two phases.

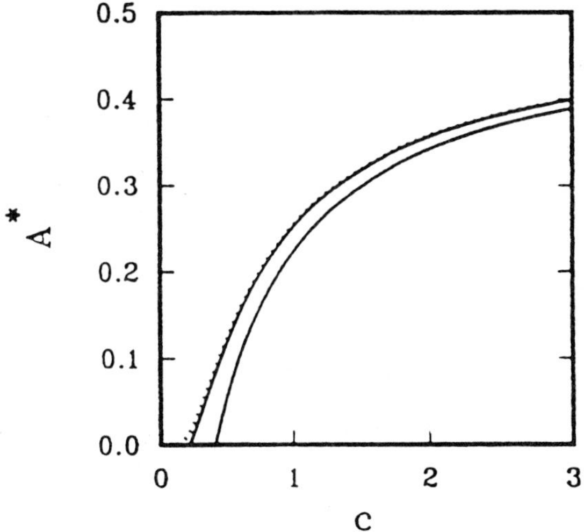

FIGURE 3 The asymptotic activity A^* versus the connectivity c for $N = 10$ and $N = 100$ according to Eq. (101) (full curves) and according to Eq. (103) (dotted curve).

The arguments used in this section were based on average values. While we would not expect fluctuations to change the qualitative picture, they might change the coefficient in a scaling law like Eq. (16). Actually they do not. The perturbative result is exact, as we see in Appendix E, where we also find the activity as a function of $c = C/N$. This activity is shown in Figure 3 for $N = 10$ and $N = 100$. In Appendix F, the systems relaxation time to the steady state is calculated for both phases, and found to diverge with mean-field exponent -1 at C_{crit}.

SUMMARY, DISCUSSION, PERSPECTIVES

For species evolving in isolation, we have obtained rigorous results to leading order in $1/N$ for the length and duration of adaptive walks in a special case of Kauffman's NK model. We found the average length scales as $\log N$, and so does the variance of the distribution of lengths. We have also obtained analytical expressions for the prefactors in these scaling laws, and found that to leading order in $1/\log N$, lengths are Poisson distributed.

For the duration of adaptive walk, we found qualitatively different results. While the average duration is proportional to N with a constant of proportionality we have found analytically, the variance of the duration is proportional to N^2, again with an analytically known coefficient. So while typical *lengths* of adaptive walks are relatively close to their average, typical *durations* vary over a range with magnitude equal to their average. We extended this result by showing analytically that in the limit $N \to \infty$, t/N has a finite distribution. Numerically, we found this distribution falls off exponentially for $t/N \geq 1$.

For coevolving species, we have shown analytically that a variant of Kauffman's *NKC* model has two phases: a *frozen* phase in which all species eventually stop evolving, because they all reach local fitness maxima, and a *chaotic* phase characterized by a balance between the number of species at local fitness maxima and the number evolving towards such maxima and changing the fitness landscape of other species in the process. As the order parameter we used the asymptotic *activity*, the fraction of species changing genetically per unit of time. We gave a closed expression determining the asymptotic activity as an implicit function of the connectivity between species. We also gave expressions for the system's relaxation time to the asymptotic activity. On the line separating the two phases in the system's parameter space, the relaxation time diverges with mean-field exponent -1.

We obtained these results in a mean-field description of the model, keeping only leading terms in an expansion in $1/N$, N being the number of genes per species. Since N typically is very large, however, our leading-order approximation in N is very good. We do not expect any qualitative differences between our leading-order $1/N$-expansion results and exact results as concerns the existence of the two phases, the location of the phase boundary, and the relaxation time. As for the exponent -1 for the divergence of the relaxation time, we have argued that it is an exact result. These results all depend on the number of species S being effectively infinite, and certainly much larger than both the number of genes N and the connectivity C.

It may well be possible to obtain other analytical results for the *NKC* model, using the methods of the present paper. For example, one may try to find the Lyapunov exponents of the chaotic phase.

As for the purpose of our investigation—the demonstration of self-organized criticality in the *NKC* model—we see no way that the maximally rugged variant studied here can be driven with perturbations from its frozen phase into a "poised," critical state, as was done in Bak et al.[3] with Conway's *Game of Life*. The maximally rugged variant cannot be "pumped up" to a "poised" state—at least not in the mean-field description—because after the model has responded to a perturbation it is back in the same state it was in before the perturbation was applied. This is not necessarily a shortcoming of the mean-field description. It willingly describes, for example, the build-up of the self-organized critical state of conservative sand pile models.[4] Rather, it is due to the maximal ruggedness of the fitness landscape. Its total absence of correlations makes any perturbation of a species wipe out all memory of the fitness that the species had acquired before the perturbation was applied. There is, so to speak, no such thing as a perturbation *of fitness* in the maximally rugged case. Genetic configurations may be perturbed by having just one or a few genes changed. But that typically results in a finite change of fitness in a maximally rugged landscape.

On the other hand, maximal ruggedness of the model's fitness landscape is crucial for our ability to derive analytical results, and these results are important in view of the difficulty of a numerical simulation of the model. So we are reluctant to abandon it. That leaves us with another, biologically appealing possibility: we can make the model more realistic (and computationally even more difficult) by treating N and C as dynamical parameters of the individual species, add criteria for their evolutionary change, and ask if evolution drives their averages onto the critical line found in the present paper. That study has yet to be done. Methods and results that appear to make such an undertaking feasible were presented here.

ACKNOWLEDGMENTS

We thank Bernard Derrida and Thordur Jonsson for discussions. PB's research was supported by the Division of Materials Science, U.S. Department of Energy, under contract DE-AC02-76CH00016. HF and BL were supported by the Danish Natural Science Research Council and the Danish Technical Research Council under contracts 11-9450-1 and 5.26-1818, respectively. PACS numbers: 87.10.+e, 02.50.+s, 05.40.+j, 64.60.Ht.

APPENDIX A: ADAPTIVE AND RANDOM WALKS

In this appendix we argue that if the dimension N of configuration space is sufficiently large compared to the length of a finite path in that space, we cannot distinguish, to leading order in N, between the path of a random walk and the path of an adaptive walk in a random fitness landscape.

Assume that the dimension N of configuration space is much larger than the length of adaptive walks in that space. Then we can neglect the fact that the adaptive walk avoids itself and all configurations previously probed by it. The reasoning goes as follows: Since mutations occur on random genes, a step is added to the walk by probing random directions in configuration space, until one leading to higher fitness is found. Then the walk is extended one step in that direction, and the procedure repeated from the new position in configuration space. By this algorithm, correlations between successive directions chosen by the walk are of order $1/N$. So to leading order in an expansion in $1/N$, successive directions are uncorrelated, and we have a *random* walk at hand. Successive directions are also *different* to leading order. Consequently, if the length of the walk is much smaller than $\sqrt{N}$, *all* directions chosen by it are different, and it obviously does not self-intersect.[6]

By assuming that the adaptive walk never probes a site in configuration space that it has probed before, we found, in Section 3, that walks have length $\sim \log N$, which *is* much smaller than $\sqrt{N}$ for N large. We conclude that our assumption that the walk is short compared to N is self-consistently correct.

We may ask whether we can find all subleading terms in an expansion in $1/N$ without knowing the entire history of an adaptive walk. The answer is negative.

An adaptive walk does not backtrack, while a random walk does with probability $\sim 1/N$ per step. We can handle a random walk without backtracking analytically. But backtracking is not the only $1/N$ effect distinguishing an adaptive walk from a random one, however. An adaptive walk also forms no closed loops, and does not visit sites in configuration space that were probed previously, but not visited for lack of fitness. Thus an adaptive walk is not only self-avoiding, but also avoids many one-mutant neighbors to itself. A short random walk visits such sites with probability $\sim 1/N$. This is seen as follows.

Self-intersection by a random walk requires the formation of a closed loop by the walk; i.e., at least two steps, of opposite orientation, must be taken in each dimension in which the loop extends. So the probability for the formation of a closed loop of length ℓ' on a random walk of length ℓ is, to leading order, suppressed by a factor $(\ell - \ell')/N^{\ell'/2}$, where $\ell' \geq 4$. Nearest neighbors to the walk can be visited in one step less, i.e., with probability $(\ell - \ell')/N^{\ell'/2-1}$. For $\ell' = 4$ this probability is $\sim \ell/N$. So to leading order in $1/N$ we can treat the adaptive walk as a random walk. We can also treat it as a random walk without backtracking, thereby describing some of the $1/N$ effects at play. But a full description of $1/N$ effects requires more information than the walk's current position in configuration space.

In summary, to leading order in $1/N$ we may add a step to the adaptive walk by treating the one-mutant neighbors to the current configurations as if they had never been visited or probed before. Consequently, the probability that M of these N neighbors are less fit than the current configuration is binomial, $B_{M,N}(F)$, where F is the fitness of the current configuration. If we take into account that the previous configuration is known to be less fit, the probability is $B_{M-1,N-1}(F)$, as given in Eq. (57).

When we forbid backtracking, our treatment is exact for a configuration space which is a Cayley tree with coordination number N. It should not be confused with an "annealed" fitness landscape, as an approximation to the "quenched" landscape we start out with. Not if "annealed" means rechoosing the fitness of a configuration every time it is probed by the adaptive walk. If we did that, we would have no maxima, since a higher fitness could always be attained by sufficiently many trials. The picture of an "annealed" fitness landscape applies only in the sense that the fitnesses of all N (or $N-1$) neighbors to a configuration are rechosen every time that configuration is visited, *and kept fixed during the visit,* thereby possibly making the visit permanent.

APPENDIX B: THE LENGTH OF WALKS

When the duration is not of interest, but the length is, the simplest quantity to work with is the probability density $p_\ell(F)$ that an adaptive walk contains (at least) ℓ steps and has fitness F after these ℓ steps. Evolution by random mutations through fitter one-mutant neighbors can be described approximately by a recursion relation:

$$p_{\ell+1}(F) = \int_0^F dF' \, \frac{1 - F'^{N-1}}{1 - F'} \, p_\ell(F') \quad \text{for } \ell = 1, 2, \ldots . \tag{17}$$

This recursion relation expresses that fitness F is acquired in $\ell+1$ evolutionary steps by acquiring any lower fitness F' in ℓ steps, and taking one more step to fitness F. Taking the last step requires that not all N one-mutant neighbors in configuration space are less fit. One is—the one that was reached after $\ell - 1$ steps. The remaining $N - 1$ neighbors have fitness less than F', each with probability F', since their fitness is random. Here we assume that they were not probed previously by the path of evolution. This assumption is only approximately true, so Eq. (17) is an approximation. Within this approximation, the probability that not all neighbors are less fit is $1 - F'^{N-1}$. When this is the case, the $(\ell + 1)$th evolutionary step will be taken, and leads to any fitness above F' with equal probability, hence to fitness F in the interval dF' with probability $dF'/(1 - F')$.

The approximation we have made with Eq. (17) relies on N being large. While the power $N - 1$ on F' excludes evolutionary backtracking the possibility, Eq. (17) does not exclude that the path of evolution *intersects* itself or visits other points in configuration space that it probed and rejected at an earlier time. Such intersections are forbidden by the dynamics, which forces the path to always higher degrees of fitness in a fixed landscape, or to stop at a local maximum. But in Eq. (17), the $N-1$ one-mutant neighbors which are not a state's immediate predecessor in evolution are all treated as if they were never probed before by the evolutionary process. Some of them may have been, in which case we know that their fitness is lower than the current one. So Eq. (17) yields an upper bound for the true value of $p_\ell(F)$, because the exact relation has a power lower than or equal to $N - 1$, where Eq. (17) has $N - 1$. This exact power depends on the entire path of evolution up to the currents state, so the approximation made with Eq. (17) causes a vast simplification of the problem. In the appendix we give arguments that this approximation is correct to leading order in an expansion in $1/N$.

In view of the further approximation considered below, all we really need are results to leading order in N. But since we can solve Eq. (17) as it stands—i.e., with backtracking forbidden and self-intersection permitted—we shall do that for definiteness.

Introducing into the monotonic function

$$H_N(F) = \sum_{k=1}^{N} \frac{1}{k} F^k \tag{18}$$

a change of variable to $H = H_{N-1}(F)$ in Eq. (17) gives

$$p_{\ell+1}(H) = \int_0^H dH'\, p_\ell(H') \quad \text{for } \ell = 1, 2, \ldots, \tag{19}$$

which is easily iterated to give

$$p_\ell(H) = \frac{1}{(\ell-1)!} \int_0^H dH'\, (H - H')^{\ell-1} p_1(H'). \tag{20}$$

For definiteness and notational convenience, we let all adaptive walks begin in the least-fit state, characterized by $F = 0$. Since there is zero probability for this state being a local maximum of fitness, the first step of the adaptive walk is always taken. For notational convenience, we let ℓ denote the number of steps taken *in excess* of this first step. Then the initial condition reads

$$p_{-1}(F) = \delta(F). \tag{21}$$

This rather eccentric choice of initial condition assures that the walk has a predecessor for all values of $\ell \geq 0$. This makes formulas look simpler, and makes Eq. (17) and Eq. (19) valid also for $\ell = -1$. They have the unique solution

$$p_\ell(F) = \frac{1}{\ell!} H_{N-1}(F)^\ell \quad \text{for } \ell = 0, 1, 2, \ldots. \tag{22}$$

Obviously, for fixed $F < 1$,

$$H_N(F) \to -\log(1 - F) \quad \text{for } N \to \infty, \tag{23}$$

while, for $F = 1$, $H_N(1)$ are the *harmonic numbers* discussed by Knuth[15]

$$H_N(1) = \sum_{k=1}^{N} \frac{1}{k} = \psi(N+1) + \gamma_E = \log N + \gamma_E + \mathcal{O}(N^{-1}), \tag{24}$$

where $\psi(x) = d \log \Gamma(x)/dx$, and $\gamma_E = 0.57721566...$ is Euler's constant. For general F we note that

$$H_N(F) = \mathrm{li}(F^{N+1}) - \log(-\log(F)) + \mathcal{O}(N^{-1}) \tag{25}$$

where li is the logarithmic integral. We shall require that

$$H_N(1 - x/N) = \log N + \gamma_E - \mathrm{Ein}(x) + \mathcal{O}(N^{-1}) \quad \text{for } x \sim \mathcal{O}(1), \tag{26}$$

where $\mathrm{Ein}(x)$ is an entire function related to the exponential integral[1]:

$$\mathrm{Ein}(x) = \int_0^x dt \frac{1 - e^{-t}}{t} = \mathrm{E}_1(x) + \log x + \gamma_E. \tag{27}$$

As stated above, Eq. (17) is the simplest relation we can write down for a probability describing the length of the adaptive walks considered here, in the approximation specified. The probability that a walk contains (at least) ℓ steps is obtained from $p_\ell(F)$ by integration over F:

$$\begin{aligned} P_\ell &= \int_0^1 dF p_\ell(F) = \frac{1}{\ell!}\int_0^1 dF H_{N-1}(F)^\ell \\ &= \frac{1}{\ell!}\sum_{k_1,\ldots,k_\ell=1}^{N-1} \frac{1}{k_1\cdots k_\ell(k_1+\cdots+k_\ell+1)} \text{for} \ell = 0,1,2,\ldots. \end{aligned} \tag{28}$$

Integration over F on both sides of Eq. (17) gives

$$P_{\ell+1} = P_\ell - \int_0^1 dF F^{N-1} p_\ell(F) \tag{29}$$

which obviously cannot be made into a closed equation for P_ℓ. The remaining integral in Eq. (29) is the probability that an adaptive walk contains *exactly* ℓ steps. This is a quantity of interest. We introduce the notation Q_ℓ for it, and $q_\ell(F)$ for the corresponding probability density that a walk stops with fitness F after exactly ℓ steps:

$$\begin{aligned} q_\ell(F) &= F^{N-1} p_\ell(F) = F^{N-1}\frac{1}{\ell!}H_{N-1}(F)^\ell \text{for} \ell = 0,1,2,\ldots \\ Q_\ell &= \int_0^1 dF\, q_\ell(F) = \frac{1}{\ell!}\int_0^1 dF F^{N-1} H_{N-1}(F)^\ell \\ &= \frac{1}{\ell!}\sum_{k_1,\ldots,k_\ell=1}^{N-1} \frac{1}{k_1\cdots k_\ell(k_1+\cdots+k_\ell+N)} \text{for} \ell = 0,1,2,\ldots. \end{aligned} \tag{31}$$

From Eq. (29) follows

$$Q_\ell = P_\ell - P_{\ell+1}. \tag{32}$$

Since Eq. (21) implies

$$Q_{-1} = 0, \quad P_0 = 1, \tag{33}$$

normalization of $q_\ell(F)$ and Q_ℓ follows trivially from Eq. (32):

$$\sum_{\ell=0}^{\infty}\int_0^1 dF\, q_\ell(F) = \sum_{\ell=0}^{\infty} Q_\ell = \sum_{\ell=0}^{\infty}(P_\ell - P_{\ell+1}) = P_0 = 1. \tag{34}$$

Here we have used $\lim_{\ell\to\infty} P_\ell = 0$, and we have set the upper limit on the sum over ℓ to infinity for convenience. Strictly speaking, this upper limit is A^N, the number of points in configuration space. We shall see below that typical values for ℓ are of order $\log N$, and much larger values of ℓ occur with probabilities that are more than exponentially suppressed. So the effect of this change in upper limit is truly negligible.

Inserting Eq. (30) in Eq. (34) and summing over ℓ, we see that normalization means

$$\int_0^1 dF F^{N-1} \exp\left(H_{N-1}(F)\right) = 1 \tag{35}$$

for any positive integer N. This identity is exact, and may also be proven directly; we leave that for the reader's entertainment.

The generating function for the probabilities Q_ℓ reads

$$\tilde{Q}(\lambda) = \sum_{\ell=0}^{\infty} \lambda^\ell Q_\ell = \int_0^1 dF F^{N-1} \exp\left(\lambda H_{N-1}(F)\right). \tag{36}$$

Despite our ability to evaluate the integral in Eq. (25), we have not been able to evaluate the integral in Eq. (36) for general λ. But as we have already neglected terms of subleading order in $1/N$, we may continue to do so with no further loss of generality. To this end we write $F = 1 - x/N$, and observe that $F^N = \exp(-x) + \mathcal{O}(x^2/N)$. Consequently, the integrand in Eq. (36) is negligible unless $x \sim 1$, and, to leading order in $1/N$, we have for $\tilde{Q}(\lambda)$, Q_ℓ, and its first moment μ_1:

$$\tilde{Q}(\lambda) = N^{\lambda-1} \int_0^\infty dx\, e^{-x-\lambda(\mathrm{Ein}(x)-\gamma_E)} \tag{37}$$

$$\tilde{Q}(1) = 1 \tag{38}$$

$$\begin{aligned} Q_\ell &= \frac{1}{N\ell!} \int_0^\infty dx\, e^{-x} \left(\log N + \gamma_E - \mathrm{Ein}(x)\right)^\ell \\ &= \frac{1}{N\ell!} \left((\log N)^\ell + \mathcal{O}((\log N)^{\ell-1})\right) \end{aligned} \tag{39}$$

$$\mu_1 = \sum_{\ell=0}^{\infty} \ell Q_\ell = \frac{dQ}{d\lambda}(1) = \log N + \mu_1^{(\text{finite})} + \mathcal{O}(N^{-1}). \tag{40}$$

Here

$$\mu_1^{(\text{finite})} = \int_0^\infty dx\, (\gamma_E - \mathrm{Ein}(x))\, e^{-x+\gamma_E-\mathrm{Ein}(x)} = 0.09913... \tag{41}$$

is a constant that we have not been able to express in terms of known constants. Our results for μ_1, and for μ_2 given in Eq. (34), agree with the two-digit numerical results given by Macken and Perelson.[16]

Equation (39) shows that to leading order in $\log N$, Q_ℓ is a Poisson distribution. This simple results has a simple explanation: the Poisson distribution is obtained because all adaptive walks terminate with essentially the same fitness F. F belongs to an interval of width $\sim 1/N$ at $F = 1$. This is seen from our rewriting $\int^1 dF F^{N-1}$ as $1/N \int_0 dx \exp(-x)$. Thus, in the interval $[0, 1]$, NF^{N-1} is almost a δ-function with support at $F \simeq 1$. If we replace it with that in the formulas above, we arrive at a Poisson distribution.

APPENDIX C: Q_ℓ'S POISSON BEHAVIOR

In this appendix we elaborate on Q_ℓ's similarity with a Poisson distribution, and compare it with such distributions for various values of N.

With the notation

$$\langle \ldots \rangle = \int_0^1 dF\, q(F) \ldots , \tag{42}$$

where

$$q(F) = \sum_{\ell=0}^{\infty} q_\ell(F) = F^{N-1}\, e^{H_{N-1}(F)} \tag{43}$$

is the probability density that an adaptive walk terminates at a local fitness maximum with fitness F, we have a positive measure on the fitness interval $[0,1]$. Equation (35) shows that this measure is normalized. We write the integral in Eq. (36) in terms of this measure and cumulant-expand it:

$$\begin{aligned}\tilde{Q}(\lambda) =& \langle e^{(\lambda-1)H_{N-1}} \rangle \\ =& \exp\left(\langle e^{(\lambda-1)H_{N-1}} - 1 \rangle_c \right) \\ =& \exp\left((\lambda-1)\mu_1 + \frac{1}{2}(\lambda-1)^2 \langle H_{N-1}^2 \rangle_c + \frac{1}{3!}(\lambda-1)^3 \langle H_{N-1}^3 \rangle_c \cdots \right)\end{aligned} \tag{44}$$

where the first cumulants are

$$\langle H_{N-1} \rangle_c = \langle H_{N-1} \rangle = \mu_1 = \log N + \mu_1^{(\text{finite})} + \mathcal{O}(N^{-1}) \tag{45}$$

$$\langle H_{N-1}^2 \rangle_c = \langle (H_{N-1} - \mu_1)^2 \rangle = 0.16733... + \mathcal{O}(N^{-1}) \tag{46}$$

$$\langle H_{N-1}^3 \rangle_c = \langle (H_{N-1} - \mu_1)^3 \rangle = -0.08370... + \mathcal{O}(N^{-1}) \tag{47}$$

$$\begin{aligned}\langle H_{N-1}^4 \rangle_c =& \langle (H_{N-1} - \mu_1)^4 \rangle - 3 \langle (H_{N-1} - \mu_1)^2 \rangle^2 \\ =& 0.03815... + \mathcal{O}(N^{-1})\end{aligned} \tag{48}$$

$$\ldots .$$

Here we have used that to leading order in $1/N$ these expectation values receive contributions only from values of F obeying $F = 1 - x/N$ with $x \sim 1$, i.e., where $H_{N-1}(F) - \mu_1 = \gamma_E - \text{Ein}(x) - \mu_1^{(\text{finite})}$. Consequently, all cumulants beyond the first are ~ 1, while the first, μ_1, is $\sim \log N$. Neglecting cumulants higher than the first in Eq. (44), we arrive at the generating function for a Poisson distribution with the same mean, μ_1, as Q_ℓ has:

$$\tilde{Q}(\lambda) = \exp((\lambda - 1)\mu_1). \tag{49}$$

Figure 1 shows Q_ℓ against ℓ for $N = 10$, 100, 1000, and 10,000 as open symbols connected by lines. The lines are only meant to guide the eye. Q_ℓ was found by numerical integration of the expressions for Q_ℓ deriving from Eq. (36). The values for the Poisson distributions with the same mean values are shown as filled circles, which in most cases fall within the open symbols. This agreement is rather striking. It is *not* just due to the central limit theorem making both Q_ℓ and the Poisson distribution well approximated by the same Gaussian distribution, hence by each other. This is illustrated in Figure 1 for the case of $N = 10$: the dashed line shows the Gaussian distribution with the same mean and variance as Q_ℓ has. Clearly, it does not approximate Q_ℓ, shown as open circles, as well as the Poisson distribution with the same mean as Q_ℓ, shown as filled circles. In addition to that, it has non-negligible support for negative values of ℓ.

We can also compare Q_ℓ's moments, μ_n, with the moments of the Poisson distribution with the same mean, μ_1:

$$\mu_1 = \langle H_{N-1} \rangle = \log N + \mu_1^{(\text{finite})} \tag{50}$$

$$\mu_2 = \mu_1 + \langle H_{N-1}^2 \rangle_c = \mu_1 + 0.16733... \tag{51}$$

$$\mu_3 = \mu_1 + \langle H_{N-1}^2 \rangle_c + \langle H_{N-1}^3 \rangle_c = \mu_1 + 0.08363... \tag{52}$$

$$\begin{aligned} \mu_4 &= \mu_1 + \langle H_{N-1}^2 \rangle_c + 3(\mu_1 + \langle H_{N-1}^2 \rangle_c)^2 + \langle H_{N-1}^3 \rangle_c + \langle H_{N-1}^4 \rangle_c \\ &= \mu_1 + 0.16733... + 3(\mu_1 + 0.16733...)^2 + 0.12215.... \end{aligned} \tag{53}$$

As expected from Eq. (44), we see that when we neglect cumulants beyond the first, the nth moment, μ_n, depends on the first moment, μ_1, as the nth moment of a Poisson distribution does. We also see that this neglect introduces an error of just a few percent in the moments shown for $N \geq 100$. We expect this error to increase with the order n of the moment μ_n, and know that it decreases as $1/\log N$.

APPENDIX D: THE DURATION OF WALKS

Let $p_{\ell,M;t}(F)$ denote the probability that an adaptive walk at time t has proceeded ℓ steps, thereby reaching a point in configuration space having fitness F and M less-fit neighbors. The time evolution of $p_{\ell,M;t}(F)$ is found as follows. As above, we neglect the fact that an adaptive walk cannot intersect itself or any site that was previously probed by its evolution and discarded for being less fit. As explained in the appendix, this is a leading order approximation in an expansion in $1/N$. Within this approximation, but explicitly forbidding backtracking, $p_{\ell,M;t}(F)$'s development in time is given by

$$p_{\ell,M;t+1}(F) = \frac{M}{N} p_{\ell,M;t}(F) + B_{M-1,N-1}(F)\Phi_{\ell-1;t}(F) \tag{54}$$

where

$$\Phi_{\ell;t}(F) = \int_0^F dF' \, \phi_{\ell;t}(F') \tag{55}$$

and

$$\phi_{\ell;t}(F) = \frac{1}{1-F} \sum_{M=0}^{N} \left(1 - \frac{M}{N}\right) p_{\ell,M;t}(F). \tag{56}$$

Equation (54) expresses that an adaptive walk has length ℓ, fitness F, and number of less-fit neighbors M at time $t+1$ for one of two mutually exclusive reasons: it was either characterized by these values at time t, and took no step between time t and time $t+1$—this happens with the probability given as the first term on the right-hand side in Eq. (54)—*or* a step *was* taken between time t and time $t+1$, and the adaptive walk arrived at values (ℓ, F, M) with that step—this happens with the probability given as the second term on the right-hand side of Eq. (54). $\Phi_{\ell-1;t}(F)$ is the transition probability density at time t to fitness F from less-fit one-mutant neighbor configurations arrived at in $\ell - 1$ steps. It is an integral over $F' < F$ of $\phi_{\ell-1;t}(F')$, the transition probability density at time t from fitness F' arrived at in $\ell - 1$ steps to any more-fit one-mutant neighbor configuration.

A configuration with fitness F, arrived at from a less-fit configuration, will have a total of M less-fit neighbor configurations, when $M-1$ of the $N-1$ new neighbor configurations are less fit. This happens with binomially distributed probability,

$$B_{M-1,N-1}(F) = \binom{N-1}{M-1} F^{M-1}(1-F)^{N-M}, \tag{57}$$

when we treat the landscape's quenched randomness as if the one-mutant neighborhood of any configuration arrived at is "annealed," thereby allowing the adaptive walk to self-intersect, with the exception that backtracking remains forbidden.

As an initial condition for Eq. (54) we choose as before, with no essential loss of generality, to let the adaptive walk start out in the least-fit configuration, at a time that is chosen to be -1 for notational convenience. We let ℓ denote the number of steps taken *in excess* to the first step, which is always taken. Then the initial condition reads

$$p_{\ell,M;-1}(F) = \delta_{\ell,-1}\, \delta_{M,0}\, \delta(F). \tag{58}$$

Inserted in Eq. (54), this initial condition gives the equivalent initial condition

$$p_{\ell,M;0}(F) = B_{M-1,N-1}(F)\, \delta_{\ell,0}. \tag{59}$$

Again our rather eccentric choice of initial condition assures that the walk has a predecessor at $t=0$, as at all later times. This makes formulas look simpler.

The simpler Eq. (17) is contained in Eq. (54): the probability that an adaptive walk reaches length ℓ and fitness F at time t is $\Phi_{\ell-1;t-1}(F)$. Consequently, the probability that it reaches length ℓ and fitness F at all, denoted $p_\ell(F)$ in Appendix B, is

$$p_\ell(F) = \sum_{t=0}^{\infty} \Phi_{\ell-1;t-1}(F) \quad \text{for } \ell \geq 0. \tag{60}$$

Using this with Eq. (54), one obtains an equation for $p_\ell(F)$, Eq. (17).

Equation (54) is a linear integro-difference equation. The fact that it is nonlocal in F does not prevent its solution, since it can be made local by appropriate differentiation after F. Introducing the generating function

$$\tilde{p}_M(\lambda, F;\tau) = \sum_{\ell=0}^{\infty} \lambda^\ell \sum_{t=0}^{\infty} \tau^t p_{\ell,M;t}(F) \tag{61}$$

and the corresponding generating functions for transition probability densities

$$\tilde{\phi}(\lambda, F;\tau) = \frac{1}{1-F} \sum_{M=0}^{N} (1 - M/N)\tilde{p}_M(\lambda, F;\tau) \tag{62}$$

and

$$\tilde{\Phi}(\lambda, F;\tau) = \int_0^F dF' \, \tilde{\phi}(\lambda, F';\tau), \tag{63}$$

the initial condition, Eq. (59), reads

$$\tilde{p}_M(\lambda, F; 0) = B_{M-1,N-1}(F) \tag{64}$$

and Eq. (54) itself, after a minor rearrangement, reads

$$\tilde{p}_M(\lambda, F;\tau) = \frac{N}{N-\tau M} B_{M-1,N-1}(F) \left(1 + \lambda\tau\tilde{\Phi}(\lambda, F;\tau)\right) \tag{65}$$

Consequently,

$$\tilde{\phi}(\lambda, F;\tau) = h_{N-1}(F;\tau)(1 + \lambda\tau\tilde{\Phi}(\lambda, F;\tau)), \tag{66}$$

where we have introduced

$$\begin{aligned} h_{N-1}(F;\tau) &= \frac{1}{1-F} \sum_{M=0}^{N-1} \frac{N-M}{N-\tau M} B_{M-1,N-1}(F) \\ &= \sum_{M=1}^{N-1} \frac{N-1}{N-\tau M} B_{M-1,N-2}(F) \\ &= \frac{N-1}{N-\tau} \frac{1}{1-F} \, {}_2F_1(2-N, 1; 2-N/\tau; F). \end{aligned} \tag{67}$$

Here ${}_2F_1$ is Gauss' hypergeometric function. For later use, we note that

$$\tau h_{N-1}(1-x/N;\tau) = (N-1)x^{-1-N(\tau^{-1}-1)} e^{-x} \gamma(1+N(\tau^{-1}-1); x) + \mathcal{O}(N^{-1}), \tag{68}$$

where γ is the incomplete gamma-function. We shall also need the function

$$H_N(F;\tau) = \int_0^F dF' h_N(F';\tau) \tag{69}$$

and make contact with Appendix B by noting that

$$H_N(F;1) = H_N(F). \tag{70}$$

For later use, we note that

$$\tau H_{N-1}(1 - x/N; (1+z/N)^{-1}) = \log N - \psi(1+z) - \gamma_E - \mathcal{I}(x;z) + \mathcal{O}(N^{-1}), \tag{71}$$

when $x \sim 1$ and $z \sim 1$, and we have introduced:

$$\mathcal{I}(x;z) = \int_0^1 dy\, y^z \frac{1 - e^{-x(1-y)}}{1-y} - \gamma_E \tag{72}$$

$$\mathcal{I}(x;0) = \text{Ein}(x) - \gamma_E. \tag{73}$$

Equation (66) is solved by

$$\tilde{\phi}(\lambda, F;\tau) = h_{N-1}(F;\tau) \exp(\lambda\tau H_{N-1}(F,\tau)) \tag{74}$$

and, consequently,

$$\tilde{p}_M(\lambda, F;\tau) = \frac{N B_{M-1,N-1}(F)}{N - \tau M} \exp(\lambda\tau H_{N-1}(F,\tau)). \tag{75}$$

In this result λ only occurs multiplied by τ. This is because in the series expansion of this result each power of λ represents one step taken in configuration space by the adaptive walk, and each such step takes one unit of time, represented by one power of τ. Powers of τ not occurring in conjunction with λ, on the other hand, represent time steps during which the adaptive walk did not progress.

The relation between the length and the duration of adaptive walks is contained in

$$p_{\ell;t} = \int_0^1 dF \sum_{M=0}^{N} p_{\ell,M;t}(F) \tag{76}$$

and therefore in

$$\tilde{p}(\lambda;\tau) = \int_0^1 dF \sum_{M=0}^{N} \tilde{p}_M(\lambda, F;\tau) = \frac{1}{\lambda\tau}\left(\frac{1}{(1-\tau)^\lambda} - 1\right) + \mathcal{O}(N^{-1}). \tag{77}$$

The generating function at time $t \geq 0$,

$$\tilde{p}_t(\lambda) = \sum_{\ell=0}^{\infty} \lambda^\ell \, p_{\ell;t} \,, \tag{78}$$

is obtained from $\tilde{p}(\lambda;\tau)$ via the relation

$$\begin{aligned} \tilde{p}_t(\lambda) =& \frac{1}{2\pi i} \oint \frac{d\tau}{\tau^{t+1}} \, \tilde{p}(\lambda;\tau) = \frac{1}{2\pi i \lambda} \oint \frac{d\tau}{\tau^{t+2}} \left(\frac{1}{(1-\tau)^\lambda} - 1 \right) + \mathcal{O}(N^{-1}) \\ =& \frac{\sin(\pi\lambda)}{\pi\lambda} B(1-\lambda, \lambda + t + 1) + \mathcal{O}(N^{-1}), \end{aligned} \tag{79}$$

where the closed path of integration in the complex τ-plane encircles $\tau = 0$ once in the positive direction. Using Cauchy's theorem, the last identity was established by moving the path so that it lies along the integrand's branch cut on the real axis, $\tau \geq 1$. The function $B(x, y)$ is the beta-function, Euler's integral of the first kind. Notice that the normalization condition

$$\tilde{p}_t(1) = \sum_{\ell=0}^{\infty} p_{\ell;t} = 1, \qquad \forall t \geq 0, \tag{80}$$

is satisfied by the result in Eq. (79). The same result gives, to leading order in $1/N$, that

$$\begin{aligned} (\bar{\ell})_t =& \sum_{\ell=0}^{\infty} \ell \, p_{\ell;t} = \frac{d\tilde{p}_t(1)}{d\lambda} = \psi(t+2) + \gamma_E - 1 \\ =& \log t + \gamma_E - 1 + \mathcal{O}(t^{-1}) \end{aligned} \tag{81}$$

and

$$\begin{aligned} \sigma_t^2(\ell) =& (\bar{\ell^2})_t - (\bar{\ell})_t^2 = \psi(t+2) + \gamma_E + \sum_{k=1}^{t+1} \frac{1}{k^2} \\ =& \log t + \gamma_E + \pi^2/6 + \mathcal{O}(t^{-1}). \end{aligned} \tag{82}$$

Thus we see our estimate confirmed: the average length of an adaptive walk grows logarithmically with time. Furthermore, we see that the variance of the length grows like the average length, like for a biased random walk. This similarity is no coincidence, since the adaptive walk in many respects resembles a simple, biased random walk.

In the last identity in Eq. (77) it was tacitly assumed that N itself was the only quantity of order N. Consequently, the time dependence found from this identity is reliable only when t is far from being of order N. This restriction need not prevent

t from being large or the asymptotic forms in Eq. (81) and Eq. (82) from being valid.

When $t \sim N$, walks reach local maxima and terminate, according to our estimate for their duration. This, of course, is an average result. For example, there is a probability $\sim 1/N$ that an adaptive walk terminates after its first step. Now let us substantiate the estimate: the probability that a walk terminates with length ℓ and fitness F at time t is

$$q_{\ell;t}(F) = p_{\ell,N;t}(F) - p_{\ell,N;t-1}(F). \tag{83}$$

Contact is made with Appendix B by observing

$$q_\ell(F) = \sum_{t=0}^{\infty} q_{\ell;t}(F) = \lim_{t\to\infty} p_{\ell,N;t}(F). \tag{84}$$

We introduce

$$Q_{\ell;t} = \int_0^1 dF\, q_{\ell;t}(F) \tag{85}$$

and

$$\tilde{Q}(\lambda;\tau) = \sum_{\ell=0}^{\infty} \lambda^\ell \sum_{t=0}^{\infty} \tau^t Q_{\ell;t} \tag{86}$$

and have

$$\begin{aligned}\tilde{Q}(\lambda;\tau) &= (1-\tau)\int_0^1 dF \tilde{p}_N(\lambda, F;\tau) = \int_0^1 dF F^{N-1} \exp(\lambda\tau H_{N-1}(F;\tau)) \\ &= e^{-\lambda(\psi(1+z)+\gamma_E)}\mathcal{F}(\lambda;z) + \mathcal{O}(N^{-1}),\end{aligned} \tag{87}$$

where Eq. (75) was used in the second identity, and $F = 1 - x/N$, $\tau = (1 + z/N)^{-1}$, $x, z \sim 1$, in the third. We have introduced the N-independent function

$$\mathcal{F}(\lambda;z) = \int_0^\infty dx\, e^{-x-\lambda\mathcal{I}(x;z)}. \tag{88}$$

Equation (87) is the time-dependent extension of Eq. (36). From the generating function in Eq. (87) we derive the average time it takes for an adaptive walk to reach a local maximum:

$$\begin{aligned}\bar{t} &= \sum_{\ell,t=0}^{\infty} tQ_{\ell;t} = \frac{\partial\tilde{Q}}{\partial\tau}(1;1) \\ &= \langle H_{N-1}(F;1) + \frac{\partial H_{N-1}}{\partial\tau}(F;1)\rangle\end{aligned}$$

$$
\begin{aligned}
&= \langle \int_0^F dF' \sum_{M=1}^{N-1} \frac{N(N-1)}{(N-M)^2} B_{M-1,N-2}(F') \rangle && (89) \\
&= N\left(\frac{\pi^2}{6} - \frac{\partial \mathcal{F}}{\partial z}(1;0)\right) + \mathcal{O}(1) \\
&= 1.22398\ldots N + \mathcal{O}(1) \\
\overline{t^2} - \bar{t}^2 &= N^2 \left(2\zeta(3) + \frac{\partial^2 \mathcal{F}}{\partial z^2}(1;0) - \left(\frac{\partial \mathcal{F}}{\partial z}\right)^2 (1;0)\right) + \mathcal{O}(N) && (90) \\
&= 1.71788\ldots N^2 + \mathcal{O}(N) && (91)
\end{aligned}
$$

where ζ is Riemann's zeta-function. We have not been able to relate the derivatives of $\mathcal{F}$ in these equations to known mathematical constants.

Comparing this appendix's results with those of Appendix B, we notice a big difference between the length and the duration of adaptive walks in a random fitness landscape: while typical lengths are relatively closer to the average length, as the system size N increases, typical durations can differ from the average by an amount the size of this average. This picture is confirmed by the following expression for Q_t, the probability that a walk has duration t:

$$
\begin{aligned}
Q_t &= \frac{}{2\pi i} \oint \frac{d\tau}{\tau^{1+t}} \tilde{Q}(1;\tau) \\
&= \frac{1}{2\pi i N} \oint dz\, e^{\frac{t}{N}z - \psi(1+z) - \gamma_E} \mathcal{F}(1;z). && (92)
\end{aligned}
$$

Here the closed path of integration in the complex τ-plane encircles $\tau = 0$ once in the positive direction, while a similar path of integration in the complex z-plane, obtained by the substitution $z = N(\tau^{-1} - 1)$, has been moved to lie along the negative real axis. That is the only place in the z-plane where $\mathcal{F}(1;z)$ is not analytic. We have not found a more closed analytical expression for Q_t in the large-N limit than Eq. (92). Equation (92) suffices, however, since it shows that for $t/N \sim 1$ we have $Q_t \sim N^{-1}$. Hence, in the limit $N \to \infty$, NQ_t is a finite function of the variable t/N. We have found this function numerically. Its graph is shown in Figure 3 as the fully drawn line. The dashed line shows the graph for the estimate in Eq. (3) with the exact value in Eq. (89) used for $\bar{t}$. From the figure it seems that for $t/N \geq 1$, Q_t is essentially an exponential function, or at least exponentially bounded, though other possibilities cannot be eliminated on the basis of the figure.

APPENDIX E: CALCULATING THE PHASE STRUCTURE

Let us denote a stationary, or fixed point, solution to Eq. (10) by $\rho^*_M(F)$. With the notation $A^* = A[\rho^*]$, $\phi^* = \phi[\rho^*]$, $\Phi^* = \Phi[\rho^*]$, and $c = C/N$, the time-independent version of Eq. (10) can be rewritten

$$\rho^*_M(F) = \frac{N}{N - M + CA^*} B_{M,N}(F)(cA^* + \Phi^*(F)). \tag{93}$$

Since A^* and Φ^* both depend on ρ^*, Eq. (93) is a nonlinear integral equation for $\rho^*_M(F)$. We can solve it, nevertheless, by temporarily treating A^* as a constant, to be determined by self-consistency in the end. This is done in the following way: By multiplying both sides in Eq. (93) with $(1 - M/N)/(1 - F)$, and summing over M, one finds

$$\phi^*(F) = g(F; cA^*)(cA^* + \Phi^*(F)), \tag{94}$$

where we have introduced the function[3]

$$g(F; x) = \frac{1}{1-F} \sum_{M=0}^{N-1} \frac{N-M}{N-M+Nx} B_{M,N}(F) = N \sum_{M=0}^{N-1} \frac{B_{M,N-1}(F)}{N-M+Nx}. \tag{95}$$

For later use we also introduce

$$G(F; x) = \int_0^F dF' \, g(F'; x) \tag{96}$$

and

$$\mathcal{G}(x) = \int_0^1 dF e^{G(F;x)}. \tag{97}$$

Since g and G have simple poles at $x = -1$, $-1 + 1/N$, $-1 + 2/N$, ..., $-1/N$, the function $\mathcal{G}$ has essential singularities at these points. The graph for $\mathcal{G}(x)$ is shown in Figure 4 for the case of $N = 10$. For $x \gg \mathcal{O}(1/N)$ or $x < -1$, $\mathcal{G}$ simplifies to

$$\mathcal{G}(x) = (1 + x) \log(1 + x^{-1}) \tag{98}$$

to leading order in $1/N$. The graph for this approximation is shown as the dotted curve in Figure 4. The approximation has a cut in the interval $[-1, 0]$ where $\mathcal{G}(x)$ has N essential singularities.

[3]Contact is made with Appendix B by observing that $g(F; 0)$ essentially is equal to the function $h_{N-1}(F; 1)$ defined there. The two functions differ only because in Appendix B we chose to account explicitly for the impossibility of backtracking in adaptive walks, though this is an effect of subleading order in $1/N$. In the present appendix, formulas are simpler when we neglect subleading terms from the start. Thus the difference between $g(F; 0)$ and $h_{N-1}(F; 1)$.

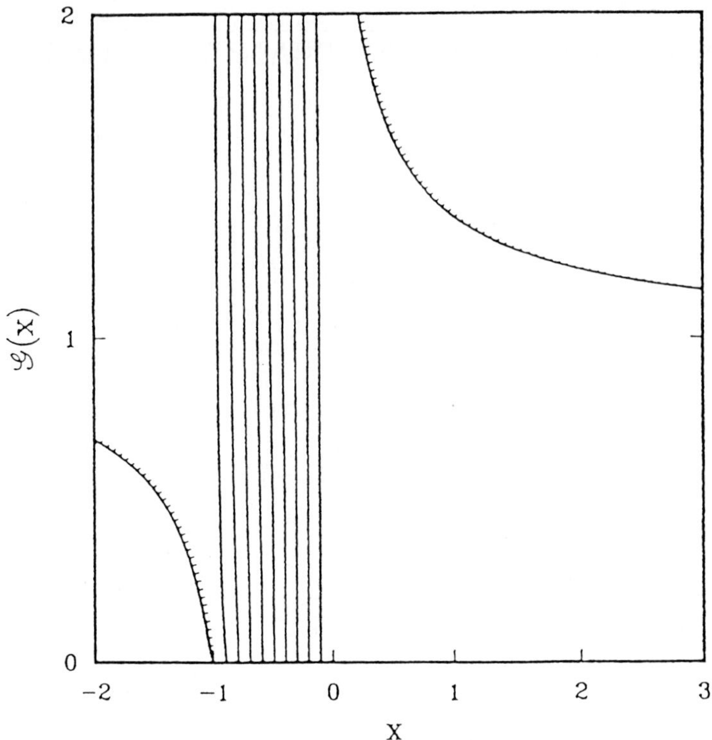

FIGURE 4 Graph of the function $\mathcal{G}(x)$ defined in Eq. (97) in the case of $N = 10$ (full curve), and its approximation given in Eq. (98) (dotted curve).

Now, remembering $\phi^*(F) = d/dF\Phi^*(F)$, we see Eq. (94) is solved by

$$\Phi^*(F) = cA^*(e^{G(F;cA^*)} - 1). \tag{99}$$

Inserting this solution in the definition Eq. (7) of the activity, we finally arrive at a self-consistency equation for A^*, given c:

$$A^* = cA^*(-1 + \mathcal{G}(cA^*)). \tag{100}$$

This equation is solved by $A^* = 0$ and by A^* satisfying

$$c^{-1} = -1 + \mathcal{G}(cA^*). \tag{101}$$

The last equation gives A^* as an implicit function of c. It has a real, positive solution A^* only for

$$c > c_{\text{crit}} = (-1 + \mathcal{G}(0))^{-1} = \mu_1^{-1} \tag{102}$$

where μ_1 is given in Eq. (15). For $cA^* \gg \mathcal{O}(1/N)$, Eq. (101) simplifies to leading order in $1/N$ to another implicit expression for $A^*(c)$,

$$c^{-1} = -1 + (1 + cA^*)\log(1 + (cA^*)^{-1}). \tag{103}$$

According to Eq. (101), $A^* \propto c - c_{\text{crit}}$ for $c \sim c_{\text{crit}}$; i.e., the critical exponent for the order parameter A^* is 1. At the other extreme, for $c \to \infty$, Eq. (101) gives $A^* = 1/2$, as we expect from Section 6. Figure 3 shows $A^*(c)$ for $N = 10$ and $N = 100$ as fully drawn curves. The approximate expression in Eq. (103) is shown as the dotted curve.

For $c > c_{\text{crit}}$, Eq. (93) then gives

$$\rho_M^*(F) = \frac{CA^*}{N - M + CA^*} B_{M,N}(F)\exp(G(F; cA^*)), \tag{104}$$

while for $c < c_{\text{crit}}$ we have

$$\rho_M^*(F) = \delta_{M,N} F^N \exp(G(F;0)). \tag{105}$$

So, as already seen in Section 6, the long-term dynamics of the coevolving species can be of two qualitatively different kinds, depending on whether the parameters C and N have values making $c = C/N$ smaller or larger than c_{crit} given above. In the first case, the activity $A(t)$ dies out because all species stop evolving as they reach local fitness maxima. This is *frozen* dynamics, characterizing the *frozen phase.* In the second case the activity converges to a nonzero value A^*, signalling *chaotic* dynamics, characterizing the chaotic phase. In this phase, species also evolve towards local maxima in fitness but, in the process of doing so, they change the fitness of other species, typically setting them back in evolution. After a transient time, a balance is reached where a certain fraction of species evolve, while another fraction remains at local fitness maxima, with individual species passing from one fraction to the other every so often.

The line $C/N = c_{\text{crit}}$ dividing the (C, N)-plane into two phases is critical in the sense that the relaxation time to asymptotic behavior diverges on this line, as shown in Appendix F.

APPENDIX F: RELAXATION TIMES

In order to find the relaxation time to asymptotic values, we linearize Eq. (10) at its fixed point solution. We write

$$\rho_M(F;t) = \rho_M^*(F) + \Delta\rho_M(F;t), \tag{106}$$

$$A(t) = A^* + \Delta A(t), \tag{107}$$

$$\Phi(F;t) = \Phi^*(F) + \Delta\Phi(F;t), \tag{108}$$

$$\phi(F;t) = \phi^*(F) + \Delta\phi(F;t), \tag{109}$$

and insert these expressions in Eq. (10). By using Eq. (93) and keeping only terms linear in Δ...., we arrive at the linearized master equation

$$\begin{aligned}\frac{\partial}{\partial t}\Delta\rho_M(F;t) = &-(1-\frac{M}{N}+cA^*)\Delta\rho_M(F;t) \\ &+c(B_{M,N}(F)-\rho^*_M(F))\Delta A(t)+B_{M,N}(F)\Delta\Phi(F;t).\end{aligned} \tag{110}$$

This equation is more easily solved by writing $\Delta\rho_M(F;t)$ as a Laplace transform:

$$\Delta\rho_M(F;t)=\int_0^\infty d\lambda\, e^{-t\lambda}\Delta\tilde{\rho}_M(F;\lambda). \tag{111}$$

$\Delta A(t)$ and $\Delta\Phi(F;t)$ are linear functionals of $\Delta\rho_M(F;t)$ and therefore commute with Laplace transformation. So with a self-explanatory notation, the inverse Laplace transform of Eq. (110) reads, slightly rewritten:

$$\Delta\tilde{\rho}_M(F;\lambda)=\frac{c\Delta\tilde{A}(\lambda)(B_{M,N}(F)-\rho^*_M(F))+B_{M,N}(F)\Delta\tilde{\Phi}(F;\lambda)}{1-M/N+cA^*-\lambda}. \tag{112}$$

By multiplying both sides of this equation with $(1-M/N)/(1-F)$, and summing over M, one finds

$$\begin{aligned}&\Delta\tilde{\phi}(F;\lambda)= \\ &\quad c\Delta\tilde{A}(\lambda)(g(F;cA^*-\lambda)-g_1(F;cA^*-\lambda))+g(F;cA^*-\lambda)\Delta\tilde{\Phi}(F;\lambda),\end{aligned} \tag{113}$$

where the function $g(F;x)$ was introduced in the previous appendix, and the function g_1 has a similar definition:

$$\begin{aligned}g_1(F;x) &= \frac{1}{1-F}\sum_{M=0}^{N-1}\frac{N-M}{N-M+Nx}\rho^*_M(F) \\ &= \frac{cA^*}{x-cA^*}(g(F,cA^*)-g(F,x))\exp(G(F,cA^*)).\end{aligned} \tag{114}$$

Equation (113) is solved by

$$\begin{aligned}&\Delta\tilde{\Phi}(F;\lambda)= \\ &\quad c\Delta\tilde{A}(\lambda)e^{G(F;cA^*-\lambda)}\int_0^F dF'\, e^{-G(F';cA^*-\lambda)}\,(g(F';cA^*-\lambda)-g_1(F';cA^*-\lambda)) \\ &= c\Delta\tilde{A}(\lambda)\left(-1+\frac{cA^*}{\lambda}e^{G(F;cA^*)}\right)\end{aligned} \tag{115}$$

where we have used the definition, Eq. (114), for g_1 to obtain the last equality. Using

$$\Delta\tilde{A}(\lambda)=\int_0^1 dF\,\Delta\tilde{\Phi}(F;\lambda), \tag{116}$$

integration over F on both sides of Eq. (115) gives an equation for $\Delta\tilde{A}(\lambda)$ that is solved by $\Delta\tilde{A}(\lambda) = 0$, as we might expect, and by

$$\frac{cA^* - \lambda}{\lambda}(\mathcal{G}(cA^* - \lambda) - \mathcal{G}(cA^*)) = 0. \tag{117}$$

The smallest value for λ solving this equation contributes the longest relaxation time

$$t_{\text{relax}}^{\text{chaotic}} = \lambda^{-1} \tag{118}$$

to $\Delta\rho_M(F;t)$ in Eq. (111). An obvious solution is

$$\lambda = cA^*. \tag{119}$$

A survey of $\mathcal{G}(x)$'s graph shows there are $N-1$ other solutions to Eq. (117), one in each interval $[cA^* + M/N, cA^* + (M+1)/N]$, where $M = 1, 2, \ldots, N-1$. So all these solutions correspond to contributions to $\Delta\rho_M(F;t)$ which decay faster in time than the mode corresponding to $\lambda = cA^*$. We conclude that the relaxation time in the chaotic phase is

$$t_{\text{relax}}^{\text{chaotic}} = \frac{1}{cA^*}, \tag{120}$$

where A^* is a function of c given implicitly by Eq. (101).

Since $A^* \sim c - c_{\text{crit}}$ for $c - c_{\text{crit}} \sim 0^+$, we see from Eq. (120) that the relaxation time diverges with exponent -1 at the critical connectivity. This typical mean-field value for the exponent comes as no surprise; it is, after all, a mean-field description we are developing. The value for this exponent is exact, however, in the limit $S = \infty$ of infinitely many species, which we are considering. The only requirement is that each species depends on a vanishing fraction of other species—i.e., $C/S = 0$—and that the species which a given species depends on were chosen at random. Whether this randomness is quenched or annealed does not matter. This point has been explained in detail by Derrida[7,8] for an, in this respect, identical problem.

In the frozen phase, where the order parameter $A^* = 0$, Eq. (117) shows

$$c^{-1} + 1 = \mathcal{G}(-\lambda) \tag{121}$$

which for a given value of $c < c_{\text{crit}}$ has N positive solutions for λ, one in each interval $[cA^* + M/N, cA^* + (M+1)/N]$, where $M = 0, 1, \ldots, N-1$. The smallest solution, which determines the relaxation time, grows from $\lambda = 0$ to $\lambda = 1/N$ for c decreasing from c_{crit} to 0. So the relaxation time grows from N to infinity when c grows from 0 to c_{crit}. This result agrees with the average relaxation time for isolated species found in Appendix D, and the expected increase in relaxation time with increased coupling.

We can summarize our results for the relaxation time in the following implicit expressions for it:

$$c^{-1} + 1 = \mathcal{G}(-t_{\text{relax}}^{-1}) \text{for} c < c_{\text{crit}}, \tag{122}$$

$$c^{-1} + 1 = \mathcal{G}(t_{\text{relax}}^{-1}) \text{for} c > c_{\text{crit}}, \tag{123}$$

where the solution for t_{relax} is obtained by using the branch of $\mathcal{G}^{-1}$ characterized by $-1/N < x < \infty$.

REFERENCES

1. Abramowitz, M., and I. A. Stegun, eds. *Handbook of Mathematical Functions*, ninth printing, page 228, footnote 3. New York: Dover Publications.
2. Bak, P., C. Tang, and K. Wiesenfeld. *Phys. Rev.* **A38** (1987): 364.
3. Bak, P., K. Chen, and M. Creutz. *Nature* **42** (1989): 780.
4. Bak, P., and H. Flyvbjerg. Unpublished paper.
5. Derrida, B. *Phys. Rev. B* **24** (1981): 2613.
6. Derrida, B., and H. Flyvbjerg. *J. Physique* **48** (1987): 971.
7. Derrida, B., and G. Weisbuch. *J. Physique* **47** (1987): 1297.
8. Derrida, B., E. Gardner, and A. Zippelius. *Europhys. Lett.* **4** (1987): 167.
9. Gardner, E., and B. Derrida. *J. Phys. A: Math. Gen.* **22** (1989): 1975.
10. Gillespie, J. *Evolution* **38** (1984): 1116.
11. Kauffman, S. A., and S. Levin. *J. Theor. Biol.* **128** (1987): 11.
12. Kauffman, S. A. *Sci. Am.* **264** (1991): 78.
13. Kauffman, S. A., ed. *Origins of Order: Self-Organization and Selection in Evolution.* Oxford: Oxford University Press, 1990.
14. Kauffman, S. A., and S. Johnsen. *J. Theor. Biol.* **149** (1991): 467.
15. Knuth, D. E. *The Art of Computer Programming*, Vol. 1. Reading, MA: Addison-Wesley, 1968.
16. Macken, C. A., and A. S. Perelson. *Proc. Natl. Acad. Sci. USA* **86** (1989): 6191.
17. Macken, C. A., P. S. Hagan, and A. S. Perelson. *SIAM J. Appl. Math.* **51** (1991): 799.
18. Raup, D. M. *Science* **231** (1986): 1528.
19. Weinberger, E. D. *J. Theor. Biol.* **134** (1988): 125.
20. Weinberger, E. D. *Phys. Rev. A* **44** (1991): 6399.

Kunihiko Kaneko and Junji Suzuki
Department of Pure and Applied Sciences, College of Arts and Sciences, University of Tokyo, Komaba 3-8-1, Meguro-ku, Tokyo 153; e-mail: chaos@tansei.cc.u-tokyo.ac.jp

Evolution to the Edge of Chaos in an Imitation Game

Motivated by the evolution of complex bird songs, an abstract imitation game is proposed to study the increase of dynamical complexity: Artificial "birds" display a "song" time series to each other, and those that imitate the other's song better win the game. With the introduction of population dynamics according to the score of the game and the mutation of parameters for the song dynamics, the dynamics are found to evolve towards the borderline between chaos and a periodic window, after punctuated equilibria. With topological chaos for complexity the importance of the edge of chaos is stressed.

1. INTRODUCTION

Increased complexity through evolution is believed to be visible in many biological systems, not only in the hierarchical organization in genotypes and phenotypes but also in animal behavior and communication. For example, a bird song increases its repertoire through evolution and developments.[1] In the bird song, two functions are believed to exist: one is defense of territory and the other is sexual attraction. A bird with a complex song (with many repertoires made from combinations of simple phrases) is stronger in defending its territory, as demonstrated by Krebs with the help of loud-speaker experiments.[1,9]

There are reports of birds that try to imitate each other's song to defend their territory.[1] Inspired by these observations, we propose an imitation game of birds with their songs, used for territorial defense. A complex song may not be imitated easily and may be powerful in territorial defense. Although here we do not claim strongly the plausibility of this hypothesis, evolution in an imitation game is interesting as a novel evolution game[12] and in the more general context of the evolution of communication or mimicry.[13] Here we consider an abstract game for imitation, where an artificial "bird" wins the game when it can imitate the other's song better than the other player or bird. If a song is simple, it may be imitated easily by others, and we may expect evolution towards a complex song.

The model we propose here has a sound basis in nonlinear dynamics. It adopts a nonlinear mapping as the generator of songs. Our birds thus can produce songs possessing the complexity of real numbers, which is one of the advantages of our model.

Another motivation of this chapter is the introduction of a simple model that realizes "evolution to the edge of chaos." Increased complexity at the edge of chaos has been discussed for cellular automata,[10] for Boolean networks,[8] and for coupled map lattices.[6] It has been suggested that in a system with mutually interacting units, this edge of chaos has potential advantages in evolution.[15] Although these studies are reasonable in their own contexts, there is no simple example that provides an evolution to the edge of chaos, in the sense of dynamical systems theory (note that chaos is defined only for dynamical systems with a continuous state and is not defined for a discrete-state system like cellular automata). Since we have adopted a nonlinear dynamical system as a song generator and an imitator, we can examine a song as it evolves towards the edge of chaos, which is a clear advantage of our model.

Indeed, we will find that the dynamics of the song time series evolves toward the edge of chaos through the imitation game. However, this edge state lies not at the onset of chaos, but at the edge between a window of a periodic cycle and chaos. The main difference is that chaotic orbits exist as a transient orbit in the windows. The importance of the existence of transient chaos will also be discussed.

2. MODELING

Our abstract model consists of the processes of song dynamics, imitation, game, and mutation.

A. SONG DYNAMICS

As a "song," we use a time series generated by a simple mapping $x_{n+1} = 1 - ax_n^2$, the logistic map. (This map is equivalent to the logistic map of the form $z_{n+1} = bz_n(1-z_n)$, by the simple transformation of variables $x_n = \frac{4}{b-2}(z_n+0.5), a = \frac{b(b-2)}{4}$. See any textbook on chaos for the bifurcation diagram of the logistic map.)

The attractor of the map shows a bifurcation sequence from a fixed point to cycles with periods $2, 4, 8, \ldots$, and to chaos as parameter a is increased.[2,11] Parameter a is assumed to be different for each individual "bird." We choose this dynamics since it is investigated in detail as the simplest generic model for bifurcation and chaos.

B. IMITATION

Each bird player i chooses an initial condition so that, by its own dynamics $x_{n+1}(i) = f_i(x) = 1 - a(i)x_n(i)^2$, the time series can imitate the song of the other player. Here we use the following imitation process for simplicity. For given transient time steps T_{imi}, a player 1 modifies its dynamics with a feedback from the other player:

$$x_{n+1}(1) = f_1[(1-\epsilon)x_n(1) + \epsilon x_n(2)]. \tag{1}$$

By this dynamics, the player 1 adjusts its value $x_n(1)$ by referring to the other player's value $(x_n(2))$. Here ϵ is a coupling parameter for imitation process. After repeating this imitation process for T_{imi} steps, the player 1 uses its own dynamics $x_{n+1}(1) = f_1(x_n(1))$. In other words, the above process is used as a choice of initial condition for the imitation of the other player's dynamics. The coupling parameter ϵ also varies by players. (However, the distribution of the parameter ϵ seems "irrelevant," judging from our simulations. We will discuss this parameter in later sections.)

C. GAME

We adopt a two-person game between "birds." After completing the imitation process, we measure the Euclidian distance $D(1,2) = \sum_{m=1}^{T} |x_m(1) - x_m(2)|^2$ over certain time steps T. By changing the roles of players 1 and 2, we measure $D(2,1)$. If $D(1,2) < D(2,1)$, player 1 imitates better than player 2 and thus is the winner of the game, and vice versa.

TOPOLOGY FOR THE PLAYERS. Here all "birds" are assumed to play the imitation game against all others with equal probability, although some simulations use a two-dimensional lattice with nearest-neighbor play and these are briefly discussed later.

After each game, the winner gets a point W, while a loser gets L ($W > L$). (Both get $(W+L)/2$ in the case of draw.) After iterating a large number of games, the population is updated proportionally to the score. This population update corresponds to reproduction, with survival of the fitter.

D. MUTATION

The parameters a ($0 < a < 2$) and ϵ ($0 < \epsilon < 1$) can vary from individual to individual, and can be changed by mutation.[5] In the population update stage, mutational errors are introduced in to the parameters: The parameters a and ϵ are changed to $a + \delta$ and $\epsilon + \delta'$, where variables δ and δ' are random numbers chosen from a suitable distribution (we use a homogeneous random distribution over $[-\mu, \mu]$ or Lorenzian distribution ($P(\delta) = 1/\{\mu(1 + (\delta/\mu)^2)\}$). The former choice inhibits a large jump of parameters, and often the parameter values are trapped at intermediate values, whereas the latter choice is often useful, since it can provide a larger variety of species.

3. RESULTS OF SIMULATIONS

First of all, we start with a game without population dynamics. The score table of the game is plotted in Figure 1. We note that the outcome of the game can depend on the initial choice of $x_n(i)$ for the players. We have computed the score of player 1 with parameter $a(1)$ against the player with parameter $a(2)$, averaged from 100 initial values chosen randomly. In Figure 1 we plot the averaged score of the player with $a(1)$ against the player with $a(2)$.

Although the score table is rather complicated, we can see some strong bands of parameters against a wide range of parameters. Generally these parameters lie around the bifurcation points: bifurcation from a fixed point to period 2, from period 2 to period 4, etc. In addition to these period-doubling bifurcation points, the borderline between chaos and windows for stable cycles is stronger, in the high nonlinearity regime with topological chaos ($a > 1.4011\ldots$).

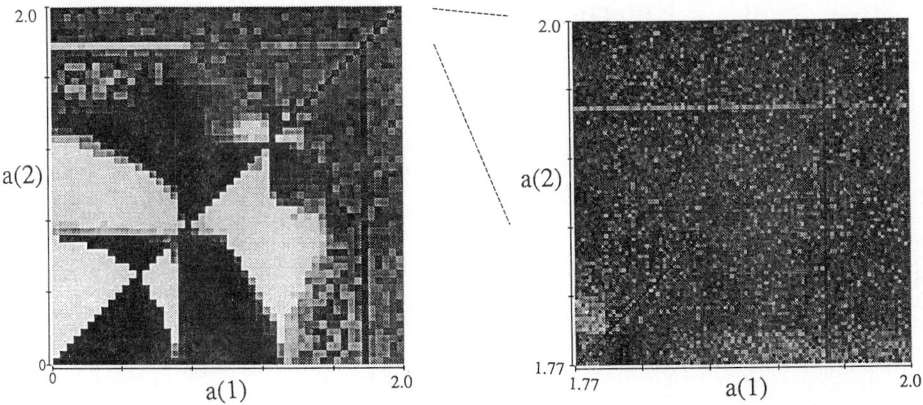

FIGURE 1 Score table of the imitation game: The score of the game between two players with parameters $a(1)$ and $a(2)$ is plotted with the help of a gray scale at the site corresponding to $(a(1), a(2))$. We have computed the score of player 1 with parameter $a(1)$ against the player with parameter $a(2)$, averaged from 100 initial values chosen randomly. Blank corresponds to 100 wins of the player with $a(2)$, while black corresponds to 100 wins of the player with $a(1)$. The darkness of the site $(a(1), a(2))$ increases as the winning ratio of the player with $a(1)$ increases. The parameter a is changed by an increment of 0.04 in the left table and 0.003 in the right one. $T_{imi} = 30$ and $T = 16$.

With the inclusion of population dynamics, the score of each player can depend on the population at the moment; we do not have a fixed fitness landscape. To see a fitness tendency, we sample the scores for each parameter range over long time steps, taking a high mutation regime where many players of different parameters exist. The average score is shown in Figure 2 with the use of bins of 0.001. The ruggedness of the landscapes is clearly seen. For example, the peaks of hills are seen at 1.25, 1.38, 1.401, 1.625, 1.77, and 1.94. They correspond to the bifurcation points from periods 2 to 4, and periods 4 to 8, the accumulation point of period doubling bifurcations, and the edges of windows of periods 5, 3, and 4, respectively.

The importance of the (schematic) rugged landscape is discussed in various biological contexts. We have to note the following points in the present example. First of all, the fitness landscape is not given in advance, in contrast with many theoretical models.[7] The rugged landscape is *emergent* through evolution. The landscape depends strongly on the population distribution of the moment. Second, the figure of the landscape is not schematic, in contrast with other landscapes, e.g., in spin glass models.[14] Indeed, the horizontal axis of Figure 2 is just the parameter of the logistic map, while the axis for the spin-glass-type rugged landscape is usually schematic.

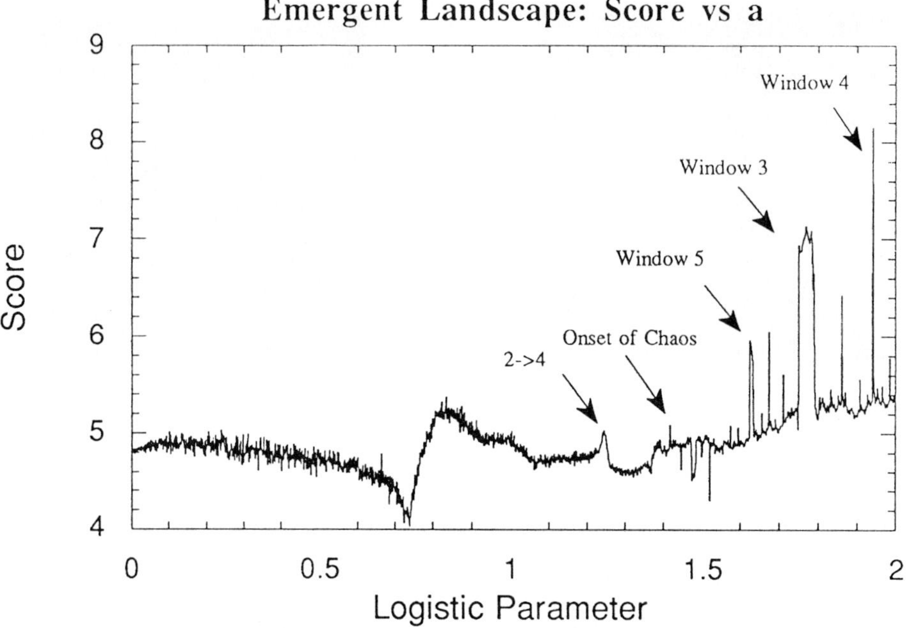

FIGURE 2 Emergent landscape: Average score for the players with parameters within $[a_i, a_i + \Delta]$ is plotted for $a_i = -1 + i \times \Delta$, with the bin size $\Delta = 0.001$. For Figures 2–4, we have adopted $W = 10, L = 1, T_{imi} = 255$, and $T = 32$. For mutational errors, random numbers δ are chosen from the Lorenzian distribution $P(\delta) = 1/\{\mu(1 + (\delta/\mu)^2)\}$, to avoid trapping at intermediate parameters. Simulation is carried out with the mutation rate $\mu = 0.1$, starting from the initial parameters $a = 0.6$ and $\epsilon = .1$. Sampled for time steps from 1000 to 1500, over all players (who always number 200).

Temporal evolution of the average of parameter a over all players is plotted in Figure 3. Plateaus corresponding to the hills in Figure 2 are observed successively, which provides an explicit example for punctuated equilibrium.[3] At the temporal domain with a plateau, the parameter values of "birds" are concentrated on the plateau value. Each plateau corresponds to the period-doubling bifurcation or to the edge of periodic windows among chaotic states. For example, the plateaus in Figure 2 are the superstable point of period 2 orbit ($a = 1$, where the derivative $f'(x_1)f'(x_2)$ changes its sign),[11,2] the bifurcation point to period 4, and then to period 8, and the edge of the period 3 window, and finally the edge of the period 4 window. As the mutation rate is decreased, plateaus have longer time intervals, and many steps corresponding to finer windows (and the accumulation point of period doublings) are observed.

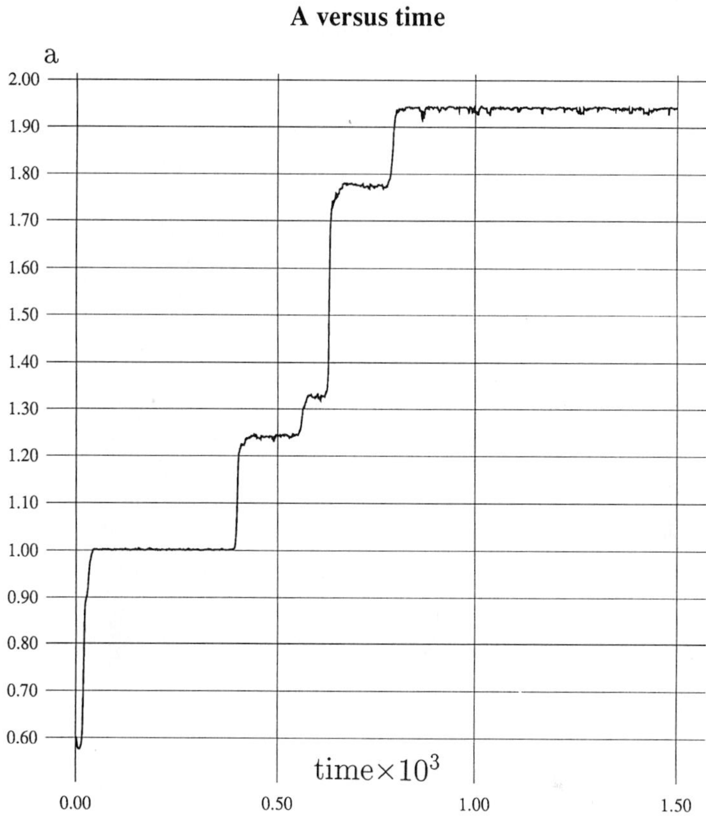

FIGURE 3 Temporal change of the average parameter a: the simulation is carried out with the mutation rate $\mu = 0.0005$, starting from the initial parameter $a = 0.6$ and $\epsilon = .1$. The average of the parameters a over all players are plotted with time. The total population is fixed at $N = 200$.

Evolution leads our system to the edge of chaos, i.e., the borderline between chaos and a window. To see clearly "the edge of chaos," we have measured the Lyapunov exponent λ. The Lyapunov exponent is a characteristic of asymptotic orbital instability, given by the small growth rate between two close orbits. The exponent is positive for a chaotic orbit, whose magnitude characterizes the strength of chaos, while it is negative for a stable periodic orbit. Thus a border between chaos and a periodic state corresponds to $\lambda = 0$. In Figure 4 we have plotted the birds' score as a function of the Lyapunov exponent λ, sampled for 200 time steps at the final stage of evolution. As shown, the score has a broad peak around $\lambda = 0$.

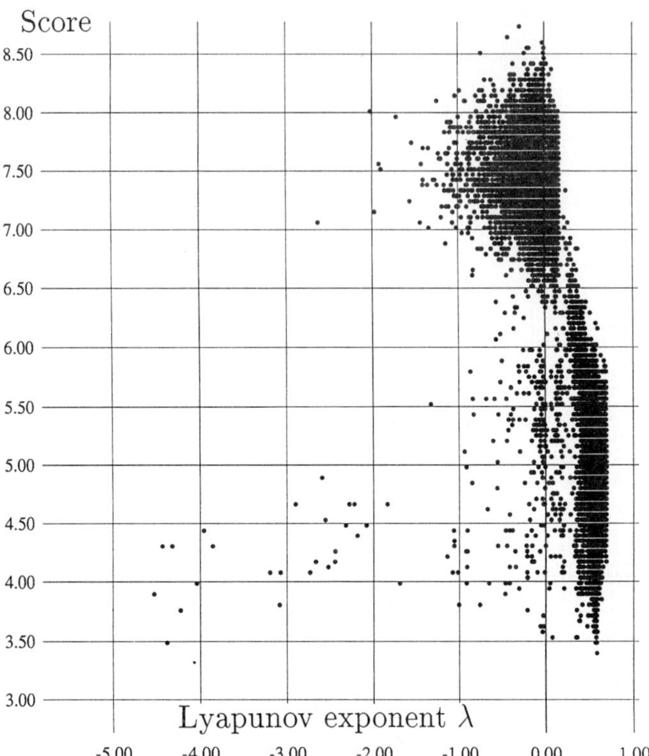

FIGURE 4 Lyapunov exponent versus score: the simulation is carried out with the mutation rate $\mu = 0.001$, starting from the initial parameters $a = 0.9$ and $\epsilon = .1$. Lyapunov exponents of the song dynamics vs. the score against all players at the moment are plotted. Sampled for time steps from 600 to 800 over all players (who always number 200), where the parameter value a is concentrated around 1.94.

Thus we have observed the evolution of a song to increase its complexity in our imitation game. The dynamics of the song time series evolves towards the edge of chaos. Although there is no "strongest" parameter in a two-player game, the borderlines between periodic windows and chaos are selected in the course of evolution. The song at the edge ($a \approx 1.94$) of the period 4 window is strong and robust.

The robustness of the parameter depends on the mutation rate. In the landscape (Figure 2), another important quantifier besides its height is the width of a hill. For example, the hill around $a \approx 1.94$ is higher than the hill around $a \approx 1.77$, but the latter hill is much wider. A wider hill is more robust against the variation of the parameter and is more advantageous in a higher mutation-rate regime. Indeed, in a simulation with a higher mutation rate, the population is concentrated around $a \approx 1.77$. The width of a hill is governed by the width of a window: thus the edge of a wider window is robust at a higher mutation rate.

We have made a large number of simulations, by changing parameters and initial conditions. Our conclusions are invariant against these changes of parameters, unless the mutation rate is too small for us to observe the successive changes within our simulation time steps (usually less than 10^5 steps).

4. MODIFICATIONS AND EXTENSIONS

There can be many generalizations to our model. Here we will briefly discuss some of these, although the results are still rather preliminary.

1. Change of the topology: Instead of the game of all-to-all, it may be interesting to study a game on a two-dimensional lattice with nearest-neighbor interaction. Most of our results in Section 3 are reproduced, although spatial differentiation and clustering are found in the lattice version for a suitable mutation rate. The coexistence, for example, of the species near the period 3 and period 4 windows is seen in this case. Needless to say, "fluctuation" is much bigger in this model than that presented in Section 3. Consequentially, some plateaus observed previously are lost depending on the value of mutational errors.
2. Symbolization of a song: Instead of judging the imitation by the Euclidian distance, we could check it after symbolizing the dynamics to few digits (here we choose 2). It is well known that the time series of the logistic map can be traced by the binary sequence (left, i.e., $x < 0$; right, i.e., $x > 0$). We check if the sign of x_m is identical between players 1 and 2, at each time step, and count the number of time steps where the symbols are not matched. In other words, we use the following function $I(1,2)$ instead of $D(1,2)$; $I(1,2) = \sum_{m=1}^{T}$ $\text{sign}(x_m(1), x_m(2))$ with the notation $\text{sign}(a,b) = 1$ if $ab < 0$ and 0 otherwise. Adopting this criterion, we have performed simulations of our game on a two-dimensional lattice. We have found that the logistic parameter a of winners often lie at a point where the binary sequence changes its pattern, the right-left sequence. Here, the temporal behavior of evolution is more complicated than that in Section 3. A punctuated equilibrium around $a = 1.77$ can be established for some intervals; however, this does not continue over a long time scale, and switchs occur to different transient states. Often species with distant parameters can coexist. Generally speaking, complexity and diversity are enhanced in the present case with symbolization. This result may give some hint to the origin of complexity and diversity in communication and language.

3. Dual dynamics for song and imitation: What happens if birds assume different parameters in singing and in imitating? This would be a reasonable question, since there is no primary reason that the two processes are governed by identical dynamics. From some simulations, the parameter for the singing process again shows punctuated equilibria among "window" values, before settling down to the border between chaos and the period 4 window value ($a \approx 1.94$). The parameter for the imitation process, on the other hand, is often settled around $a = 0.75$, the bifurcation point from a fixed point to period 2.
   If we adopt the symbolization dynamics as in (2), the dynamics is again more complicated. Again, the parameter for the imitation process shows punctuated equilibria among "window" values. However, the parameter does not settle to a window value (e.g., at $a \approx 1.94$), but wanders upwards and downwards forever. On the other hand, the parameter for singing often increases up to the value for fully developed chaos, although there is a small amplitude variation, in synchronization with the change of the parameter for imitation.
4. Changes in the choice of mapping: This would be the most immediate idea for most of us. It would be far more interesting to include evolution, not only within the logistic maps with a variable parameter, but also towards different types of maps.

5. DISCUSSIONS

In this chapter we have presented a "minimal" model showing evolution to the edge of chaos and punctuated equilibrium. Indeed, it gives the first explicit example for this type of evolution in the sense of dynamical systems, as is demonstrated by evolution to a state with a Lyapunov exponent close to zero. In addition, we note that our edge state lies between a window and chaos. A logistic map has topological chaos at the window and can show chaotic transients before the dynamics settles down to a stable cycle. The existence of transient chaos is useful in the imitation of different dynamics. In the window regime, the logistic map includes a variety of transient orbits, some of which are close to a periodic orbit, while others are chaotic. A window at a higher nonlinearity regime includes a variety of unstable cycles, as coded by Sharkovskii's ordering,[2,11] and can provide a larger variety of dynamics as transients. This is the reason why the edge of the window is strong in our imitation game. The above speculation suggests the importance of transient chaos, besides the edge of it, for the adaptation to a wide range of external dynamics.

Our imitation process is based on the synchronization of a player's dynamics to the other's song. This synchronization process may not be physiologically unrealistic. Indeed, there have been recent extensive studies on the entrainment of oscillation. So far the studies on synchronization is mostly focused on the visual cortex,[4] but there is no reason to suspect the importance of synchronization in auditory systems.

Evolution of complexity to escape from imitation may play a key role in many fields. Our imitation game provides one route to the evolution to complexity. We may expect that this route can be seen in some examples of biological evolution, besides our original motivation of a bird song. One is the evolution of a communication code only within a given group: "secret code." When there are two groups, it is often important to send some messages only within the same group, so that they are not encoded by the other. Another example is the Batesian mimicry.[12] Here again, we have two groups. One of the groups can survive better by imitating the pattern of the other's group, while the other group's advantage in survival is lost if it is not distinguished well from the other. Complexity in the patterns of butterflies in the mimicry relationship, for example, may be increased through the imitation game process. Our simulation gives the first conceptual model for the evolution to complexity by imitation, while modeling with two distinct species with different roles will be necessary in the future.

ACKNOWLEDGMENTS

The authors would like to thank T. Ikegami, K. Tokita, and S. Adachi for useful discussions. The work is partially supported by Grant-in-Aids for Scientific Research from the Ministry of Education, Science, and Culture of Japan.

REFERENCES

1. Catchpole, C. K. *Vocal Communication in Birds* [city]: Edward Arnold, 1979.
2. Collet, P., and J. P. Eckmann. *Iterated Maps on the Intervals as Dynamical Systems.* Birkhasuer: Boston-Basel-Stuttgart, 1980.
3. Eldredge, N., and S. J. Gould. In *Models in Paleobiology*, edited by T. J. M. Schopf. New York: Freeman, 1972.
4. Gray, C. M., P. Koenig, A. K. Engel, and W. Singer. *Nature* **338** (1989): 334.
5. Holland, J. "Escaping Brittleness." In *Machine Learning II*, edited by R. S. Mishalski, J. G. Carbonell, and T. M. Miechell. Los Altos, CA: Morgan-Kaufmann, 1986.
6. Kaneko, K. *Physica* **34D** (1989): 1.
7. Kauffmann, S. A., and S. Levin. *J. Theor. Biol.* **128** (1987): 11.
8. Kauffmann, S. A., and S. Johnson. *J. Theor. Biol.* **149** (1991): 467.
9. Krebs J. R., R. Ashcroft, and M. Weber. *Nature* **271** (1978): 539–542.

10. Langton, C. *Physica* **42D** (1990): 12.
11. May, R. *Nature* **26** (1976): 459.
12. Maynard Smith, J. *Evolution and the Theory of Games.* Cambridge, MA: Cambridge University Press, 1982
13. Maynard Smith, J. *Evolutionary Genetics.* Oxford: Oxford University Press, 1989.
14. Mezard, M., G. Parisi, and M. A. Virasoro, eds. *Spin Glass Theory and Beyond.* Singapore: World Scientific, 1988.
15. Packard, N. H. In *Dynamic Patterns in Complex Systems*, edited by J. Kelso, A. J. Mandell, M. F. Shlesinger, 293–301. Singapore: World Scientific, 1988.

Peter J. Angeline and Jordan B. Pollack
Laboratory for Artificial Intelligence Research, Computer and Information Science Department, The Ohio State University, Columbus, Ohio 43210
e-mail: pja@cis.ohio-state.edu and pollack@cis.ohio-state.edu

Coevolving High-Level Representations

Several evolutionary simulations allow for a dynamic resizing of the genotype. This is an important alternative to constraining the genotype's maximum size and complexity. In this paper, we add an additional dynamic to simulated evolution with the description of a genetic algorithm that coevolves its representation language with the genotypes. We introduce two mutation operators that permit the acquisition of modules from the genotypes during evolution. These modules form an increasingly high-level representation language specific to the developmental environment. Experimental results illustrating interesting properties of the acquired modules and the evolved languages are provided.

INTRODUCTION

A central theme of artificial life is to construct artifacts that approach the complexity of biological systems. To accomplish this, the benefits and limitations of current methods and tools must be considered. The representation employed in a genetic algorithm, for example, must permit an appropriate level of expression in order for the genotypes to evolve.[18,22] Artificial life researchers favor genetic algorithms both for their proven ability in a wide range of environments and for their obvious association with natural evolution. However, in most genetic algorithms, the representation's length is fixed *a priori*, placing an ad hoc restriction on the developing genotypes.

Other genetic algorithms emphasize a dynamic approach to developing genotypes. Their representations expand and contract according to environmental requirements providing a flexible alternative when simulating evolution. In this paper, such systems are described as *dynamic genetic algorithms.* While these representations allow the size of the genotypes to vary with the environment, their expressiveness remains static.

In this paper, a dynamic genetic algorithm is described that not only alters a genotype's size but also coevolves a high-level representational language tailored to the environment. This is accomplished with the addition of two specialized genetic operators that create stable modules from portions of the developing genotypes. Each created module serves as an abstract addition to the representational language reflecting some specifics of the environment. We demonstrate this technique and illustrate some of the properties of the evolved modules. The following section provides a review of genetic algorithm theory and related genetic algorithm implementations.

SIMPLE AND DYNAMIC GENETIC ALGORITHMS

A genetic algorithm[14,17] is a search method analogous to natural selection that is surprisingly adept in ill-formed environments. Genetic algorithms evaluate a population of genotypes with respect to a particular environment. The environment includes a *fitness function* that rates the genotype's viability. Genotypes reproduce proportionally to their relative fitness using a variety of genetic operators. One operator, termed *crossover*, uses the recombination of two parents to construct novel genotypes. The *mutation* operator creates new genotypes from a single parent with a probabilistic alteration.

To analyze how genetic algorithms process genotypes, Holland defines a class of genotypes that share specific attributes to be a *schema.*[17] A schema's *defining length* is the number of places where a crossover operator when applied to any member of the schema creates a genotype that is no longer in the schema. The *order* of a schema is similarly the number of positions where the application of a mutation operator removes a member from the schema. Holland's *schema theorem*[17] shows that a schema with a consistently above-average fitness, short defining length, and low-order propagates through the population exponentially fast under ideal conditions. The *building block hypothesis*[14] suggests that the power of genetic algorithms originates in the recombination of good schemata into better schemata. For a complete description of the schema theorem and the building block hypothesis, see Goldberg.[14]

SIMPLE GENETIC ALGORITHMS

Simple genetic algorithms use a fixed-length string representation. Two tacit assumptions that accompany simple genetic algorithms limit their evolutionary abilities. First, the representation's fixed length places a strict upper-limit on the number of distinct genotypes and hence the number of phenotypes. A second, more subtle limitation occurs when converting a genotype into its associated phenotype. Typically, the individual positions of the genotype's string are interpreted as distinct features to be included in the phenotype. This *positional encoding* of the genotype reduces its phenotype to a simple concatenation of features, reducing the potential complexity available to simulated evolution.

The limitations of simple genetic algorithms can be overcome by a variety of techniques. Jefferson et al.[18] allow limited variability by defining an extremely long binary string as the genotype that is converted into either a finite-state automata (FSA) or a neural network as the phenotype. The content of the fixed-length genotype dictates the actual size of the phenotype. While this interesting idea holds promise, it still dictates a maximum size for both the genotype and the phenotype. Others[5,7] have opted to increase the complexity of the conversion process to construct more interesting phenotypes. But, this approach begs the question of evolving a complex artifact since sufficient representational complexity is provided *a priori* in the conversion function.

DYNAMIC GENETIC ALGORITHMS

In dynamic genetic algorithms, the representation of the genotype is a variable size structure, typically a tree or graph. The requirements of the environment dictate the final size and shape of the genotypes. Various dynamic structures have been used to represent genotypes in these genetic algorithms including the linear "assembly code" programs of Tierra,[25] the hierarchical LISP expression trees of the genetic programming paradigm,[20,21] and the FSA representations of early evolutionary programming.[10,11,12] In each of these systems, the genotype is expressed as an executable structure with a language of primitives suited for the environment. The phenotype is then the genotype's simulated *behavior*. Typically, the genotype's behavior is of interest in these systems instead of the specification of the behavior altered by the genetic operators. For clarity, we refer to all such executable structures as "programs."

Given that a dynamic genetic algorithm relies on the evaluation of a program for the expression of the phenotype, the design of the primitive language is crucial. First, the syntax of the language should be chosen to minimize the number of programs that are syntactically invalid.[18,25] A good primitive language reduces the probability of constructing nonviable offspring during evolution. Second, the relationship between the primitive language's semantics and the environment determines how difficult a desired phenotype is to evolve.[18,19] For instance, given a high-level language where the individual primitives are tailored to the environment, small programs exhibit sufficient behaviors. With a lower level language, the minimal size of a program that exhibits an equivalent behavior is proportionally larger and thus much more difficult to evolve. Jefferson et al. provide some excellent additional guidelines to consider when evolving dynamic representations.[18]

While the designable aspect to the primitive languages in dynamic genetic algorithms is convenient when constructing applications, it leaves dynamic genetic algorithms susceptible to the same over-design problems as in simple genetic algorithms. A researcher can easily avoid the problem of extracting the complexity from an environment entirely by providing a suitably designed representational language.

THE GENETIC LIBRARY BUILDER (GLIB)

The Genetic Library Builder (GLiB) is a dynamic genetic algorithm modeled after Koza's genetic programming paradigm (GPP).[20,21] Primitive languages in both GPP and GLiB employ a LISP expression tree syntax to represent genotypes in the population. Both also use a crossover operator that exchanges randomly selected subtrees between parents to create the novel genotypes. The principal difference between GPP and GLiB is the addition of two novel mutation operators. First, the *compression* operator extracts environment-specific additions to the primitive language from the genetic material of the population. Each compressed portion of a genotype, called a *module,* increases the expressiveness of the language and decreases the average size of the genotypes in the population. The usage of the module by subsequent generations provides a measure of a module's worth. A second operator, called expansion, replaces a compressed module by its original definition. As with all mutation operators, compression and expansion alter the structure of the genotype. Unlike other mutation operators, they never directly or indirectly alter the phenotype.

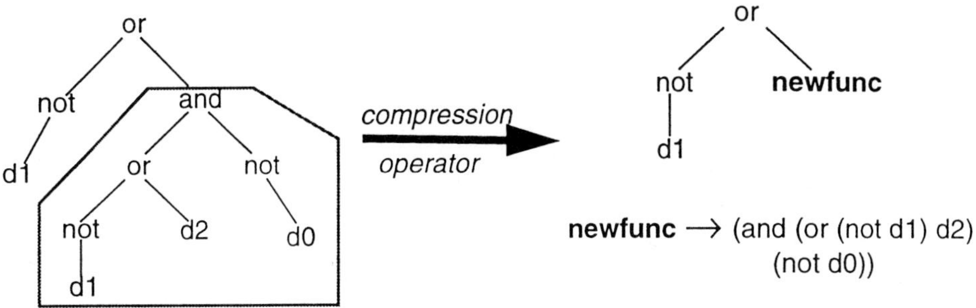

FIGURE 1 Creation of a compressed module from a randomly selected subtree of a genotype. **Newfunc's** definition is added to the library.

CREATION OF ENVIRONMENT-SPECIFIC MODULES

A compression operation begins with the selection of a random position in the genotype. This position identifies the root of the subtree to be compressed. When the module is defined, the subtree is extracted from the genotype and replaced with a unique symbolic name. GLiB binds the extracted subtree to the module's symbolic name by entering the pair into the "genetic library." This library is the collection of all created modules and serves only as a symbol reference for GLiB. During either a genotype's execution or the expansion of a compressed module, GLiB retrieves the definitions of symbols from the genetic library.

The *depth compression* method of module creation extracts the subtree beginning at the randomly selected position of the program and clips off any branch exceeding a maximum depth. The maximum depth for each new module is a random number selected uniformly from a user-defined range. Figure 1 shows the result of a compression when each branch of the subtree is within the maximum depth. Here, compression removes the entire subtree from the genotype, assigns it a unique module name, and defines it as a new LISP function with no parameters. A call to the newly defined module replaces the original subtree as shown in Figure 1. When GLiB evaluates the genotype's fitness, it retrieves the compressed module's definition instead of directly executing the original subtree.

When one or more subtree branches exceed the maximum depth, the defined module has a very different definition. Rather than extracting the entire subtree, only the portions within the selected depth appear in the new module's definition. The branches exceeding the depth are replaced with unique variable names. GLiB then defines the new module with the variables defined as parameters to the module.

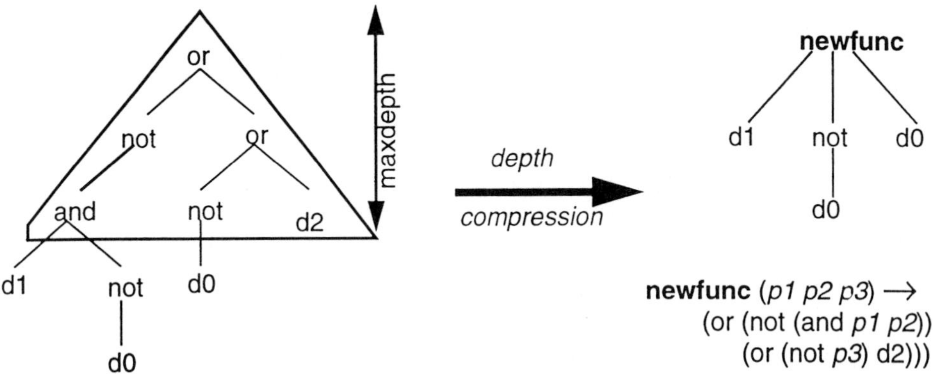

FIGURE 2 Depth compression applied to a subtree larger than the maximum depth. The module's definition has parameters as a result.

When GLiB replaces the subtree in the genotype with the call to the module, the sections of the original subtree below the selected depth become the parameter bindings. Figure 2 shows this instance of module creation.

Depth compression is the simplest method GLiB currently employs and is the form of compression used in the experiments described in the next section. Other compression methods based on more involved computations are also under investigation. Future research will explore their associated benefits and weaknesses.

GLiB's compression operator introduces a new dynamic into the evolutionary process: the evolution of the representational language. Each compressed module extends the representational language of the genotypes, like a library of functions extends a particular programming language. Since the new modules will be constructed from the primitives and modules already in the language, they will tend to be at a higher level of abstraction. The genotypes and the representational language coevolve until the language is sufficiently expressive to succinctly encode appropriate behaviors.

EXPANSION OF COMPRESSED MODULES

One drawback to the compression operator is that each compression removes genetic material that otherwise would be available to construct novel genotypes in subsequent generations. After only a few generations of moderately applying the compression operator, the available genetic material in the population is insufficient to significantly improve the performance of the population. The loss of genetic diversity in a population leading to premature convergence to a suboptimal solution is a well-known problem in genetic algorithms research.[8,9,14] Typical solutions include altering the scaling of the fitness values or altering the frequency of crossovers and mutations.[14] Neither of these solutions are appropriate in this situation.

To offset the loss of diversity in the population due to the application of the compression operator, GLiB provides an additional mutation operator. The *expansion* mutation randomly selects a compressed module from the genotype and expands it back into the original subtree, reintroducing the previously removed genetic material. The expansion process is exactly the reverse of compression. A copy of the module's definition from the genetic library is expanded by replacing all occurrences of parameters with the bindings specified in the genotype's call to the module. GLiB then splices the expanded subtree back into the genotype. One could literally reverse the directions of the arrows in Figures 1 and 2 as illustrations.

The complementary nature of the compression and expansion operators provides an iterative refinement mechanism for the evolving modules. The random selection of a subtree for compression provides no guarantee that a compressed module is beneficial to the developing genotypes in its initial form. It could be that the acquired module contains only a portion of a useful subtree, a useful subtree along with some extraneous additions, or simply nothing of import. The original subtree's replacement provides the chance to capture a better version of the module in a later compression.

MODULE EVALUATION BY THE ENVIRONMENT

Randomly compressing subtrees from the genotypes does not guarantee useful additions to the representation language that benefit the evolutionary process. It is not enough to define modules without an appropriate mechanism to judge and discard them if necessary. One appropriate criterion for judging an evolved module is to examine the number of calls made to it in the subsequent generations.[3] If the population uses a module frequently, then it must have provided some consistent beneficial behavior in previous generations. When a module provides such a consistent advantage, we say it is *evolutionarily viable*, meaning, literally, the module can survive the evolutionary process. Given the criterion of evolutionary viability as an appropriate evaluation metric, GLiB must copy beneficial modules into the offspring of the next generation while inhibiting the proliferation of inconsequential modules. Goldberg's "enlightening" guidelines for genetic algorithms design cautions that one should never be too clever when designing genetic algorithms as a "frontal assault" usually defeats the inherent nonlinear interactions.[15]

Appropriately enough, the evolutionary process which propagates and inhibits the reproduction of the genotypes evaluates the worth of every compressed module without any additional intervention. The primary effect of the compression operator is to protect potentially useful schema within the modules. Because the compressed modules are examples of schemata, the genetic algorithm evaluates them automatically. Consider that modules appear in only a single genotype when created. If this genotype has an above average phenotype, then it will be used to create several offspring in the next generation. As stated above, the schema theorem suggests that a schema appearing inconsistently above average genotypes with a short defining length and low order will proliferate exponentially through the population under ideal conditions. By definition, a compressed module has a defining length of zero and is of order one, making it a perfect candidate for quick proliferation. After several reproductive cycles, a significant portion of the population will rely on a beneficial module and its number of calls per generation will be high. Likewise, if a module consistently appears in below-average genotypes, it is removed from the population, resulting in a proportional decrease in its number of calls. When no member of the population contains a call to a particular module, it is no longer in the representational language.

THE EMERGENCE OF ENVIRONMENT-SPECIFIC MODULES

In this section, we present experiments illustrating some of the properties of GLiB's evolved programs and their associated modules. In the first set of experiments, designed to provide initial insights into the nature of an evolved high-level representation, GLiB created programs to solve the familiar Tower of Hanoi problem. These results permit analysis of the properties of evolved modularity. The second experiment takes a more ambitious track. A general primitive language is used to evolve programs to play and beat an automated Tic Tac Toe opponent. The experiments show GLiB preserves important schemata in the created modules while highlighting some additional unexpected features of evolved modular programs.

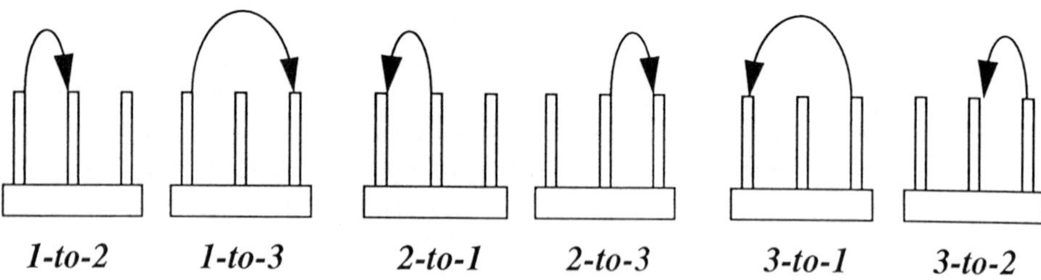

FIGURE 3 Primitive language for Tower of Hanoi experiments. Each command moves a single disk between the specified pegs.

TOWER OF HANOI EXPERIMENTS

Our initial experiments with GLiB evolved solutions for the classic Tower of Hanoi problem. This puzzle consists of three pegs, labeled by position, and several uniquely sized disks. The initial state has all disks on the first peg such that no disk is on top of a smaller disk. The goal is to reconstruct the tower of the initial configuration on the third peg by moving the disks, one at a time, such that a larger disk never rests on a smaller disk.

The primitive language for the genotypes included the six possible moves from peg to peg shown in Figure 3. Each primitive moves the top disk from one peg and places it on the other peg. A solution then is a sequence of single disk moves in an order that reconstructs the tower on the third peg. Additionally, we assumed four disks in the initial tower and permitted a maximum of 32 moves to complete the task. This does not imply a limit of the 32 steps in a program, only that the evaluation function stopped executing a program after it had executed 32 primitive operations. The fitness function gave points only for the set of disks an evolved program correctly moved to the third peg. The points awarded for a disk were proportional to the size of the disk. Illegal moves, i.e., ones that would leave a larger disk on top of a smaller disk, were not executed. As a penalty, programs lost one time step for every illegal move executed.

Several trials of this experiment were executed with 10 percent of the population undergoing compressions and expansions each generation. The population size was set to 1000 and 50 generations of decendents were evolved. In each run, GLiB performed as GPP has on similar problems.

A consistent property appearing in the modules evolved during these runs is the ability to improve upon a variety of intermediate problem configurations. As an example, the steps preserved within one evolved module are pictured in Figure 4. Note that step 4 moves no disk in the configuration shown. While this module encodes neither an optimal nor a completely legal subsequence from certain configurations, it is very versatile given the penalty for using illegal or unnecessary commands is so mild, i.e., only the loss of a time step. For instance, given the 27 possible intermediate configurations of the problem with the largest disk still on the first peg, the evolved module shown in Figure 4 moves the largest disk to the third peg in 4 of them. This module also does well from many other configurations.

While broad applicability is clearly an advantage in an evolutionary setting, it is unclear whether or not the flexibility found in the evolved modules is a direct reflection of developmental usage or simply a feature that guarantees survival. In either case, it is evident that the modules capture some of the limitations and constraints of the environment.

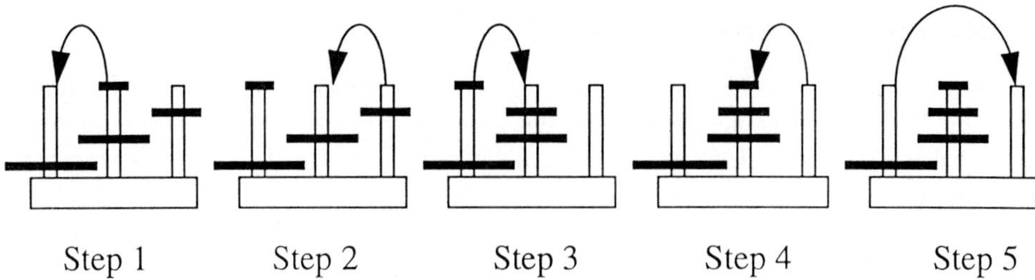

FIGURE 4 Sequence of moves implemented by an evolved module. The configuration of the disks in Step 1 is the configuration at which the module is called in the evolved program.

EXPERIMENTS WITH TIC TAC TOE

The second set of experiments investigate the attributes of modular programs evolved to play Tic Tac Toe (TTT) against a player of our own design. Figure 5 shows the primitive language for the TTT genotypes. The primitives ***pos00*** to ***pos22*** are the data points representing the nine positions on the TTT board. For the remaining primitives, the return value is either one of these positions or NIL.[1] For instance, the binary operators ***and*** and ***or*** each take two arguments. When both arguments are non-NIL, and returns the second. If either argument is NIL, then it returns NIL. ***Or*** returns the first non-NIL argument and returns NIL otherwise. The ***play-at*** primitive takes a single argument. If the argument is a position and no player has placed a mark there, then the current player's mark is placed at that position and the turn is halted. Otherwise, ***play-at*** will return whatever it is passed. Finally, the operators ***mine***, ***yours***, and ***open*** take a position and return it when the mark in that position fits the test. Otherwise, they too return NIL.

This language is general enough to cover any number of games playable on a 3-by-3 board. Consequently, there is no guarantee that a random program in this language will observe the rules of TTT or even place a single mark on a TTT board. If the program does not make a valid move, then its turn is forfeited, providing a significant advantage for its opponent. We consider the rules of play to be an additional component of the environment and consequently should be induced by GLiB.

[1]NIL represents FALSE in LISP, GLiB's current implementation language.

pos00 .. pos22 - board positions
and - binary LISP "and"
or - binary LISP "or"
if - if <test> then <arg1> else <arg2>
open - returns <arg> if unplayed else NIL
mine - returns <arg> if player's else NIL
yours - returns <arg> if opponent's else NIL
play-at - places player's mark at <arg>

pos00	*pos01*	*pos02*
pos10	*pos11*	*pos12*
pos20	*pos21*	*pos22*

FIGURE 5 Primitives used to evolve modular programs to play Tic Tac Toe.

How well a genotype plays a pre-programmed "expert" TTT algorithm determines its viability. The expert is constructed such that it will not lose a game unless forked by an opponent. A *fork* is any board configuration where the addition of a single mark results in more than one possible winning play on the player's next turn. The expert also adapts over a series of consecutive games to individual opponents by selecting positions the opponent frequents when no better move is available. Adaptability of this sort increases the diversity of the expert's strategy and forces the evolving programs to be more robust. Given the generality of the primitives combined with the expert's level of play, this is a difficult environment for learning TTT.

The fitness function used in this experiment averages a genotype's score over a total of four games played against the expert. Individual moves by the genotype receive varying point values. First, because there is only a small possibility of a random program making a legal play, a program receives 1 point for every legal move it makes. It also receives 1 additional point if that move blocks the expert from winning on the next turn. If the game ends in a draw or a win, the scoring function increases the accumulated score by 4 or 12 points respectively. It is important to note that the score for a genotype is a lump sum and provides no indication of which actions are being rewarded. The same results should be achieved if only the final state of the board is scored.

In this experiment, the population contained 1000 programs and the environment consisted of the described expert and scoring method. GLiB applied the compression and expansion operators to 10 percent of the population each generation. All other parameters were as set in Koza's "ant" experiment.[20]

```
(FUNC26677
  (FUNC43031
    (PLAY-AT (PLAY-AT (MINE (MINE (MINE (FUNC42966))))))
    (IF (FUNC20328) (POS20) (POS20))
    (PLAY-AT
      (AND
        (FUNC46557
          (FUNC12959)
          (AND (POS12) (PLAY-AT (OR (POS22) (POS20))))
          (FUNC20917)
          (PLAY-AT
            (YOURS (YOURS (OR (PLAY-AT (OPENP (FUNC16153))) (POS21)))))
          (PLAY-AT
            (PLAY-AT
              (IF (FUNC20917)
                (PLAY-AT (OR (PLAY-AT (FUNC23500)) (FUNC20671)))
                (AND
                  (PLAY-AT (IF (FUNC4076) (MINE (POS01)) (POS10)))
                  (POS20)))))
          (FUNC36830))
        (AND (OR (POS12) (POS12)) (POS12))))
    (FUNC68808)
    (AND (PLAY-AT (FUNC70469 (POS00))) (POS22))))
```

FIGURE 6 Best evolved modular program from described experiment. Evolved modules are named FUNC####.

RESULTS AND DISCUSSION

The best evolved program after 200 generations, shown in Figure 6, had an average score of 16.5 points for the four games, a maximum depth of 13, and 15 evolved modules in its top level. As expected, expanding the definition of these modules back into their original subtrees revealed additional modules in their definitions. In all, this evolved program used a total of 43 distinct modules in 89 calls. The "unrolled" size of the program is 477 nodes with a maximum depth of 39. Two of the modules had a total of nine separate calls each.

In Figure 7 we show the first game played between the evolved program and the expert. There is an interesting point to be made about the apparent strategy of the evolved program. Notice that it establishes a fork on its third move (Figure 7(c)) but did not win the game until two turns later (Figure 7(e)). While this seems an odd strategy, recall that the evolved program gets points for every move it makes and

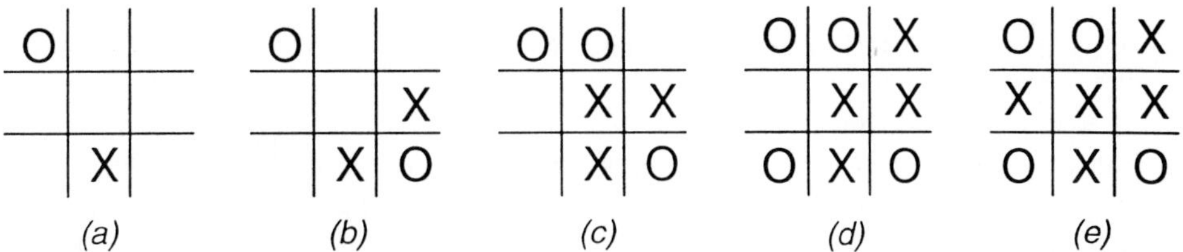

FIGURE 7 A sample game from the described run. The evolved program is "X" and is playing first. It sets up the first fork in (b) and completes it in (c). The second fork is completed in (d) and leads to a win in (e). The evolved program received a score of 20 points for this game.

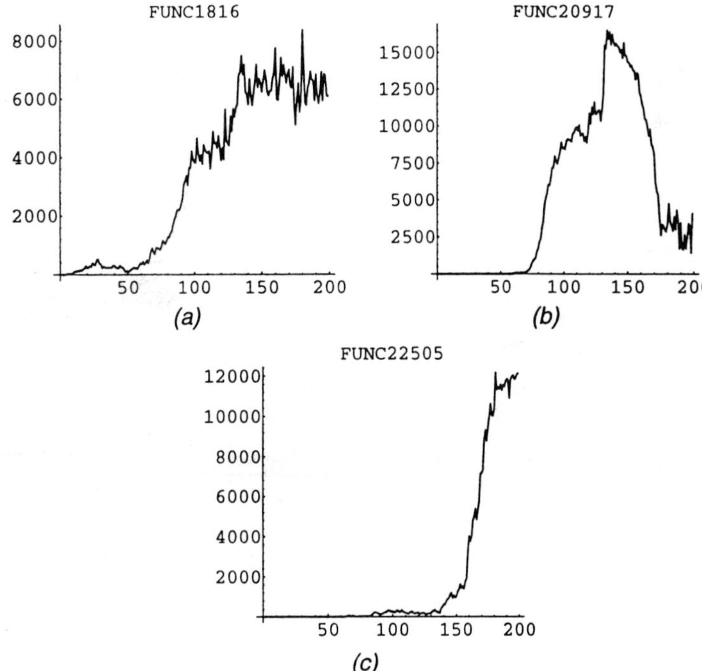

FIGURE 8 Graphs showing number of calls (y-axis) per generation (x-axis) for three evolved modules. Note the scales are different in each.

extra points when it blocks the expert. Its strategy is to maximize its total point score by forking the expert not once but *twice* in the same game. If the program had ended the game on its fourth move, it would have received 3 fewer points. By extending the game it increases its score without the possibility of losing. It remains an open research question whether environments can be designed that encourage good sportsmanship.

The analysis of the evolved modules is equally interesting. The compression operator created 16,852 modules during the experiment with only 257 in use during the final generation. In Figure 8 we show the number of calls per generation for three of the evolved modules. Each of these has a distinct period where its number of calls increases rapidly. In Figure 8(a), an initial increase occurs soon after the module is defined, showing it provided an immediate advantage to the population. This module's usage quickly rises again a short period later, showing its application to additional facets of the environment. In Figure 8(b) we show a module that was extremely useful shortly after its creation but whose use fell off dramatically toward the end, possibly due to a preferred module appearing in the population. Finally, In Figure 8(c) we show a module that was present for almost 100 generations before the population recognized its usefulness, resulting in its explosive proliferation. Such rapid changes are indicative of "phase transitions" which form the basis for inductions in dynamic learning systems.[3,24]

Discovering general properties of the modules constructed in the Tic Tac Toe experiment was not as simple as in the Tower of Hanoi experiments. Obviously, the broad application of the modules seen in the Tower of Hanoi experiment is apparent from the description of the evolved Tic Tac Toe program and the high usage of individual modules evident from the graphs of Figure 8. Unfortunately, more specific properties have been extremely difficult to identify. One observation is that the evolved modules do not employ the typical conceptual breakdown for playing Tic Tac Toe. For instance, we found no modules that test for either an opportunity to win on the current turn or the possibility of the opponent winning on the subsequent turn. Second, the program modularity that results is not amenable to normal analysis techniques. As expected from an evolutionary process, the usage and semantics of the evolved language is extremely non-standard. For instance, several modules encode numerous side effects in the test position of a conditional. In short, the resulting programs, while containing multiple modules, are not examples of modular programming but share more commonality with the *distributed representations* of procedural knowledge found in connectionist networks.[16]

CONCLUSIONS

It is apparent from Figure 8 and the statistics of the runs that GLiB was successful in capturing modules advantageous to the construction of complex programs and that reflect the idiosyncracies of the developmental environment. While the number of modules created is large, the computation expended to capture them is insignificant compared to the evolutionary search process. In addition, the emergence of a useful module reduces the size of the genotypes and, consequently, the number of available crossover and mutation points. This focuses the evolutionary process on improving the evolving programs at a higher level of abstraction rather than destroying previously discovered building blocks. GLiB's construction of useful high-level modules as a by-product of purely local interactions identifies yet another emergent property of genetic algorithms.[13]

The Tower of Hanoi experiments illustrate that the evolved modules are often applicable to multiple environmental situations given the penalty for such flexibility is not too extreme. This flexibility was also evident in the Tic Tac Toe experiments. Additionally, the modularity that resulted from the more complex Tic Tac Toe environment was significantly more difficult to analyze. This shows our technique is free from the extremely restricted form of programming humans practice. It also suggests a trade-off between complexity of behavior and interpretability of structure when evolving complex modular programs. David Fogel identifies this same trade-off for different reasons.[10,11]

Extracting subtrees from the programs and placing them into GLiB's genetic library is a convenience rather than a necessity for our technique. The genetic library lends nothing to the development process but is very beneficial in reducing the stored size of the developing genotypes. Exactly the same results would be achieved if the subtrees were not extracted from the genotypes but were still equally protected from alteration.

Although the techniques of compression and expansion encoded in GLiB were illustrated using only one type of dynamic genetic algorithm, they are universally applicable to any method employing an analogy to natural selection. For instance, simple genetic algorithms could designate protected positions for each individual with parents propagating their protected regions to their offspring. Evolutionary programming, which relies on mutation as its only method of constructing novel population members, could restrict changes to unsettled or unprofitable sections of the representation in a similar manner. Previous experiments have demonstrated that a simplified form of compression produces a speed-up in the acquisition of Finite State Machines by an evolutionary program.[2]

This technique also holds promise for removing the limitations of human designers from the creation of complex structures. We discuss elsewhere how GLiB's operators aid in the construction of arbitrarily complex programs.[3] Similarly, modular designs for artificial organisms in the spirit of Brooks[6] could be evolved rather than hand-constructed, possibly leading to wholly new design principles for complex artificial organisms. We are currently exploring this possibility by extending previous work to evolve modular neural networks for the control of artificial organisms.[1,26]

We feel that an evolutionary process necessarily requires the ability to evolve environment-specific abstractions in order to be open-ended or at the very least scalable. Patching up inadequate representations with complex interpretations dodges the issue, as does a representation language that assumes overly powerful primitives. GLiB's technique for coevolving environment specific representations with the population offers a third, universally applicable alternative that preserves the integrity of simulated evolution.

ACKNOWLEDGMENTS

We owe many thanks to Greg Saunders, John Kolen, Tilmann Wendel, and David Fogel for commenting on various drafts of this paper. This research is supported by the Office of Naval Research under contract number N00014-92-J1195.

REFERENCES

1. Angeline, P., G. Saunders, and J. Pollack "An Evolutionary Algorithm that Constructs Recurrent Neural Networks." *IEEE Trans. on Neur. Net.*, 1993: in press.
2. Angeline, P. and J. Pollack "Evolutionary Module Acquisition." In *Proceedings of the Second Annual Conference on Evolutionary Programming*, edited by D. Fogel, in press. Palo Alto, CA: Morgan Kaufmann, 1993.
3. Angeline, P. and J. Pollack "The Evolutionary Induction of Subroutines." In *Proceedings of the Fourteenth Annual Conference of the Cognitive Science Society,* edited by John K. Kruschke, 236–241. Hillsdale, NJ: Lawrence Erlbaum, 1992.
4. Bedau, M., and N. Packard. "Measurement of Evolutionary Activity, Teleology, and Life." In *Artificial Life II,* edited by C. Langton, C. Taylor, J. Farmer and S. Rasmussen. Santa Fe Institute Studies in the Sciences of Complexity, Proc. Vol X, 431–462. Redwood City, CA: Addison-Wesley, 1991.
5. Belew, R., J. McInerney, and N. Schraudolph. "Evolving Networks: Using the Genetic Algorithm with Connectionist Learning." In *Artificial Life II,* edited by C. Langton, C. Taylor, J. Farmer and S. Rasmussen. Santa Fe Institute Studies in the Sciences of Complexity, Proc. Vol. X, 511–548. Redwood City, CA: Addison-Wesley, 1991.
6. Brooks, R. "Intelligence Without Representation." *Art. Int.* **47** (1991): 139–159.
7. Dawkins R. *The Blind Watchmaker* New York: W. W. Norton, 1987.
8. Davis, L. *The Handbook of Genetic Algorithms.* New York: Van Nostrand Reinhold, 1991.
9. DeJong, K. "An Analysis of the Behavior of a Class of Genetic Adaptive Systems." Ph.D. Dissertation, University of Michigan, 1975.
10. Fogel, D. "Evolving Artificial Intelligence." Ph.D. Dissertation, University of California at San Diego, 1992.
11. Fogel, D. "A Brief History of Simulated Evolution." In *Proceedings of the First Annual Conference on Evolutionary Programming,* edited by D. Fogel and J. Atmar, 1–16. Palo Alto, CA: Morgan Kaufmann, 1992.

12. Fogel, L., A. Owens, and M. Walsh *Artificial Intelligence Through Simulated Evolution* New York: John Wiley & Sons, 1966.
13. Forrest, S. "Emergent Computation: Self-Organizing, Collective, and Cooperative Phenomena in Natural and Artificial Computing Networks." In *Emergent Computation,* edited by S. Forrest, 1–11. Cambridge, MA: MIT Press, 1991.
14. Goldberg, D. *Genetic Algorithms in Search, Optimization, and Machine Learning.* Reading, MA: Addison-Wesley, 1989.
15. Goldberg, D. "Zen and the Art of Genetic Algorithms." In *Proceedings of the Third International Conference on Genetic Algorithms,* edited by J. Schaffer, 80–85. Palo Alto, CA: Morgan Kaufmann, 1989.
16. Hinton, G., J. McClelland, and D. Rumelhart. "Distributed Representations." In *Parallel Distributed Processing: Explorations in the Microstructure of Cognition,* edited by D. Rumelhart and J. McClelland, Vol 1., 77–109. Cambridge, MA: MIT Press, 1986.
17. Holland, J. *Adaptation in Natural and Artificial Systems.* Ann Arbor, MI: University of Michigan Press, 1975.
18. Jefferson, D., R. Collins, C. Cooper, M. Dyer, M. Flowers, R. Korf, C. Taylor, and A. Wang "Evolution as a Theme in Artificial Life: The Genesys/Tracker System." In Langton et al.,[23] 549–578.
19. Kolen, J. and J. Pollack "Apparent Computational Complexity in Physical Systems." In *Proceedings of the Fifteenth Annual Conference of the Cognitive Science Society*, 1993, in press.
20. Koza, J. "Genetic Evolution and Co-Evolution of Computer Programs." In Langton et al.,[27] 431–462.
21. Koza, J. *Genetic Programming* Cambridge, MA: MIT Press, 1992.
22. Langton, C. "Introduction." In Langton et al.,[23] 3–23.
23. Langton, C., C. Taylor, J. Farmer, and S. Rasmussen *Artificial Life II* Santa Fe Institute Studies in the Sciences of Complexity, Proc. Vol X. Redwood City, CA: Addison-Wesley, 1991.
24. Pollack, J. "The Induction of Dynamical Recognizers." *Mach. Learning* **7** (1991): 227–252.
25. Ray, T. "An Approach to the Synthesis of Life." In Langton et al.,[23] 371–408.
26. Saunders, G., J. Kolen, P. Angeline, and J. Pollack. "Additive Modular Learning in Preemptrons." In *Proceedings of the Fourteenth Annual Conference of the Cognitive Science Society*, edited by John K. Kruschke, 1098–1103. Hillsdale, NJ: Lawrence Erlbaum, 1992.

Kristian Lindgren†‡ and Mats G. Nordahl†
†Santa Fe Institute, 1660 Old Pecos Trail, Suite A, Santa Fe, New Mexico 87501, USA
‡Institute of Physical Resource Theory, Chalmers University of Technology, S-412 96 Göteborg, Sweden

Artificial Food Webs

Artificial ecologies with explicit resource flows from an external environment are constructed, and the resulting food webs are studied. The interactions between species are modeled using game theory. In addition to a strategy, in some of the models, genomes also include preferences for whom to play, and a tag gene on which other organisms base this choice. This allows a higher degree of specialization, and the evolution of camouflage mechanisms and mimicry.

1. INTRODUCTION

One of the most important aspects of ecological organization is the flow of resources in an ecosystem. The aim of this article is to create simple artificial ecologies where the interactions between species are based on game theory, and where resources are explicitly included. This allows us to observe who is eaten by whom, and to extract observables such as food webs. The statistical properties of these artificial food webs can be compared to regularities observed or conjectured for real ecological communities.

This can also be viewed as an attempt to build coevolutionary models which, to a larger extent, take ecological structure into account. Coevolution is an area of biology where even present work on artificial life could yield some insights. To understand the dynamics of many coevolving species, we need to reduce the byzantine complexity of the real world to something manageable, and make a simplified model of the world and the interactions between species. Models of coevolutionary systems (or systems that could be viewed as such) can be found on a spectrum from the very simple to the quite complex; a few examples are the following:

- Stenseth and Maynard Smith[43] considered two coupled ordinary differential equations for the number of species and the mean evolutionary lag of the population. In this way, the structure of the world is summarized in six real numbers.
- In the absence of detailed information about the interaction between species, a reasonable approach is to make some statistical assumptions and to study *generic* properties of the resulting class of systems. Models of gene regulation,[11] stability of ecological systems,[24,25] and fitness landscapes[12] have been constructed in this spirit. Similar models of coevolution have also been introduced.[13,30]
- Others have modeled the interaction between species using game theory.[27] One particular model of this kind based on the iterated Prisoner's Dilemma was studied by Lindgren,[18] and showed a number of interesting evolutionary phenomena such as periods of stasis, large extinctions, and evolution of symbiosis.
- Models where individual creatures explore complex model worlds, and search for food, fight, reproduce, etc., have been considered by a number of Alife researchers (see, e.g., these and earlier Artificial Life proceedings[15,16]).

A complex model world could allow for more interesting behavior and a larger range of emergent behaviors, but it is then hard to avoid introducing a high degree of arbitrariness in the model. This might make it harder to argue that the model is of biological relevance, and it also becomes difficult to perform simulations on truly evolutionary time scales. For example, the simulations of Yeager's PolyWorld[44] span up to 500 generations (using a week of CPU time on a Silicon Graphics workstation), while a simplier game-theory-based model[18] allows simulations over several million generations under the same circumstances. However, the choice of a very simple model may restrict the types of phenomena that can be observed.

When constructing such models, it is clearly important to ask what quantities could actually be compared to what is observed in nature. We are particularly interested in areas where a *quantitative* comparison could be made, at least in principle, disregarding, e.g., limitations on the availability of experimental data.

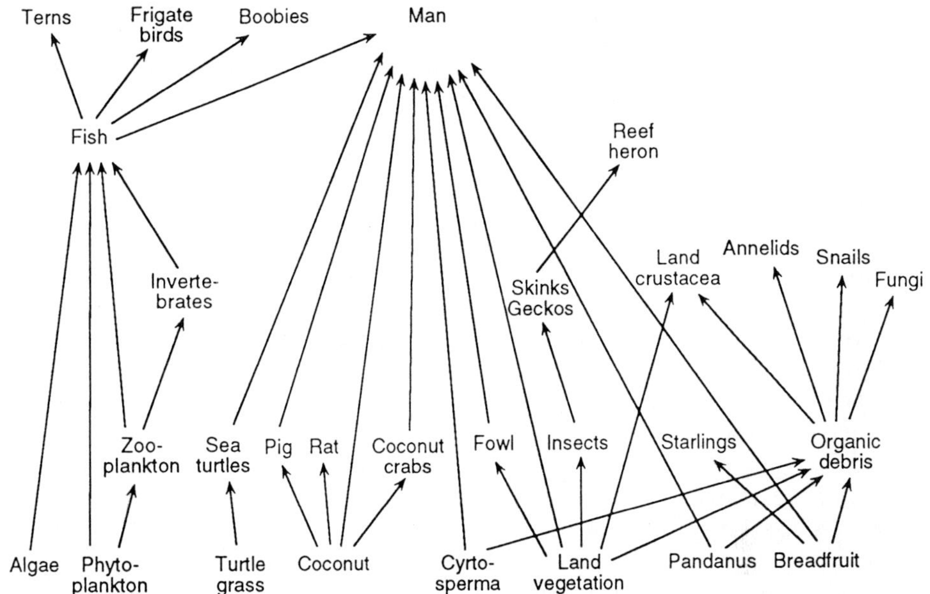

FIGURE 1 Food web from the Kapingamarangi atoll in Micronesia, drawn using the data of Niering.[29]

One possibility could be to consider the temporal behavior of evolutionary systems: Does evolution proceed gradually or intermittently[6] (a question which we believe should be asked and addressed in a quantitative way)? In the absence of external perturbations, does the system reach a fixed point, or does it enter a never-ending evolutionary arms race, as suggested by the Red Queen hypothesis by Van Valen?[45] Can one make statements about the size distribution of extinction events[41] (a power law distribution has been suggested by Kauffman[13])?

Another approach would be to ask questions about properties of the system at some particular time; i.e., one could study the patterns of organization in biological systems, and how these are determined by evolution. Patterns of organization obviously exist at many different levels in biological systems—this article is primarily concerned with *ecological* organization (for a more general approach to biological organization, see Fontana and Buss.[8])

Examples of statistical regularities in ecological systems can be found in the structure of food webs. A *food web* is a representation of who eats whom in an ecological system, often compiled in a qualitative rather than quantitative fashion by ecologists. An example of an experimental food web from the literature is shown in Figure 1 (redrawn from Niering[29]). This is not a complete representation of the system, or even of the resource flow in the system. However, food webs may still be useful observables to consider, both for real and artificial ecologies.

From a comparison of a number of food webs in the ecological literature, a number of statistical regularities have been suggested[4,36]:

- The average number of trophic linkages L is approximately proportional to the number of species S, $L \sim kS^{1+\epsilon}$, where ϵ is small.
- The fractions of top predators, intermediate species, and basal species are approximately constant across webs. This is also the case for the fractions of trophic linkages of different types: top-intermediate, top-bottom, intermediate-intermediate, and intermediate-bottom.
- Averages of the maximal chain length are typically fairly short (approximately 3–4) and do not appear to depend on the productivity of the environment.[35] They do, however, seem to depend on the dimensionality of the environment; three-dimensional environments, such as pelagic water columns or forest canopies, appear to have longer food chains than two-dimensional environments such as grasslands and intertidal zones.
- Cycles are rare (food webs in the literature typically do not include the cycles that arise from cannibalism or from the presence of decomposers).
- Food webs can often be represented as overlap graphs of a set of subintervals of the real line (interval graphs).

Several examples of criticisms of food web theory can be found in the literature.[32,38] Most published food webs contain a fairly limited number of species, and trophically similar species are often considered as groups. Experimental studies that to a larger extent reflect the complexity of real ecosystems[46,38] could conceivably change some of the above conclusions. The study of artificial ecologies might be useful in illuminating these issues, since data are more easily available in this case, and the artificial ecosystems can be freely manipulated.

Examples of other types of statistical regularities in ecological systems can also be found in the literature. One of the most well known is the area-species law,[23] a power law for the number of species on an island as a function of its area, $S \sim c \cdot A^z$, with an exponent $z \approx 0.3$ (the exponent does not appear to be universal). Under certain assumptions, this power law can be related to a log-normal distribution of population abundances.[39,26] Regularities in the patterns of resource flows have also been suggested.[28]

Many important properties of ecological systems involve resource flow. In the evolutionary models studied by Lindgren,[18] where cooperation is modeled using the Prisoner's Dilemma, there are no explicit notions of resources or an environment. In this contribution, we build models where resources are explicitly represented. This allows the study of a number of phenomena involving resource flow. Our previous models might still be equally relevant to biology—this depends on the relative importance in nature of cooperative interactions compared to predation and competition.[14,33]

The main purpose of this paper is to build a class of models where explicit resource flows are included, and where the system can be interpreted as an open system. Explicit resources do appear in other artificial life models, for example in the work by Holland,[9] and Rasmussen and coworkers,[40] but we are not aware of any more systematic studies of the patterns of resource flow and utilization. This paper is a first step towards attempting to make *quantitative* comparisons with biological systems, e.g., via various scaling laws. One should note that the models we construct are not models of food webs *per se* (as are the random feedforward graphs considered, e.g., by Cohen et al.[4]). Instead we attempt to build simple artificial ecologies, and food webs are used as examples of observable quantities.

Two different models are considered in this paper. In Section 2, we introduce a model (sometimes refered to as model 1 below) where the system lives in and derives resources from an external environment, and the resources are distributed according to the result of a game between organisms. The dynamics and the resulting food webs are analyzed for an environment represented by a single strategy, as well as more diverse environments consisting of a large number of randomly chosen strategies. In Section 3, a model (model 2) is introduced where genomes also include preferences for whom to play, and a tag gene on which other organisms base this choice. This allows a higher degree of specialization, and the evolution of camouflage mechanisms and mimicry.

2. INCLUDING RESOURCE FLOWS IN GAME-THEORETIC EVOLUTIONARY MODELS

In this section, we build a model where the interaction between individuals is based on the infinitely terated Prisoner's dilemma with noise. The score in the game determines how a certain resource (energy) is transfered from the environment and distributed among the organisms in the population. The fitness of a species, and thus its rate of reproduction, is then proportional to the amount of energy accumulated per individual.

In the model considered in this section (model 1), the details of genomes and mutations are identical to those used in an earlier model of coevolving strategies.[18] These are described in Section 2.1. Section 2.2 explains how the dynamics of the model is constructed to take resource flows into account, and Section 2.3 contains results from simulations.

2.1 REPRESENTATION AND ADAPTIVE MOVES

The Prisoner's Dilemma is a two-person nonzero-sum game where each player has a choice of two moves—C for cooperate and D for defect—and where the payoff matrix has the following form:

$$\begin{array}{c} \\ C \\ D \end{array} \begin{array}{cc} C & C \\ \left(\begin{array}{c} (R,R) \\ (T,S) \end{array} \right. & \left. \begin{array}{c} (S,T) \\ (P,P) \end{array} \right) \end{array}$$

where $T > R > P > S$ and $2R > T + S$. In the simulations below, we use the common parameter values $R = 3$, $S = 5$, $T = 0$, and $P = 1$. The iterated Prisoner's Dilemma has been extensively studied as a model of the evolution of cooperation, in particular by Axelrod.[2] If the game is only played once, or a fixed finite number of times against a rational opponent, a rational player defects. Both players would, however, gain by establishing a pattern of cooperation, which constitutes the dilemma.

In the present model, we introduce a quantity e which is redistributed according to the score difference in the game. We call this quantity energy, though a more plausible thermodynamic analogy might be free energy. In this way, the game becomes equivalent to a zero-sum game. Cooperative effects can, however, be reintroduced by letting the dissipation of energy for a certain species be determined by its absolute score in the game.

We consider the infinitely iterated game, where the actions of the players are influenced by noise, so that with probability p_{err} the action performed is opposite to that intended. The introduction of noise makes the game more complex, and allows

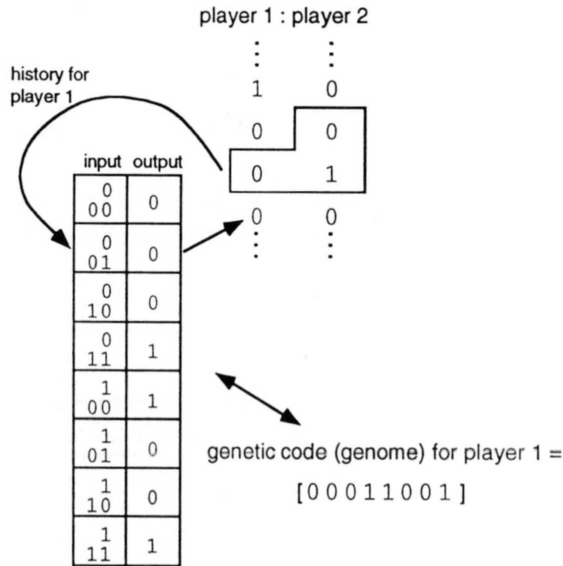

FIGURE 2 The representation of strategies as binary strings is illustrated for the memory 3 strategy 00011001. The symbols 0 and 1 stand for defect and cooperate, respectively.

the evolution of longer memory strategies with protocols for correcting occasional noise-induced mistakes.[18]

Genomes in our model represent strategies in the game. A strategy is a rule (deterministic or stochastic) for determining the next move of a player given the history $h = ((x_0, y_0), \ldots, (x_t, y_t))$ of the game, where the pair (x_t, y_t) consists of the moves of the player and the opponent, respectively, at time t. We consider only deterministic strategies of finite memory m, where $m \geq 1$ is a parameter such that for m even, the strategy takes into account the last $m/2$ moves of both players. For m odd, the strategy depends on the last $[(m+1)/2]$ moves of the opponent and the last $[(m-1)/2]$ moves of the player herself. If we write 1 for the action C, and 0 for the action D, a strategy of memory m can be represented as a binary string σ of length 2^m, as shown in Figure 2. The symbol at position $i = (b_1, b_2, \ldots, b_m)_2$ in the string gives the action when the history truncated to the m last moves is given by $\bar{h} = ((b_1, b_2), \ldots, (b_{m-1}, b_m))$, assuming that m is even. If m is odd, the history taken into account is $\bar{h} = (b_1, (b_2, b_3), \ldots, (b_{m-1}, b_m))$, so that at the most distant past considered only the move of the opponent affects the strategy. For a finite game, the definition of a strategy should also include initial conditions for the first time steps when the history is shorter than the maximal memory of the strategy. However, in our case this is irrelevant, since we only consider an infinitely iterated game with noise, where initial conditions do not affect the average score.

As stated above, the interaction between organisms is based on the result in the infinitely iterated game with noise. We denote the average score for a strategy i when playing strategy j as s_{ij}. Since the noisy game can be viewed as an ergodic Markov chain of memory $[(m+1)/2]$, with symbols (CC), (CD), (DC), and (DD), we can solve for the stationary probability distribution on the states of the Markov chain and calculate the average score.[18] Solving the game analytically speeds up simulations by a significant factor compared to performing explicit simulations of the iterated game.

The mutations that may occur in the model are of three different types: point mutations, gene duplications, and split mutations. Point mutations flip single bits in the genome; the frequency of point mutations in the simulations was set to $p_{\text{mut}} = 2 \cdot 10^{-5}$ per site (which gives a mutation rate per genome slightly lower than that of a bacterium such as *Salmonella typhimurium*[5]). The gene duplication operator increases the memory of a strategy from m to $m+1$ while leaving the actual strategy unchanged. This corresponds to attaching a copy of the genome to itself, so that $01 \rightarrow 0101$, for example. A mutation frequency of $p_{\text{dupl}} = 1 \cdot 10^{-5}$ per genome was used in this case. The gene duplication is a neutral mutation, which increases the size of the evolutionary search space without immediately affecting the phenotype. Additional point mutations may then give rise to new strategies that cannot be represented using shorter memory. The importance of gene duplication and unequal crossover for increasing the complexity of real genomes has been emphasized, e.g., by Ohno.[31] We have included a neutral gene duplication operator and a growing genome in an application of evolutionary methods to recurrent neural networks.[19] The split mutation splits a genome into two parts of equal size; one of these is chosen at random and kept in the population. The frequency of split mutations was $p_{\text{split}} = 1 \cdot 10^{-5}$ per genome.

When statistical comparisons with biology are attempted, one would like to remove the arbitrariness associated with the choice of a particular game, and study the generic properties of models with stochastically generated interactions between species. One possible interaction between genomes of length 2^{k_1} and 2^{k_2} ($k_1, k_2 \leq N$) which allows a neutral gene duplication operator is given by the following expression[30]:

$$s_{ij} = \sum_{\alpha,\beta=1}^{2^N} m_{\alpha\beta} v^{\alpha}[x_i] v^{\beta}[y_j] \tag{1}$$

where x_i and y_j are the two genomes, $m_{\alpha\beta}$ is a random $2^N \times 2^N$ matrix with elements drawn from a zero-symmetric distribution, and the length 2^N vector $v^{\alpha}[x_i]$ is obtained by mapping each element x to $2x - 1$ (which gives us a vector of ± 1), and then concatenating 2^{N-k_1} copies to give us $v^{\alpha}[x_i]$ (and similarly for $v^{\beta}[y_j]$). This example is similar to the Sherrington-Kirkpatrick spin-glass Hamiltonian[42]; more generally the degree of ruggedness of the coevolutionary landscape can be tuned by replacing $m_{\alpha\beta}$ by a random tensor that contracts n_1 copies of $v^{\alpha}[x_i]$ and n_2 copies of $v^{\beta}[y_j]$ (for fixed genome length, similar models were considered by Kauffman and Johnsen[13]). Statistical results from simulations of these models will be reported elsewhere[21]; in this contribution all models are based on the iterated Prisoner's Dilemma.

2.2 DYNAMICS OF THE MODEL

The evolutionary process is modeled using a dynamical system with a variable number of degrees of freedom. Strategies change in abundance according to the population dynamics of the system, while new strategies are added through mutations (a stochastic process), and strategies are removed from the system when their abundance goes below a threshold ϵ. This approach to the modeling of a potentially infinite-dimensional system was previously used in certain models of autocatalytic sets.[7,3]

If we label the genomes present in the population at some particular time t by an index $i = 1 \ldots N$, then genome i is associated with a binary string σ_i representing a strategy as described above, and also a population size x_i and energy e_i (both real numbers).

The population lives in an external environment consisting of N_{env} environment strategies, which are determined at the outset of the simulation. In the simulations shown below, the environment consists either of a single strategy such as Tit-for-Tat, or of a larger number (typically ranging from 8 to 64) of randomly chosen strategies with maximal memory 4 or 5. Energy flows into the system via the environment strategies; these are characterized by variables σ_l, x_l, and e_l (where $l = 1 \ldots N_{\text{env}}$) analogous to the population variables introduced above. In some cases we also need indices that run both over the environment strategies and the population; for these indices we use capital letters I and J.

The simplest imaginable population dynamics for the system considers discrete nonoverlapping generations. In each generation all of the energy e_i is converted into organisms. With a suitable choice of units, we can then assume that

$$x_i = e_i \,, \tag{2}$$

and the variable x_i disappears from further consideration. For a certain strategy i, the dynamics of the energy variable e_i takes into account the redistribution of energy due to the interaction with other strategies in the population and the environment. It also contains a dissipation term.

Let us first consider the energy transfer. The energy f_{ij} transfered from j to i in the interaction between two strategies i and j in the population is taken to be proportional to the score difference $s_{ij} - s_{ji}$ in the game between i and j, and also to the population sizes e_i and e_j (imagine a number of individuals of each kind, where each plays against all others, and a small fraction of the energy of the loser is transfered to the winner in each interaction). We also include a functional response for predators, i.e., a function that limits the amount of energy a predator can absorb per unit time. Let P_i be the total amount of energy stored by those organisms from which i receives energy (including the environment):

$$P_i = \sum_{J: f_{iJ} > 0} e_J \tag{3}$$

and let

$$r(P_i) = \frac{1}{1 + bP_i} \tag{4}$$

(a functional response of this form, i.e., linear at low prey densities, and gradually approaching a constant at large densities, is sometimes refered to as being of Holling's type II[10]). This would give us the following expression for the energy transfer from strategy j to strategy i: If $s_{ij} - s_{ji} > 0$, let

$$f_{ij}^{(0)} = \alpha(s_{ij} - s_{ji}) e_i e_j r(P_i); \tag{5}$$

otherwise, let $f_{ij}^{(0)} = 0$. However, we also need to introduce a cutoff so that not more energy than is available to strategy j is actually transfered. In this way we obtain the following expression for the energy transfer f_{ij}:

$$f_{ij} = \begin{cases} f_{ij}^{(0)} & \text{if } \sum_i f_{ij}^{(0)} \le e_j; \\ \dfrac{f_{ij}^{(0)}}{\sum_i f_{ij}^{(0)}} e_j & \text{otherwise.} \end{cases} \tag{6}$$

The energy transfer from the environment to a certain species i is treated in a somewhat different way; here we take the energy transfer from an environment strategy l to be proportional to a transfer function $t(s_{il})$—the choice is here somewhat arbitrary, in the simulations we have used $t(s_{il}) = s_{il} - 2.0$ for $s_{il} > 2.0$, $t(s_{il}) = 0.0$ otherwise. In this way we have

$$f_{il}^{(0)} = \alpha_{\text{env}} t(s_{il}) e_i e_l r(P_i), \tag{7}$$

and

$$f_{il} = \begin{cases} f_{il}^{(0)} & \text{if } \sum_i f_{il}^{(0)} \le e_l; \\ \frac{f_{il}^{(0)}}{\sum_i f_{il}^{(0)}} e_l & \text{otherwise.} \end{cases} \tag{8}$$

The energy e_i of a strategy in the population is then updated as follows: first let

$$e_i' = e_i + \sum_{J:f_{iJ}>0} f_{iJ} - \sum_{j:f_{ji}>0} f_{ji} \,. \tag{9}$$

We then introduce a dissipation term $d(s_i)$ for each strategy i that depends on its average score $s_i = \sum e_J s_{iJ} / \sum e_J$. A simple choice for this term is a linear function, with a cutoff to keep the dissipation non-negative; in the simulation we used $d(s_i) = d_0\,(1 - s_i/3)$ for $s_i \le 3$, and $d(s_i) = 0.0$ for $s_i > 3$, with a dissipation constant of $d_0 = 0.2$ (we did not observe any sensitive dependence on the exact value of d_0, except when d_0 was set exactly to 0). The updated energy is then given by

$$e_i(t+1) = e_i' - d(s_i) e_i' \,. \tag{10}$$

This dissipation term can be viewed as due to organisms dying of other causes than being killed by a predator. One can also introduce a dissipation term proportional to the energy flow, which would describe the limited efficiency of organisms. In the simulations described here, all organisms were assumed to be maximally efficient.

Finally, energy flows into the system via the environment strategies, which obey the following dynamics (where e_{flow} and e_{max} are constants):

$$e_l(t+1) = e_l - \sum_{j:f_{jl}>0} f_{jl} + e_{\text{flow}}(1 - \frac{e_l}{e_{\text{max}}}). \tag{11}$$

The update procedure performed each time step also includes removal of strategies whose population fall below a threshold ϵ (a value of $\epsilon = 0.001$ was used below, which means that a unit of energy corresponds to 1000 individuals), and a mutation step. The mutations were described above; if a genome not previously present in the population is created through a mutation, it is added to the population with an initial energy $e_i = \epsilon$ taken from its parent.

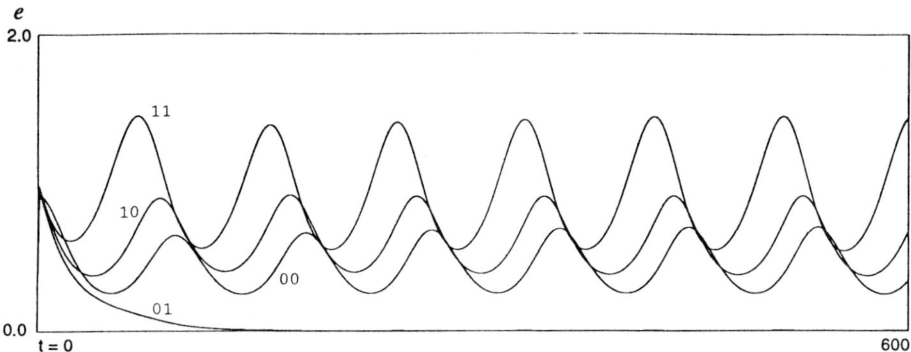

FIGURE 3 Species densities as a function of time for the first 600 generations of a simulation with a Tit-for-Tat environment. Prey-predator oscillations involving the species 11, 10, and 00 are seen.

In this article, species (as opposed to genotypes) will be defined as *trophic species*[34]; i.e., classes of genomes that interact in the same way with all other genomes within some tolerance limit ϵ_1. A distance $d(i,j)$ between the genomes i and j can be defined in terms of the matrix of energy flows, e.g., as

$$d(i,j) = \sum_K |\bar{f}_{iK} - \bar{f}_{jK}|(\bar{f}_{iK} + \bar{f}_{jK}) + \sum_K |\bar{f}_{Ki} - \bar{f}_{Kj}|(\bar{f}_{Ki} + \bar{f}_{Kj}), \qquad (12)$$

where the relative energy flow $\bar{f}_{iJ}$ is defined by $\bar{f}_{iJ} = f_{iJ}/\sum_K f_{iK}$. The clustering algorithm used to identify trophic species starts with the most frequent genome, groups everything within a distance ϵ_1 of that genome into a trophic species, and repeats this procedure with the remaining genomes as long as there are genomes left. Species are labeled by their most frequent genotype.

2.3 EXAMPLES OF SIMULATIONS

In the first of our simulation examples, the environment consists only of the strategy Tit-for-Tat with genotype 01. The initial state consists of all four memory 1 strategies: 00 (always defect), 01 (TfT), 10 (ATfT), 11 (always cooperate), with equal population sizes given by $e_i = 1.0$. For the parameters α, α_{env}, b, e_{flow}, and $e_{\max}$ the following values were used: $\alpha = 0.005$, $\alpha_{\text{env}} = 0.04$, $b = 0.1$, $e_{\text{flow}} = 0.1$, and $e_{\max} = 10.0$. The first 600 generations of this simulation are shown in Figure 3. Prey-predator oscillations involving the three species 00, 10, and 11 are observed. The corresponding food web is shown to the left in Figure 4, where we can see that the always cooperating strategy 11 feeds only on the environment, while 10 feeds on 11 and the environment, and the always defecting top predator 00 feeds on 11 and 10. Our convention for drawing food webs is that environment strategies are located at the bottom of the web, and a connection from one species to another below it indicates that the first feeds on the second. Links representing less than a fraction ϵ_2 of the energy inflow for a species are not included.

The system then typically follows one of two possible evolutionary pathways. Both alternatives are shown in Figure 4. Either the intermediate species 10 is replaced by 1110, which feeds mainly on the environment but also on 11, or alternatively the basal species 11 is replaced by 1011. The second of these cases occurs in the simulation of Figure 5, where 96,000 generations of the evolutionary process are shown. In this case, the second prey-predator oscillation period is interrupted by the appearance of the strategy 00111011, which exploits the environment more efficiently than 1011. This particular simulation was continued for several hundred thousand generations beyond those shown here without the system reaching a stable state. A Tit-for-Tat environment in general appears to generate very unstable

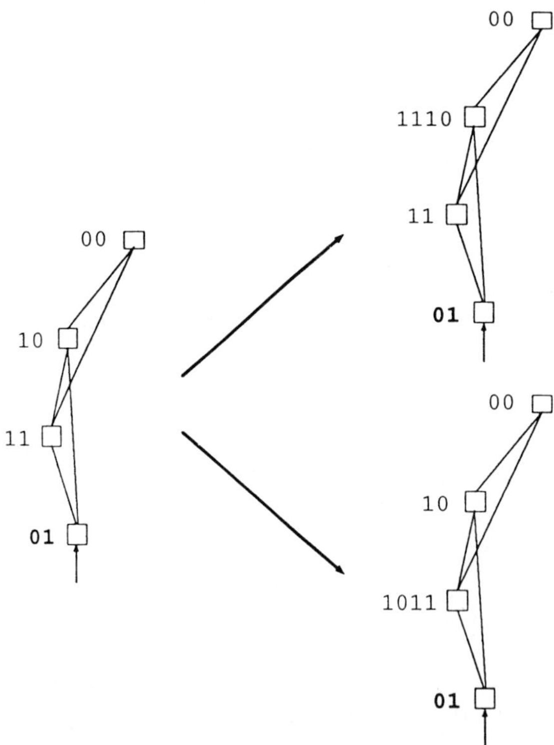

FIGURE 4 Food chains for two common initial evolutionary paths in a Tit-for-Tat environment. In the simulation of Figure 5, the lower alternative is realized.

Red Queen dynamics, even though single species may persist for long periods of time. Notice the species 1111100101011001 which dominates the population most of the time between generations 50,000 and 80,000. A possible explanation for the instability could be that a large number of longer memory strategies play about equally well against Tit-for-Tat. While these species are cooperative and easily exploited by predators, the degeneracy makes it easy for them to evolve and still use the environment efficiently, forcing the predators to evolve as well, and so on. For more cooperative environments, this kind of degeneracy typically does not exist—in these cases a single uncooperative strategy can both exploit the environment and defend itself against predators in a more or less optimal way.

The food chains seen in the simulation of Figure 5 are typically quite simple, which is not too surprising when the environment consists only of a single resource. They are, however, more complex than suggested by stability based arguments relating the number of species to the number of resources.[17,22]

If the diversity of the environment is increased, more complex food webs evolve. Typical environments that we have used consist of from 8 to 64 randomly chosen strategies. The degree of cooperativity of the environment can be adjusted by generating the background strategies at random with a density p_C of 1's and $1 - p_C$ of 0's. As mentioned above, in more cooperative environments the ecologies generated often become less complex since a small number of uncooperating species dominate. On the other hand, in very harsh uncooperative environments few species (or none at all) may be able to survive. Intermediate values of p_C, but biased somewhat towards non-cooperation, tend to give the most complex ecologies. In the simulations below, the parameter p_C was set to $p_C = 0.4$.

Two simulations with more diverse environments are shown in Figures 6 and 8. In Figure 6 the environment consists of 16 randomly generated strategies (the list of environment strategies can be found in the food webs of Figure 7). The initial state and other parameters are the same as in the simulation with a Tit-for-Tat environment; 72,000 generations of the simulation are shown in Figure 6. As is often the case in random environments (which typically contain some strategies that are very easy to exploit), the generalist species 00 dominates at first, and is then replaced by 0001. More complex food webs with predators and more specialized basal species then develop. Figure 7(a) shows the food web and the matrix of energy flows at $t = 33600$ in more detail.

The food web of Figure 7(a) was extracted from a rather unstable period. Certain species manage to survive for extended periods of time—note, for example, the basal species 0011101100011101, which is present at least from $t = 24000$ until it goes extinct at the beginning of the stable period which begins around $t = 47000$. This particular run reached a very stable state around $t = 56000$; the corresponding food web is shown in Figure 7(b). This web is in fact quite similar to that of Figure 7(a); both top predators of Figure 7(a), 0011001000011100 and 1001000100011100, appear as predators in Figure 7(b), and the species 0001001010010111 also appears in both webs, though it has lost its prey 1111101000000100 and instead is eaten by 1111100001000100.

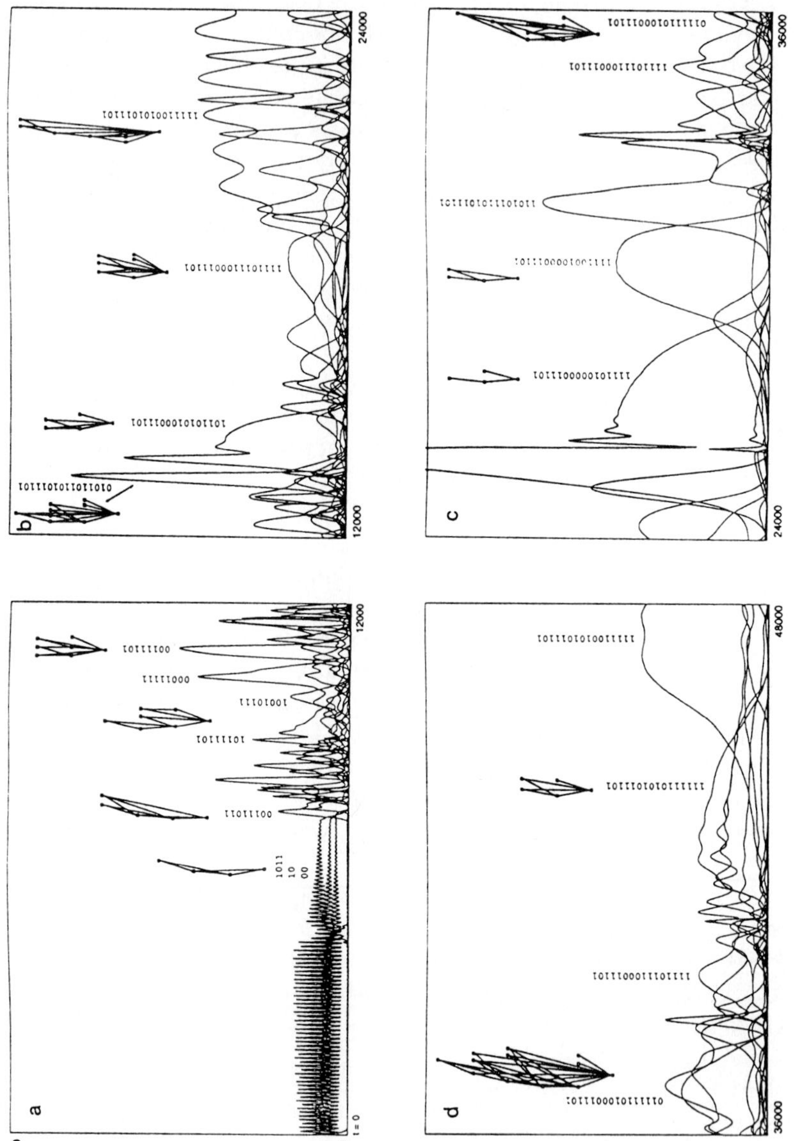

FIGURE 5 (a–h) Densities of different species when a population in a Tit-for-Tat environment is followed for 96,000 generations. Food webs and genomes of dominating strategies are shown at selected times.

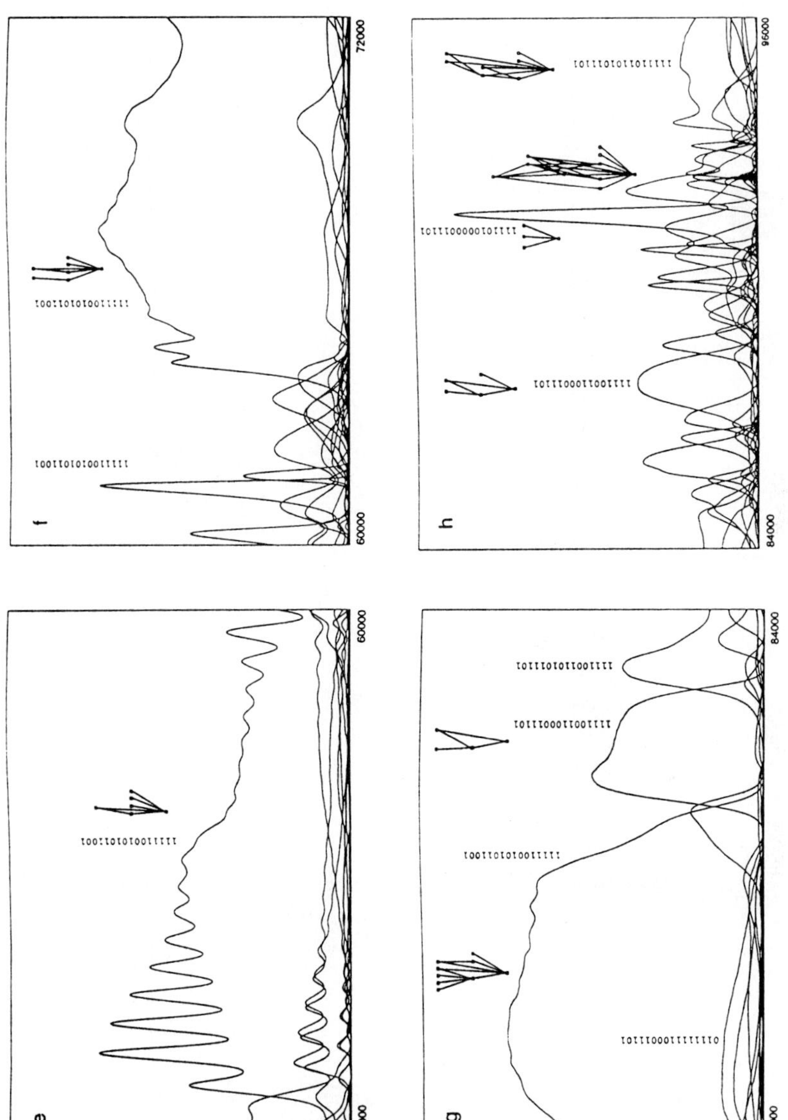

FIGURE 5 (continued)

FIGURE 5 (continued)

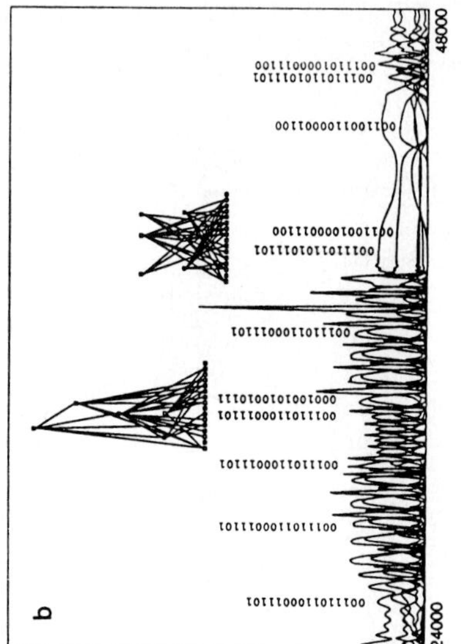

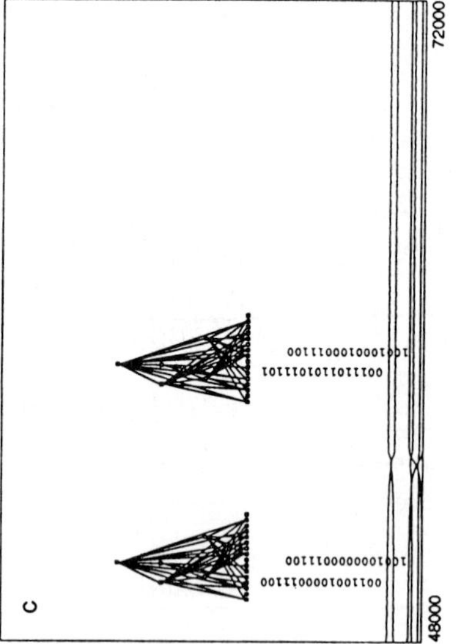

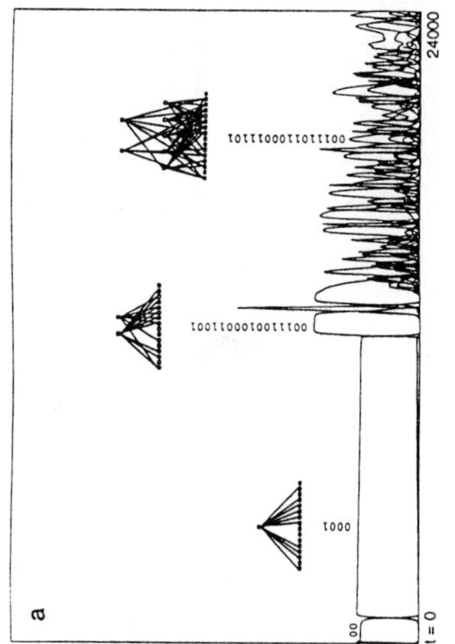

FIGURE 6 (a–c) Species densities as a function of time for a simulation where the environment consists of 16 randomly generated strategies ($p_C = 0.4$). Food webs and genotypes of dominating strategies are shown at selected points in time.

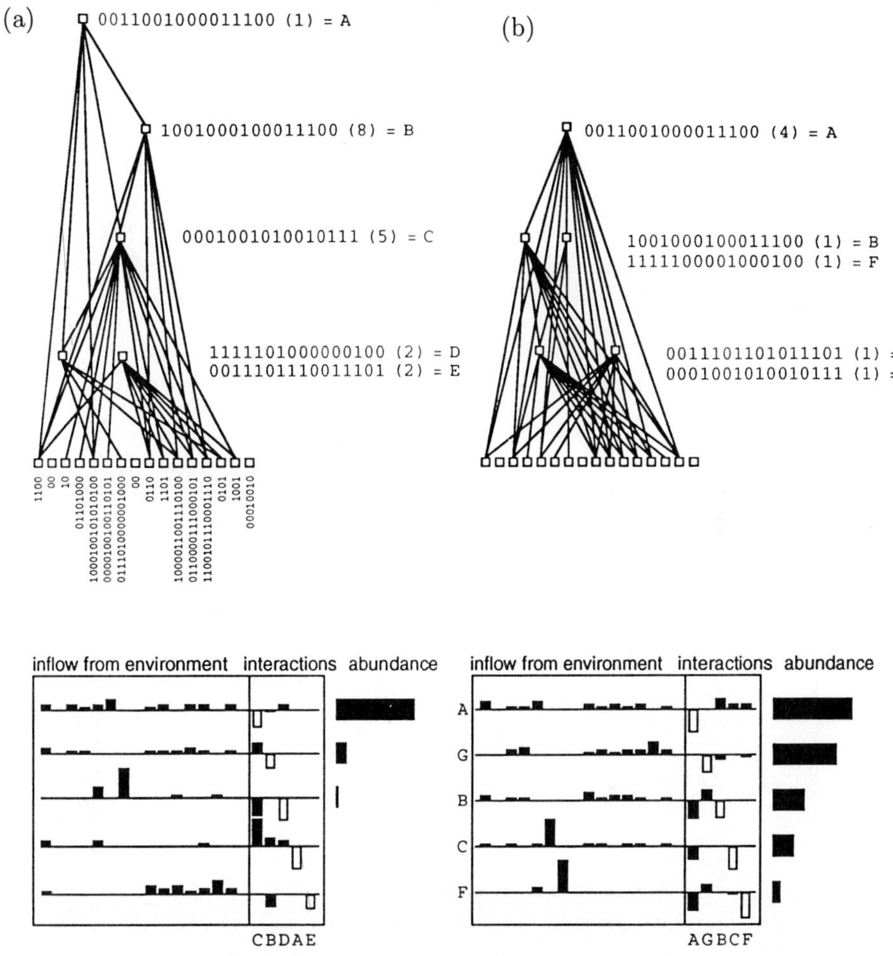

FIGURE 7 (a) The food web and energy flow matrix at $t = 33600$ in the simulation of Figure 6. The first part of the histogram represents energy inflows from the environment (always positive); the second energy flows between species (signs depend on the directions of the flows). The open bars show the energy dissipation for each species. The magnitude of the flow is shown relative to the total energy flow for the species in question. The parameters ϵ_1 and ϵ_2 were both set to 0.05. (b) The food web and energy flow matrix at $t = 72000$ in the same simulation.

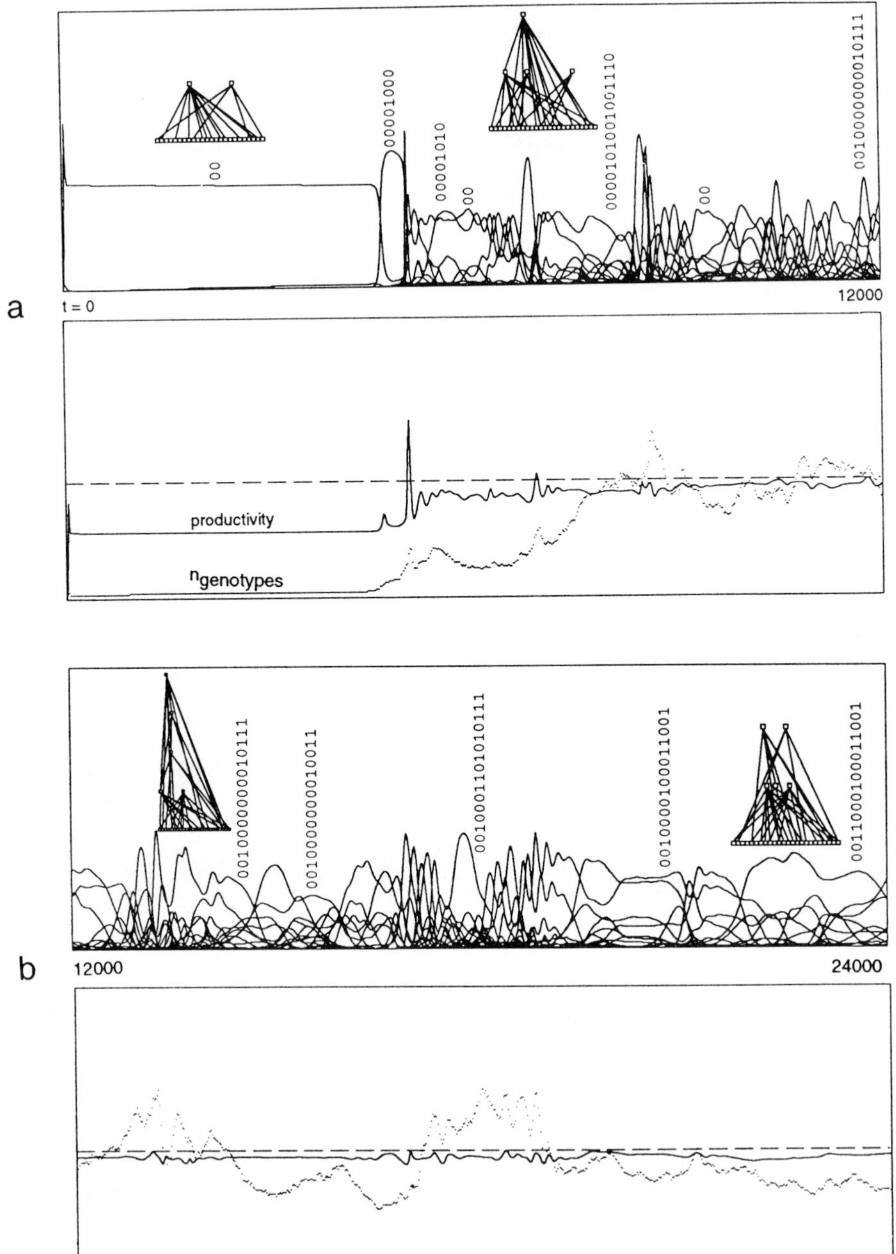

FIGURE 8 (a–d) A simulation with an environment consisting of 24 randomly generated strategies. ($p_C = 0.4$). The upper diagrams show species densities as a

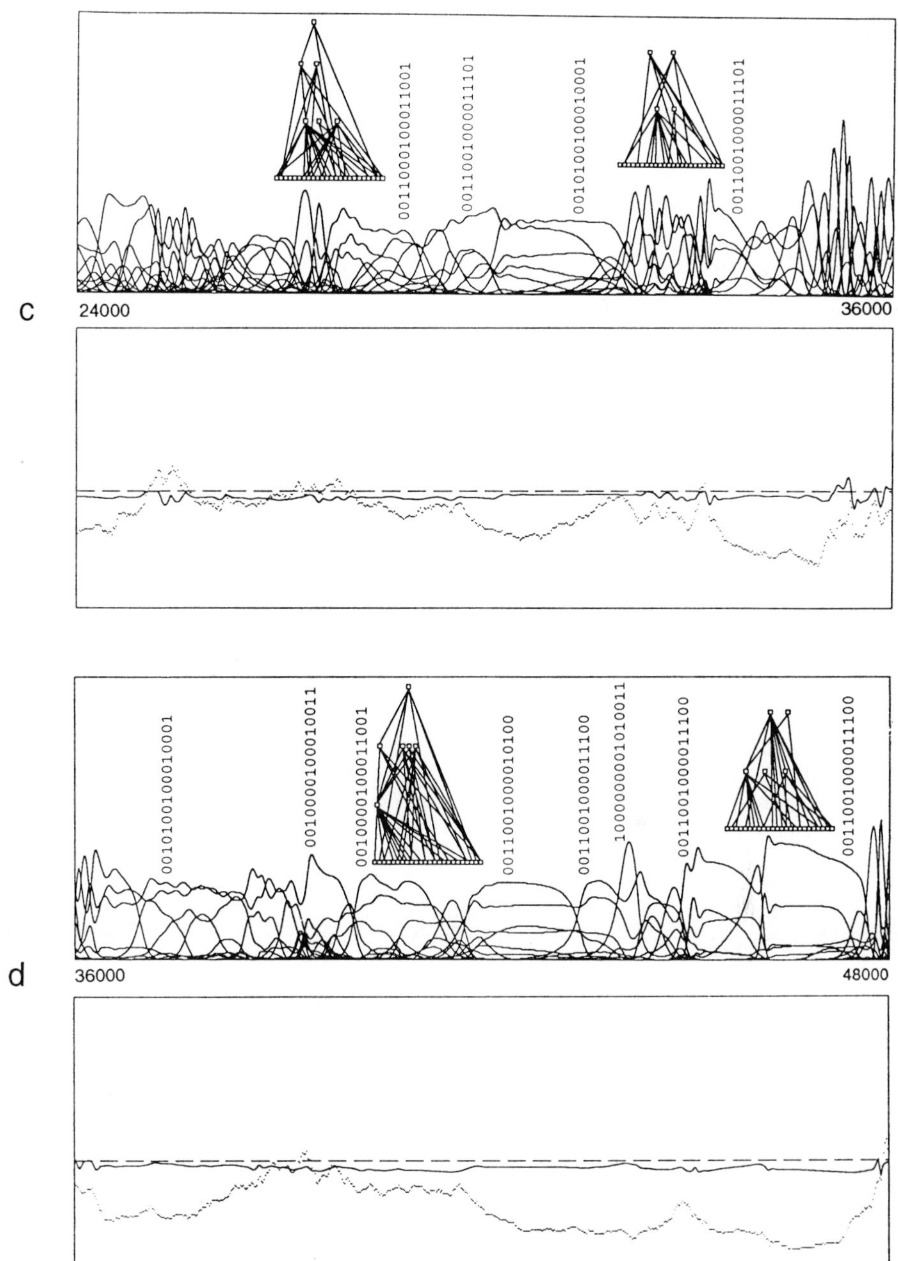

FIGURE 8 (cont'd.) function of time; the lower show the productivity (solid line) and number of genotypes (dotted line). For the number of genotypes the scale on the y-axis ranges from 0 to 200. The maximal average productivity is shown as a horizontal line.

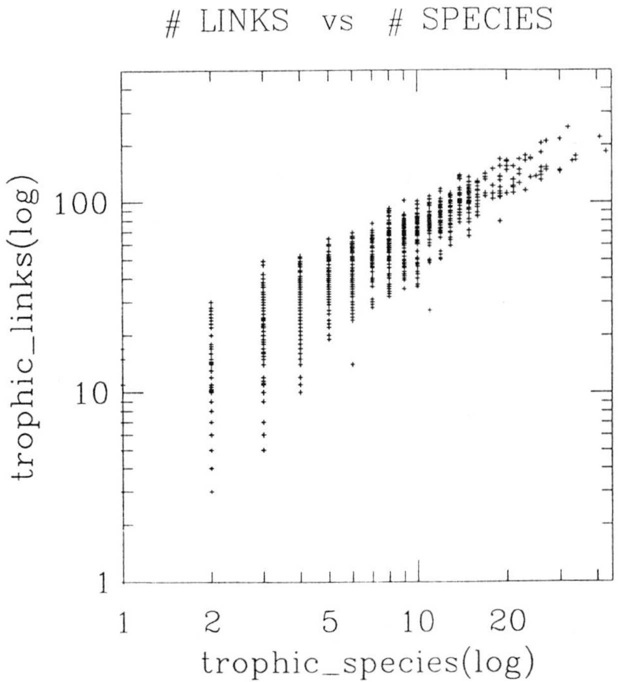

FIGURE 9 Number of trophic links versus number of trophic species for a set of 250 communities with between 8 and 64 environment strategies.

An example of more unstable dynamics is shown in Figure 8, where the number of environment strategies is given by $N_{\text{env}} = 24$. The time evolution of the productivity of the system, defined as the total amount of energy transfered from the environment to the population per unit time, is also shown in Figure 8. The horizontal line indicates the maximal average productivity, adjusting for the fact that four of the environment strategies in this simulation were equivalent to 00, which cannot be exploited at all. Also in Figure 8 are shown the number of genotypes as a function of time. A considerable genetic diversity within trophic species is observed. The number of genotypes appears to increase in unstable periods.

Most of the time, the productivity fluctuates between 90% and 100% of its maximum, with occasional excursions above 100%, which are possible since part of the energy inflow may be temporarily stored by increasing the population of one or more environment strategies. No systematic increase in the productivity is visible except at the beginning of the simulation.

A number of statistical measurements have also been carried out for the Prisoner's Dilemma model. These will be reported in more detail elsewhere[21]; here we only consider one of these, namely the number of trophic links as a function of the number of trophic species. In Figure 9, we have plotted the number of links versus the number of species for 250 communities chosen at random points in time from simulations with 8 to 64 environment strategies.

A linear relationship of the form $N_{\text{links}} \sim 8.0 \cdot N_{\text{species}}$ fits the data well. A linear functional form for this relation agrees with the observations for real food webs by Cohenet al.[4] The proportionality constant is significantly higher than that found for real ecological communities,[4,36] but this constant is affected by the choice of cut-off both in simulations and experiment,[38] so one cannot make a direct comparison.

3. A MODEL WITH EVOLVING PREFERENCES AND APPEARANCES

In the model studied in the previous section, each strategy plays against every other strategy. This means that an energy flow exists between a pair of strategies unless they play equally well in the game. In this way, one might suspect that food webs could become too densely connected. In this section, we introduce a model where the choice of whom to play is genetically determined. The game between two genotypes is played when at least one so desires (a reasonable assumption for the case of predation). The model also includes a tag gene, which determines this choice for other organisms. This allows the possibility of evolving a higher degree of specialization, and also allows phenomena such as camouflage and mimicry to evolve.

3.1 DETAILS OF THE MODEL

In the model considered in this section, the genomes consists of three parts: a strategy gene, a gene that indicates preferences for whom to play, and a tag gene on which other organisms base their choice. In this way a more distinct difference between genotype and phenotype is introduced (even in the earlier model such a distinction could be made, since gene duplications did not change the actual strategy) — the tag gene can be regarded as the "visual appearance" of an organism.

The strategy σ_i is represented in the same way as in the previous section; the mutations on the strategy gene are also identical. The preference gene consists of two binary strings p_i and m_i, where p_i can be viewed as an appearance that i attempts to match, and m_i consists of care (represented by 1) and don't care (0) symbols, which indicate the positions of p_i actually matched against the tag gene t_j of another species. If we denote the maximal memory allowed as M, the matching is done by expanding all genes involved periodically to the reference length 2^M, and checking whether

$$\sum_\alpha m_i^\alpha (p_i^\alpha t_j^\alpha + (1-p_i^\alpha)(1-t_j^\alpha)) = \sum_\alpha m_i^\alpha \tag{13}$$

where $\alpha = 1 \ldots 2^M$. This corresponds to requiring perfect matching only for the positions indicated by 1's in the expansion of m_i. We write a genome as a concatenation of three strings separated by a *: the strategy σ_i; a preference gene p_i' where the sites can take values $0 \ldots 3$, so that $p_i' = p_i$ if $m_i = 1$, and $p_i' = p_i + 2$ if $m_i = 0$; and finally the tag gene t_i. In this way, the genome 01*03*11 represents an organism that plays the strategy 01, looks like 11 to others, and prefers to play organisms with a tag gene that matches 0# (where # stands for don't care).

The mutations on preference genes and tag genes are identical to those on strategy genes, except that for simplicity we use only a single gene duplication operator which acts simultaneously on m_i, p_i, and t_i (and also a simultaneous split operator). Identical mutation frequencies were used for the different parts of the genome.

The dynamics is then identical to that of model 1, except that the energy flow now is obtained by first computing

$$f_{ij}^{(0)} = \alpha(s_{ij} - s_{ji})F(i,j)e_i e_j R(i), \tag{14}$$

where the function $F(i,j)$ determines whether at least one of i and j wants to play; i.e, $F(i,j) = 1$ if $\sum_\alpha m_i^\alpha (p_i^\alpha t_j^\alpha + (1-p_i^\alpha)(1-t_j^\alpha)) = \sum_\alpha m_i^\alpha$ or $\sum_\alpha m_j^\alpha (p_j^\alpha t_i^\alpha + (1-p_j^\alpha)(1-t_i^\alpha)) = \sum_\alpha m_j^\alpha$, and $F(i,j) = 0$ otherwise, and $R(i)$ stands for the response function for genotype i. The energy flow f_{ij} is then obtained from $f_{ij}^{(0)}$ in the same way as above. The environment strategies have an appearance identical to their strategy, and do not have any preferences for playing against other species. Thus, to gain energy from the environment, an individual in the population must have an explicit preference for playing against some environment strategy.

We have used several different forms for the response function $R(i)$; in the first simulation shown below the sum over prey only includes those genotypes j for which i has a preference. This can be viewed as a cost associated with actively hunting a prey species, rather than hiding passively and waiting for prey to pass by (like certain fish such as *Thaumatichthys*, for example). This means that the response function is given by

$$R(i) = r_1(P_i') = \frac{1}{1 + bP_i'} \tag{15}$$

where $P_i' = \sum_J f_{iJ}\ \delta(\sum_\alpha m_i^\alpha (p_i^\alpha t_J^\alpha + (1-p_i^\alpha)(1-t_J^\alpha)) - \sum_\alpha m_i^\alpha)$; i.e., the sum is over those J that i wants to play against. In the second of the simulations of this section, we instead use a response function identical to that of model 1, which associates a cost to consumption.

3.2 SIMULATIONS

In our first example, shown in Figures 10 and 11, the environment consists only of the fairly cooperative memory 2 strategy 0111 ($p_C = 0.75$). The initial state consists of the four genomes 00*22*00, 01*23*01, 10*32*10, and 11*33*11; i.e., all individuals have tag genes identical to their strategies, and none have any special preferences for whom to play, so that at first everyone plays against everyone.

Initially, we can see that 10*32*10 dominates the population, which is due to the fact that 10 is an excellent strategy to play against 0111. In a game between 10 and 0111, 0111 cooperates and 10 defects, and an occasional cooperation by 10 is ignored by 0111, while a single defection by 0111 results in a sequence of actions that takes both players back to the original pattern, as shown in Figure 12. Strategy 10 is, however, easily exploited in turn by 00.

The first specialized species that enters is 00*02*00. The specialization here serves the purpose of being able to avoid hunting for 10*32*10. Recall that in this case an energetic cost is associated with hunting. Instead the totally nondiscriminating 10*32*10 will itself choose to play with the predator 00*02*00. At the time marked (a) in Figure 10, the partially specialized predator 00*12*00 enters and replaces 00*02*00, and also reduces the population of its prey 10*32*10 significantly.

At (c), the first species (10*3212*0000) that attempts to tune its preferences to the structure of the environment enters. Up to this point, all predators have had the appearance 00; now the slightly camouflaged species 00*0212*0010 also enters,

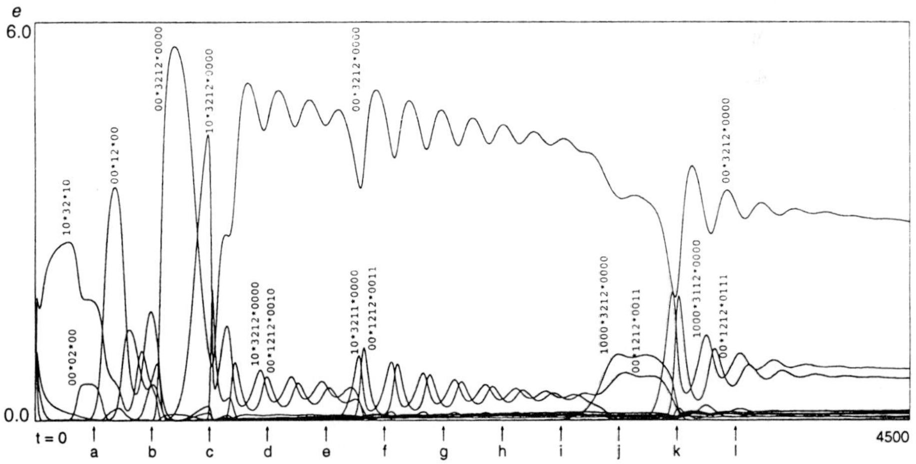

FIGURE 10 Species densities as function of time for a simulation of model 2 in an environment consisting of the strategy 0111.

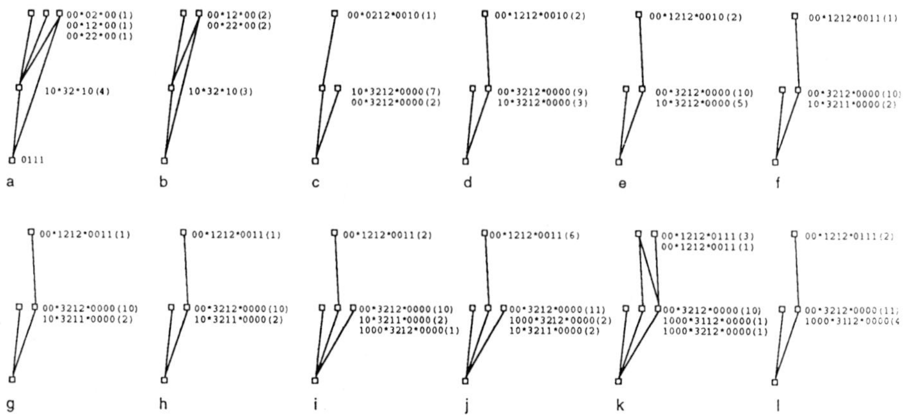

FIGURE 11 Food webs at selected points in time (marked as (a) – (l) in Figure 10) in the previous simulation.

10: • • • D D D C* D D D C C D D D • • •
0111: • • • C C C C C C D* D C C C C • • •

→ *t*

FIGURE 12 An excerpt from an iterated game between the strategies 10 and 0111. Mistakes by either player are marked by *.

but is soon replaced by the mutant 00*1212*0010, which avoids playing against itself. An explicit preference for cannibalism is unfavorable for a species due to the response function.

Between (e) and (f), the basal species 10*3211*0000 enters the population. Its success is due to the fact that it manages to avoid the predator 00*1212*0010, which results in the extinction of that predator. However this success is only temporary, since the improved predator 00*1212*0011 soon appears.

Between (i) and (j), 1000*3212*0000 enters and is able both to exploit the environment more efficiently and to protect itself better against 00 predators. From (j) to (l) we first see the appearance of the basal species 1000*3112*0000, which is characterized by an improved search strategy. This is followed by the appearance of a predator 00*1212*0111, which has a tag gene identical to the environment, and thus has evolved a perfect camouflage. We have only shown 4500 generations of this particular simulation. On longer time scales, the cooperative 0111 environment usually gives rise to very stable states dominated by basal species with strategies such as 00001000. These are both capable of exploiting the environment efficiently and defending themselves very well against predators.

In our second example, shown in Figures 13 and 14, the environment consists of the strategy Tit-for-Tat, and the dynamics now contains the same predator response function as in model 1. A succession of oscillatory periods (some remarkably complex) is seen. We shall not describe these in detail here; instead we restrict our attention to a few selected food webs.

At (a), we find a food web with some general features that are commonly seen in the simulations. The basal species 01*31*10, 11*31*10, and 0111*31*10 have all evolved a partial preference for playing against the environment 01; the intermediate species 10*30*01 has developed a preference for the basal species with appearance 10, and is at the same time camouflaged as the environment 01. The top predator 00*31*10 has evolved a partial preference for playing against the environment, and is thus able to exploit 10*30*01 which attempts to hide in the environment. A predator 00*30*11, which 11*31*10 cannot distinguish from the environment, also exists in the web.

In the transition between (b) and (c), we first see the appearance at (b) of a top predator 00*10*00, which exploits the insufficiently specific preferences of the intermediate species 10*30*01. Shortly thereafter at (c), a basal species 0111*21*00, with the same appearance 00 as this predator, enters. This is an example of how

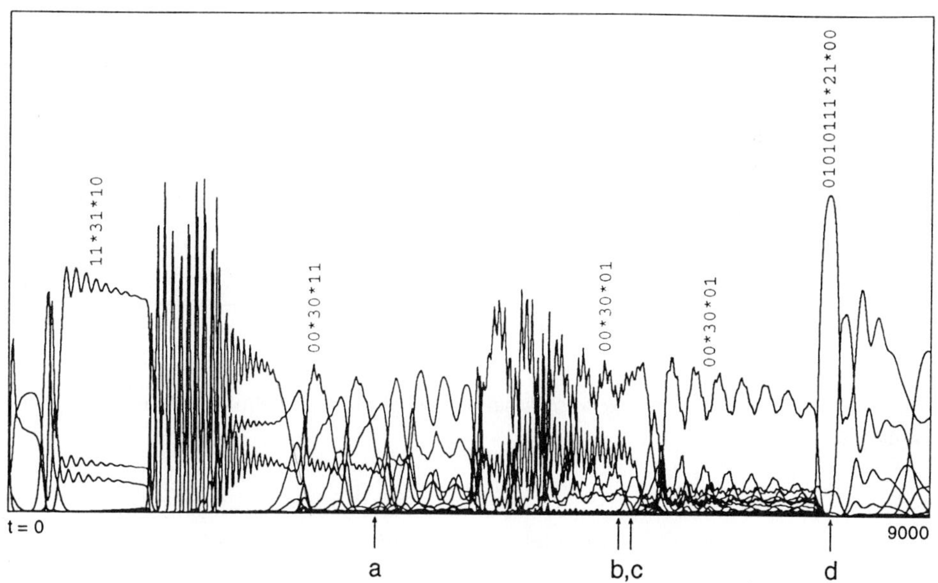

FIGURE 13 A simulation of model 2 with a Tit-for-Tat environment and a response function independent of preferences.

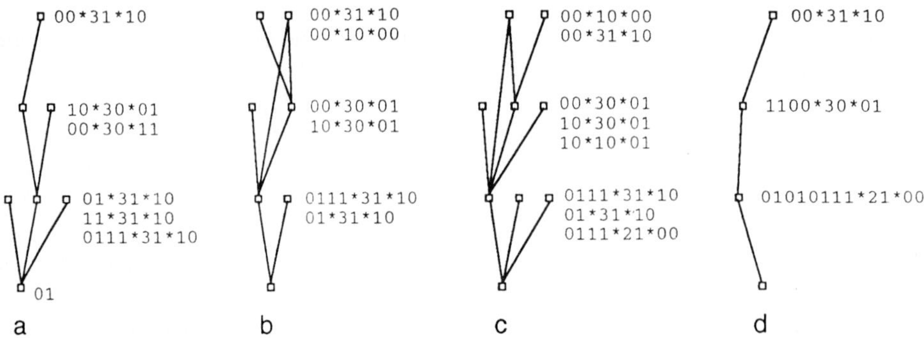

FIGURE 14 Food webs extracted at selected points in time (marked as (a)–(d) in Figure 13) from the simulation above.

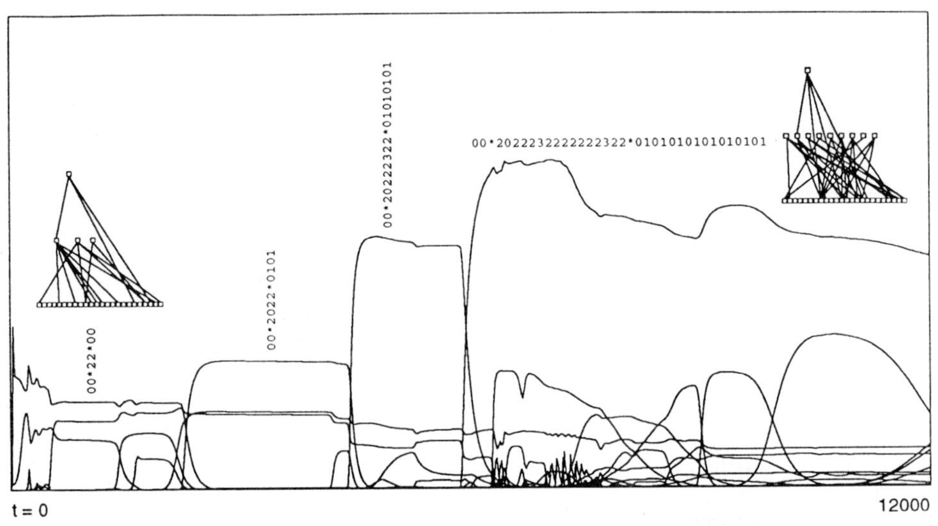

FIGURE 15 Species densities from a simulation of model 2 with 24 environment strategies ($p_C = 0.4$).

mimicry can be seen in this model. Simulations on longer time scales with complex environments to encourage more complex appearances are, however, needed to demonstrate that this effect appears also in a more general setting.

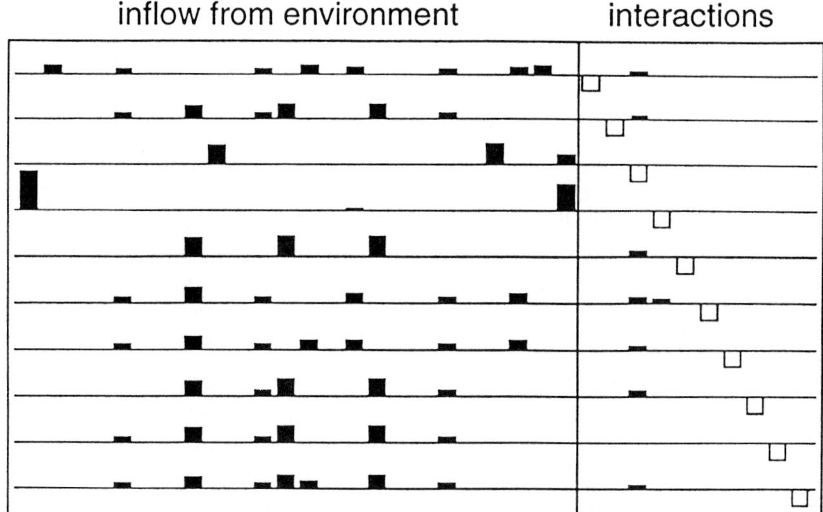

FIGURE 16 The energy flow matrix at $t = 12000$ in the simulation above.

In Figure 15, the first 12,000 generations of a simulation of model 2 with 24 environment strategies generated at random with $p_C = 0.4$ is shown. One can see that the preference genes increase in length with time and match the environment strategies more accurately. In Figure 16 we show the energy flow matrix at $t = 12000$ in this simulation. In random environments, model 2 appears to generate a higher degree of specificity, while typical lengths of food chains remain comparable to those of model 1. More extensive statistical studies of model 2 have not yet been carried out.

4. CONCLUSIONS AND DISCUSSION

Artificial ecologies can be useful in developing our understanding of real ecosystems. Systems can be studied on time and length scales not accessible in real ecological experiments,[37] data can easily be collected, and systems can be manipulated without restrictions. Furthermore, an artificial ecosystem containing simple organisms with lifelike behavior may have fewer built-in theoretical prejudices than a simple mathematical model based, say, on the Lotka-Volterra equations, and could be more likely to surprise us and provide new insights.

This approach assumes some degree of universality in community properties, so that communities consisting of simple artificial organisms in some respects are similar to biological communities. Such similarities could, for example, show up in various statistical regularities and scaling relations. This probably requires taking some of the physical constraints on real ecosystems into account, such as conservation of energy and matter, and the dissipation of energy in irreversble processes.

The simulations of the previous sections illustrate how the introduction of resource flows into very simple game-theory-based evolutionary models allows us to observe a number of phenomena connected with who eats whom in an ecology. The food webs observed are qualitatively similar to those found in nature; for environments consisting only of a few resources, quite simple structures are found, while more diverse environments give rise to more complex webs. In the more complex webs, most of the productivity of the system is typically found at low trophic levels. Food chain lengths are typically comparable to those of real food chains.

These models, or closely related models with more generic interactions, could also serve as a basis for quantitative comparisons with biological data. For the example of food webs, results from artificial ecosystems might be useful in attempting to determine whether various conjectured regularities in food web structure are real or caused by limitations in the available experimental data (the number of experimentally observed food webs reported in the literature is fairly small, at most a few hundred; see Cohen et al.[4] for a compilation). For example, the statistical results found for model 1, e.g., for the number of trophic links as a function of the number of species, do appear to be in reasonable agreement with proposed structural properties of food webs. Such a project would require much more extensive computer simulations than those reported here.

In our second model, where a less trivial genotype to phenotype mapping was used, the evolution of camouflage mechanisms and mimicry was seen. Mimicry effects in particular should be further studied in simulations of larger systems where organisms can develop more complex appearances. The statistical properties of this model, which encourages a higher degree of specialization, need to be further investigated.

These models could also be extended in several directions. One obvious extension would be to consider more than one resource (this has also been proposed by Holland[9]). Another would be to introduce spatial degrees of freedom. In a spatially extended model of the coevolving strategy,[18] we have observed a rich variety of spatiotemporal dynamic behavior. Introducing resources in a spatial model would allow us to treat heterogenous environments in a natural way.

A simplification that appears in the models above is the treatment of energy dissipation, which corresponds to simply removing dead organisms from the system. In this way, no cyclic energy flows due to decomposition can appear. While the same approximation is often made by ecologists when reporting food webs and in many theoretical food web models, it would be interesting to construct a model which incorporated decomposition and cyclic resource flows in a natural fashion.

Finally, to further develop quantitative comparisons between real and artificial ecologies, one should give some thought to what quantities should be measured. This could for example, involve developing observables that more accurately characterize the patterns of resource flows in an ecosystem, e.g., in terms of information theory.

ACKNOWLEDGMENTS

K.L. thanks Doyne Farmer for stimulating discussions on evolutionary models, the Santa Fe Institute for financial support for two extended visits, and the Center for Nonlinear Studies at Los Alamos National Laboratory for supporting the first of these visits. This research was also supported by the Swedish Natural Science Research Council.

REFERENCES

1. Ackley, D., and M. Littman. "Interactions Between Learning and Evolution." In *Artificial Life II*, edited by C. G. Langton, J. D. Farmer, S. Rasmussen, and C. Taylor, 487–509. Redwood City CA: Addison-Wesley, 1991.
2. Axelrod, R. *The Evolution of Cooperation.* New York: Basic Books, 1984.
3. Bagley, R. J., and J. D. Farmer, "Spontaneous Emergence of a Metabolism." In *Artificial Life II*, edited by C. G. Langton, J. D. Farmer, S. Rasmussen, and C. Taylor, 93–158. Redwood City CA: Addison-Wesley, 1991.
4. Cohen, J. E., F. Briand, and C. M. Newman. *Community Food Webs.* Berlin-Heidelberg: Springer-Verlag, 1990.
5. Drake, J. W. "The Role of Mutation in Microbial Evolution." In *Evolution in the Microbial World*, edited by M. J. Carlile and J. J. Skehel, 41–58. Cambridge: Cambridge University Press, 1974.
6. Eldredge, N., and S. J. Gould. "Punctuated Equilibria: An Alternative to Phyletic Gradualism." In *Models in Paleobiology*, edited by T. J. M. Schopf, 82–115. San Fransisco: Freeman Cooper, 1972.
7. Farmer, J. D., S. A. Kauffman, and N. H. Packard. "Autocatalytic Reproduction of Polymers." *Physica D* **22** (1986): 50–67.
8. Fontana, W., and L. W. Buss. "The Arrival of the Fittest: Toward a Theory of Biological Organization." *Bull. Math. Biol.* in press.
9. Holland, J. H. *Adaptation in Natural and Artificial Systems*, 2nd ed. Cambridge, MA: MIT Press, 1992.
10. Holling, C. S. "The Functional Response of Predators to Prey Density and Its Role in Mimicry and Population Regulation." *Mem. Ent. Soc. Can.* **45** (1965): 1–60.
11. Kauffman, S. A. "Metabolic Stability and Epigenesis in Randomly Constructed Genetic Nets." *J. Theor. Biol.* **22** (1969): 437–467.
12. Kauffman, S. A., and S. Levin. "Towards a General Theory of Adaptive Walks on Rugged Landscapes." *J. Theor. Biol.* **128** (1987): 11–45.
13. Kauffman, S. A., and S. Johnsen. "Coevolution to the Edge of Chaos: Coupled Fitness Landscapes, Poised States, and Coevolutionary Avalanches." In

Artificial Life II, edited by C. G. Langton, J. D. Farmer, S. Rasmussen, and C. Taylor, 325–369. Redwood City CA: Addison-Wesley, 1991.

14. Kawanabe, H., J. E. Cohen, and K. Iwasaki, eds. *Mutualism and Community Organization.* Oxford: Oxford University Press, 1993.
15. Langton, C. G., ed. *Artificial Life.* Santa Fe Institute Studies in the Sciences of Complexity, Proc. Vol. VI. Redwood City, CA: Addison-Wesley, 1989.
16. Langton, C. G., J. D. Farmer, S. Rasmussen, and C. Taylor, eds. *Artificial Life II.* Santa Fe Institute Studies in the Sciences of Complexity, Proc. Vol. X. Redwood City, CA: Addison-Wesley, 1989.
17. Levin, S. A. "Community Equilibria and Stability, an Extension of the Competitive Exclusion Principle." *Am. Nat.* **104** (1970): 413–423.
18. Lindgren, K., "Evolutionary Phenomena in Simple Dynamics." In *Artificial Life II*, edited by C. G. Langton, J. D. Farmer, S. Rasmussen, and C. Taylor, 295–312. Redwood City, CA: Addison-Wesley, 1991.
19. Lindgren, K., A. Nilsson, M. G. Nordahl, and I. Råde, "Regular Language Inference Using Evolving Neural Networks." In *COGANN-92, International Workshop on Combinations of Genetic Algorithms and Neural Networks*, edited by L. D. Whitley and J. D. Schaffer, 75–86. Los Alamitos CA: IEEE Computer Science Press, 1992.
20. Lindgren, K., and M. G. Nordahl. "Evolutionary Dynamics of Spatial Games." *Physica D* to appear.
21. Lindgren, K., and M. G. Nordahl. In preparation.
22. MacArthur, R. H., and R. Levins. "Competition, Habitat Selection and Character Displacement in a Patchy Environment." *Proc. Natl. Acad. Sci. U.S.* **51** (1964): 1207–1210.
23. MacArthur, R. H., and E. O. Wilson. *The Theory of Island Biogeography.* Princeton: Princeton University Press, 1967.
24. May, R. M. "Will a Large Complex System be Stable?" *Nature* **238** (1972): 413–414.
25. May, R. M. *Stability and Complexity in Model Ecosystems.* Princeton: Princeton University Press, 1973.
26. May, R. M. "How Many Species are There on Earth?" *Science* **241** (1988): 1441–1449.
27. Maynard Smith, J. *Evolution and the Theory of Games.* Cambridge: Cambridge University Press, 1982.
28. McNaughton, S. J., M. Oesterheld, D. A. Frank, and K. J. Williams. "Ecosystem-Level Patterns of Primary Productivity and Herbivory in Terrestrial Habitats." *Nature* **341** (1989): 142–144.
29. Niering, W. A. "Terrestrial Ecology of Kapingamarangi Atoll, Caroline Islands." *Ecol. Mono.* **33** (1963): 131–160.
30. Nordahl, M. G. "A Spin-Glass Model of Coevolution." Preprint.
31. Ohno, S. *Evolution by Gene Duplication.* Berlin-Heidelberg: Springer-Verlag, 1970.

32. Paine, R. T. "Food Webs: Road Maps of Interactions or Grist for Theoretical Development?" *Ecology* **69** (1988): 1648–1654.
33. Paine, R. T. "Food-Web Analysis Through Field Measurement of Per Capita Interaction Strength." *Nature* **355** (1992): 73–75.
34. Pimm, S. L. *Food Webs.* London: Chapman & Hall, 1982.
35. Pimm, S. L., and R. L. Kitching. "The Determinants of Food Chain Lengths." *Oikos* **50** (1987): 302–307.
36. Pimm, S. L., J. H. Lawton and J. E. Cohen. "Food Web Patterns and Their Consequences." *Nature* **350** (1991): 669–674.
37. Pimm, S. L. *The Balance of Nature?: Ecological Issues in the Conservation of Species and Communities.* Chicago: The University of Chicago Press, 1991.
38. Polis, G. A. "Complex Trophic Interactions in Deserts: An Empirical Critique of Food-Web Theory." *Am. Nat.* **138** (1991): 123–155.
39. Preston, F. W. "The Canonical Distribution of Commonness and Rarity: Part I, *Ecology*, **43** (1962) 185–215; Part II, *Ibid.* **43** (1962) 410–432.
40. Rasmussen, S., C. Knudsen, and R. Feldberg. "Dynamics of Programmable Matter." In *Artificial Life II*, edited by C. G. Langton, J. D. Farmer, S. Rasmussen, and C. Taylor, 211–254. Redwood City CA: Addison-Wesley, 1991.
41. Raup, D. M. "Biological Extinction in Earth History." *Science* **231** (1986): 1528–1533.
42. Sherrington, D., and S. Kirkpatrick. "Solvable Model of a Spin Glass." *Phys. Rev. Lett.* **35** (1975): 1792–1796.
43. Stenseth, N. C., and J. Maynard Smith. "Coevolution in Ecosystems: Red Queen Evolution or Stasis?" *Evolution* **38** (1984): 870–880.
44. Yeager, L. "Computational Genetics, Physiology, Metabolism, Nural Systems, Learning, Vision, and Behavior, or PolyWorld: Life in a New Context." This volume.
45. Van Valen, L. "A New Evolutionary Law." *Evol. Theory* **1** (1973): 1–30.
46. Winemiller, K. O. "Spatial and Temporal Variation in Tropical Fish Trophic Networks." *Ecol. Mono.* **60** (1990): 331–367.

Alan R. Johnson
Department of Biology, University of New Mexico, Albuquerque, NM 87131;
e-mail: ajohnson@algodones.unm.edu

Evolution of a Size-Structured, Predator-Prey Community

Many biological and ecological variables are related to an organism's body size or mass. These relations can often be expressed as allometric functions of the form:

$$X = aM^b = cL^d$$

where X is the variable of interest, M is the body mass of the organism, L is a measure of the linear size of the organism, and a, b, c, and d are constants. This study constitutes an effort to mimic the assembly of an ecological community with certain body-size constraints explicitly taken into account, and to investigate the resulting community structure and dynamics.

The model simulates organisms on a lattice. Organisms are assigned to discrete size classes. The smallest size class, 0, consists solely of autotrophs. All organisms of larger size classes are heterotrophs. Heterotrophs can eat only organisms smaller than themselves. Both an organism's size class and its dietary specificity (which size classes it eats) are genetically encoded in bit strings, which can mutate during reproduction.

Heterotrophs move randomly in search of food. Movement rate is specified as a function of body mass. When a predator encounters a suitable prey, it consumes the prey and augments its internal energy reserves. Metabolic energy costs are subtracted from the energy reserves at a rate that is an allometric function of body mass. If an organism's energy reserves exceed an upper threshold, it reproduces. If energy reserves fall below a lower threshold, the organism dies of starvation.

Starting with only autotrophs, simulations exhibited the evolution of heterotrophs, the establishment of typical oscillatory predator-prey dynamics, periods of stable community composition punctuated by abrupt shifts, a succession of food web structures, and convergence to a small subset within the much larger set of possible community structures and size distributions. The abundance, biomass, population growth rate, and biomass production rate of dominant heterotrophs exhibited power-law scaling with body mass, establishing new allometric functions that had not been explicitly programmed in beforehand. The current model leads to relatively simple communities with few species, and the emergent allometric relationships differ quantitatively from those reported in the ecological literature, indicating that some features of actual ecosystems may not be adequately represented. Nonetheless, the basic approach promises to be a valuable tool in the elucidation of size-related constraints in ecological systems.

INTRODUCTION

Biological organisms are amazingly diverse. Their diversity in form is readily apparent—as a simple example, consider the variety of colors and shapes exhibited in the reproductive organs of vascular plants (i.e., flowers). However, there is another factor in the diversity of organisms which is so obvious that it is frequently overlooked. Organisms differ greatly in terms of their body size (or mass). It is becoming increasingly clear that organismal size can have profound consequences, both for the organism itself and for the ecological community of which it is a member.

Modern biological organisms have masses spanning a range of 21 orders of magnitude, from unicellular, bacterialike mycoplasmas ($< 10^{-13}$ g) to blue whales ($> 10^{8}$ g).[33] Species are not uniformly distributed over this range. Numerous researchers have compiled size distributions of species within selected taxa. Van Valen estimated size distributions of flowering plants, birds, and mammals.[39] May compiled distributions for a variety of groups, and synthesized an estimated size distribution for all terrestrial animals.[20]

Attempts have been made to explain the observed distribution of body sizes among species. Most empirical data display a unimodal distribution. Species richness drops relatively slowly over larger-than-modal logarithmic size classes, whereas the decrease is more rapid, and the distribution truncated, at smaller sizes. Explanations of this distribution have generally focused on the differential partitioning of the environment into niches and rates of resource utilization by organisms of varying body size. A combinatorial model proposed by Hutchinson and MacArthur produces a qualitatively plausible size distribution.[16] As May[20] points out, this model implies that at the large end of the spectrum, the number of species is roughly proportional to L^{-2}, where L is the length or characteristic linear dimension of an organism. Much of the empirical data conforms reasonably well to this prediction. A more recent theoretical approach explains the size distribution of species in terms of reproductive power, the rate at which assimilated energy is converted to offspring.[9] Reproductive power depends upon the rate at which energetic resources can be gathered from the environment, and upon the physiologically determined rate at which offspring can be produced. The former is the primary constraint for large organisms, whereas the latter constrains small organisms.

Not only the number of species, but also the number of or total biomass of individuals varies with body size. Based on data for pelagic marine ecosystems, Sheldon and his colleagues have argued that the amount of biomass in logarithmic size classes is constant.[35] In other words, organism biomasses follow an exponential distribution. Lurie and Wagensberg demonstrate that an exponential probability density function maximizes the biomass diversity (as measured by Shannon's entropy) subject to the constraint of a constant total biomass.[19] Data for other systems is scarce, but this relationship may have wider applicability.[30] It has even inspired one fanciful calculation of the number of monsters that could be supported by Loch Ness.[34] The underlying reason for the observed constancy of biomass over logarithmic size classes remains obscure, although Pahl-Wostl has recently argued that this may result from a dynamic self-similar structure in the matter and energy flows of a trophic network.[27]

Many important biological variables are related to an organism's body size. The relationship often takes the form of a power law:

$$X = aM^b = cL^d \tag{1}$$

where X is the biological variable of interest, M is the mass of the organism, L is the length of the organism (or a characteristic part of the organism), and a, b, c, and d are constants. Some variables are directly proportional to the linear size of the organism, such that $d = 1$ in Eq. (1). Such variables are said to display isometric scaling with body size. More frequently, however, a power-law relationship with $d \neq 1$ is observed, in which case the variable is said to display allometric scaling.

Allometric scaling of biological structures is often related to physical constraints. For instance, in large trees, trunk diameter is not linearly related to height, but rather to the 3/2 power of height. The reason for this appears to be that the critical diameter at which buckling of a cylinder occurs is proportional to the 3/2 power of height.[22] A similar relationship applies to the dimensions of bones in many animals.[22]

Allometric relationships apply not only to structural features of organisms, but to functional aspects as well. The metabolic rate, or energetic requirements of an organism, is largely determined by its size. Most studied is basal metabolic rate, the amount of energy expended per unit time by a resting organism in a favorable thermal environment (representing a near-minimal energy requirement for the maintenance of life). In 1932 Kleiber[17] demonstrated that for a variety of homeothermic[1] animals, basal metabolic rate varies as the 3/4 power of body mass. The same scaling exponent has since been shown to apply to standard (not necessarily basal) metabolic rates in poikilotherms and in unicellular organisms.[30] In homeotherms, maximal metabolic rates (under conditions of physcial exertion) are approximately 10 times basal rates, and some studies suggest that its scaling exponent may be somewhat greater than 3/4.[30,38] The average realized metabolic rate for free-ranging homeotherms is typically 2–3 times the basal rate and, although the data are limited, there is an indication that this quantity scales with mass to an exponent somewhat less than 3/4.[30]

Metabolic rates are sometimes expressed on a per-unit body mass basis, in which case they are referred to as specific rates. If a metabolic rate scales with mass to an exponent b, the corresponding specific rate must scale with the exponent $b-1$. Thus, specific basal metabolic rates scale with an exponent of $-1/4$. The decrease of specific metabolic rates with increasing body mass reflects the greater energetic efficiency of large organisms. This is one of the advantages of being big.

The power that must be expended in support of 1 g of elephant biomass (5.8×10^{-4} Watt) is less than that required for support of 1 g of mouse biomass (1.1×10^{-2} Watt). However, due to the larger body mass of the elephant (2600 kg) than the mouse (20 g), the power requirements on a per-individual basis are larger for the elephant (1.5×10^{3} Watt) than for the mouse (2.2×10^{-1} Watt). Thus, as would be expected, larger animals require greater amounts of food. Since food resources are spatially distributed within the environment, larger animals must search larger areas to obtain the food they require.

Searching larger areas requires movement over greater distances per unit time. Numerous studies have been conducted on the effect of body size on locomotion. Whether swimming, flying, walking, or running, it is generally observed that the rate of movement increases with increasing body size. Allometric exponents for maximal velocities are in the range of 0.14 to 0.38.[30] Energetic costs associated with

[1]The terms *homeothermic* and *poikilothermic* are roughly synonymous with the less technical terms warm-blooded and cold-blooded, respectively, The terms warm-blooded and cold-blooded can be misleading since, in some environments, poikilotherms may have body temperatures that are higher than those of homeotherms. The proper contrast is between the relatively constant and physiologically controlled body temperature of homeotherms, as opposed to the more variable and environmentally controlled temperature of poikilotherms.

movement also scale allometrically with body size. Exponents for transportation costs are in the range of 0.52 to 0.86.[30]

For animals that occupy an identifiable home range, the area of the home range scales allometrically with body size. Among birds and mammals, carnivores have larger home ranges than herbivores of similar size, and home range area increases more rapidly as a function of body size for carnivores than herbivores. One study of reptiles indicates that they have smaller home ranges than homeotherms of corresponding size. These trends are what would be expected based on energetic considerations.[30]

Since a larger individual requires a larger portion of the environment and the available resources to meet its metabolic requirements, a finite environment can support fewer large animals than small animals. By this reasoning, population density (the number of individuals per unit area) should vary inversely with body size. Indeed, based on data for a variety of mammals, birds, reptiles, amphibians, and invertebrates, Peters has estimated that population density scales as $M^{-0.98}$, although there is substantial scatter around the regression line.[30] Debate has arisen, however, concerning the exponent in this scaling relationship. Recent work, focusing on insects dwelling on plant surfaces, has suggested that the exponent may depend on the fractal dimension of the habitat.[24,42] More fundamentally, Brown and Mauer[8] argue that the data used by Peters and others may be biased, primarily representing densities of common species from scattered localities. Their analyses focus on faunal assemblages at a single site, including rare species. They find that, on average, densities vary as $M^{-0.30}$ for North American land birds, as $M^{+1.17}$ for nine Chihuahaun desert rodents, and as $M^{0.0}$ for marine fish.

Population growth rates also vary with body size. The intrinsic rate of natural increase, $r_{\max}$, is the per-capita number of offspring produced per unit time under conditions of unconstrained (i.e., exponential) population growth. This quantity has been observed to be nearly proportional to $M^{-1/4}$, reminiscent of the scaling of specific metabolic rates.[3,14] Population growth can also be quantified as a specific production rate, measured as the ratio of gross annual production of new biomass (P) to the time-averaged population biomass for the year (B). A detailed analysis focusing on invertebrates concluded that P/B scales with an exponent of -0.37, significantly different than $-1/4$.[2] These results are regarded as at least paradoxical, if not contradictory.[30] Studies of marine and freshwater pelagic communities have demonstrated shallow slopes for the community as a whole (exponents around -0.18), but steeper slopes (about -0.39) within taxonomic groups.[5,12] Clearly, the exponent value depends upon the taxonomic grouping of the data, and may reflect a combination of physiological and ecological effects.[5,12]

Body size is also an important determinant of community structure. Trophic relationships are size-related, since most predators consume prey that are smaller than themselves.[2] Studies of terrestrial vertebrates indicate that not only does

[2] Exceptions, such as predation on moose by wolves, are often due to social organization and cooperation among individual predators (e.g., the wolf pack).

mean prey size increase with increasing predator size, but so does the diversity of prey.[15] The distribution of prey sizes utilized by a predator is approximately log-normal and skewed to smaller sizes. Insectivores consume a wider range of prey sizes than carnivores, and piscivores are intermediate.[41]

Size-selective predation can have cascading effects through trophic levels, as has been particularly well illustrated in aquatic systems.[6,23,40] Consider a typical pelagic food chain consisting of phytoplankton, which are grazed by zooplankton, which are eaten by small planktivorous fish, which are consumed by large piscivorous fish. An increase in the number of top predators (i.e., piscivores) will increase predation pressure on the planktivores, leading to a lower biomass at that trophic level, and a shift toward smaller body sizes. The reduction in planktivores will reduce predation pressure on the zooplankton leading to a greater biomass at that trophic level, and a shift toward larger species and larger individuals within species. Finally, due to their increased numbers and larger sizes, the zooplankton will be capable of grazing on phytoplankton more efficiently, leading to decreased biomass, smaller organism sizes, and shifts in species composition at the primary producer level.

The size of an organism affects its ability to compete for environmental resources. Among plants, larger individuals may be at a competitive advantage for light (shading out smaller plants) and for water (with their more extensive root systems). Large size also frequently confers a competitive advantage to animals, although the underlying mechanisms are subject to debate.[29] Moreover, the work by Brown and Mauer[8] indicates that in a variety of ecosystems, larger organisms monopolize a greater proportion of the available resources.

Given the large number of ways in which body size affects an organism and its interaction with the environment, it should come as no surprise that body size is subject to natural selection. Again, given the large number of consequences of body size, it is easy to imagine that there might be many different, and often conflicting, selective pressures operating on organismal body size. It may therefore be somewhat surprising that paleontologists have identified a systematic trend toward evolution of organisms with larger body size over time. This empirical generalization is known as Cope's rule. This trend has attracted considerable attention, although no generally accepted explanation is available at this time.[18]

In considering evolutionary trends in body size, it is important to distinguish between two types of trend—anagenetic and cladogenetic.[21] Anagenetic trends are those found within a single lineage, such as evolutionary changes in the heights of humans. Many studies of anagenetic change demonstrate fluctuations, both up and down, in body size over time, although long-term trends ($> 10^6$ yr) seem to favor size increase.[21] Cladogenetic trends are those that involve changes within a taxonomic group, from the level of a genus, family, or order up to the entirety of life forms on earth. Cladogenetic change can involve speciation and extinction events—the initiation and termination of lineages—which are not a part of anagenetic processes. It was cladogenetic changes within vertebrate lineages that led

Cope in the last century to draw attention to the trend that now bears his name, and the same trend has since been documented in invertebrate lineages.[4,25]

In many respects the long-term anagenetic trend toward size increase is the most difficult to explain. It could be a result of sampling bias, reflecting a preference among researchers to work on lineages showing size increase, or, taxonomic uncertainty could lead cladogenetic changes to be misclassified as anagenetic.[21] If the trend is real, it presumably reflects some overall advantage of large body size. Numerous advantages of large size are evident: greater metabolic efficiency, avoidance of predation, greater competitive ability. However, there are also disadvantages to being large: lower reproductive rate, greater resource requirements, lower population density. The selective forces operating on body size would seem to depend to a large extent on details of the biology and ecology of the particular species. It seems unlikely that a simple, single explanation can be advanced for anagenetic trends in a wide variety of life forms.

McKinney[21] has offered a general scenario based partly on earlier work of Stanley,[36] for explaining cladogenetic trends toward increasing body size. First, it is argued that most clades (groups linked by descent from a common ancestor) originate at a relatively small body size. In part, this is a simple consequence of the fact that relatively small-bodied species are more common than large-bodied species. Moreover, many clades originate after major extinction events. For a variety of reasons (lower reproductive rate, greater resource requirements, lower population density) large-bodied species may be more susceptible to such extinction events,[11,31] leaving an even greater preponderance of small-bodied species among the survivors. These small organisms are well suited for colonization of new habitats, and rapid diversification and speciation ensues. The subsequent increase in maximum or mean body size within the clade can then be regarded as a simple diffusion process. Individual lineages are envisioned to execute a random walk in "body size" space, with a reflecting boundary at some physiologically determined minimum size. The trend for the clade is set by those lineages that evolve to large body size, often by species that reoccupy those niches vacated by the extinction of previous large-bodied organisms.

When the entire history of life on earth is considered, another perspective on the evolution of body size can be obtained. The earliest forms of life for which we have fossil records were unicellular organisms. Single cells are necessarily small, as surface-to-volume considerations place an upper limit on how large a cell can be and still efficiently exchange materials with its environment. Multicellular organisms are not only capable of being larger than unicellular organisms, but by allowing for specialization of different cell types within the organism, they can become more complex. Thus, the overall trend toward larger organisms may be a necessary corollary of the evolution of more complex biological entities.[4]

THE MODEL

MOTIVATION

The empirical data attest to the large number of biological and ecological variables that depend upon body size. A number of scaling relationships were reviewed in the Introduction, and many more have been reported in the literature.[30,33] No doubt many of these relationships are interdependent. For instance, given the scaling of metabolic energy requirements with body size, and some assumptions about the distribution of resources in the environment, the scaling of population density with body size may follow as a necessary consequence.

Similarly, paleontological studies document trends in the evolution of body size. But any particular evolutionary trajectory can be thought of as a single realization of an underlying stochastic, dynamical process. Given a single realization, it is difficult (if not impossible) to discern those features that are general and those that are idiosyncratic. In other words, to what extent (or in which aspects) is evolutionary history robust to small changes in initial conditions, and to what extent (or in which aspects) is it sensitive to historical contingencies?

Adequately addressing these mechanistic questions requires an experimental approach. Some ecological and evolutionary experiments in actual biological systems are possible, but the complexity of natural ecosystems and the long time scale of evolutionary change limit what can be done. Artificial Life (ALife) offers a complementary approach, in which the evolution and ecological implications of body size (or its analogue) can be studied in artificial organisms in artificial environments. Principles elucidated for the role and evolution of body size in an appropriate set of artificial systems should form a general theoretical framework, within which the principles governing actual biological systems could be regarded as a special case.

The model presented here is a preliminary attempt to use the paradigm of ALife to study the ecology and evolution of body size. The field is still some distance away from being able to realize artificial life forms in which the ecology and evolution of body size (or its analogue) can be studied directly.[13,28] We are, however, able to simulate the ecology and evolution of body size. In the spirit of ALife, a minimalistic approach was used in developing the simulation model, in the sense that relatively few biological details were explicitly specified, with the hope that additional phenomena would emerge as a result of the dynamics of the system.

DESCRIPTION OF THE MODEL

The model focused on the level of individual organisms. The model explicitly tracked the movement, metabolism, foraging, reproduction, and death of individual organisms within a lattice environment. Population- and community-level dynamics arose as a consequence of the behavior of individuals.

Each organism had two genes, represented in the model as bit strings. One gene controlled the organism's body size. The size gene is bit string containing a single 1; when interpreted as a binary number, the size gene gave the body mass of the organism. The size class of the organism was defined as the log (base 2) of the size gene. For instance, the size genes ...0001, ...0010, and ...1000 would have indicated size class 0, 1, and 3 organisms, with body masses of 1, 2, and 8 units, respectively. The other gene specified the organism's dietary specificity. This gene could contain multiple 1's, the locations of which indicated size classes that could serve as prey for the organism. For instance, a prey gene of ...0101 would have indicated that the organism could eat only size class 0 or size class 2 prey. The restriction was imposed that an organism could only eat prey that were smaller than itself.

Organisms can be divided into two broad categories: autotrophs and heterotrophs. Autotrophs are organisms that fulfill their metabolic energy demands using inorganic sources. In natural systems this role belongs to plants and other photosynthetic organisms, plus certain microorganisms that derive energy from the oxidation of inorganic chemicals. In the model, all autotrophs were size class 0 organisms, and all size class 0 organisms were necessarily autotrophs (since there were no smaller organisms that could serve as their prey). Heterotrophs are organisms that derive energy from the consumption of other organisms. Heterotrophs that eat only autotrophs will be called herbivores, and heterotrophs that eat only other heterotrophs will be called carnivores. The terms predator and prey will be used generically to refer to the consumer and the consumed, respectively, whether the predator is consuming an autotroph or a heterotroph. Note that size class 1 organisms were obligate herbivores, since only autotrophs are smaller than they are.

The model kept track of the internal energy reserves of each organism, on a scale of essentially -10 to 100 (although these bounds could be somewhat exceeded, as explained below). For autotrophs, energy reserves were augmented each time step due to photosynthetic activity. For heterotrophs, energy reserves were augmented by consumption of prey, and diminished by normal metabolism.

Each iteration of the simulation consisted of separate metabolism, foraging, mortality, and reproduction phases. Organisms were initially placed at random locations on a square lattice. In the simulations reported here, lattices of 20×20 and 50×50 sites were used, and the initial community consisted solely of size class 0 organisms (autotrophs).

During the metabolism phase, the internal energy reserves of the organism were adjusted to reflect normal metabolic processes. For autotrophs, energy reserves were augmented to reflect net photosynthesis. In the simulations reported here, net photosynthetic rates were set at 20 energy units per iteration. For heterotrophs, energy reserves were decreased by an amount that depended allometrically on body mass. Note that since the energy scale was normalized to the same range for organisms of all size classes, the corresponding allometric relationship pertained to a specific metabolic rate.

During the foraging phase, heterotrophs searched for and consumed prey; autotrophs were stationary. Heterotrophs executed a random walk, with steps being between nearest-neighbor sites (in the four orthogonal directions of the lattice, with periodic boundary conditions). The movement rate, and therefore area searched, was related to body size. During each iteration of the simulation, each heterotroph was allowed to take a number of steps equal to its size class. If, at any step during its walk, a heterotroph moved to a site occupied by a suitable prey, it ate the prey organism and occupied that site; no further steps were taken during that iteration. If a heterotroph moved to a site occupied by an organism that was its predator, it was eaten. Only one organism at a time could occupy a site, so moves that would have placed a heterotroph on a site with an organism that was neither its predator nor its prey were disallowed.

Upon consuming a prey organism, the internal energy reserves of the predator were augmented by an amount that depended on the relative body masses of predator and prey. In general, if the body mass of the prey was M_{prey} and that of the predator was M_{pred}, the predator received an energetic benefit of $100(M_{\text{prey}}/M_{\text{pred}})$ units. Thus, if a size class S predator consumed a size class $S-1$ prey, it received 50 energy units whereas, if it consumed a size class $S-2$ prey, it received 25 units, and so on.

If an organism's internal energy reserves ever fell to -10 units or less, it died from starvation. During the mortality phase of the simulation, organisms whose energy reserves had fallen below this threshold were removed from the lattice.

If an organism's energy reserves exceeded 100, it reproduced. Reproduction was asexual. A parent with energy reserves of X units was replaced with a parent and an offspring, with energy reserves of $(X - 100)/2$ units each. The offspring was placed at a randomly selected site on the lattice. If the site contained an organism that was a prey for the offspring, the offspring ate it and occupied the site. If the site contained an organism that was a predator of the offspring, the offspring was eaten. If the site contained an organism that was neither predator nor prey, the offspring was killed due to overcrowding. This competition for space introduced density dependence in the realized reproductive rates of the organisms.

Whenever an organism reproduced, there was a possibility of mutation in either (or both) of its genes. The organism could mutate to an adjacent size class (either larger or smaller). Note that autotrophs (size class 0) were allowed to mutate into size class 1 heterotrophs. Mutations could also occur in the dietary specificity gene. These consisted of bit flips (i.e., converting 0's to 1's or vice versa) in any of the allowable positions of the bit string (i.e., those represent prey size classes smaller than the organism). Heterotrophs with dietary specificity genes consisting of all zeros arose, but were nonviable, since they lost energy via metabolism and were unable to replenish this loss via consumption of prey. Since autotrophs had prey genes consisting of all zeros, two mutations were required to convert an autotroph into a viable heterotroph: both the size gene and the dietary specificity gene had to change. The program allows probabilities for size increase, size decrease, and dietary gene bit-flip mutations each to be specified separately. In the simulations presented in this paper, all three probabilities were set at 0.01 (per reproductive event).

SIMULATION RESULTS

Many features of the dynamics of the model can be seen by examination of output from a typical simulation. We take as an example a simulation on a 20 × 20 lattice, starting with an initial population of 63 size class 0 autotrophs. Autotrophs gained 20 energy units per time step, and specific metabolism for heterotrophs was governed by an allometric relationship (Eq. (1)) with constants of $a = 10$ and $b = -0.25$.

In the absence of herbivores, the autotroph population grows rapidly (see Figure 1). In simulations in which the appearance of herbivores was sufficiently delayed, autotrophs filled the entire lattice, their population approaching carrying capacity in the familiar sigmoid fashion. In this simulation, however, mutation gave rise to an herbivore around time step 25, which established a population of size class 1 heterotrophs that expanded rapidly due to the abundant food resources (see Figure 1). As the heterotrophs eventually depleted their resource base, their numbers also

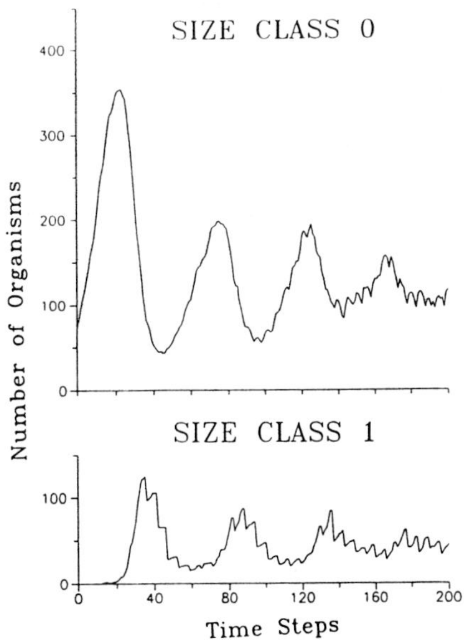

FIGURE 1 Population dynamics during the first 200 time steps of a typical simulation (20 × 20 lattice).

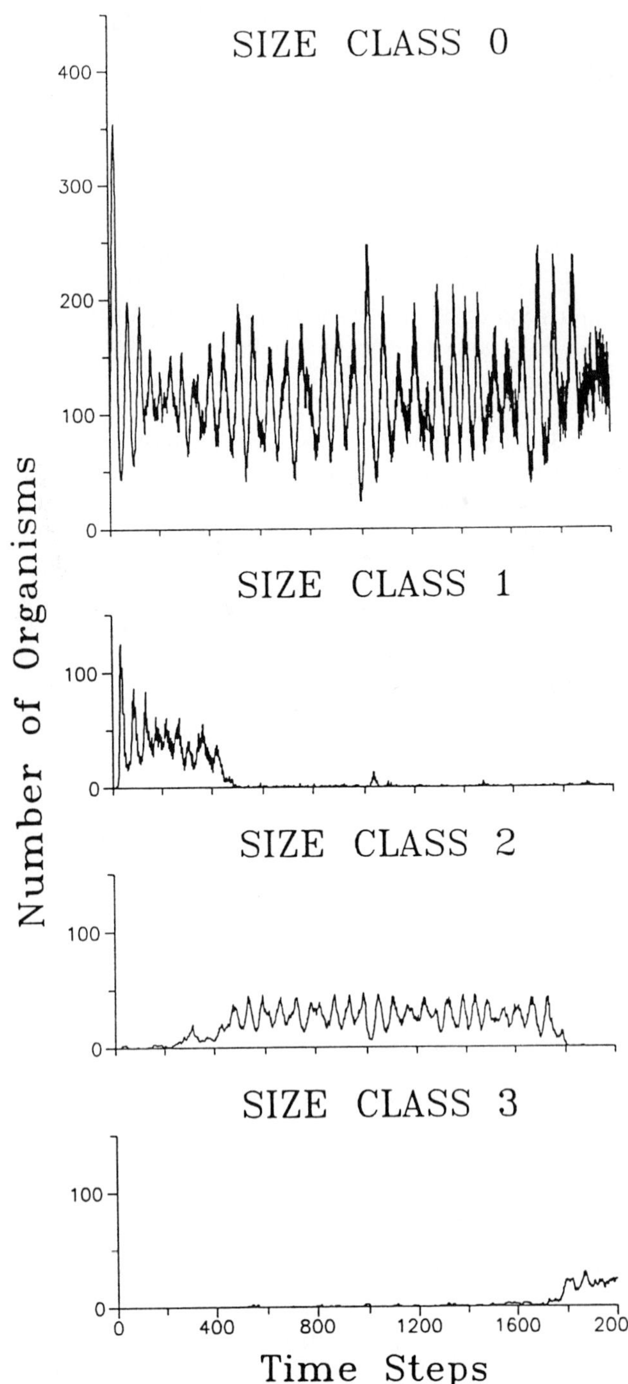

FIGURE 2 Population dynamics during the first 2,000 time steps of a typical simulation (20×20 lattice).

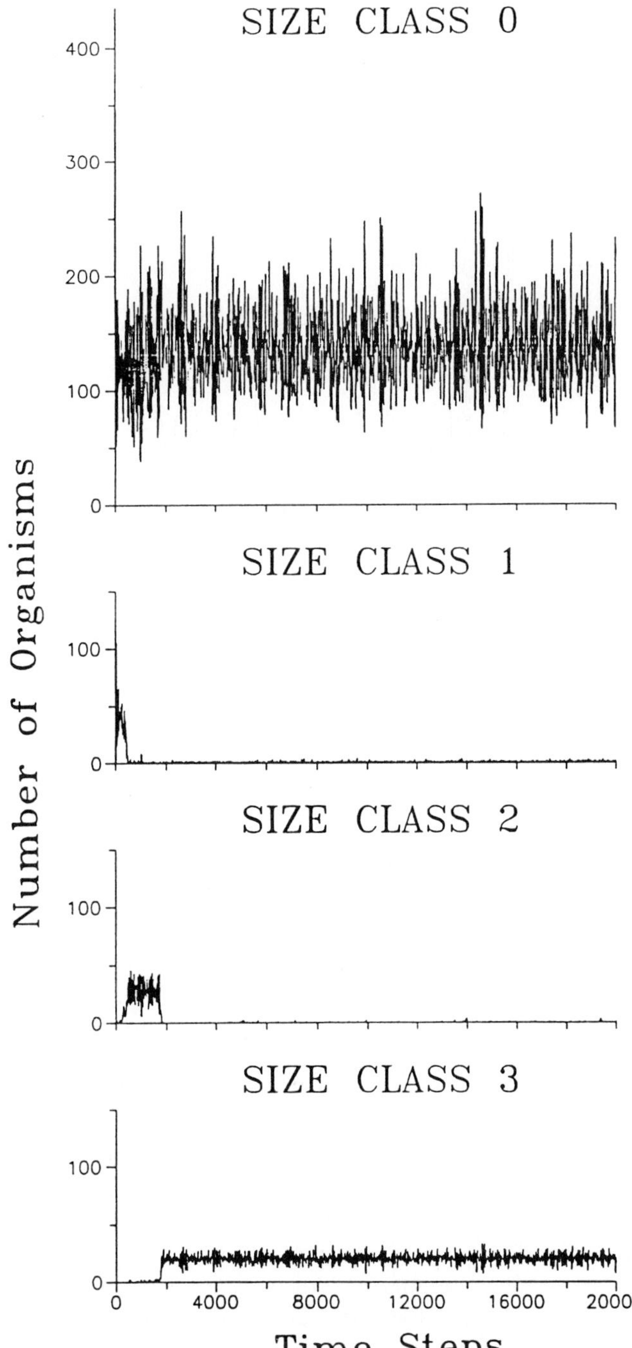

FIGURE 3 Population dynamics during the first 20,000 time steps of a typical simulation (20×20 lattice).

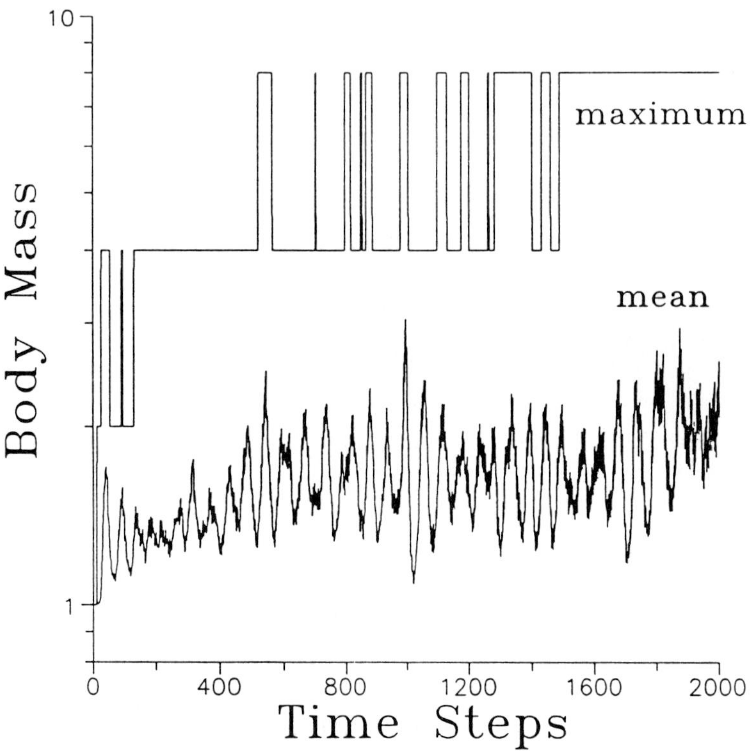

FIGURE 4 Average and maximum body mass for the community as a function of time in a typical simulation (20×20 lattice).

declined, and a typical predator-prey oscillation was established (see Figure 1). In addition, a high-frequency, low-amplitude oscillation superimposed on the primary oscillation became apparent (see Figure 1).

Examining the dynamics over a longer time scale, other features become apparent (see Figure 2). Autotroph populations continued to oscillate around a fairly stable mean, although the amplitude of the oscillations varied. Around time step 450, size class 2 organisms became the dominant heterotrophs, and size class 1 organisms decreased in abundance. At about time step 1800, size class 3 heterotrophs attained dominance, size class 2 organisms became extinct, and size class 1 organisms were occasionally present at very low abundance. The behavior exhibited here is typical of nearly all simulations conducted. Heterotrophs are dominated by one size class at a time. Most of the time, the autotrophs and the dominant heterotrophs exhibit coupled predator-prey oscillations. This is punctuated, at irregular intervals, by relatively rapid shifts in which heterotroph size class is dominant.

Examination of the dynamics over an even longer time scale shows that size class 3 organisms were able to retain dominance for an extended period (see Figure 3). The early evolution of the system gives the appearance of evolving through a succession of metastable states. It is not known whether the community dominated by size class 3 organisms was completely invulnerable to other invasions, or whether potentially successful invaders had simply not evolved over the time scale simulated.

Since only organisms of the smallest size class were present at the beginning of the simulation, evolution—if it were to occur at all—had to lead to an initial increase in average body size. In our example simulation, both mean and maximum body mass (calculated over all organisms in the community) exhibited a generally increasing trend (see Figure 4). Superimposed on this general trend, mean body mass exhibits oscillations reflecting predator-prey dynamics, and maximum body mass exhibits irregularly spaced, brief fluctuations reflecting the sporadic appearance of size class mutants that fail to establish a persistent population (see Figure 4). The steplike character of the maximum body mass curve is due to the use of discrete size classes.

It was possible to reconstruct detailed information about the composition of the community, beyond size information, by looking at the dietary specializations of the heterotrophs. An organism's prey gene specified what organisms it could eat, and its size gene determined what organisms could eat it. Taken together, the two genes specified a particular trophic role for the organism, and thus defined a "trophic species."[10] To refer to a particular trophic species, the following notational convention will be adopted. Each trophic species will be identified by a pair of numbers, the first of which is the size class ($= \log_2$[size gene]), and the second of which is the decimal representation of the prey gene. Thus, a size class 3 organism that could eat organisms of all smaller size classes would have ...01000 as its size gene, and ...00111 as its prey gene, and would be a member of trophic species 3–7.

A food web is a directed graph representing trophic relations (i.e., who eats who) in an ecological community. The sequence of food webs that developed during the first 2000 time steps of the example simulation are shown in Figure 5. Since numerous mutants briefly entered the community and died out without establishing viable, reproducing populations, only trophic species that occupied more than 1% of the lattice sites (i.e., ≥ 5 organisms) were included in the reconstructed food webs. The community started with autotrophs (trophic species 0–0). Herbivores (trophic species 1–1) entered the food web at time step 21. This two-species web persisted until a competing herbivore (trophic species 2–1) came in at time step 255. Once established, both herbivores coexisted until a new trophic species, 2–3, entered the food web at time step 432. This trophic species not only competed with the herbivores for the consumption of autotrophs, but was also a predator on 1–1. Species 1–1 was quickly eliminated from the web (time step 481). This was followed by coexistence of 2–3 and 2–1, with 2–1 occasionally being driven below the 5-organism threshold, until 2–1 was ultimately eliminated at time step

1001. Except for brief fluctuations, the two-species system of 0–0 and 2–3 persisted until another new trophic species, 3–7, entered the web at time step 1769. Being both a competitor and predator of 2–3, trophic species 3–7 was able to quickly eliminate 2–3 from the web (time step 1797). The two-species system of 0–0 and 3–7 persisted until time step 2000 and, except for occasional fluctuations, prevailed for the remainder of the simulation (up to time step 20,000).

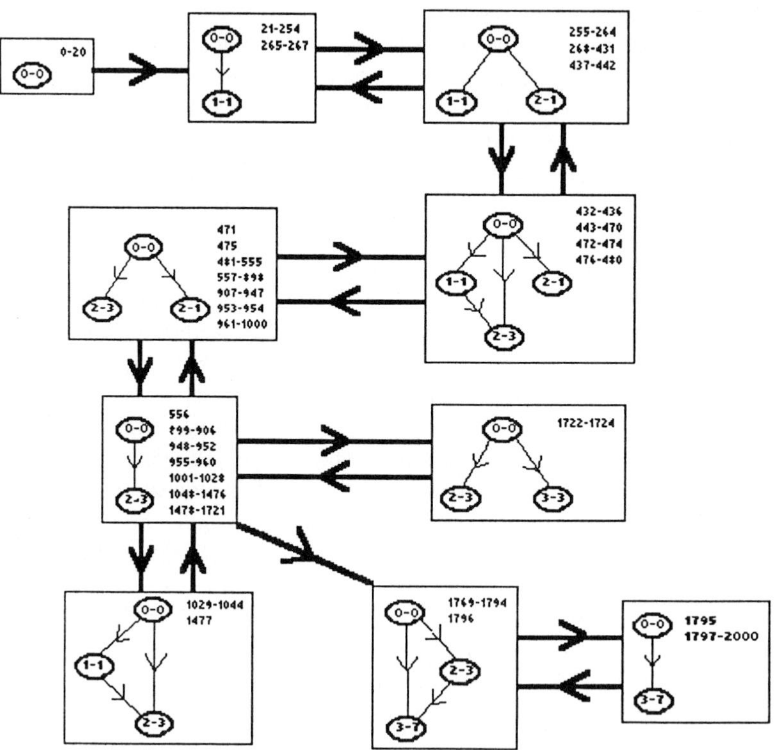

FIGURE 5 Succession of food web structures in a typical simulation (20×20 lattice). Species are represented by a two-number code indicating their size class and dietary specificity. Arrows in food webs show the direction of energy flow (i.e., from prey to predator). Bold arrows indicate transitions that occurred between food web structures. Numerical ranges in each box indicate the time periods the community exhibited a given food web. Food webs do not include rare organisms which occupied $\leq$ 1% of the lattice sites (i.e., <5 individuals).

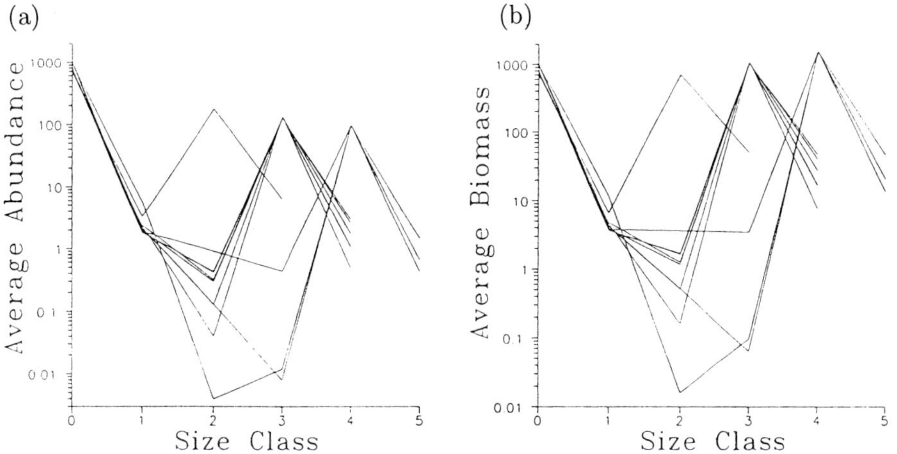

FIGURE 6 Distribution of (a) organisms and (b) biomass among size classes, averaged over time steps 1750 to 2000, for ten replicate simulations (50×50 lattice).

A notable feature of the ecological communities that evolved in this model is that they are dominated, both numerically and in terms of biomass, by autotrophs and a single heterotroph size class. In a single simulation, there is typically a succession of community structures. A particular size class structure may persist for many time steps, and then be replaced with a relatively rapid transition to a community dominated by a different heterotroph size class. Which heterotroph size class dominates the community may vary from simulation to simulation, but the qualitative pattern of relatively stable size class structures punctuated by transitions to new structures is common to all simulations.

Size class structures that evolved in a set of ten simulations on a 50×50 lattice are shown in Figure 6. The simulations were run for 2000 time steps; all other parameters were the same as the example simulation discussed above. The graphs in Figure 6 show abundances and biomasses in each size class averaged over the final 250 time steps. Size class 2 heterotrophs were dominant in one simulation, size class 3 heterotrophs in six simulations, and size class 4 organisms in three simulations. Among simulations with the same dominant heterotroph, the distribution of abundances and biomasses is quite repeatable between simulations.

The abundance and biomass distributions shown in Figure 6 do not exhibit power-law scaling. However, if the abundance, biomass, population growth rate, or biomass production were computed for each heterotroph size class during a period when it was dominant, power-law relationships appear (see Figure 7). Such power laws might also arise in the following situation. Consider a landscape composed of a mosaic of internally homogeneous, relatively isolated patches. Within each patch, communities would evolve that are dominated by a single heterotroph size class,

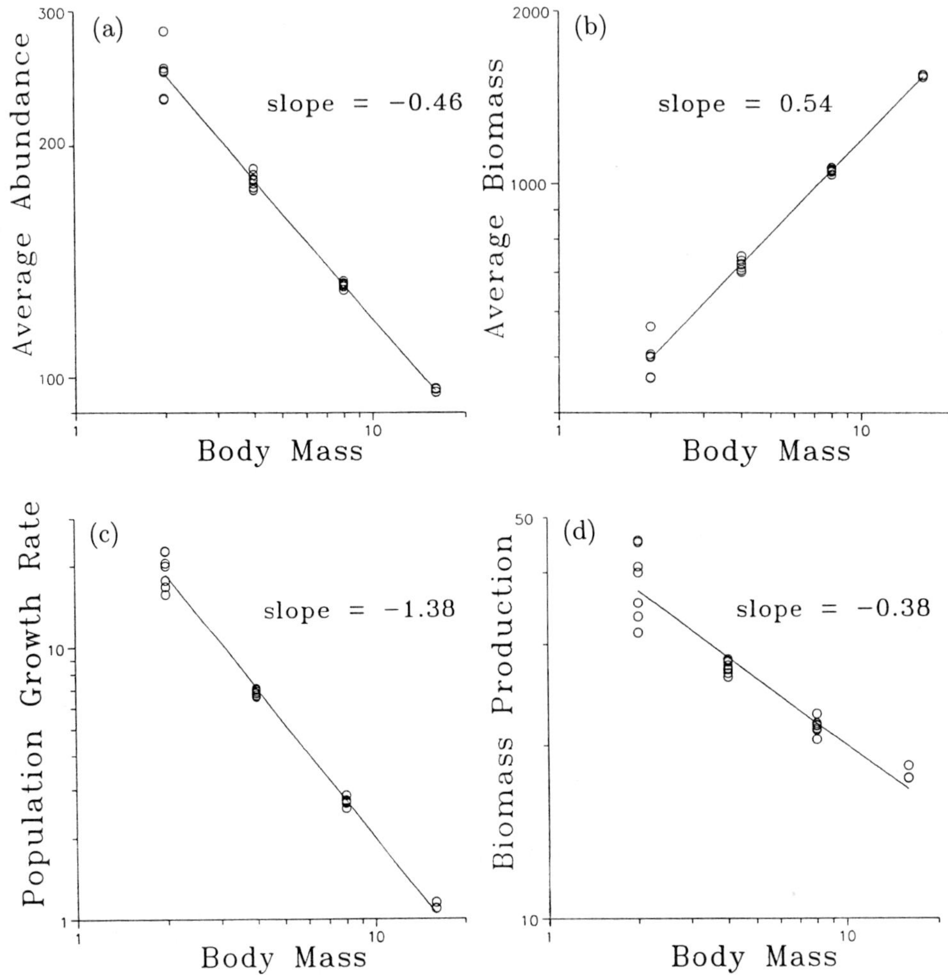

FIGURE 7 (a) Heterotroph abundance, (b) biomass, (c) population growth rate, and (d) biomass production rate, averaged over time periods of stable size class structure and plotted as a function of body mass of the dominant heterotroph.

and that display local abundance and biomass distributions like those in Figure 6. However, different patches would be dominated by different heterotrophs, so that the landscape as a whole might display patterns like those in Figure 7.

The abundance of dominant heterotrophs scaled with body mass as $M^{-0.46}$ (see Figure 7(a)). Since the area of the lattice was constant, the same scaling exponent applies to population density. This differs considerably from the scaling of $M^{-0.98}$ reported by Peters,[30] but comes closer to the scaling of $M^{-0.30}$ reported by

Brown and Mauer for North American land birds.[8] Direct comparisons, however, are complicated by methodological considerations. Brown and Mauer were careful to use only data for species that simultaneously occupied a common environment. The data in Figure 7 is not of that nature. The data used by Peters was culled from various literature sources, but if, as Brown and Mauer assert, it emphasizes common species from a variety of locations, in fact it may be methodologically more comparable to Figure 7.

The total biomass in each size class was simply the number of organisms in each size class multiplied by the body mass of an individual of that size class. Thus, total biomass scaled as $M^{0.54}$, where the exponent was 1 greater than the exponent for abundance (see Figure 7(b)). This increase in biomass with size class is contrary to the constancy of biomass over logarithmic size class intervals predicted by theoretical arguments[19] and observed in marine ecosystems.[35] Again, however, methodological differences complicate such direct comparisons.

In the model, population growth rate was measured as the number of new individuals produced per time step, and production as the biomass of those individuals. These quantities were observed to scale as $M^{-1.38}$ and $M^{-0.38}$, respectively (see Figures 7(c) and 7(d)). As with abundance and total biomass, the exponents necessarily differ by 1. Since production scaled as $M^{-0.38}$, and biomass scaled as $M^{0.54}$, specific production (P/B) must scale as $M^{-0.92}$. This departs strongly from the expectation commonly advanced that specific production should scale in the same manner as specific metabolic rate (i.e., $M^{-0.25}$).[30] Since this argument is based solely on individuals, and not on a consideration of the ecological community in which they are embedded, methodological considerations should not hamper direct comparisons in this case. The simulations used a scaling of $M^{-0.25}$ for specific metabolism at the individual level. These results demonstrate that ecological factors can substantially modify the scaling of specific production from what would be expected from individual metabolic considerations alone.

Autotroph biomass varied according to the dominant heterotroph size class in the community. When size class 1 was dominant, average autotroph biomass was 743.21 ± 49.57 (mean $\pm$ std dev., $N = 7$). When size class 2 was dominant, average autotroph biomass dropped slightly to 693.51 ± 7.44 ($N = 9$). Average autotroph biomass increased in communities dominated by size class 3 or 4 heterotrophs, to 789.02 ± 7.97 ($N = 9$) and 1000.78 ± 16.89 ($N = 3$) mass units, respectively. Trends in autotrophic production mirrored those in average biomass. New autotroph biomass was produced at rates of 116.79 ± 12.07 ($N = 7$), 102.97 ± 1.27 ($N = 9$), 117.74 ± 1.41 ($N = 9$), and 153.17 ± 3.20 ($N = 3$) mass units per time step in communities dominated by size class 1, 2, 3, and 4 heterotrophs, respectively.

Ratios of autotroph-to-heterotroph biomass and production for a typical simulation (50×50 lattice) are given in Table 1. These values were computed based on total heterotroph biomass or production, including nondominant size classes. Ratios for actual ecosystems computed from data in the ecological literature[26] are given for comparison. The ratio of heterotroph to autotroph biomass was relatively high; indeed, when size class 2 or larger heterotrophs were dominant, heterotroph biomass exceeded autotroph biomass. This situation is uncommon in terrestrial ecosystems, but occurs frequently in aquatic ecosystems, and is exemplified by the pond data in Table 1.

The ratio of heterotroph to autotroph production was considerably higher in the simulations than in any of the real-world ecosystems (see Table 1). Since all heterotroph biomass is ultimately derived from the consumption of autotrophs, conservation of mass implies that the heterotroph-to-autotroph production ratio cannot exceed 1. Indeed, since some consumed food must be used to meet metabolic needs rather than to produce new biomass, the ratio must be strictly less than 1. This ratio measures the efficiency at which autotrophic production is converted to heterotrophic production. It is interesting to note that as the community became dominated by larger heterotrophs, this efficiency decreased. Evolutionary dynamics can be regarded as an optimization process, but at an ecosystem level it is often not clear what system property is being optimized. It is clear that, in these simulations at least, whatever is being optimized, it is not production efficiency.

TABLE 1 Ratios of heterotroph to autotroph biomass and production in simulations and actual ecosystems. Ratios for ecosystems calculated from data compiled and reported by R.V. O'Neill.[26]

SYSTEM	HETEROTROPH-TO-AUTOTROPH BIOMASS RATIO	HETEROTROPH-TO-AUTOTROPH PRODUCTION RATIO
SIMULATION		
size class 1 dominant	0.772	0.379
size class 2 dominant	1.14	0.290
size class 3 dominant	1.40	0.179
size class 4 dominant	1.57	0.116
ECOSYSTEMS		
tundra	0.00400	0.00404
tropical forest	0.00926	0.0268
deciduous forest	0.0156	0.0133
salt marsh	0.0268	0.0373
pond	5.44	0.0220

DISCUSSION

The model presented here simulated the evolution of artificial ecological communities subject to three explicitly imposed constraints related to body size: (1) Metabolic energy costs for heterotrophs were a power-law function of body mass. In accordance with empirical data, metabolic rates were assumed to scale as $M^{3/4}$, and specific metabolic rates as $M^{-1/4}$. (2) Movement rates of heterotrophs during foraging were a function of body size. The number of steps in a random walk a heterotroph executed per model iteration were equal to its size class ($= \log_2 M$). The area searched by a random walker (i.e., the number of distinct sites visited) scales as $N/\log N$, where N is number of steps taken. (3) Predators were only allowed to eat prey that were smaller than themselves.

As a result of these imposed constraints, the communities that evolved exhibited a number of other patterns related to body size. The distribution of abundance or biomass over size class routinely displayed one of a few, repeatable patterns. In every case, these distributions were dominated by autotrophs plus a single heterotroph size class, although which heterotroph was dominant varied, both across replicate simulations, and over time within a simulation.

Abundance, biomass, population growth rate, and biomass production of dominant heterotrophs displayed a power law dependence on body mass. This power law describes the upper limit of the values of these variables attainable by heterotrophs within these artificial communities. This upper limit is reached by the dominant size class, whereas nondominant heterotrophs are suppressed to lower values. Particularly in data encompassing broad spatial scales, plots of ecological variables as a function of body mass frequently show considerable scatter, but with sharp boundaries at upper or lower limits.[7] This has led to the concept of a "constraint envelope" which sets body-size related bounds on ecological variables. The power-law functions described here appear to form part of such a constraint envelope for the artificial communities produced by the model.

One striking feature of the communities that evolved under this model was their relative simplicity and the small number of coexisting trophic species. Food webs frequently consisted solely of autotrophs and herbivores. Occasionally an organism that could consume both autotrophs and heterotrophs entered the food web, but it would typically drive its heterotrophic prey to extinction. Strict carnivores were never observed to become established members of the community. On the basis of the simulations done so far, the reasons for the small number of coexisting species cannot be determined with certainty, but likely factors are the small size and spatial homogeneity of the environment. It is well established that the number of species contained in an environment increases with increasing area.[1] There is also strong theoretical evidence to support the contention that spatial heterogeneity enhances species coexistence.[38] In contrast, the model developed here lacks refuges for prey, lacks spatial barriers which would impede diffusive mixing of predators and prey, and even promotes mixing by placing offspring at random locations on the lattice rather than localizing reproduction near the parent.

Despite the simple nature of the ecological communities, a trend toward increasing body size (both mean and maximum) over time was apparent. This analogue of Cope's rule was clearly an example of a cladogenetic trend, since the fixed definition of size classes precluded anagenetic changes in body size within a lineage. However, the heterotrophs which became successfully established in the community were substantially smaller than the maximum allowable size (the size gene would accommodate up to size class 15 organisms). It is not clear whether the communities that existed at the end of long simulations were, in fact, invulnerable to invasion by larger size classes, or if the appropriate invader had simply failed to appear during the time period simulated. This question could be resolved by perturbation experiments in which potential invaders are deliberately introduced into an apparently stable community.

In addition to the simulations reported in detail here, factorial sets of simulations were performed in which the rate of autotrophic energy gain was varied between 5, 10, or 20 units per time step, and the exponent governing the specific metabolic rate for heterotrophs was varied between -0.05, -0.25, and -0.45. Two simulations lasting 20,000 time steps, plus two simulations lasting 2,000 time steps were conducted on 20×20 lattices for each combination of parameters. It was clear that the maximum size of heterotroph that a community could support increased as the rate of autotrophic energy gain increased. The specific metabolic rate exponent had no clearly noticeable effect. These results suggest that larger organisms might be successful if the rate of autotrophic energy gain were increased yet further.

The focus of this study has been on the simulation of ecological community dynamics, with body size as a state variable, and certain dynamical constraints related to body size being explicitly represented in the model. Body size is an important variable in the real world because organisms and ecosystems are nonequilibrium thermodynamic entities which process matter and energy, and are governed by physical constraints. It is interesting to speculate upon what analogous constraints may obtain to realizations of artificial organisms or ecosystems in a computer. In most attempts at computer realizations (as opposed to simulations) of artificial life, the artificial life forms compete for memory space and CPU time. An instructive example is the Tierra system developed by Ray in which the life forms are self-replicating programs.[32] The number of machine instructions in such a program provides a natural analogue to the size of an organism. The Tierran operating system includes a time slicer that doles out small slices of CPU time to the programs. The number of instructions executed per time slice is proportional to the program length raised to an exponent. If the exponent is greater than one, there is a selection pressure for larger programs. If it is less than one, smaller programs are favored. An exponent equal to one is size-neutral. However, as Ray points out,[32] there is no analogue of metabolic costs in the Tierran system. Inclusion of such metabolic costs, and comparison of the size-distribution and community patterns that evolve with and without such costs would provide valuable information about those aspects of ecosystems that are a result of such physical (thermodynamic) constraints. Elucidation of the consequences of physical laws for ecosystem structure and dynamics is a major problem in theoretical ecology. By allowing for experimentation in worlds with alternative physical laws, the artificial life paradigm provides a valuable tool for addressing this problem.

ACKNOWLEDGMENTS

Several of the ideas for the model were developed in the course of discussions with Doyne Farmer. My interest in the role of body size in structuring ecological systems was stimulated by Pablo Marquet, Jim Brown, and Bruce Milne. Research was supported in part by NSF grant no. BSR-9107339.

REFERENCES

1. Arrhenius, O. "Species and Area." *J. Ecol.* **9** (1921): 95–99.
2. Banse, K., and S. Mosher. "Adult Body Mass and Annual Production/ Biomass Relationships of Field Populations." *Ecol. Monogr.* **50** (1980): 355–379.
3. Blueweiss, L., H. Fox, V. Kudzma, D. Nakashima, R. Peters, and S. Sams. "Relationships Between Body Size and Some Life History Parameters." *Oecologia* **37** (1978): 257–272.
4. Bonner, J. T. *The Evolution of Complexity.* Princeton: Princeton University Press, 1988.
5. Boudreau, P. R., L. M. Dickie, and S. R. Kerr. "Body-Size Spectra of Production and Biomass as System-Level Indicators of Ecological Dynamics." *J. Theor. Biol.* **152** (1991): 329–339.
6. Brooks, J. L., and S. I. Dodson. "Predation, Body Size, and Composition of Plankton Assemblages." *Science* **150** (1965): 28–35.
7. Brown, J. H., and B. A. Mauer. "Macroecology: The Division of Food and Space Among Species on Continents." *Science* **243** (1989): 1145–1150.
8. Brown, J. H., and B. A. Mauer. "Body Size, Ecological Dominance and Cope's Rule." *Nature* **324** (1986): 248–250.
9. Brown, J. H., P. A. Marquet, and M. A. Taper. "Evolution of Body Size: Consequences of an Energetic Definition of Fitness." *Am. Nat.* (in press).
10. Cohen, J. E., F. Briand, and C. M. Newman. *Community Food Webs: Data and Theory.* Berlin: Springer-Verlag, 1990.

11. Dial, K. P., and J. Marzluff. "Are the Smallest Organisms the Most Diverse?" *Ecology* **69** (1988): 1620–1624.
12. Dickie, L. M., S. R. Kerr, and P. R. Boudreau. "Size-Dependent Processes Underlying Regularities in Ecosystem Structure." *Ecol. Monogr.* **57** (1987): 233–250.
13. Farmer, J. D., and A. d'A. Belin. "Artificial Life: The Coming Evolution." In *Artificial Life II*, edited by C. G. Langton, C. Taylor, J. D. Farmer and S. Rasmussen, 815–840. Santa Fe Institute Studies in the Sciences of Complexity, Proc. Vol. X. Redwood City, CA: Addison-Wesley, 1991.
14. Fenchel, T. "Intrinsic Rate of Natural Increase: The Relationship with Body Size." *Oecologia* **14** (1974): 317–326.
15. Gittleman, J. L. "Carnivore Body Size: Ecological and Taxonomic Correlates." *Oecologia* **67** (1985): 540–554.
16. Hutchinson, G. E., and R. H. MacArthur. "A Theoretical Ecological Model of Size Distributions Among Species of Animals." *Am. Nat.* **93** (1959): 117–125.
17. Kleiber, M. "Body Size and Metabolism." *Hilgardia* **6** (1932): 315–353.
18. LaBarbera, M. "The Evolution and Ecology of Body Size." In *Patterns and Processes in the History of Life*, edited by D. M. Raup and D. Jablonski, 69–98. Berlin: Springer-Verlag, 1986.
19. Lurie, D., and J. Wagensberg. "An Extremal Principle for Biomass Diversity in Ecology." In *Thermodynamics and Regulation of Biological Processes*, edited by I. Lamprecht and A. I. Zotin, 257–273. Berlin: Walter de Gruyter, 1984.
20. May, R. M. "The Dynamics and Diversity of Insect Faunas." In *Diversity of Insect Faunas*, edited by L.A. Mound and N. Waloff, 188–204. Oxford: Blackwell, 1978.
21. McKinney, M. L. "Trends in Body-Size Evolution." In *Evolutionary Trends*, edited by K.J. McNamara, 75–118. Tucson: University of Arizona Press, 1990.
22. McMahon, T. A., and J. T. Bonner. *On Size and Life.* New York: Scientific American Books, 1983.
23. Mills, E. L., J. L. Forney, and K. J. Wagner. "Fish Predation and Its Cascading Effect on the Oneida Lake Food Chain." In *Predation: Direct and Indirect Impacts on Aquatic Communities*, edited by W. C. Kerfoot and A. Sih, 118–131. Hanover: University Press of New England, 1987.
24. Morse, D. R., J. H. Lawton, M. M. Dodson, and M. H. Williamson. "Fractal Dimension of Vegetation and the Distribution of Arthropod Body Lengths." *Nature* **314** (1985): 731–733.
25. Newell, N. D. "Phyletic Size Increase, an Important Trend Illustrated By Fossil Invertebrates." *Evolution* **3** (1949): 103–124.
26. O'Neill, R. V. "Ecosystem Persistence and Heterotrophic Regulation." *Ecology* **57** (1976): 1244–1253.

27. Pahl-Wostl, C. "The Hierarchical Organization of the Aquatic Ecosystem: An Outline How Reductionism and Holism May Be Reconciled." *Ecological Modelling* **66** (1993): 81–100.
28. Pattee, H. H. "Simulations, Realizations, and Theories of Life." In *Artificial Life*, edited by C. G. Langton, 63–77. Santa Fe Institute Studies in the Sciences of Complexity, Proc. Vol. VI. Redwood City, CA: Addison-Wesley, 1988.
29. Persson, L. "Asymmetrical Competition: Are Larger Animals Competitively Superior?" *Am. Nat.* **126** (1985): 261–266.
30. Peters, R. H. *The Ecological Implications of Body Size.* Cambridge: Cambridge University Press, 1983.
31. Pimm, S. L., H. L. Jones, and J. Diamond. "On the Risk of Extinction." *Am. Nat.* **132** (1988): 757–785.
32. Ray, T. S. "An Approach to the Synthesis of Life." In *Artificial Life II*, edited by C. G. Langton, C. Taylor, J. D. Farmer and S. Rasmussen, 371–408. Santa Fe Institute Studies in the Sciences of Complexity, Proc. Vol. X. Redwood City, CA: Addison-Wesley, 1991.
33. Schmidt-Nielsen, K. *Scaling: Why Is Animal Size So Important?* Cambridge: Cambridge University Press, 1984.
34. Sheldon, R. W., and S. R. Kerr. "The Population Density of Monsters in Loch Ness." *Limnol. Oceanogr.* **17** (1972): 796–798.
35. Sheldon, R. W., A. Prakash, and W. H. Sutcliffe, Jr. "The Size Distribution of Particles in the Ocean." *Limnol. Oceanogr.* **17** (1972): 327–340.
36. Stanley, S. M. "An Explanation for Cope's Rule." *Evolution* **27** (1973): 1–26.
37. Taylor, A. D. "Metapopulations, Dispersal, and Predator-Prey Dynamics: An Overview." *Ecology* **71**(1990): 429–433. This paper provides a good introduction to the large literature on the role of spatial heterogeneity in promoting persistence of predator-prey systems.
38. Taylor, C. R., N. C. Heglund, T. A. McMahon, and T. R. Looney. "Energetic Cost of Generating Muscular Force During Running: A Comparison of Large and Small Animals." *J. Exp. Biol.* **86** (1980): 9–18.
39. Van Valen, L. "Body Size and Numbers of Plants and Animals." *Evolution* **27** (1973):27–35.
40. Vanni, M. J. "Effects of Nutrients and Zooplankton Size on the Structure of a Phytoplankton Community." *Ecology* **68** (1987): 624–635.
41. Vézina, A. F. "Empirical Relationships Between Predator and Prey Size Among Terrestrial Vertebrate Predators." *Oecologia* **67** (1985): 555–565.
42. Williamson, M. H., and J. H. Lawton. "Fractal Geometry of Ecological Habitats." In *Habitat Structure: The Physical Arrangement of Objects in Space*, edited by S. S. Bell, E. D. McCoy and H. R. Mushinsky, 69–86. London: Chapman & Hall, 1991.

E. Ann Stanley,† Dan Ashlock,† and Leigh Tesfatsion†‡
†Department of Mathematics, Iowa State University, Ames, IA 50011
‡Department of Economics, Iowa State University, Ames, IA 50011

Iterated Prisoner's Dilemma with Choice and Refusal of Partners

This paper studies the effects of partner selection on cooperation in an artificial ecology. Agents, represented by finite automata, interact with each other through an iterated prisoner's dilemma—inxxprisoner's dilemmaiterated (IPD) game with the added feature that players choose and refuse potential game partners on the basis of continually updated expected payoffs. Analytical studies reveal that the subtle interplay between choice and refusal in N-player IPD games can result in various long-run player interaction patterns: e.g., mutual cooperation; mixed mutual cooperation and mutual defection; parasitism; and/or wallflower seclusion. Simulation studies indicate that choice and refusal can accelerate the emergence of cooperation in evolutionary IPD games. More generally, however, choice and refusal can result in the emergence and persistence of multiple payoff bands, reflecting the possible existence of ecological attractors characterized by play behavior that is not entirely cooperative. The existence of a spectrum of payoff bands, in turn, leads to the emergence of new ecological behaviors such as band spiking and band tunneling.

1. INTRODUCTION

This paper investigates cooperative behavior in an artificial ecology in which egoistic agents interact with each other through a tournament of iterated games. The tournament played is a variant of Axelrod's[2] iterated prisoner's dilemma (IPD) game, modified to permit players to choose and refuse potential game partners. In each iteration of our modified game, players use expected payoffs updated on the basis of past encounters to make offers to a limited number of preferred players and to refuse offers from unacceptable players.

The introduction of choice and refusal fundamentally modifies the way in which players interact in the IPD game, and the characteristics which result in high payoff scores. Choice allows players to increase their chances of encountering other cooperative players. Refusal gives players a way to protect themselves from defections without having to defect themselves. Ostracism of defectors occurs endogenously as an increasing number of players individually refuse their game offers. But choice and refusal also permit clever ripoff players to home in quickly on exploitable players and form parasitic relationships.

Following Miller,[17] player strategies for the IPD game with choice and refusal (or "IPD/CR game" for short) are represented by means of finite automata.[1] This representation has a number of advantages over the original code-based formulation by Axelrod.[2] It permits the same modeling of complex strategic behavior, but it is simpler and cleaner to program. Moreover, behavior modification in response to endogenously occurring events can be readily incorporated in the form of genetic algorithms.[2] Consequently, the basic single-tournament IPD/CR game is easily generalized to an "evolutionary" IPD/CR game, i.e., a multiple-tournament IPD/CR game in which the strategies of the players evolve between tournaments.

The choice/refusal mechanism is characterized by six parameters: initial expected payoff; the minimum tolerance (expected payoff) level below which game offers will be refused; the maximum number of game offers that can be made in each iteration; the rejection payoff received when a game offer is refused; the wallflower payoff received when game offers are neither made nor accepted; and a memory weight that determines the relative importance of distant to recent payoffs in the calculation of updated expected payoffs. Analytical parameter sensitivity studies are undertaken for a variety of single-tournament, two-player IPD/CR games. To simplify and systematize the analysis, we restrict our attention to six illustrative

[1] A "finite automaton" is a system specified by a finite collection of internal states together with a state transition function, driven by input, which gives the next internal state the system will enter. See Section 5, below, for a detailed discussion of the specific finite automaton representation used in this paper.

[2] A "genetic algorithm" uses a Darwinian selection principle to optimize a solution to a problem with respect to various selected problem features. Genetic algorithms are powerful tools for evolving high-performance strategies from simple representative strategy types; see Holland.[11]

player types that roughly span the range from uncooperative to cooperative behavior. For each pair of player types, we determine the precise conditions under which refusal first occurs, and the average payoff scores that are achieved, as a function of the choice/refusal parameters.

We then extend the analysis to single-tournament N-player IPD/CR games. The interplay between choice and refusal can be quite subtle for such games. We illustrate this with a detailed study of a single-tournament, five-player IPD/CR game in which the ultimate pattern of player interactions can include the formation of successful long-term parasitic relationships unless the minimum tolerance level is set suitably high.

We also report on simulation experiments carried out for the evolutionary 30-player IPD game studied by Miller,[17] modified to allow for choice and refusal of game partners. The experimental results indicate that, in comparison to Miller's findings, the emergence of cooperation is accelerated over much of the choice/refusal parameter space. However, high values of the minimum tolerance level and the wallflower payoff can result in a "wallflower trap" ecology consisting primarily of antisocial hermits.

Moreover, for nonextreme settings of the choice/refusal parameters, our simulation studies reveal an interesting clustering effect presaged by our analysis of single-tournament, N-player IPD/CR games. In Miller's simulation experiments, most ecologies evolve to a set of players whose average payoff scores are near the mutual cooperation payoff. For players who are not ultimately cooperative, the average payoff scores will end up scattered between the mutual defection and mutual cooperation payoffs. In contrast, in our studies with choice and refusal, we typically observe the ultimate formation of two or more distinct tight bands of average payoff scores reflecting the emergence of stable behavioral patterns that are not entirely cooperative. The existence of this payoff band spectrum, in turn, leads to the emergence of new ecological behaviors. For example, we see "band spiking" in which an ecology abruptly moves from one payoff band to another and then back again, and "band tunneling" in which an ecology that has long resided in one payoff band suddenly traverses to another payoff band and remains there.

The addition of choice and refusal to the Axelrod/Miller IPD game is motivated by an interest in human interactions, particularly in the sexual-partner selection process which leads to the spread of AIDS and other sexually transmitted diseases.[12] Not only do rates of sex and new partner acquisition influence the spread of this epidemic, but the structure of the contacts determines who becomes infected. Current models of the spread of AIDS typically assume that behavior is predetermined and responds in exogenously determined ways to changing circumstances. This makes it difficult to understand the impact of intervention strategies and to predict the conditions under which particular types of sexual marketplaces will arise.

The choice/refusal ecology studied in this paper is a first step towards a model of AIDS transmission in the context of an endogenously evolving social milieu. This is a long-term development project in which elements of realism will be introduced one step at a time. The current model focuses on a key feature essential for the final ecology: the possibility of choice and refusal in social interactions.

The relation of this paper to previous work on cooperation in IPD games is outlined in Section 2. Section 3 reviews the essential features of the basic IPD game set out in Axelrod,[2] and extends this framework to include a choice/refusal mechanism. Section 4 briefly describes some simulation experiments with an IPD game studied in Miller[17] which illustrate the potential sensitivity of evolutionary IPD game outcomes to the introduction of choice and refusal. An analytical study of single-tournament IPD/CR games is undertaken in Sections 5 and 6. These results are used in Section 7 to provide a more careful interpretation of various simulation results obtained for evolutionary IPD/CR games. Concluding comments are given in Section 8.

2. RELATION TO PREVIOUS WORK

In a series of pathbreaking studies, Axelrod[1,2,4] has explored the initial emergence and viability of cooperative behavior in the absence of either altruism or binding commitments, using the IPD game as a paradigm for social interactions. In each iteration, each player plays one prisoner's dilemma game with every player in a fixed pool of N players. The only possible choice for each player in each two-player game is either to cooperate or to defect, and both players must choose simultaneously.

As discussed by Axelrod[2] and by May,[15] the cooperative Tit-for-Tat strategy is a collectively stable strategy for the IPD game if the number of iterations is either uncertain or infinite and the probability that any two players meeting in a current iteration will meet again in a future iteration is sufficiently high.[3] In an IPD game with a known finite number of iterations, however, cooperation is hard to sustain. Mutual defection occurs in the final iteration of the game because no player foresees any future gains to cooperation, and this typically leads by backwards recursion to mutual defection in every iteration.

A number of modifications have been proposed that permit the viability of cooperative strategies in IPD games even if the number of iterations is a known finite number. Kreps et al.[14] establish that mutual cooperation can be sustained in every iteration up to some iteration close to the end of the game if one player assigns positive probability to the possibility that the strategy followed by the other player is Tit-for-Tat. Thompson and Faith[21] and Hirshleifer,[9] among others, have

[3]The *Tit-for-Tat* strategy is defined as follows: cooperate initially and, thereafter, do whatever the other player did in his previous move. A single mutant strategy introduced into a pool of identical native strategies is said to *invade* the native strategy if the newcomer receives a higher payoff from playing against a native strategy than a native strategy receives from playing against another native strategy. A native strategy is said to be *collectively stable* if no mutant strategy can invade it.

shown that cooperation can be sustained if players can credibly commit themselves to use retaliatory strategies in response to defections by opposing players. Finally, Hirshleifer and Rasmusen[10] use the possibility of group ostracism to sustain mutual cooperation in all but the last iteration. Empirical support for the cooperation-inducing effects of group ostracism can be found in a case study by Barner-Barry.[5]

Although such modifications enhance the viability of cooperative strategies in IPD games, a major difficulty remains. As pointed out by Axelrod and Hamilton,[3] cooperative strategies cannot successfully invade a population of defectors playing the basic IPD game unless the initial frequency of interactions between cooperative strategies is sufficiently large. Consequently, it is difficult for cooperation to emerge spontaneously from noncooperation in the basic IPD framework.

One limitation of many iterated game studies of social interaction that hampers the emergence of cooperation is the implicit assumption that individual players have no control over which opponents they play; see, for example, the models reviewed by Maynard Smith.[16] Players either engage in a round-robin tournament—i.e., each player in each iteration plays one game with every player in a pre-determined set of players—or games occur through random encounters. In actuality, however, social interactions among organisms are typically characterized by the choice and refusal of partners rather than by a random or deterministic matching mechanism. How do herds form for foraging and protection? How do animals choose mates? How do family and social structures protect cooperative players from noncooperative players? And how do humans choose their friends and sexual partners? The question thus arises whether the long-run viability of cooperation in the IPD game would be enhanced if players were more realistically allowed to choose and refuse their potential game partners.

Conjectures along these lines have been explored by a number of previous researchers. For example, in the context of a Darwinian fitness model, Eshel and Cavalli-Sforza[7] show that full cooperativeness is the only evolutionarily stable strategy if encounter probabilities are sufficiently biased in favor of meeting an individual using the same strategy. Feldman and Thomas[8] investigate conditions under which multiple IPD strategies can coexist in a stable equilibrium, assuming the probability that a player stays in the IPD game depends either on his own current play or on the current play of his opponent. Dugatkin and Wilson[6] examine the ability of a roving always-defect player to invade a Tit-for-Tat player population that is partitioned into "patches" of different, externally specified size and duration. Kitcher[13] argues that altruistic play (giving weight to the payoffs of other players) can evolve more readily in contexts where game play is optional and the possibility exists for taking actions whose fitness effects are independent of the actions of others.

The present paper complements and extends this work by allowing players to choose and refuse potential game partners on the basis of continually updated expected payoffs. Player encounters are thus determined by anticipated rewards rather than by a probability of encounter biased towards cooperative behavior per se. Moreover, ostracism (end of game play) for noncooperative players occurs endogenously as an increasing number of players individually refuse their game offers.

3. THE BASIC IPD GAME WITH CHOICE AND REFUSAL

In this section we first review the essential features of the basic single-tournament IPD game set out in Axelrod.[2] We then extend this framework to include a choice/refusal mechanism.

3.1 THE BASIC IPD GAME

The prisoner's dilemma (PD) game is a game with two players. Each player has two possible moves, "cooperate" or "defect," and each player must move without knowing the move of the other player. If both players defect, each receives a payoff D. If both cooperate, each receives a payoff C that is strictly greater than D. Finally, if one defects and the other cooperates, the cooperating player receives the lowest possible payoff B and the defecting player receives the highest possible payoff S, where $B < D < C < S$. For reasons clarified below, the payoffs are also restricted to satisfy $(S + B)/2 < C$.

The dilemma is that, if *both* players defect, both do worse than if both had cooperated; yet there is always an incentive for an individual player to defect. More precisely, the payoffs (C, C) achieved with mutual cooperation are higher than the payoffs (D, D) achieved with mutual defection. Nevertheless, defection is the best response to any move an opponent might make. The best response to defection is to defect, because this avoids the lowest possible payoff B; and the best response to cooperation is to defect, because this achieves the highest possible payoff S.

The *iterated prisoner's dilemma (IPD) game* is a tournament consisting of the repeated (iterated) play of a round-robin of PD games at discrete time intervals. In each iteration, each player plays one PD game with all other players in a fixed pool of N players. The only information a player has about another player is the history of payoffs achieved in previous game plays with that player. The restrictions on payoffs guarantee that the players cannot escape their dilemma by taking turns exploiting each other. For any two players, the average (per game) payoffs (C, C) achieved with mutual cooperation over the course of the IPD game are higher than either the average payoffs (D, D) achieved with mutual defection or the average payoffs $((S+B)/2, (S+B)/2)$ achieved with alternating plays of cooperation against defection and defection against cooperation.

3.2 INTRODUCTION OF CHOICE AND REFUSAL

The *IPD game with choice and refusal*, henceforth abbreviated by IPD/CR, is an IPD game with the added feature that players can choose and refuse game partners in each iteration. A fixed pool of N players engages in a tournament consisting of indefinitely many iterations. Each iteration, in turn, consists of five stages: (1) a *choice stage* in which each player makes PD game offers to a limited number of potential game partners with high expected payoffs; (2) a *refusal stage* in which each player refuses PD game offers with unacceptably low expected payoffs; (3) a *play stage* in which nonrefused PD game offers are played out as PD games; (4) a *cleanup stage* in which payoffs are calculated and recorded for both active and inactive players; and (5) an *update stage* in which expected payoffs are updated for active players on the basis of newly received payoff information.

The logical progression of the IPD/CR game will now be more fully described. At the beginning of the first iteration 1, all players are assumed to assign the same *initial expected payoff* π^0 to each possible play of a game. Then, in each iteration $i = 1, 2, \ldots$, the five component stages take the following form:

- *Choice Stage:* Each player determines which other players are tolerable game partners. Given any player n, a player $m \neq n$ is *tolerable* for player n in iteration i if
$$\pi^{i-1}(m|n) \geq \tau, \tag{1}$$
where $\pi^{i-1}(m|n)$ denotes the the expected payoff to n from playing a PD game with player m in iteration i, and τ denotes an exogenously given *minimum tolerance level.* If the number of tolerable players for player n is no greater than an exogenously given *upper choice bound* K, where $1 \leq K \leq N-1$, then player n makes a PD game offer to each tolerable player. If the number of tolerable players exceeds K, player n makes a PD game offer to the K tolerable players m for whom his expected payoff $\pi^i(m|n)$ is highest. Ties are settled by a random draw.
- *Refusal Stage:* Each player then examines the PD game offers he has received. Any offer coming from an intolerable player is refused and any offer coming from a tolerable player is accepted. Thus, a player gets to reject odious offers, but he must accept offers from anyone he has judged to be tolerable in the Choice Stage.
- *Play Stage:* All nonrefused PD game offers are played out as PD games. Even if there are mutual offers between two players, only one PD game is played.
- *Clean-Up Stage:* A player receives a *rejection payoff* R for each PD game offer he made in iteration i that was refused, and a PD payoff (depending on the actual play of the game) for each PD game offer he made or received in iteration i that was not refused. An inactive player—i.e., a player who neither made nor accepted offers in iteration i—is assigned a *wallflower payoff* W. Note that a player makes no offers if and only if he judges all other players to be intolerable.
- *Update Stage:* Consider any two players n and m. If n neither made nor accepted a PD game offer from m in the current iteration i, then n's expected payoff $\pi^{i-1}(m|n)$ for the play of a PD game with m in iteration i is trivially updated to
$$\pi^i(m|n) = \pi^{i-1}(m|n) \tag{2}$$
for play in the next iteration $i+1$. On the other hand, suppose player n either made a PD game offer to m (who subsequently either accepted or rejected it)

or accepted a PD game offer from m. In the former case, the payoff to n is either a PD payoff or the rejection payoff R; in the latter case, the payoff to n is a PD payoff. In either case, let this payoff be denoted by U. Then player n's updated expected payoff for making a PD game offer to player m in the next iteration $i+1$ takes the form of a weighted average over player n's payoff history with player m,

$$\pi^i(m|n) = \omega\pi^{i-1}(m|n) + (1-\omega)U, \tag{3}$$

where the *memory weight* ω controls the relative weighting of distant to recent payoffs.[4] Note that an increase in ω implies an increase in the weight put on past payoffs relative to current payoffs, which in turn leads to more inertia in the partner selection process.

In summary, the choice and refusal mechanism for the IPD/CR game is characterized by the following six parameters:

Initial Expected Payoff:	π^0
Minimum tolerance level:	τ
Upper choice bound:	K
Rejection payoff:	R
Wallflower payoff:	W
Memory weight:	ω

At the end of the IPD/CR game, the overall success of each player is measured by his average payoff score, calculated as the total *sum* of his payoffs divided by the total *number* of his payoffs.[5]

It is assumed that players in IPD/CR games do not anticipate future conditions under which PD game play with a current game partner could end, implying that they play each PD game with another player as if the total number of PD games to be played with that player is indefinite. Since the choice/refusal mechanism itself is exogenously determined, this implies that players can be distinguished from one another on the basis of the rules they use to play an indefinite sequence of PD games with an arbitrary opposing player, i.e., on the basis of their *PD rules*. As will be seen in Section 5, a rich class of PD rules can be represented as finite automata.

[4]This mechanism for updating expected payoffs is a special case of "criterion filtering"—i.e., the direct updating of expected return functions on the basis of past return observations. Criterion filtering is an operationally feasible alternative to the indirect updating of criterion functions via Bayes' rule which, given an appropriate specification for the filter weights, can yield a strongly consistent estimate for the true expected return function; see Tesfatsion.[20]

[5]In an IPD/CR game, the players have some degree of control over the number of PD games they play—equivalently, over the number of moves they make—and players not participating in PD games can still receive wallflower and rejection payoffs. Consequently, average payoff per payoff made is used as a measure of overall success rather than total payoff or average payoff per game.

Note that the IPD/CR game reduces to the basic Axelrod IPD game in two cases: either (i) $K = N - 1$ and $\tau = 0$ or (ii) $K = N - 1$, $\omega = 1$, and $\tau < \pi^0$. In either case, each player ends up playing one PD game with all other players in each iteration, i.e., each iteration reduces to a round-robin among the N players.

4. IPD WITH CHOICE AND REFUSAL: AN ILLUSTRATION

An intriguing artificial life experiment is reported in an evolutionary IPD game study by Miller.[17] A population of thirty finite automata playing an IPD game was allowed to evolve by means of a genetic algorithm that used high payoff scores as its selection principle. To illustrate the potential sensitivity of evolutionary outcomes to the introduction of choice and refusal, this section briefly describes some simulation results obtained for Miller's evolutionary IPD game after the introduction of a choice/refusal mechanism. A more careful discussion of these results is given in Section 7, following a preliminary analytical investigation of single-tournament IPD/CR games.

We first implemented our own version of Miller's experiment, without noise. As in Miller, 50 successive tournaments were conducted. Each tournament was separated from the next by a genetic step in which only the 20 most successful players among the 30 automata constituting the current player set were allowed to reproduce, resulting in a modified player set of 30 automata for the next tournament. The 50-tournament run thus resulted in an "ecology" consisting of an evolved population of 30 automata.

Each tournament, in turn, consisted of 150 iterations of round-robin PD games among 30 automata. The four possible per-game payoffs were $B = 0$ (for cooperating against a defecting player), $D = 1$ (for mutual defection), $C = 3$ (for mutual cooperation), and $S = 5$ (for defecting against a cooperating player). The entire run of 50 tournaments separated by genetic steps was repeated 40 times to obtain 40 distinct ecologies.

For each of the 50 tournaments, we show the average payoff score obtained by the 30 automata over all 40 ecologies (see Figure 1). These results generally conform to the results obtained by Miller. Note the initial "dip" in the average payoff score. This dip reflects the exploitation of some players by other opportunistic players, until the implacable forces of evolution eliminate the chumps. The upward progress thereafter is the result of cooperative but relatively unexploitable players—of which Tit-for-Tat is a sterling example—beating out the opportunists whose victims have died out.

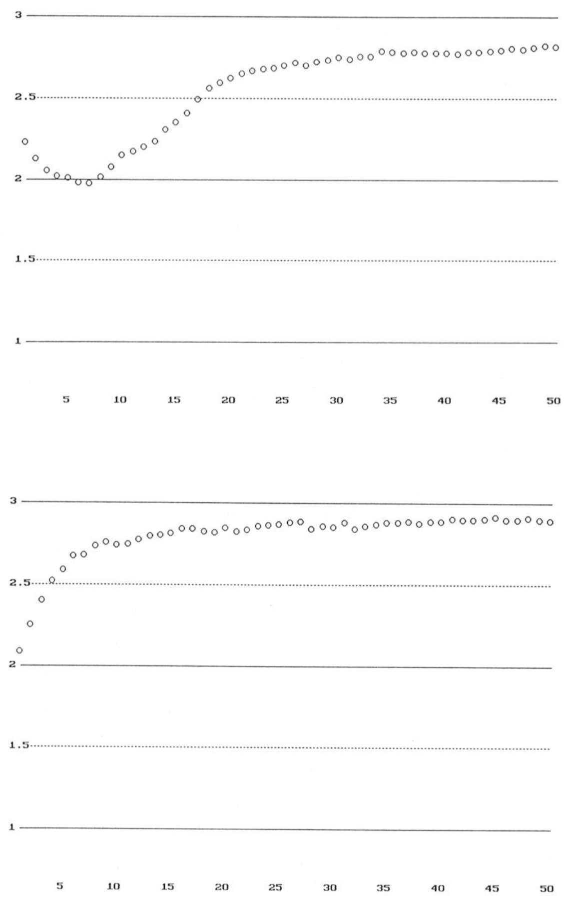

FIGURE 1 Average ecology payoffs for the replicated Miller game.

FIGURE 2 Average ecology payoffs for the IPD/CR game with $\tau = 1.6$.

We then incorporated our choice/refusal mechanism into Miller's experimental set up. In Figure 2 we see the analogous diagram to Figure 1 for simulations of the resulting modified Miller IPD game in which, in each iteration: (i) players have initially rosy payoff expectations prior to play ($\pi^0 = 3$); (ii) each player makes a PD game offer to at most one other preferred player ($K = 1$); (iii) received offers with expected payoffs greater than or equal to $\tau = 1.6$ are played out, and received offers with expected payoffs less than this value are refused; (iv) a player whose offer is refused receives a rejection payoff $R = 1$; (v) a player neither making nor accepting any offers receives a wallflower payoff $W = 1.6$; and (vi) players have reasonable but not excellent memories of their past payoff outcomes ($\omega = 0.7$).

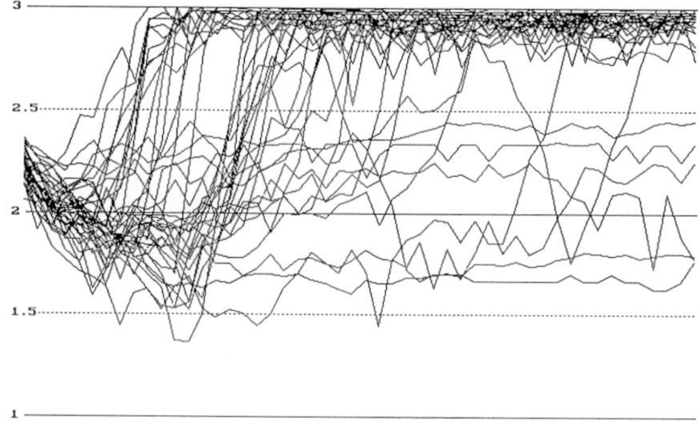

FIGURE 3 Individual ecology payoffs for the replicated Miller game.

It is interesting to note in Figure 2 that the initial dip in average payoff score in Miller's experimental data has vanished. The Miller dip occurred because various nice players were fatally exploited by predatory players, which lowered the average payoff score. In contrast, in an IPD/CR game, it sometimes happens that players who are exploitable in the basic IPD game can protect themselves by refusing offers from predatory players and, hence, remain in the ecology indefinitely. One type of highly exploitable player is a player whose PD rule closely approximates "always cooperate." Such players can attain high payoff scores when playing one another and, hence, can dominate the genetic step reproduction process when protected from predatory players by a refusal mechanism. As a result, unlike Miller, an initial dip in average payoff score need not occur in an evolutionary IPD/CR game.

While replicating Miller's work, we became curious about the behavior of individual ecology payoffs. To this end, we plotted the average tournament payoffs for each of the 40 ecologies in our replication of Miller's experiment and obtained Figure 3. As one can see, 9 of the 40 Miller ecologies did not end in the cluster of essentially cooperative ecologies; hence, the standard deviation of average tournament payoffs across ecologies is rather large.

In Figure 4 we give the average tournament payoffs for each of the forty ecologies in the choice/refusal experiment depicted in Figure 2. These payoffs display an interesting behavior when compared to the noiselike outlier payoffs obtained for Miller's experiment in Figure 3. Notice that *two* distinct payoff bands have now emerged, one close to 3 and the other just below 2.8. The ecologies corresponding to the payoff band near 3 ultimately consist of a single large group within which essentially random partner selection takes place. In contrast, the ecologies corresponding to the payoff band just below 2.8 ultimately consist of several small player groups whose members only choose to play one another.

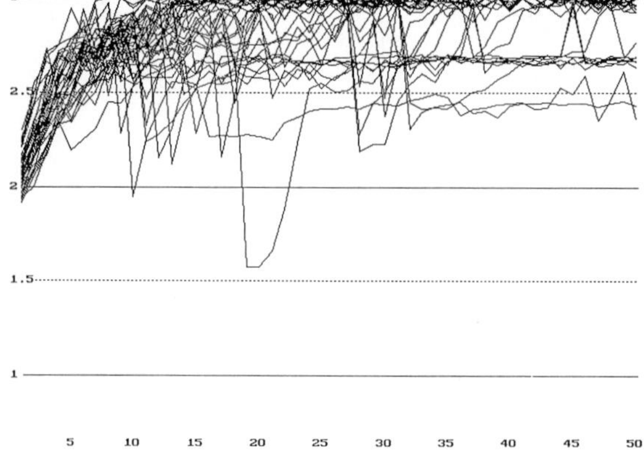

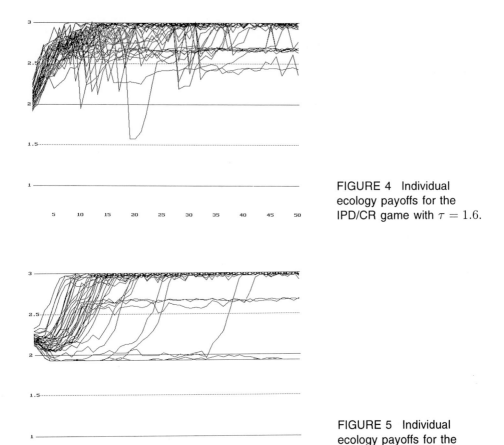

FIGURE 4 Individual ecology payoffs for the IPD/CR game with $\tau = 1.6$.

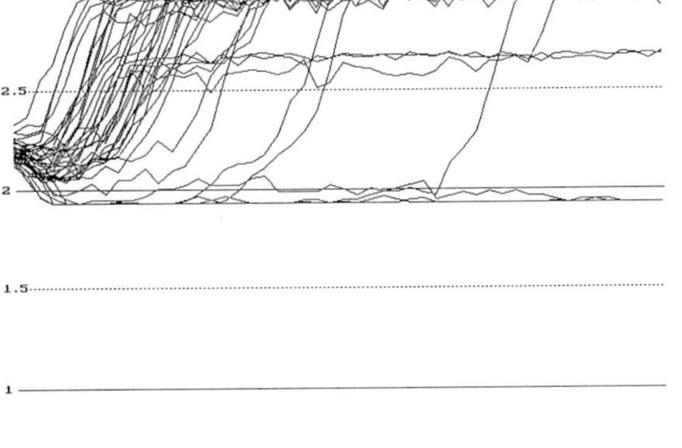

FIGURE 5 Individual ecology payoffs for the IPD/CR game with $\tau = 2.5$.

We then raised both the minimum tolerance level and the wallflower payoff from 1.6 to 2.5. As depicted in Figure 5, this change results in the gain of a third tight payoff band. The ecologies associated with this band ultimately consist of many players who neither make nor accept game offers, choosing instead to survive on the wallflower payoff.

As these examples indicate, the addition of choice and refusal to Miller's original IPD experiment induces the formation of an interesting new ecological feature—the formation of multiple distinct payoff bands. Somewhat at odds with our initial expectations, choice and refusal do not merely speed up the emergence of cooperation; in some parameter regimes they also allow the appearance of stable player interaction patterns that are not entirely cooperative.

5. SINGLE-TOURNAMENT, TWO-PLAYER IPD/CR GAMES

The simulation results presented in the previous section for an evolutionary IPD/CR game are suggestive but preliminary. This section undertakes a more careful analysis of the role of choice and refusal in the simpler context of a single-tournament, two-player IPD/CR game. For such games, the upper choice bound $K \in \{1, \ldots, N-1\}$ is forced to equal $1 = N - 1$. In Section 6 we take up the more general case of a single-tournament, N-player IPD game with $K = 1$. The results of Sections 5 and 6 are used in Section 7 to provide a more careful interpretation of the simulation results obtained for evolutionary IPD/CR games.

Since even a single-tournament, two-player IPD/CR game can be very complicated to analyze, we first describe six illustrative player types. We then analyze single-tournament, two-player IPD games using various combinations of these player types.

5.1 ILLUSTRATIVE PLAYER TYPES

As discussed in Section 3.2, players in IPD/CR games can be identified with the PD rules they use in playing an indefinite sequence of PD games with an arbitrary opposing player. We will study the pairwise interactions of six types of players (PD rules): (1) Always Defect (AllD); (2) Ripoff-Artist (Rip); (3) Gentle Ripoff (GRip); (4) Tit-for-Tat (TFT); (5) Tit-for-Two-Tats (TFTT); and (6) Always Cooperate (AllC).

These six players roughly span the range from uncooperative to cooperative behavior. With the exception of GRip, all of these players have previously been used in studies of IPD games. GRip was invented in order to have a relatively subtle opportunistic player who initially appears cooperative but who repeatedly sneaks in defections after an opposing player has built up a rosy payoff expectation.

Finite automaton representations for these six players are depicted in Figure 6, ordered by their complexity. All six players make an opening move, either cooperate (c) or defect (d), and then enter state 1; this opening move is indicated next to the arrow pointing to state 1. Thus, GRip, TFT, TFTT, and AllC initially cooperate, while AllD and Rip initially defect.

After one of the six players, say n, has arrived at a current state, his next move is conditioned on the previous move of the opposing player. This move sequence then determines a transition to a new state. A transition to a new state is indicated by an arrow, and the move sequence(s) which result(s) in this transition are indicated beside the arrow in a move-slashmark-move format. The previous move of the opposing player appears to the left of the slashmark and the next move of player n appears to the right of the slashmark; i.e., moves are time-sequenced from left to right.

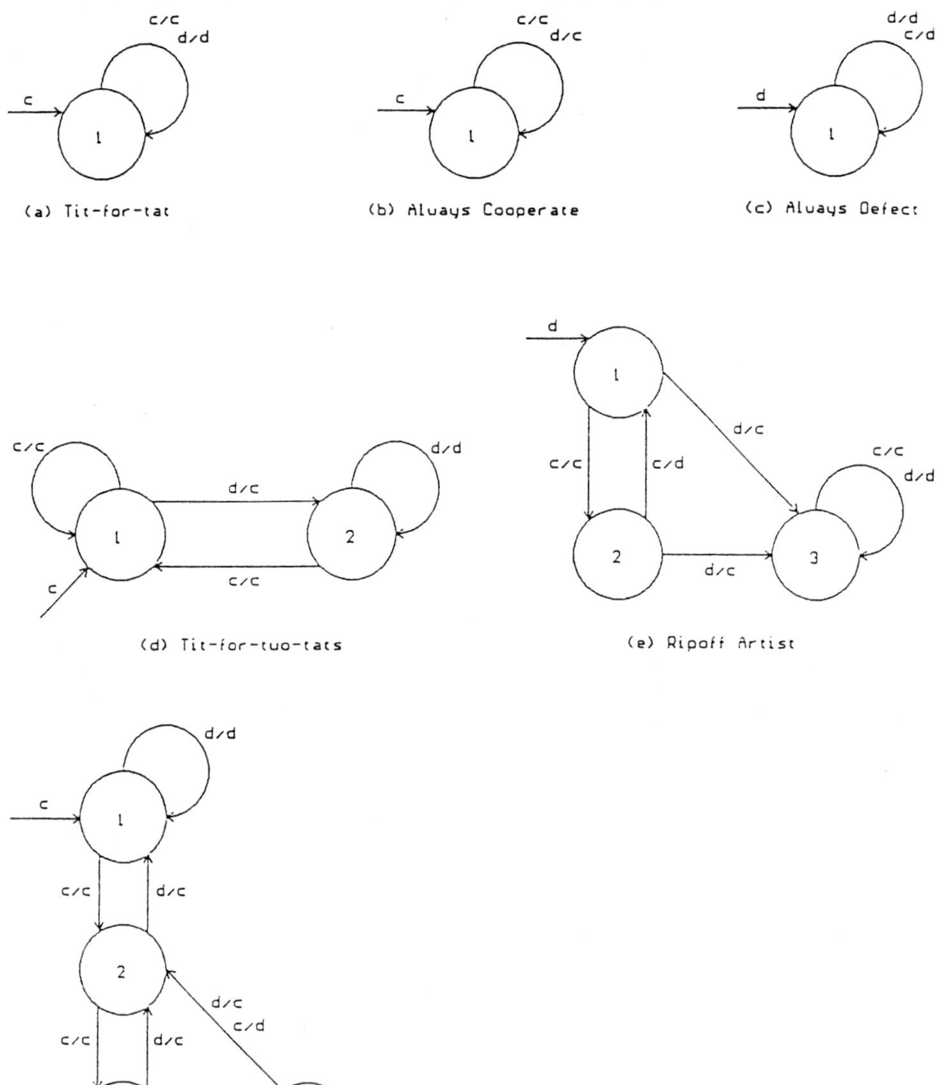

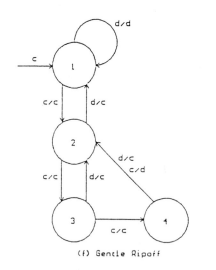

FIGURE 6 Illustrative Player types: (a) Tit-for-Tat; (b) Always Cooperate; (c) Always Defect; (d) Tit-for-Two-Tats; (e) Ripoff-Artist; and (f) Gentle Ripoff.

In particular, TFT begins by cooperating and then mimics whatever move his opponent made in the previous PD game. Consequently, once the player is in the initial state 1, two move sequences are possible for TFT, either $c|c$ or $d|d$, and each move sequence results in a transition back to state 1; see Figure 6(a). The finite automaton representations for AllC and AllD are similarly straightforward; see Figures 6(b) and 6(c).

TFTT begins by cooperating, subsequently defects only if his opponent defects twice in a row, and quickly reverts back to cooperation as soon as his opponent cooperates. TFTT's behavior is depicted in Figure 6(d) by a finite automaton with two states. Initially, TFTT cooperates and enters state 1; TFTT will continue to cooperate and to return to state 1 as long as his opponent cooperates. The first time his opponent defects, if ever, TFTT cooperates but enters a new state 2. If his opponent then defects again, TFTT defects and returns to state 2; but if his opponent cooperates, TFTT cooperates and returns to state 1. Thus, TFTT is less provocable than TFT, requiring two successive defections before retaliating with a defection; but TFTT is equally quick to forgive as soon as the other player is nice.

Rip is more complex than TFT or TFTT, requiring three states for its finite automaton representation in Figure 6(e). Rip evolved in an evolutionary IPD game which included TFT, TFTT, AllC, and AllD in the initial player pool. Consequently, it is not surprising that Rip takes advantage of TFT, TFTT, and AllC, while protecting itself from AllD. More precisely, Rip first tests the mettle of an opponent with a defection and enters state 1. If the opponent ever defects, Rip cooperates and enters the "TFT" state 3, resulting in TFT play for all further moves against this player. Thus, after first defecting and then cooperating, Rip attains mutual cooperation with TFT, with other Rips, and with GRip, and mutual defection with AllD. However, against opponents such as AllC and TFTT which never retaliate after a single defection, Rip alternately cooperates and defects, moving from state 1 to state 2 and back again. Rip thus takes advantage of the low provocability of AllC and TFTT—full advantage in the case of TFTT.

GRip is even more complex, requiring four states for its finite automaton representation in Figure 6(f). GRip initially cooperates, but eventually GRip tries out a defection against "nice" players. Against TFTT and AllC, GRip cooperates for his first four moves; but then GRip enters a three-move cycle (d, c, c) while TFTT and AllC continue to cooperate. Thus, like Rip, GRip rips off TFTT and AllC, although not as frequently. Against another GRip, each GRip immediately enters into a five-move cycle (c, c, c, c, d). Against TFT and Rip, GRip ultimately enters a five-move cycle (d, c, c, c, c) against which TFT and Rip play (c, d, c, c, c).

5.2 ANALYTICAL RESULTS FOR TWO-PLAYER IPD/CR GAMES

An analytical study was undertaken of single-tournament, two-player IPD/CR games between all possible pairs of players described in Section 5.1. In Table 1 (see Appendix) we present the key results from these studies, and in Figures 7 and 8 we highlight some of these results.

Reading the columns of Table 1 from left to right, we have: (1) the player pairs; (2) the players' (updated) expected payoffs at the end of iteration i; (3) sufficient conditions for PD game play to come to a halt in finitely many iterations; (4) the maximum (possibly infinite) number i^* of iterations during which PD games will be played (i.e., the last iteration before either a refusal of a PD game offer occurs, or no PD game offers are made); (5) the player(s) who would refuse any PD game offer made by the opposing player in iterations subsequent to i^*, if any; and (6) the average payoff score of each player in each iteration through i^*. The quantities in columns two through four and in column six are determined as functions of the PD game payoffs and the parameters describing the choice/refusal mechanism.

The results in Table 1 are presented in normalized form. More precisely, the PD game payoffs $\{B, D, C, S\}$ and the initial expected payoff π^0 are normalized by subtracting B and dividing by $D-B$. The resulting normalized values are indicated below as hatted quantities:

$$\hat{B} = 0;\ \hat{D} = 1;\ \hat{C} = \frac{C-B}{D-B};\ \hat{S} = \frac{S-B}{D-B};\ \hat{\pi}^0 = \frac{\pi^0 - B}{D-B}. \tag{4}$$

Thus, after normalization, the lowest possible PD game payoff B becomes 0, the mutual defection payoff D becomes 1, and the nonoscillation condition $(S+B)/2 < C$ becomes $\hat{S}/2 < \hat{C}$. This normalization decreases the number of parameters by two and demonstrates that the signs of the payoffs are not important for the IPD/CR game. However, since it is not possible to normalize $\hat{C}$ and $\hat{S}$ any further, changing the relative distance between the payoffs can have an effect. For expositional simplicity, the hats are dropped from the normalized payoffs and initial expected payoff in Table 1 and throughout the remainder of this paper.

In obtaining the results presented in Table 1, we assumed that the initial expected payoff π^0 was not unrealistically high, but that it was high enough to guarantee that at least one game would always be played. Specifically, we assumed that $\tau < \pi^0 \leq S$. We also assumed that the minimum tolerance level τ was not unrealistically high or low, i.e., $0 \leq \tau \leq C$, and that the memory weight ω satisfied $0 < \omega < 1$. Finally, we dropped the conditional player indexing on expected payoffs, since it is generally obvious who is playing whom when there are only two players; subscripts are used for clarification when necessary.

Before we discuss the specific results in Table 1, we demonstrate how they were obtained by describing a couple of the calculations.

AllD VERSUS AllD. Suppose two AllD players are playing a two-person IPD/CR game. Since $\tau < \pi^0$ by assumption, both players make PD game offers to each other in the first iteration. In the subsequent play of the PD game, both players defect and receive a payoff of $D = 1$. Their expected payoffs for the next iteration then take the form $\pi^1 = \omega\pi^0 + (1-\omega)$.

Suppose the players play a PD game in i successive iterations, $i \geq 1$. They will then each have an expected payoff of

$$\begin{aligned} \pi^i &= \omega\pi^{i-1} + (1-\omega) \\ &= \omega^i\pi^0 + (1-\omega^i). \end{aligned} \tag{5}$$

PD game play stops at the end of iteration i if and only if the expected payoff π^i drops below τ, i.e., if and only if

$$\omega^i \pi^0 + (1 - \omega^i) < \tau. \tag{6}$$

If $\tau \leq 1$, this inequality is not satisfied for any i, implying that $i^* = \infty$. Suppose $1 < \tau$. Then $1 < \pi^0$ must hold as well. Solving the inequality for i then yields

$$\ln\left(\frac{\tau - 1}{\pi^0 - 1}\right) \frac{1}{\ln(\omega)} < i. \tag{7}$$

Letting i^* denote the smallest integer value of i for which this inequality is true, i^* is the last iteration during which a PD game is played between AllD and AllD. For example, if $\tau = 2$, $\pi^0 = 3$, and $\omega < 0.5$, then $i^* = 1$, implying that exactly one PD game is played.

If $i^* < \infty$, the expected payoff which AllD associates with another play of a PD game with AllD remains frozen at $\pi^{i^*} < \tau$ for all $i \geq i^*$, and no more PD game offers are made. Each player then simply receives the wallflower payoff in all subsequent iterations.

Note that an optimistically high initial expected payoff π^0 will result in many PD game plays (large i^*) unless the memory weight ω is small. That is, the specification of a high initial expected payoff tends to encourage repeated play.

Rip VERSUS Rip. Both Rips defect against each other on the first iteration, which results in an updated expected payoff $\pi^1 = \omega\pi^0 + (1-\omega)$ for each player. If $\pi^1 < \tau$, they will stop playing without rejection. On the other hand, if $\pi^1 \geq \tau$, they will continue playing forever, because they will cooperate with each other on all PD games except the first. Thus, play between two Rip players will stop after a finite number of iterations if and only if the first PD game results in an updated expected payoff $\pi^1 < \tau$, in which case $i^* = 1$.

Rip VERSUS AllC OR TFTT. AllC and TFTT always cooperate with Rip, while Rip alternately defects and cooperates. Without loss of generality, consider TFTT. TFTT receives C in even iterations and 0 in odd iterations, while Rip receives C in even iterations and S in odd iterations. Given any payoffs E and F, the expected payoffs when receiving alternating payoffs of E in even iterations and F in odd iterations take the form

$$\pi^{2i} = \omega\pi^{2i-1} + (1 - \omega)E; \qquad \pi^{2i+1} = \omega\pi^{2i} + (1 - \omega)F. \tag{8}$$

Solving these coupled difference equations gives

$$\pi^{2i} = \omega^{2i}\pi^0 + \frac{1 - \omega^{2i}}{1 + \omega}(E + \omega F) \tag{9}$$

and

$$\pi^{2i+1} = \omega^{2i+1}\pi^0 + \omega\frac{1-\omega^{2i}}{1+\omega}(E+\omega F) + (1-\omega)F. \tag{10}$$

If $(C\omega)/(1+\omega) \geq \tau$, then TFTT and Rip never refuse each other because their expected payoffs remain greater than τ in each iteration i. Consequently, $i^* = \infty$. If $(C\omega)/(1+\omega) < \tau$ and $\pi^0 \leq C/(1+\omega)$, the expected payoff of TFTT falls below τ after the first PD game play with Rip: $\pi^1 = \omega\pi^0 \leq (C\omega)/(1+\omega) < \tau$. Thus $i^* = 1$.

Finally, suppose $(C\omega)/(1+\omega) < \tau$ and $C/(1+\omega) < \pi^0$. In this case the expected payoff of TFTT falls below τ in the first odd iteration $2i+1$ that satisfies

$$\ln\left(\frac{\tau(1+\omega) - C\omega}{\pi^0(1+\omega) - C}\right)\frac{1}{\ln(\omega)} < 2i+1. \tag{11}$$

By using the difference equation (9), together with the assumption that $\tau \leq C$, it can be shown that the expected payoff of TFTT in all previous even iterations is still greater than τ. Moreover, for all previous iterations through this odd iteration, the expected payoff of Rip exceeds τ, implying that Rip has never refused TFTT. Consequently, this odd iteration gives i^*, the iteration of last PD game play between Rip and TFTT.

Once TFTT begins to refuse Rip, Rip receives the rejection payoff R. The expected payoff for Rip is updated according to

$$\pi_R^i = \omega^{i-i^*}\pi_R^{i^*} + (1-\omega^{i-i^*})R. \tag{12}$$

If $R \geq \tau$, π_R^i never falls below τ, and Rip will receive the rejection payoff in each subsequent iteration. If $R < \tau$, Rip will stop choosing to play TFTT when π_R^i declines below τ. Thereafter, both players will receive the wallflower payoff.

PLAY STOPPAGE CONDITIONS. Examining the third column in Table 1, we see that the minimum tolerance and memory parameters τ and ω play a crucial role in determining whether or not play eventually stops for our six player types. Figures 7(a) through 7(c) show the regions of the $\omega - \tau$ plane in which play stoppage occurs for AllD, Rip, and GRip, respectively, in IPD/CR game play with other player types. These regions are determined for the particular parameter specifications $\pi^0 = C = 3$ and $S = 5$.

Note that some results are determined by τ alone, regardless of all other parameter and payoff values:

$\tau > 1$: Play stops for all two-player IPD/CR games involving AllD;

$\tau \leq 1$: Play never stops for AllD v. AllD, Rip v. Rip, or GRip v. GRip; and

$\tau > 0$: AllC eventually refuses AllD.

When memory fades quickly, i.e., when ω is small, defection can trigger refusal by fairly tolerant players despite a long history of mutual cooperation or a high initial expected payoff. For example, as seen in Figure 7(c), Rip eventually refuses GRip for small enough ω (if GRip has not already refused Rip) even for minimum tolerance levels τ less than the mutual defection payoff $D = 1$. Note that the line demarking when Rip refuses GRip is independent of π^0.

If defections occur early and then stop, play behavior depends critically on simple relationships among the minimum tolerance level τ, the memory weight ω, and the initial expected payoff π^0. A high initial expected payoff π^0 tends to decrease a player's sensitivity to early defections, whereas a low memory weight ω tends to increase his sensitivity to early defections by downgrading the importance of the initial expected payoff in all subsequent updated expected payoffs. Thus, for example, a defection on the first iteration will result in a refusal in the next iteration by any opposing player who: (i) cooperated in the first iteration, if (ω, τ) lies above the line $\tau = \omega\pi^0$ in the $\omega - \tau$ plane; or (ii) defected in the first iteration, if (ω, τ) lies above the line $\tau = \omega[\pi^0 - 1] + 1$ in the $\omega - \tau$ plane. Consequently, the lower the initial expected payoff π^0, the larger is the (admissible) region in the $\omega - \tau$ plane over which Rip, TFT, and GRip will play only one PD game with any player who defects in the first iteration.

The fourth column of Table 1 highlights another interesting characteristic of two-player IPD/CR games: the maximum (possibly infinite) number of iterations during which PD game play will take place, denoted by i^*. In Figure 8 we depict regions of the $(1 - \omega) - \tau$ plane where the iteration number i^* takes on a constant value, conditional on $\pi^0 = C = 3$ and $S = 5$, for various player pairs. In each case, the region where i^* takes on its smallest value is the largest region, and the regions get ever smaller as i^* increases. In general, the regions where i^* takes on large but finite values are very small indeed; that is, for most player pairs, intolerance (play stoppage) either occurs after only a few iterations, or it does not occur at all.

Recall that, given $K = N - 1$, the IPD/CR game reduces to the IPD game if either $\tau = 0$ or $\omega = 1$ and $\tau < \pi^0$. The analytical results of this section indicate that, near these boundaries in parameter space, the IPD/CR game with $K = N - 1$ will behave much like the IPD game: either players will play each other forever without refusal, or they will play many PD games before refusal. However, as τ increases and/or ω decreases, refusal becomes more likely, and we move away from the IPD game.

GENERIC RESULTS. Certain parameter regimes can be directly associated with particular types of play behavior. In Section 7, these parameter regimes will be used to predict and to explain the ultimate outcomes in our evolutionary (multi-tournament) IPD/CR game simulations. The most important parameter regime is when any defection against a cooperation results in future refusal by the cooperating player. The next most important regime is when any defection results in immediate play stoppage.

(a)

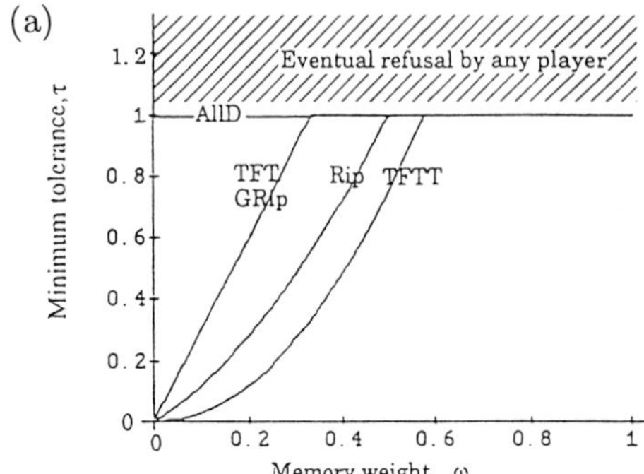

(b)

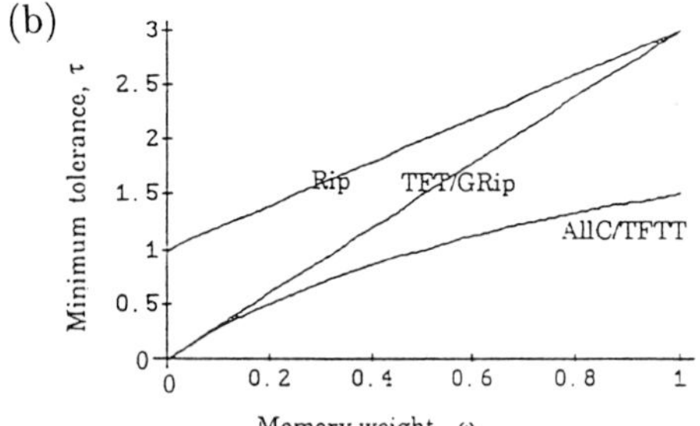

(c)

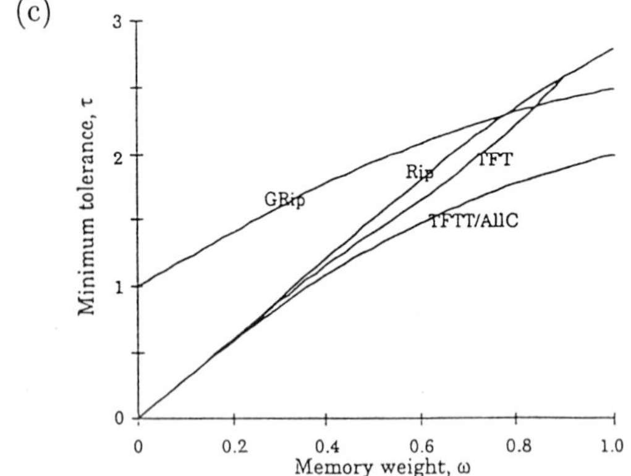

FIGURE 7 (a) A player eventually finds AllD intolerable in an IPD/CR game play if τ and ω lie in the region above the curve indicated for this player's type. (AllD is always eventually refused by AllC.) (b) A player eventually finds Rip intolerable in an IPD/CR game play if τ and ω lie in the region above the curve indicated for this player's type. Play stops between two Rips by mutual assent, but, in the other indicated cases, Rip is eventually refused. (c) A player eventually finds GRip intolerable in an IPD/CR game play if τ and ω lie in the region above the curve indicated for this player's type. The parameter values for each plot are $B = 0$, $D = 1$, $C = 3$, $S = 5$, and $\pi^0 = 3$.

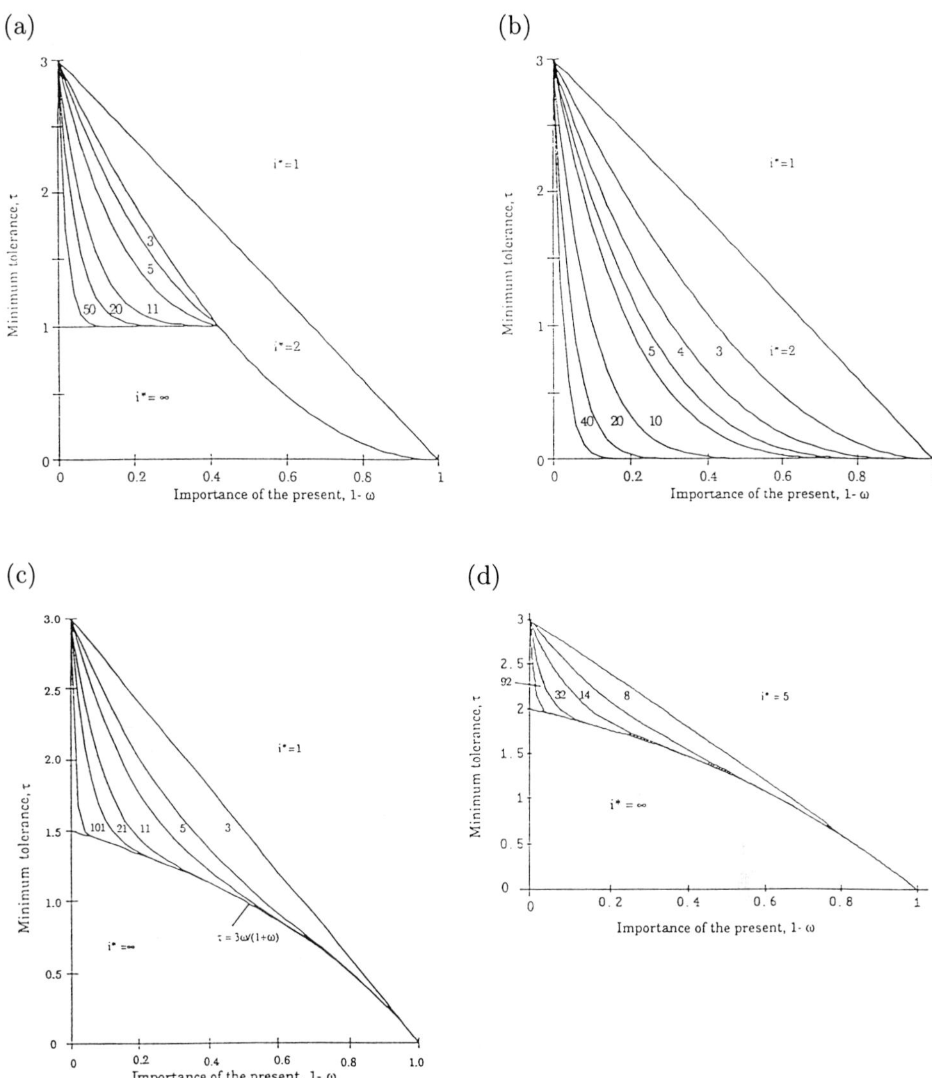

FIGURE 8 (a) The total number of PD games played in a two-player IPD/CR game between AllD and TFTT before play stoppage, as a function of τ and ω. The number i^* gives the last iteration during which a PD game takes place in a region. The other numbers indicate the maximum number of games played in a region. Each region includes its upper boundary line but not its lower boundary line. Except for the top boundary line, all boundary lines cross each other exactly once, at $\omega = 1/\sqrt{3}$ and $\tau = 1$. The d/c move sequence in the first two iterations provokes refusal by TFTT in a large region of this plane. (b) The total number of PD games played in a two-player IPD/CR game between AllD and AllC before play stoppage, as a function of τ and ω. (continued)

FIGURE 8 (continued) The number i^* gives the last iteration during which a PD game takes place in a region. The other numbers indicate the maximum number of games played in a region. Each region includes its upper boundary line but not its lower boundary line.(c) The total number of PD games played in a two-player IPD/CR game between Rip and either AllC or TFTT before play stoppage, as a function of τ and ω. The number i^* gives the last iteration during which a PD game takes place in a region. The other numbers indicate the maximum number of games played in a region. Each region includes its upper boundary line but not its lower boundary line. The boundary lines cross the line $\tau = 3\omega/(1+\omega)$ only at the axis. (d) The total number of PD games played in a two-player IPD/CR game between GRip and either AllC or TFTT before play stoppage, as a function of τ and ω. The number i^* gives the last iteration during which a PD game takes place in a region. The other numbers indicate the maximum number of games played in a region. Each region includes its upper boundary line but not its lower boundary line. The boundary lines are tangent to the line dividing eventual play stoppage ($i^* < \infty$) from no stoppage ($i^* = \infty$), and they meet at each axis.

If the first defection in a two-player tournament occurs on iteration i, with one player cooperating and the other defecting, the cooperating player will refuse further play if

$$\tau > \omega^i \pi^0 + \omega(1 - \omega^{i-1})C. \tag{13}$$

Examining this inequality, we see that if $\tau > \max\{\omega\pi^0, \omega C\}$, then any defection at any time against a cooperation results in immediate refusal to play. For example, in the evolutionary simulations illustrated in Section 4 and detailed in Section 7, we set $\pi^0 = C = 3$ and $\omega = 0.7$; for these fixed parameter values, defections against cooperation are not tolerated for any minimum tolerance level τ greater than 2.1.

If, instead, the first defection is mutual, both players will stop choosing each other if

$$\tau > \omega^i \pi^0 + \omega(1 - \omega^{i-1})C + 1 - \omega. \tag{14}$$

Comparing Eq. (14) with Eq. (13), we see that *any* initial defection results in refusal to play if $\tau > (1-\omega) + \max\{\omega\pi^0, \omega C\}$. In particular, for $\pi^0 = C = 3$ and $\omega = 0.7$, any initial defection results in immediate play stoppage if $\tau > 2.4$.

6. SINGLE-TOURNAMENT, N-PLAYER IPD/CR GAMES

The analysis in Section 5 describes the characteristics of various illustrative two-player IPD/CR games with $K = 1$. This two-player analysis can be extended in a straightforward way to analyze any N-player IPD/CR game in which $K = N - 1$. Given this specification of K, each player chooses to play a PD game with every tolerable player in each iteration, implying that the N-player IPD/CR game decomposes into a collection of $N(N-1)/2$ two-player IPD/CR games of the type analyzed in Section 5. Consequently, setting $K = N - 1$ trivializes the role of choice in the determination of PD game partners; it is refusal which is critical.

On the other hand, when the number N of players is greater than 2 and K is less than $N - 1$, players will not necessarily choose to play all tolerable players during each iteration. Choice then becomes more important, and one would expect to see choice and refusal working together in a more subtle fashion. In fact, as we shall see, in many interactions choice is much more important than refusal.

The outcomes of N-player IPD/CR games can be characterized in a number of different ways, including the average payoff scores both for individual players and for the entire ecology, the fraction of game plays of each type, and the formation of long-run player interaction patterns. For example, who ends up playing whom when a steady state is reached, if ever? To what extent is the ultimate steady-state ecology characterized by a hierarchy of social cliques, ranging from high-status player groups which everyone else would like to join, all the way down to ostracized individual players whom all other players avoid?

In order to see how play in a multiple-player tournament proceeds and how long-run player networks form, consider an illustrative five-player IPD/CR game with three TFTT players and two Rip players, played for infinitely many iterations. As in Section 5.2, let payoffs be normalized so that $B = 0$ and $D = 1$. For simplicity, suppose that each player can choose at most one potential game partner in each iteration ($K = 1$), that the initial expected payoff, common to all players, is $\pi_0 = C$, that the wallflower payoff W coincides with the minimum tolerance level τ, and that the rejection payoff R does not exceed 1. Finally, assume that when a player needs to choose among partners with the same expected payoff, the choice is made by a random draw. Since the initial expected payoff π^0 is the same for all potential partners, this implies in particular that each player chooses each of the other four players as a potential first partner with probability 1/4.

Recall from Section 5 that two TFTTs always cooperate and that a TFTT receives a payoff sequence $(0, C, 0, C, ...)$ in an IPD game with a Rip. Moreover, a Rip receives a payoff sequence $(S, C, S, C, ...)$ in an IPD game with a TFTT, and a payoff sequence $(1, C, C, ...)$ in an IPD game with another Rip. Observe, also, that $\pi^0 = C$ implies $C\omega/(1 + \omega) < \omega\pi^0 + (1 - \omega)$. Consequently, there are three possible parameter regimes for this five-player IPD/CR game. If $\tau \leq C\omega/(1 + \omega)$, all players remain tolerable to all other players. If $C\omega/(1+\omega) < \tau \leq \omega\pi^0 + (1-\omega)$, Rips are eventually refused by TFTTs but remain tolerable to each other forever. And if $\tau > \omega\pi^0 + (1 - \omega)$, Rips are refused by TFTTs after only one PD game and also find each other mutually intolerable after only one PD game. What happens in each of these three parameter regimes will now be described in more detail.

CASE I: $\tau \leq C\omega/(1+\omega)$

In this case, no player ever refuses to play another player in repeated PD game plays. Play eventually stabilizes in one of two possible player interaction patterns. In both cases, each TFTT chooses randomly in each iteration between the other two TFTTs, and each Rip chooses to play one TFTT repeatedly. But in one case the two Rips both repeatedly play the *same* TFTT player, while in the other they each repeatedly play *different* TFTT players. Which network forms is determined by the particular realizations of the random draws used to break expected payoff ties in the first few iterations.

We will now look in greater detail at the sequences of plays. First, note that the probability that the two Rips ever play each other is 7/16, the probability that they play each other in the first iteration. This occurs because both Rips defect if they play each other in the first iteration, implying that their expected payoffs from playing each other decrease below C. Since each Rip always has an expected payoff of at least C from any TFTT, each Rip then chooses a TFTT in all future iterations. On the other hand, if the two Rips do not play each other in the first iteration, each Rip plays a PD game with a TFTT and receives a payoff of S. The expected payoff for playing additional PD games with this TFTT is then larger than C, and the Rip prefers this TFTT to all other players, including the other Rip. In any subsequent PD game with this TFTT, it receives either C or S, and its expected payoff from this TFTT remains above C. From then on, it always prefers the TFTT that it played in the first iteration to the other Rip; hence, while it may eventually learn to prefer a different TFTT to the first one it played, it always chooses a TFTT. Since neither Rip chooses the other Rip after the first iteration, they never play each other unless they do so in the first iteration.

The expected payoff that a TFTT has from another TFTT stays at C. The expected payoff that a TFTT has from a Rip is also C until the TFTT first plays that Rip, after which it is less than C no matter how many times the TFTT plays that Rip again. A TFTT therefore chooses randomly among all players until he encounters a Rip (either because he makes a PD game offer to a Rip or because he receives a PD game offer from a Rip). If enough games are played, a TFTT eventually plays both Rips at least once. Once a TFTT plays a Rip, it never chooses that Rip again, but the Rip may choose the TFTT on future plays.

Rips may oscillate among different TFTT game partners for several iterations, but eventually each Rip settles into playing only one particular TFTT. Which TFTT each Rip ends up playing is a matter of chance.

For example, suppose that Rip #1 by chance plays TFTT #1 in the first iteration and gets S. He then chooses TFTT #1 again in iteration 2 and gets C. If TFTT #2 by chance chooses Rip #1 in iteration 2, Rip #1 gets S from TFTT #2, and so chooses TFTT #2 in iteration 3. After getting C from TFTT #2 in iteration 3, Rip #1 is indifferent between TFTT #1 and TFTT #2, but prefers either of these TFTTs to the remaining TFTT #3. If TFTT #3 by chance chooses Rip #1 in iteration 3, Rip #1 then prefers TFTT #3 in iteration 4 and subsequently chooses randomly among all three TFTTs in iteration 5. Since Rip #1 receives S from the TFTT that it chooses in iteration 5, and the other TFTTs never choose Rip #1 again, Rip #1 chooses this iteration 5 TFTT in all future iterations. This particular sequence of events is depicted in Figure 9.

Note that each Rip in the long-run player interaction pattern has a higher average payoff than any TFTT. The average payoff of a Rip eventually approaches $(S + C)/2 > C$, as a result of the successful long-run parasitic relationship established with a TFTT. On the other hand, the average payoff for any TFTT who is playing only other TFTTs in the final network approaches C from below as the number of iterations increases; the average payoff for TFTTs playing one Rip are lower than this; and TFTTs playing two Rips do the worst.

In summary, in Case I, the low minimum tolerance level allows each opportunistic Rip player to find and exploit a nice TFTT player while avoiding the other Rip. Moreover, not all TFTT players achieve the same average payoff: at least one TFTT player will manage by chance to avoid Rip parasites, and at least one will not.

CASE II: $C\omega/(1 + \omega) < \tau \leq \omega\pi^0 + (1 - \omega)$

In this case a TFTT eventually refuses a Rip in repeated PD game plays, but Rips remain tolerable to each other. The players thus eventually break into two groups, one consisting of the three TFTTs and the other consisting of the two Rips. Average payoffs approach the mutual cooperation payoff C for all players.

How long the final network takes to form depends on the magnitude of τ. If τ is close to the lower bound $C\omega/(1 + \omega)$, then the pattern from Case I holds until the payoffs that TFTTs expect from Rips all drop below τ. Once this occurs, Rips begin receiving rejection payoffs R from TFTTs. Eventually the Rips stop choosing the TFTTs and settle on each other. On the other hand, if τ is large enough (see Section 5), TFTTs refuse Rips after only one PD game and the long-run player interaction pattern gels after only a few iterations. The reader is referred to Stanley et al.[19] for further details.

Note that the TFTTs primarily use refusal to protect themselves from Rips and use choice to rotate among each other. Rips are able to take advantage of TFTTs only for a limited number of iterations. Ultimately, if the tournament lasts long enough, each Rip is refused by all TFTTs, and the Rips end up choosing to play only with each other.

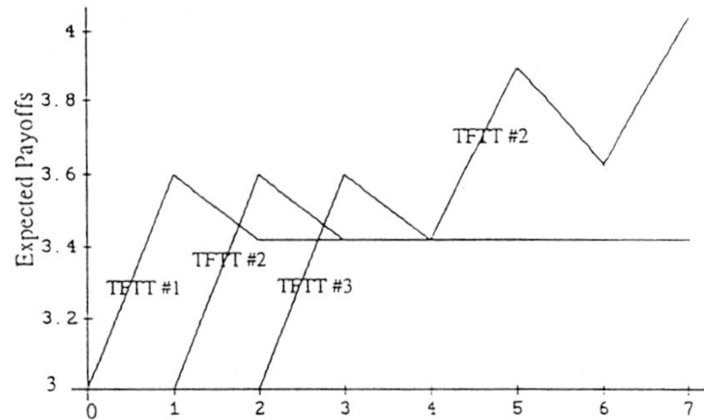

FIGURE 9 A schematic of the expected payoffs that a Rip anticipates from three TFTT (or AllC) players when $K = 1$ and $\pi^0 = C$. In this sample scenario, Rip plays TFTT #1, TFTT #2, and TFTT #3 for the first time in iterations 1, 2, and 3, respectively. Before iteration 5, Rip has the same expected payoff for all three TFTTs and chooses TFTT #2 at random. It is assumed that $i^* \geq 5$, i.e., that τ and ω are set so that at least five PD games are played between Rip and a TFTT (or AllC) player before the TFTT (or AllC) player refuses Rip.

CASE III: $\omega\pi^0 + (1 - \omega) < \tau$

In this case, recalling the analysis of two-player IPD/CR games in Section 5.2: (i) a Rip never refuses a TFTT in repeated PD game plays; (ii) a TFTT refuses a Rip after only one PD game play (since $\pi^1 = \omega\pi^0 < \tau$); and (iii) a Rip refuses another Rip after only one PD game play. The final player interaction pattern is the group of TFTTs and two ostracized Rip players. Average payoffs approach the mutual cooperation payoff C for the TFTTs and the wallflower payoff W for the Rips. Choice plays almost no role, and refusal works so strongly that Rips never discover that, in all but the first PD game play, they could receive C from plays with each other.

THE ROLE OF π^0

Suppose the initial expected payoff π^0 in the five-player IPD/CR game described above is less than C. Then, once a TFTT choses to play another TFTT, he never chooses to play a Rip in any future iteration. Lowering the initial expected payoff thus cools the system in the sense that exploration is reduced. Rips behave the same as before.

If the initial expected payoff π^0 is greater than C, the long-run player interaction patterns are unchanged. However, each TFTT chooses to play every other player at least once by iteration 5, and plays more PD games with Rip. Morever, the region in the $\omega - \tau$ plane where Rips refuse each other is smaller than before. Thus, as one would anticipate, increasing the initial expected payoff results in increased exploration.

7. EVOLUTIONARY IPD/CR GAMES

This section describes the software used to run the evolutionary IPD/CR game simulations briefly reported on in Section 4 and provides a more careful discussion of the simulation findings.

7.1 SOFTWARE AND HARDWARE

The software used to replicate the experimental results of Miller,[17] and to modify Miller's framework to include a choice/refusal mechanism, was developed in Turbo Pascal 5.5. We used the random number generator included with the Pascal compiler. The actual runs were done on a number of machines, all 80386-based PC-compatibles with 80387 coprocessors.

Our algorithm is initialized with 30 finite automata of the sort described and illustrated in Section 5.1. Each automaton possesses 16 states and a uniformly distributed random assignment of state transitions, indicated by arrows, together with an initial move and allowable move sequences appearing as arrow labels. Each automaton has access to the values for the parameters $(\pi^0, \tau, K, R, W, \omega)$ characterizing the choice/refusal mechanism and has a memory comprised of (i) its play history with every other automaton and (ii) its current expected payoff for a PD game play with every other automaton.

The 30 automata take part in an evolutionary IPD/CR game consisting of 50 successive single-tournament IPD/CR games separated by genetic steps. Each single-tournament IPD/CR game (or "tournament" for short) consists of 150 iterations, where each iteration constitutes one pass through the five-stage choice/refusal mechanism described in Section 3.2. In the genetic step at the end of each tournament, the current population of 30 automata is replaced via reproduction by a new population of 30 automata. Each run of the evolutionary IPD/CR game thus results in a distinct "ecology" consisting of an evolved population of 30 automata.

The genetic step separating successive tournaments proceeds as follows. First, the automata are sorted by the average (per payoff) payoff score they achieved during the previous tournament. Second, as in Miller,[17] the 10 automata with the worst scores are discarded and the 20 automata with the best scores are retained. Finally, the 10 emptied slots are filled by "sexual reproduction" of the best 30 automata, with "mutation" of the resulting offspring. The resulting 30 finite automata then constitute the genetically altered population of players for the next tournament.

This "sexual reproduction" is accomplished as follows. First, take the initial move (one bit), 32 arrows (each arrow represented by four bits to describe the "next state"), and 32 arrow labels (one bit each) of each automata and turn them into a bit string 161 bits long.[6] Second, select two automata to be parents, where each automaton's probability of selection is directly proportional to the average payoff score it received in the preceding tournament. Third, generate a random variable q distributed uniformly over the discrete range $1, 2, \ldots, 161$. Fourth, exchange the bits in positions q through 161 of the parental bit strings to obtain the bit strings for two offspring. Finally, repeat the second and third stages for five pairs of parents, thus obtaining five pairs of offspring.

"Mutation" simply consists of flipping the value of each of the bits of each offspring's bit string with probability 5 in 1,000. Once the final bit string for any offspring is obtained, reverse the original conversion process from automaton to bit string to obtain the finite automaton representation for the offspring.

7.2 SIMULATION RESULTS

Based on the analysis in Section 5.2, we decided to focus on the role of the minimum tolerance level τ. In the experiments reported below, we use the following fixed PD game payoff specifications

$$B = 0,\ D = 1,\ C = 3,\ S = 5; \tag{21}$$

and we set

$$\pi^0 = 3,\ K = 1,\ R = 1,\ \omega = 0.7. \tag{22}$$

The wallflower payoff W is always set equal to τ. Forty runs are carried out for each tested τ value to allow for stochastic variability.

In Figures 10 through 13 we display some of the simulation results obtained for a 50-tournament evolutionary IPD/CR game as the minimum tolerance level τ was set at 0.5, 1.5, 1.6, and 2.4, respectively. More precisely, each figure displays the following information for the indicated τ value: (a) individual ecology payoff trajectories obtained for 40 different runs of the 50-tournament game; (b) the ecology payoffs obtained for each of the 50 tournaments, averaged across the 40 runs; and (c) the fraction of various types of play behavior (mutual cooperation CC, successful defection CD, mutual defection DD, and rejection RJ) exhibited in each of the 50 tournaments, averaged across the 40 runs. In Figure 14 we highlight characteristics of the ecologies associated with the lowest payoff band appearing in Figure 13(a).

As will be clarified in the following discussion, In Figures 10 through 14 we illustrate many of the interesting results of our simulation study. A more detailed

[6] A bit can take on only two values, 0 or 1. A string of four bits therefore can represent $2^4=16$ different states. The arrow labels require only one bit, which encodes the current move to be made in response to a previous move of an opponent, because the information about the previous move of an opponent is coded by position in the string. The finite automaton and bit string representations used in this paper differ from the representations used in Miller.[17]

discussion of these results can be found in Stanley et al.[19] A complete data set, together with the PC-compatible software needed to view it, is available; please contact the second author for information.

Consider, first, the payoff band patterns exhibited by the payoff trajectories for individual ecologies in Figures 10(a) through 13(a). Given $\tau = 0.5$ [Figure 10(a)], we see many of these trajectories clustered into three payoff bands at heights of roughly 3.0, 2.8, and 2.3, as well as a fairly large number of trajectories in no apparent band. This is typical of the runs with low τ values. Band formation seems to be primarily a choice-moderated phenomenon. Payoffs with no apparent band association become more common with decreases in τ, hence with decreases in rejections; but payoff bands still form even at $\tau = 0$ where all rejections cease. Notice that the fraction of rejections RJ is effectively zero for $\tau = 0.5$ (see Figure 10(c)).

As τ increases to 1.5 and to 1.6 (see Figures 11(a) and 12(a)), two of the three payoff bands persist, but the lowest payoff band at level 2.3 is only faintly apparent at best. Fewer ecology payoffs lie outside of a payoff band, and those that do lie outside appear to be either on their way into or out of a band. As τ further increases to 2.4 (see Figure 13(a)), the payoff band at 2.3 disappears altogether, the payoff band at 2.8 becomes extremely sparse, and a new payoff band arises in the vicinity of 1.9. The ecologies with payoffs lying in the latter "wallflower isolation" band consist of players that engage in mutual defection until they are rejected by all other players, and that thereafter collect the wallflower payoff $W = 2.4$. (Recall that the wallflower payoff W is always set equal to the minimum tolerance level τ.) A moment's reflection suggests that such play behavior is perfectly reasonable, given the high value for the wallflower payoff. Characteristics of these wallflower isolation ecologies are separately displayed in Figure 14.

Preliminary analysis of the genetic diversity of ecologies in the terminal (50th) tournament has been undertaken in an attempt to obtain a better understanding of this payoff band phenomenon. As noted above, payoff bands occur at levels 2.3 and 2.8 for relatively low τ values, and the terminal ecologies associated with each of these payoff bands appear to split into two genetically distinct groups. One group has a relatively low genetic diversity; that is, each ecology in this group is dominated by players with identical or closely related finite automaton structures, leading to highly synchronized play behavior. The stable average payoff behavior exhibited by this group is intuitively understandable; for a synchronized pattern of play can result in fixed fractions of distinct play behaviors, and hence in a roughly constant average payoff. The other group of ecologies has a relatively high genetic diversity, and the stable average payoff behavior exhibited by this group is harder to understand. The terminal ecologies associated with the mutually cooperative payoff band 3.0 and the wallflower isolation band are not as strongly bimodal as the ecologies associated with the middle bands 2.3 and 2.8.

The payoff trajectories for individual ecologies exhibit other interesting features as well. For $\tau < 2.1$, a number of ecologies exhibit an abrupt payoff collapse, a feature also present in the Miller replication results. For example, one of the

(a)

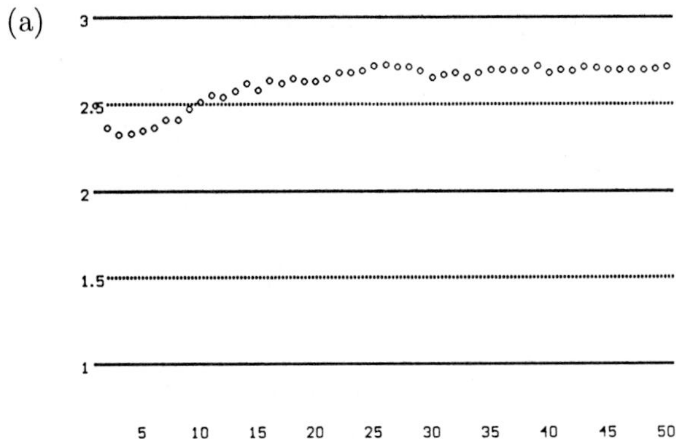

(b)

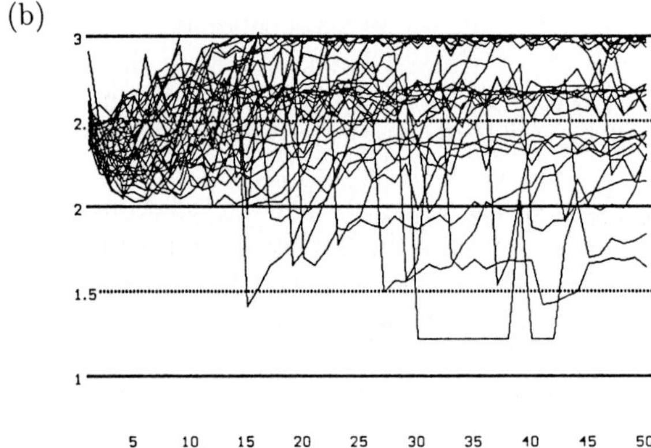

(c)

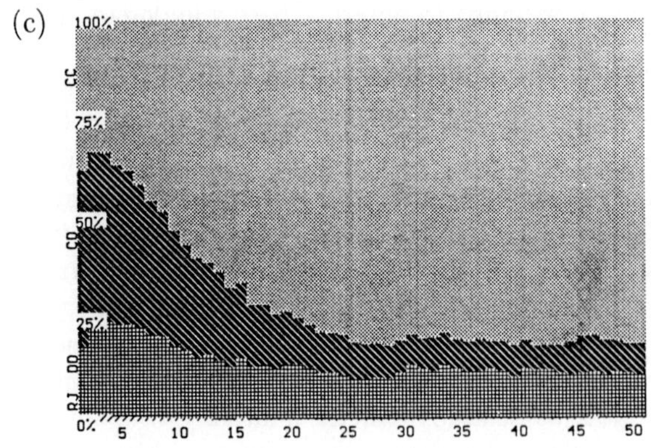

FIGURE 10 For the IPD/CR game with τ=0.5, we show (a) individual ecology payoffs, (b) average ecology payoffs, and (c) play behavior fractions.

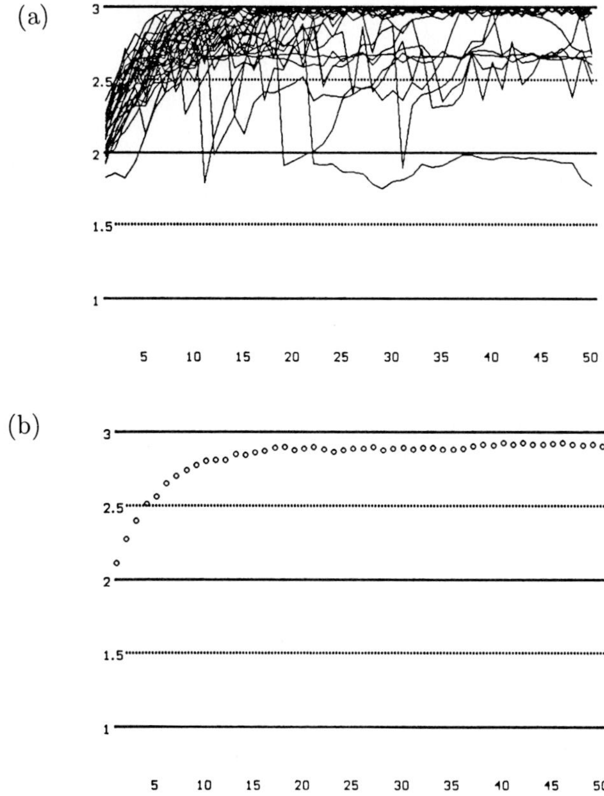

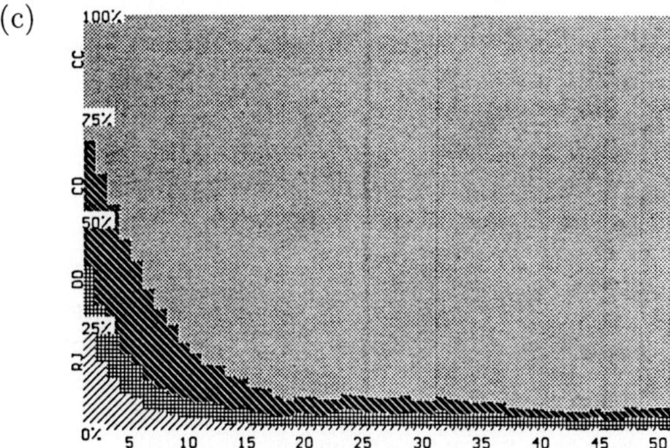

FIGURE 11 For the IPD/CR game with $\tau = 1.5$, we show (a) individual ecology payoffs, (b) average ecology payoffs, and (c) play behavior fractions.

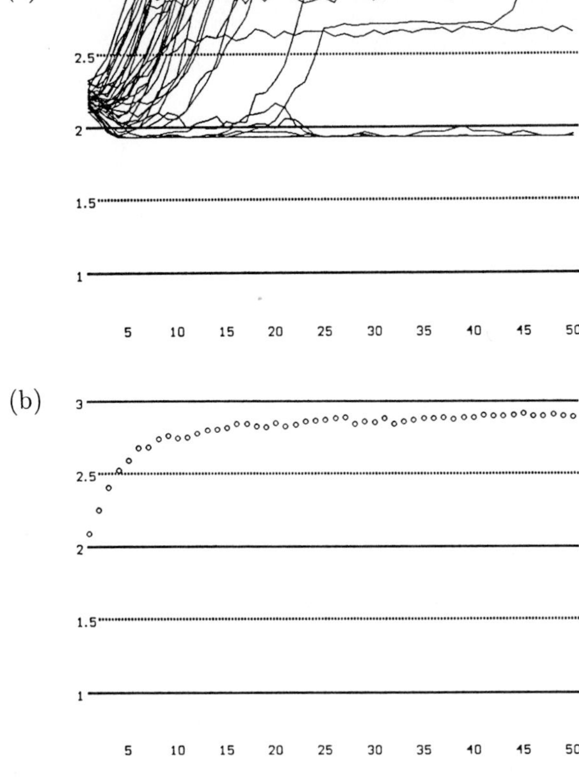

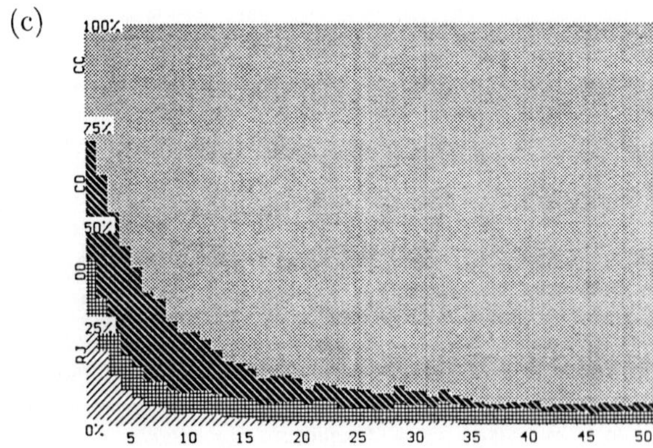

FIGURE 12 For the IPD/CR Game with $\tau = 1.6$, we show (a) individual ecology payoffs, (b) average ecology payoffs, and (c) play behavior fractions.

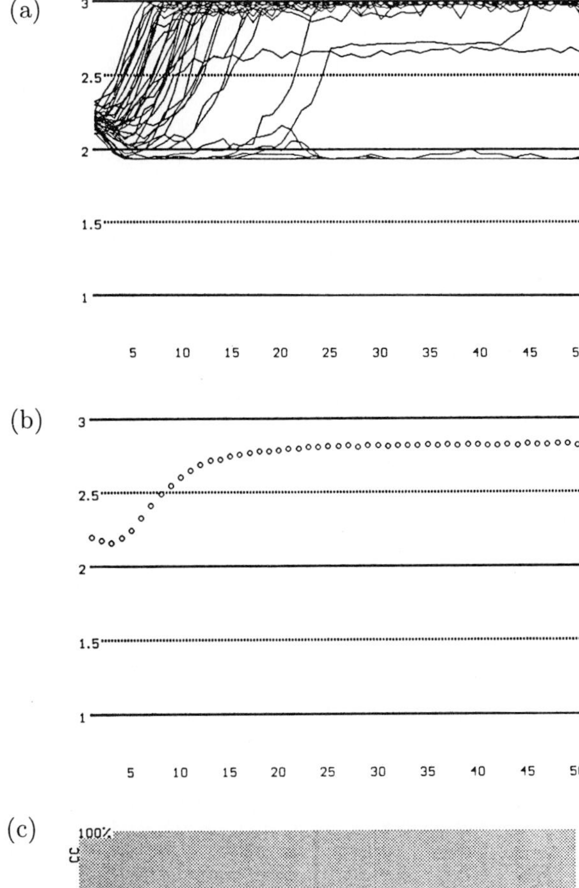

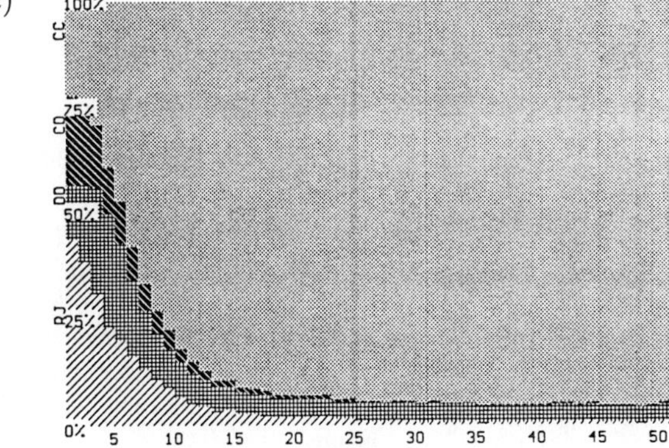

FIGURE 13 For the IPD/CR game with $\tau = 2.4$, we show (a) individual ecology payoffs, (b) average ecology payoffs, and (c) play behavior fractions.

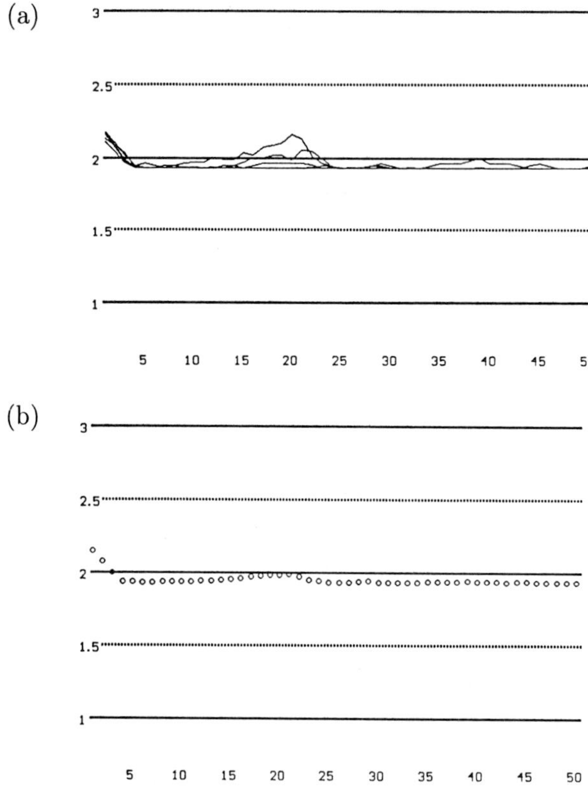

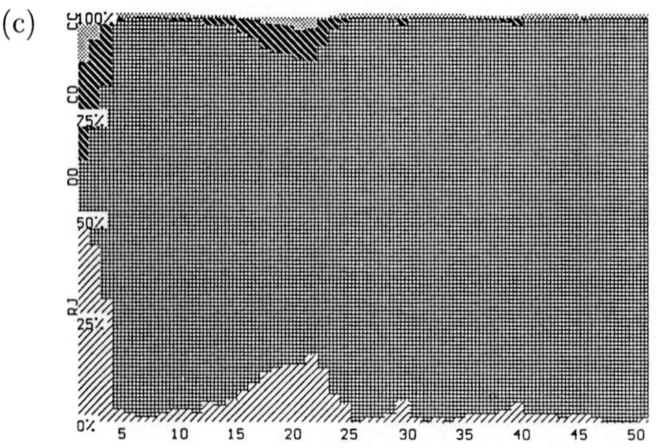

FIGURE 14 For the IPD/CR game with $\tau = 2.4$, we show (a) individual ecology payoffs for the wallflower band, (b) average ecology payoffs for the wallflower band, and (c) play behavior fractions for the wallflower band.

payoff trajectories depicted in Figure 10(a) abruptly collapses in generation 15 to a level well below that of all other trajectories, after which it slowly recovers. Examining in detail the ecology associated with this particular payoff trajectory, we saw that a spike occurred in the fraction of successful defections on the downward leg of the collapse, mutual defection became the dominant mode of play at the bottom of the collapse, and successful defection again became frequent during the subsequent payoff recovery. Note that the payoff of this ecology is roughly stable over generations 24 to 30 at the level 2.5, the average payoff level obtained by IPD players engaging in alternating *cd* and *dc* play; but the ecology payoff level subsequently climbs to the mutually cooperative payoff level 3.0.

It is also possible for ecologies to collapse to, and remain at, low payoff levels for an extended period of time. A dramatic example appears in Figure 11(a), where an ecology suffers a payoff collapse at about generation 22 and remains below the payoff level 2.0 for the remainder of the simulation.

"Band spiking" is another interesting feature exhibited by the payoff trajectories for individual ecologies for $\tau < 2.1$. An ecology sometimes exhibits an abrupt jump from one payoff band to another, followed by an abrupt jump back to its original payoff band. Preliminary evidence suggests that this spiking might be caused in part by the following type of population dynamics. Suppose the average payoff achieved by a type A player is higher than that of a type B player as long as type A players are rare, but that the average payoff of type A players drops below that of type B players as type A players become more prevalent in the population due to reproduction or fortuitous crossover. The ecology payoff then might systematically oscillate between the type A and type B payoff levels. When an ecology of this type was saved and used to "seed" a set of 40 new runs, the resulting payoff trajectories exhibited a great deal of spiking.

In addition, the payoff trajectories for individual ecologies exhibit "band tunneling," particularly for high τ values. That is, ecologies that have been long-term residents of one payoff band suddenly exhibit an abrupt change in the gradient of their payoff score, followed by a rapid and smooth traverse to another payoff band. For example, in Figure 13(a) a number of ecologies tunnel directly from the wallflower isolation band to the mutually cooperative payoff band at 3.0; and one ecology tunnels from the wallflower isolation band to the payoff band at 2.8, remains at this payoff band for fifteen generations, and then tunnels to the mutually cooperative payoff band. These tunneling events appear to be correlated with the fraction of rejections. That is, as the fraction of rejections increases for an individual ecology, the payoff level of the individual ecology increases; and the fraction of rejections recedes as the ecology merges into a new payoff band. In addition, partial tunneling events also take place. For example, in Figure 14(a) an ecology rises out of the wallflower isolation band between generations 15 and 20, but then rapidly recedes back into it.

The properties displayed by *average* ecology payoffs are also interesting. Comparing Figure 10(b) with the replicated Miller results reported in Figure 1, the average ecology payoffs obtained for the low minimum tolerance level $\tau = 0.5$ display less of an initial dip than in the Miller replication, and do not attain as high an ultimate payoff level. These differences can be traced to the emergence of new payoff bands. As seen in Figure 10(a), these new bands are at low enough levels, and contain enough ecology payoffs, to decrease markedly the average payoff across ecologies. While a substantial fraction of the individual ecology payoff trajectories appear to converge—indeed, as many as in the Miller replication—many of these trajectories do *not* converge to the mutually cooperative payoff level. For low τ values, the evolution of cooperation is, in fact, retarded as compared to the Miller replication.

When τ is increased to 1.5 and to 1.6, the initial dip in average ecology payoffs completely disappears and the evolution of cooperation is substantially accelerated [Figures 11(b) and 12(b)]. Moreover, the fraction of CC play behavior markedly increases [Figures 11(c) and 12(c)]. The minimum tolerance level τ is now high enough to give the refusal mechanism some teeth. As τ continues to increase, however, the wallflower isolation payoff band eventually appears. The initial average ecology payoffs begin to flatten out again and, by $\tau = 2.4$, an initial dip has reappeared (see Figure 13(b)). This dip is smaller than for the Miller replication and has an entirely different cause. Rather than representing the destruction of chump players by rapacious players, the dip results from the early formation of wallflower isolation ecologies which drag down the initial average ecology payoffs.

Finally, the fractions of distinct play behavior exhibited under different τ values are also of interest (see Figures 10(c)–13(c)). These fractions appear to undergo a phase change around $\tau = 2.1$. For $\tau < 2.1$, the simulation runs settle to a state where, on average, three types of play behavior in the PD game persist. The most common behavior is mutual cooperation CC, followed by mutual defection DD, and then by successful defection CD. For $\tau > 2.1$, however, successful defection effectively vanishes after the first few generations of players. This finding is predictable; as noted in Section 5.2, for $\tau > 2.1$, any successful defection results in the refusal of all future PD game offers. Mutual defection also becomes less common, appearing primarily in those ecologies constituting the wallflower isolation payoff band (see Figure 14(c)).

8. CONCLUDING REMARKS

The incorporation of a choice/refusal mechanism into the standard IPD framework has resulted in the emergence of several new ecological features. First, even without evolution, choice and refusal permit a broader range of interactions between mutually cooperative players, between potential hosts and would-be parasites, and between like-minded opportunists: for example, internally generated ostracism, persistent parasitism, and assortative partner selection. Second, with evolution, choice and refusal can accelerate the emergence of cooperation. More generally, however, choice and refusal can lead to the emergence of multiple tight payoff bands, reflecting the possible existence of multiple ecological attractors. The existence of a spectrum of payoff bands, in turn, permits the emergence of interesting new ecological behaviors, such as band spiking and band tunneling.

We hope to obtain a better quantitative understanding of these new ecological features as we undertake more elaborate studies of our choice/refusal mechanism. For example, it now seems clear that the genetic diversity of an ecology should be tracked throughout each simulation run, and that more detailed records should be maintained concerning which types of ecologies exhibit band spiking, band tunneling, and payoff collapse. At present, we have examples of ecologies that belong to the various payoff bands, but only the mutual cooperation and wallflower isolation payoff bands are well understood. The emergence of the middle payoff bands remains somewhat mysterious. Currently we are conducting extensive sensitivity studies to ascertain the conditions under which these middle payoff bands form.

In addition, a number of revisions to our model and methods are currently under consideration. For example, the minimum tolerance level τ and the initial expected payoff π^0—both currently set as fixed exogenous constants—instead should be allowed to evolve over time along with the structure of the PD rules. As seen in both the analytical and simulation studies reported here, higher levels of τ result in increased resistance to parasitic relationships, and differing resistance to parasites has been conjectured as one reason why populations divide into smaller reproductive groups.[18] Moreover, π^0 turns out to be a critical parameter that affects both the willingness of players to tolerate defections and the willingness of players to seek out new potential partners.

Also, using a fixed convex combination of past expected payoff and current payoff to obtain an updated expected payoff is too simplistic, even in the present context where players know nothing a priori about other player types. For example, surely a player could learn, through observation, that the current move of an opponent is functionally dependent on the previous move he made against that opponent. The memory weights used to obtain updated expected payoffs could reflect this dependence. For example, if a player's previous move against a current opponent was c, the updated expected payoff for this opponent could be a weighted average only over all past payoffs obtained under the same circumstance, i.e., obtained when the previous move against this opponent was c. Moreover, once a player learns that his current moves affect the later moves of other players, deliberate attempts to modify these later moves become possible.

Finally, the analysis of individual finite automata is difficult and time consuming. We are currently attempting to develop automatic methods for detecting unusual or interesting play behavior. We are also attempting to adapt some of the techniques of systematics from evolutionary biology to develop better descriptive methods for studying our artificial ecologies.

APPENDIX

TABLE 1 Pairwise interactions prior to the first refusal

Players in 2×2 games	π^i for $i \leq i^*$	Conditions for PD games to stop
AllD v. AllD	$\omega^i \pi^0 + (1 - \omega^i)$	$1 < \tau$
Ripoff v. Ripoff	for $i = 1$: $\omega \pi^0 + (1 - \omega)$; for $i > 1$: $\omega^i \pi^0 + \omega^{i-1}(1 - \omega) + C(1 - \omega^{i-1})$	$\omega \pi^0 + (1 - \omega) < \tau$
Any pair of AllC, TFT, or TFTT	$\omega^i \pi^0 + C(1 - \omega^i)$	never
AllD v. Ripoff	Let: $X = S$ for AllD and 0 for Ripoff for $i = 1$: $\omega \pi^0 + (1 - \omega)$ for $i = 2$: $\omega^2 \pi^0 + \omega(1 - \omega) + (1 - \omega)X$ for $i \geq 3$: $\omega^i \pi^0 + \omega^{i-2}(1 - \omega)X + (1 - \omega^{i-2} + \omega^{i-1} - \omega^i)$.	$\omega^2 \pi^0 + \omega(1 - \omega) < \tau$ or $1 < \tau$
AllD v. TFT or GRip	Let: $X = S$ for AllD and 0 for TFT and GRip for $i = 1$: $\omega \pi^0 + (1 - \omega)X$ for $i \geq 2$: $\omega^i \pi^0 + \omega^{i-1}(1 - \omega)X + (1 - \omega^{i-1})$.	$\omega \pi^0 < \tau$ or $1 < \tau$
AllD v. AllC	Let: $X = S$ for AllD and 0 for AllC $\omega^i \pi^0 + X(1 - \omega^i)$.	$0 < \tau$
AllD v. TFTT	Let: $X = S$ for AllD and 0 for TFTT for $i = 1$: $\omega \pi^0 + (1 - \omega)X$; for $i = 2$: $\omega^2 \pi^0 + (1 - \omega^2)X$; for $i \geq 3$: $\omega^i \pi^0 + \omega^{i-2}(1 - \omega^2)X + (1 - \omega^{i-2})$.	$\omega^2 \pi^0 < \tau$ or $1 < \tau$

i^*	Refuser	Average payoff for $i \le i^*$
$\left\lceil \ln\left(\frac{\tau-1}{\pi^0-1}\right)\frac{1}{\ln(w)}\right\rceil$	both players	1
if $\omega\pi^0 + (1-\omega) < \tau$: $i^* = 1$ otherwise: $i^* = \infty$.	both players	for $i = 1$: 1; for $i > 1$: $\frac{1}{i} + \frac{i-1}{i}C$.
$i^* = \infty$	neither	C
if $\omega\pi^0 + (1-\omega) < \tau$: $i^* = 1$; else, if $\omega^2\pi^0 + \omega(1-\omega) < \tau$: $i^* = 2$; else, if $1 < \tau$: $i^* = \left\lceil \ln\left(\frac{\omega^2(\tau-1)}{\omega^2\pi^0 - (1-\omega+\omega^2)}\right)\frac{1}{\ln(\omega)}\right\rceil$; otherwise: $i^* = \infty$.	if $i^* = 1$: both players; if $i^* \ge 2$ Rip only.	for $i = 1$: 1; for $i = 2$: $(1+X)/2$; for $i \ge 3$: $((i-1)+X)/i$.
if $\omega\pi^0 < \tau$: $i^* = 1$; else, if $1 < \tau$: $i^* = \left\lceil \frac{1}{\ln(\omega)} \ln\left(\frac{\omega(\tau-1)}{\omega\pi^0 - 1}\right)\right\rceil$; otherwise: $i^* = \infty$.	TFT or GRip only	for $i = 1$: X; for $i \ge 2$: $(X + (i-1))/i$.
if $0 < \tau$: $i^* = \left\lceil \ln\left(\frac{\tau}{\pi^0}\right)\frac{1}{\ln(\omega)}\right\rceil$.	AllC only	X
if $\omega\pi^0 < \tau$: $i^* = 1$; else, if $\omega^2\pi^0 < \tau$: $i^* = 2$; else, if $1 < \tau$: $i^* = \left\lceil \frac{1}{\ln(\omega)} \ln\left(\frac{\omega^2(\tau-1)}{\omega^2\pi^0 - 1}\right)\right\rceil$; otherwise: $i^* = \infty$.	TFTT only	$i = 1, 2$: X; $i \ge 2$: $\frac{2X + (i-2)}{i}$.

TABLE 1 Pairwise interactions prior to the first refusal (continued)

Players in 2×2 games	π^i for $i \leq i^*$	Conditions for PD games to stop
Ripoff v. AllC or TFTT	Let: $X = S$ for Ripoff and 0 for AllC and TFTT for $i = 2k$: $\omega^{2k}\pi^0 + \frac{1-\omega^{2k}}{1+\omega}(C + \omega X)$; for $i = 2k+1$: $\omega^{2k+1}\pi^0 + \omega\frac{1-\omega^{2k}}{1+\omega}(C + \omega X) + (1-\omega)X$.	if $\pi^0(1+\omega) > C$ (π^{2k+1}_{TFTT} declines): $\frac{C\omega}{1+\omega} < \tau$; otherwise, if $\pi^0(1+\omega) \leq C$ (π^{2k+1}_{TFTT} increases): $\omega\pi^0 < \tau$.
Ripoff v. TFT	Let: $X = S$ for Ripoff and 0 for TFT $Y = 0$ for Ripoff and S for TFT for $i = 1$: $\omega\pi^0 + (1-\omega)X$ for $i = 2$: $\omega^2\pi^0 + \omega(1-\omega)X + (1-\omega)Y$; for $i \geq 3$: $\omega^i\pi^0 + \omega^{i-1}(1-\omega)X + \omega^{i-2}(1-\omega)Y + C(1-\omega^{i-2})$.	$\omega\pi^0 < \tau$
GRip v. TFTT or AllC	Let: $X = S$ for GRip and 0 for TFTT and AllC for $i \leq 4$: $\omega^i\pi^0 + (1-\omega^i)C$; for $i = 4+3j$, $j = 0, 1, 2, \ldots$: $\omega^{3j}\pi^4 + \frac{1-\omega^{3j}}{1+\omega+\omega^2}(\omega^2 X + (1+\omega)C)$; for $i = 5+3j$, $j = 0, 1, 2, \ldots$: $\omega^{3j+1}\pi^4 + \frac{\omega(1-\omega^{3j})}{1+\omega+\omega^2}(\omega^2 X + (1+\omega)C) + (1-\omega)X$; for $i = 6+3j$, $j = 0, 1, 2, \ldots$: $\omega^{3j+2}\pi^4 + \frac{\omega^2(1-\omega^{3j})}{1+\omega+\omega^2}(\omega^2 X + (1+\omega)C) + (1-\omega)(\omega X + C)$.	$\frac{\omega(1+\omega)C}{1+\omega+\omega^2} < \tau$

i^*	Refuser	Average payoff for $i \leq i^*$
if $\pi^0(1+\omega) > C$ and $\frac{C\omega}{1+\omega} < \tau$: $i^* =$ The smallest odd integer greater than $\ln\left(\frac{\tau(1+\omega) - C\omega)}{\pi^0(1+\omega) - C}\right)\frac{1}{\ln(\omega)}$; else if $\pi^0(1+\omega) \leq C$ and $\omega\pi^0 < \tau$: $i^* = 1$; otherwise: $i^* = \infty$.	AllC or TFTT only	$i = 2k$: $\frac{C+X}{2}$; $i = 2k+1$: $\frac{(k+1)X + kC}{2k+1}$.
if $\omega\pi^0 < \tau$: $i^* = 1$; otherwise: $i^* = \infty$.	TFT only	$i = 1$: X; $i \geq 2$: $\frac{S + (i-2)C}{i}$.
if $\frac{\omega(1+\omega)C}{1+\omega+\omega^2} < \tau$: $i^* = 2 + 3j$, where j is the smallest positive integer such that $3j > \frac{\ln\left(\frac{\tau - \omega(1+\omega)(C-\tau)}{\omega^2(1+\omega+\omega^2)(\pi^0 - C) + C}\right)}{\ln \omega}$; otherwise: $i^* = \infty$.	TFTT or AllC	$i \leq 4$: C. for $j = 0, 1, 2, \ldots$; $i = 4 + 3j$: $\frac{(4+2j)C + jX}{4+3j}$; $n = 5 + 3j$: $\frac{(4+2j)C + (j+1)X}{5+3j}$; $i = 6 + 3j$: $\frac{(5+2j)C + (j+1)X}{6+3j}$.

TABLE 1 Pairwise interactions prior to the first refusal (continued)

Players in 2×2 games	π^i for $i \leq i^*$	Conditions for PD games to stop
Rip v. GRip	Let: $X = S$, $Y = 0$ for GRip, and $X = 0$, $Y = S$ for Rip $Z = Y + \omega X + (1 + \omega + \omega^2)\omega^2 C$ $q = 1 + \omega + \omega^2 + \omega^3 + \omega^4$ for $i = 1$: $\omega\pi^0 + (1 - \omega)Y$; for $i = 2$: $\omega^2\pi^0 + (1 - \omega)(\omega Y + X)$; for $i = 2 + 5j$, $j = 0, 1, 2, \ldots$: $\omega^{5j}\pi^2 + \dfrac{1 - \omega^{5j}}{q} Z$; for $i = 2 + m + 5j$, $m = 1, 2, 3$ and $j = 0, 1, 2, \ldots$: $\omega\pi^{2+5j} + (1 - \omega^m)C$; for $i = 6 + 5j$, $j = 0, 1, 2, \ldots$: $\omega^4\pi^{2+5j} + (1 - \omega^3)\omega C + (1 - \omega)X$; Special case: $\pi^6_{\text{Rip}} = \omega^6\pi^0 + (1 - \omega)[(\omega + \omega^2 + \omega^3)C + \omega^5 S]$.	$\omega\pi^0 < \tau$ or $\pi^0 \geq C$ and $\pi^6_{\text{Rip}} < \tau$ or (Rip's expected payoffs are decreasing each cycle of 5 games) $\pi^2_{\text{Rip}} > Z_{\text{Rip}}/q$ and (eventually becomes less than τ) $\dfrac{(\omega + \omega^2 + \omega^3)C + \omega^4 S}{q} < \tau$
GRip v. TFT	Let: $X = S$, $Y = 0$ for GRip, and $X = 0$, $Y = S$ for TFT $Z = Y + \omega X + (1 + \omega + \omega^2)\omega^2 C$ $q = 1 + \omega + \omega^2 + \omega^3 + \omega^4$ for $i = 1$: $\omega\pi^0 + (1 - \omega)C$; for $i = 5j + 1$, $j = 0, 1, 2, \ldots$: $\omega^{5j}\pi^1 + \dfrac{1 - \omega^{5j}}{q} Z$; for $i = 5j + 1 + m$, $m = 1, 2, 3$ and $j = 0, 1, 2, \ldots$: $\omega\pi^{5j+1} + (1 - \omega^m)C$; for $i = 5j + 5$, $j = 0, 1, 2, \ldots$: $\omega^4\pi^{5j+1} + (1 - \omega^3)\omega C + (1 - \omega)X$.	$\omega^5\pi^0 + C(1 - \omega^4)\omega < \tau$ or (π^{5j}_{TFT} is decreasing with j) $S - C < \omega\pi^0 + \omega^2(1 + \omega + \omega^2 + \omega^3)(\pi^0 - C)$ and (eventually becomes less than τ) $\dfrac{(\omega + \omega^2 + \omega^3)C + \omega^4 S}{q} < \tau$
GRip v. GRip	for $i = 1, 2, 3, 4$: $\omega^i\pi^0 + (1 - \omega^i)C$; for $i = 5j$, $j = 0, 1, 2, \ldots$: $\omega^{5j}\pi^0 + (1 - \omega^{5j})A$; for $i = 5j + m$, $m = 1, 2, 3, 4$ and $j = 0, 1, 2, \ldots$: $\omega^m\pi^{5j} + (1 - \omega^m)C$.	$\tau > A$ where $A = \dfrac{(1 - \omega) + \omega(1 - \omega^4)C}{1 - \omega^5}$

i^*	Refuser	Average payoff for $i \leq i^*$
if $\omega\pi^0 < \tau$: $i^* = 1$; else if $\pi^6_{\mathrm{Rip}} < \tau$: $i^* = 6$; else if $\frac{\omega^4 S + (1+\omega+\omega^2)\omega C}{1+\omega+\omega^2+\omega^3+\omega^4} < \tau,$ and $\frac{(1-\omega)S + (\omega^2+\omega^3+\omega^4)C + \omega^6 S}{q} < \pi^0,$ $i^* = 1 + 5j$, j the smallest positive integer such that $5j > 1 + \frac{\ln\left(\frac{\tau-\omega C+\omega^4(C-Z_{\mathrm{Rip}}/q)}{\pi^2_{\mathrm{Rip}}-Z_{\mathrm{Rip}}/q}\right)}{\ln(\omega)}.$	if $i^* = 1$, GRip only otherwise Rip only	$i = 1$: Y; $i = 2$: $S/2$; $i = 2 + 5j + m$, $m = 1, 2$ and $j = 0, 1, 2, \ldots$: $\frac{(j+1)S + (3j+m)C}{2+m+5j};$ $i = 5j$, $j = 1, 2, \ldots$: $\frac{jS + 3jC}{5j};$ $i = 5j + 1$, $j = 1, 2, \ldots$: $\frac{jS + 3jC + X}{5j+1}.$
if $\omega^5\pi^0 + C(1-\omega^4)\omega < \tau$: $i^* = 5$; else if $S - C < \omega\pi^0 + \omega^2(1+\omega+\omega^2$ $+\omega^3)(\pi^0 - C)$ and $\frac{\omega^4 S + (1+\omega+\omega^2)\omega C}{1+\omega+\omega^2+\omega^3+\omega^4} < \tau,$ $i^* = 5j + 5$, j the smallest postive integer such that $5j + 4 > \frac{\ln\left(\frac{q\tau-(\omega^4 S+(1+\omega+\omega^2)\omega C}{\omega\pi^0+\omega^2(1+\omega+\omega^2+\omega^3)(\pi^0-C)-S+C}\right)}{\ln(\omega)}$	TFT only	$i = 1, 2, 3, 4$: C; $i = 5j$, $j = 1, 2, \ldots$: $\frac{(j-1)S + (3j+1)C + X}{5j};$ $i = 5j + m$, $m = 1, 2, 3, 4$ and $j = 1, 2, \ldots$: $\frac{jS + (3j+m+1)C}{m+5j}.$
if the condition holds: $i^* = 5j$ where j is the smallest postive integer greater than $\frac{\ln\left(\frac{\tau-A}{\pi^0-A}\right)}{5\ln\omega};$ otherwise: $i^* = \infty$.	Both players	$i = 1, 2, 3, 4$: C; $i = 5j + m$, $m = 0, 1, 2, 3, 4$ and $j = 1, 2, \ldots$: $\frac{j + (4j+m)C}{5j+m}.$

ACKNOWLEDGMENTS

This work was partially supported by ISU University Research Grant No. 430-17-03-91- 0001. A preliminary version of this paper was presented at the ALifc III Conference sponsored by the Santa Fe Institute, Santa Fe, New Mexico, June 13–19, 1992. The authors are grateful to conference participants and to three anonymous referees for helpful comments.

REFERENCES

1. Axelrod, R. "The Emergence of Cooperation Among Egoists." *Am. Pol. Sci. Rev.* **75** (1981): 306–318.
2. Axelrod, R. *The Evolution of Cooperation.* New York: Basic Books, 1984.
3. Axelrod, R., and W. D. Hamilton. "The Evolution of Cooperation." *Science* **211** (1981): 1390–1396.
4. Axelrod, R., and D. Dion. "The Further Evolution of Cooperation." *Science* **242** (December 1988): 1385–1390.
5. Barner-Berry, C. "Rob: Children's Tacit Use of Peer Ostracism to Control Aggressive Behavior." *Ethol. & Sociobiol.* **7** (1986): 281–293.
6. Dugatkin, L. A., and D. S. Wilson. "Rover: A Strategy for Exploiting Cooperators in a Patchy Environment." *Amer. Natur.* **138** (1991): 687–701.
7. Eshel, I., and L. L. Cavalli-Sforza. "Assortment of Encounters and Evolution of Cooperativeness." *Proc. Natl. Acad. Sci. U.S.A.* **79** (February 1982): 1331–1335.
8. Feldman, M., and E. A. C. Thomas. "Behavior-Dependent Contexts for Repeated Plays of the Prisoner's Dilemma II: Dynamical Aspects of the Evolution of Cooperation." *J. Theor. Biol.* **128** (1987): 297–315.
9. Hirshleifer, J. "On the Emotions as Guarantors of Threats and Promises." In *The Latest on the Best: Essays on Evolution and Optimality*, edited by J. Dupre. Cambridge: MIT Press, 1987.
10. Hirshleifer, D., and E. Rasmusen. "Cooperation in a Repeated Prisoners' Dilemma with Ostracism." *J. Econ. Behav. & Org.* **2** (1989): 87–106.
11. Holland, J. *Adaptation in Natural and Artificial Systems.* Cambridge: MIT Press, 1992.
12. Hyman, J. M., and E. A. Stanley. "Using Mathematical Models to Understand the AIDS Epidemic." *Math. Biosci.* **90** (1988): 415–473.
13. Kitcher, P. "Evolution of Altruism in Repeated Optional Games." Working Paper, Department of Philosophy, University of California at San Diego, July 1992.

14. Kreps, D., P. Milgrom, J. Roberts, and R. Wilson. "Rational Cooperation in the Finitely Repeated Prisoner's Dilemma." *J. Econ. Theory* **27** (1982): 245–252.
15. May, R. M. "More Evolution of Cooperation." *Nature* **327** (May 1987): 15–17.
16. Maynard Smith, J. *Evolution and the Theory of Games*. Cambridge: Cambridge University Press, 1982.
17. Miller, J. H. "The Coevolution of Automata in the Repeated Prisoner's Dilemma." Revised Working Paper No. 89-003, Santa Fe Institute, July 1989.
18. Rennie, J. "Trends in Parisitology: Living Together." *Sci. Am.* **266** (January 1992): 122–133.
19. Stanley, A., D. Ashlock, and L. Tesfatsion. "Iterated Prisoner's Dilemma with Choice and Refusal." Economic Report No. 30, Iowa State University, July 1992.
20. Tesfatsion, L. "Direct Updating of Intertemporal Criterion Functions for a Class of Adaptive Control Problems." *IEEE Trans. Sys., Man, & Cyber.*. **SMC-9** (1979): 143–151.
21. Thompson, E., and R. Faith. "A Pure Theory of Strategic Behavior and Social Institutions." *Am. Econ. Rev.* **71** (1981): 366–380.

Yukihiko Toquenaga,* Masanori Ichinose, Tsutomu Hoshino,** and Koichi Fujii***
*Institute of Biological Sciences, University of Tsukuba
**Institute of Engineering Mechanics, University of Tsukuba

Contest and Scramble Competitions in an Artificial World: Genetic Analysis with Genetic Algorithms

Field observations of bean weevils (Coleoptera: Bruchidae) suggest that contest and scramble competition are favored in small and large beans, respectively. To test this hypothesis, using two artificial systems we examined the mode of competition in bean weevil species. The mode of competition in these species was examined with two artificial systems. First we conducted multiple-generation experiments with laboratory populations of three bruchid species. In interspecific competitions with *Callosobruchus analis* (contest type) and *C. phaseoli* (scramble type), there was no genetic interactions between the two species, and experimental results supported the hypothesis. To study intraspecific competition, we used *C. maculatus*. This species has contest and scramble strains whose competition types were controlled by additive genetic mechanisms, and hybridization occurred between the two strains. The results of the intraspecific competition experiment with these two strains supported the hypothesis in terms of population dynamics. However, predicted divergence in competition type did not occur at the genetic level.

We then made artificial systems in the computer to investigate the discrepancy between the genetic and the population levels. Competition types were decomposed into three larval parameters (developmental rate, interference behavior, and centripetal tendency) and one adult parameter (egg dispersion). We applied a diploid genetic algorithm (GA) system at the genetic level and coded the four life history parameters into GA bit strings. The results reproduced well the population dynamics observed in laboratory experiments. Evolutionary divergence in life history parameters between the two competition types also corresponded well to the difference in those parameters observed in real systems.

INTRODUCTION

The complexity of ecological systems has discouraged numerous attempts to analyze their underlying rules and mechanisms. The complexity increases at higher levels, from populations, to communities, to ecosystems. The hierarchical approach, i.e., analyzing phenomena at each level, has encouraged ecologists to promote many useful concepts to describe level-specific phenomena.[21]

The terms "contest" and "scramble" were first coined by Nicholson[20] to describe competition phenomena at the population level. In contest competition, by interfering with subordinates, dominant individuals monopolize the limited resource so that a constant number of individuals can survive. Thus, the number of survivors plotted against initial density becomes a saturated curve as shown in Figure 1(a). The survivors always get enough of the resource, and the survivors' individual intake remains constant irrespective of initial density (see Figure 1(a)). In contrast, with scramble competition, all competitors share the limited resource. The number of survivors initially increases as the density increases. When all or some of the competing individuals cannot get the minimal amount of the resource needed to complete their developmen, the density begins to decrease. Survivors' individual intake decreases monotonically as the initial density increases (see Figure 1(a)).

Many researchers have classified field animal populations into either contest or scramble type. However, such categorization per se is not enough to understand the population or community dynamics. These are just phenomena observed at the population level, and we should investigate the mechanisms of these phenomena at lower levels, such as individual, physiological, and genetic levels. Several theoretical models, such as kinetic and quantitative genetic models, have provided useful insights. Nevertheless, the curse of hierarchy has prevented us from applying these models in nature.[24]

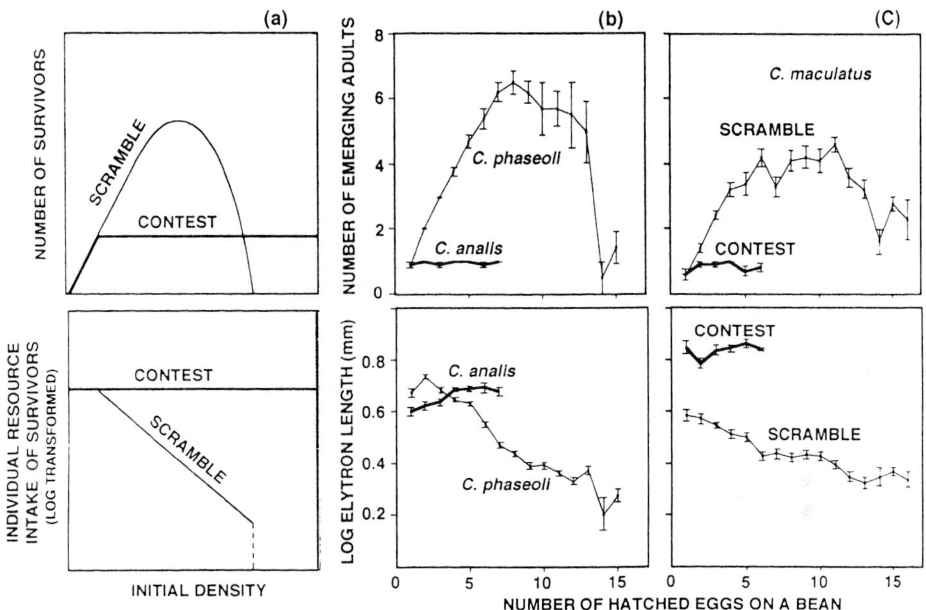

FIGURE 1 Competition curves (above figures) and resource sharing patterns (bottom figures) of contest and scramble competitions: (a) hypothetical, (b) variation between bruchid species, and (c) variation within *C. maculatus* strains.

During the first half of this century, some animal ecologists have tried to attack the problem using "semi-artificial" systems in laboratory experimental conditions (e.g., Gause,[3] Park,[22] Pearl,[23] and Utida[41]). These artificial "bottle experiments" provided excellent probes to abstract significant interactions from natural systems.[18] We have analyzed contest and scramble competition at three levels—population, individual, and genetic—using laboratory insect populations.[34,35,36,37,38,39] Our bottle experiments have successfully led us down the hierarchy steps to the individual level. At the genetic level, however, we found it helpful to study another artificial system using GAs to deduce the mechanism underlying the two competition types.

In the following sections, we present the three worlds of contest and scramble competition. First we review the contest and scramble competition in nature. Next we review our laboratory experiments that simulated evolution in the real world and show the nonlinear relationship among genetic, individual, and population levels. Finally we introduce an artificial world *in silico* to investigate such nonlinear relationships among multiple levels.

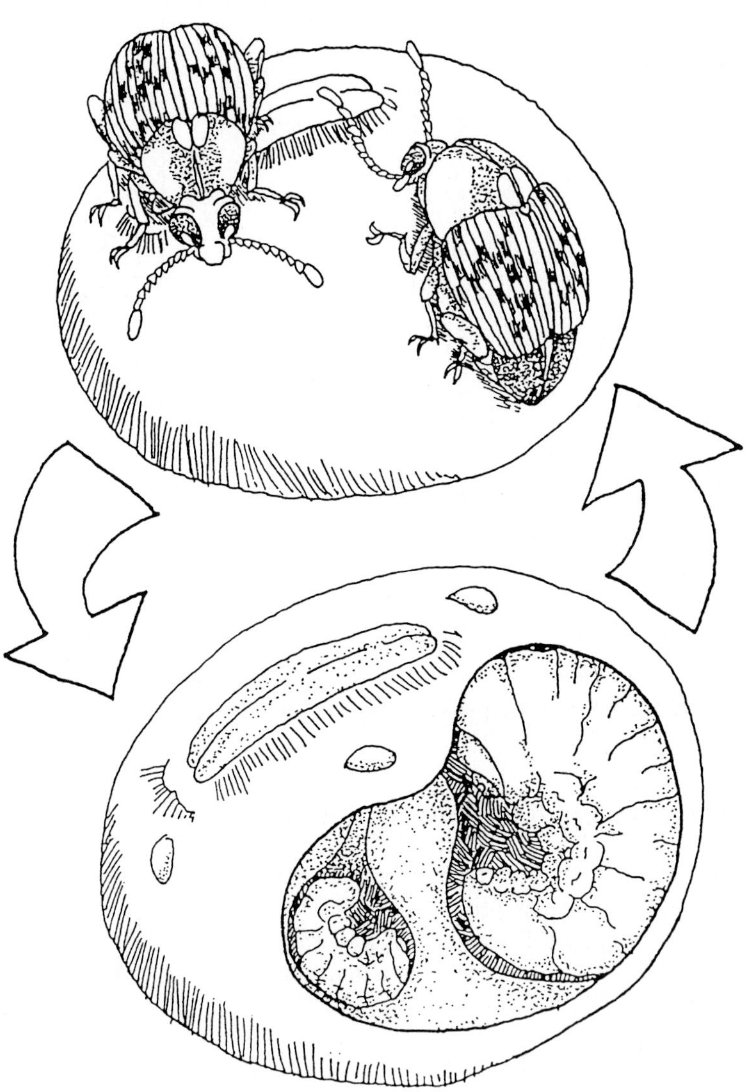

FIGURE 2 Biology of bruchid species.

THE WORLD IN NATURE

Our material insects are bean weevils (Coleoptera: Bruchidae), including notorious pests for cultivated and stored legumes worldwide.[29] In the field, female adults attach their eggs to bean pods. In stored or laboratory conditions, females lay their eggs directly on the bean surface. Larvae hatched from the eggs chew the pod shell and testa and go into the beans. They complete their four larval stages and pupal stage inside the beans before emerging again (see Figure 2).

Most pest bruchids feeding on domestic beans are scramble type; i.e., several adults can emerge from a single bean. In contradistinction, most bruchids feeding on wild small beans adopt contest competition; i.e., a constant number of adults (only one in most cases) can emerge from a single bean.[12] The divergence of contest

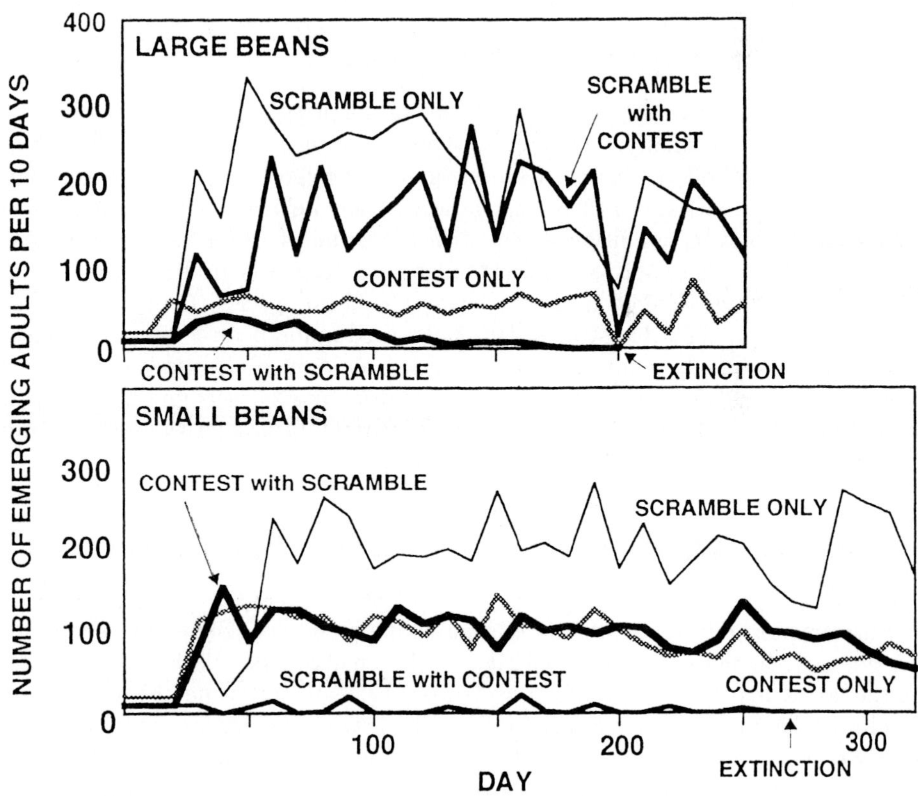

FIGURE 3 Multiple-generation competition between *C. analis* and *C. phaseoli.*

and scramble competition in bruchid species has been explained in terms of coevolution between legumes and bean predators.[10] Legumes have two alternative tactics to avoid seed predation: decreasing seed size accompanied by increasing seed number, or acquiring toxicity without seed size reduction. Contest-type competition by bruchids should be advantageous in wild small seeds because it ensures a bruchid a monopoly of the limited resource.

Cultivation of legumes by human beings may have reversed the coevolutionary process in which legumes reduced their seed size and bruchids adopted a contest strategy. Humans have selected larger beans, thereby unintentionally selecting for scramble-type bruchids which may have been selected against in wild small beans. Supporting this view is the fact that only 3 out of 20 pest bruchid species have been reported to show contest competition in cultivated or stored beans.[12,36]

THE WORLD OF BOTTLE EXPERIMENTS

Field observations have suggested that bruchids adopting contest competition have been favored in the small beans whereas those adopting scramble competition have been favored in the large beans. To investigate this hypothesis we conducted competition experiments using three pest bruchids that belong to the genus *Callosobruchus* in laboratory conditions (30°C, 70% R.H., and 24L/0D). *C. analis* adopts a contest strategy where only one individual emerges from a bean. In contrast adults of *C. phaseoli* may emerge from one bean. The third species, *C. maculatus*, has both contest and scramble strains.[31,34,38] We used mung beans, *Vigna radiata*, as the resource for the bruchid species. The mean developmental periods (from egg to adult emergence) were about 26 days for *C. analis* and 24 days for *C. phaseoli* in our laboratory condition. Those of *C. maculatus* were 23 days for the contest strain and 21 days for the scramble strain, respectively.

CONTEST AND SCRAMBLE COMPETITIONS BETWEEN THE TWO SPECIES

We first examined interspecific contest and scramble competition with *C. analis* and *C. phaseoli*. In the upper panel of Figure 1(b) we show the relationships between initial larval density (number of hatched eggs) per bean and the number of emerging adults in the two *Callosobruchus* species. The lower panel shows adult body size, which reflects individual resource intake during the larval period, against the initial larval density. The contrast between contest and scramble types is easily seen in both competition curves and the resource-sharing patterns.

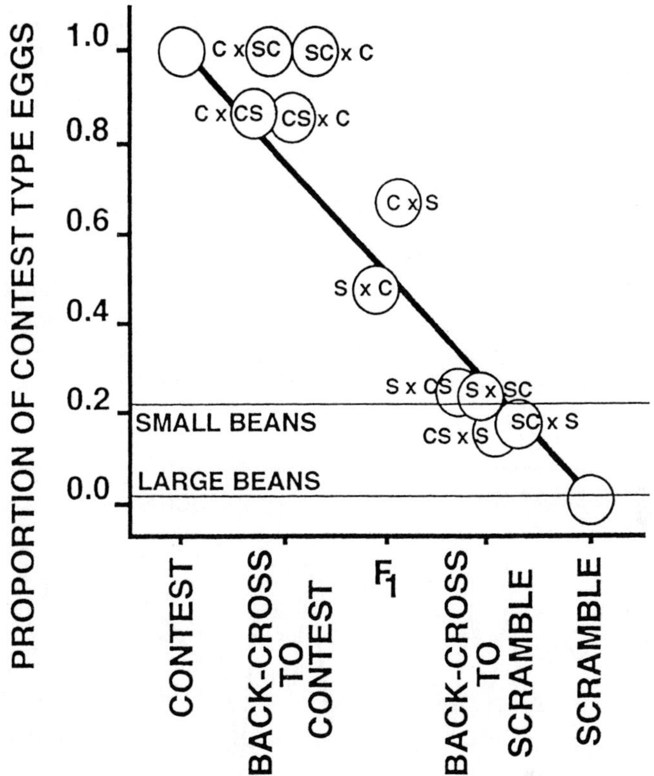

FIGURE 4 Genetic analysis of the competition types.

The competition between *C. analis* and *C. phaseoli* was an iterated two-strategy game because hybridization between the two species was impossible. We conducted semi-continuous multiple-generation competition experiments to simulate contest and scramble competition on an evolutionary time scale. We started the competition systems with 5 pairs of each species and 5 g of either large (about 70 pulses) or small (about 140 pulses) mung beans. For the controls, we started the systems with 10 pairs of either species. For the following 30 days, we introduced 5 g of beans every 10 days with the same number of adult bruchids as that at the setup time. Afterward we continued resource renewal every 10 days without adding bruchid adults. We counted the number of live and dead adults in the systems at each resource renewal. The live adults were returned to the systems but the dead ones were removed.

In Figure 3 we show the results of multiple-generation competition experiments between *C. analis* and *C. phaseoli*. In the large beans, *C. phaseoli* kept the same equilibrium level in the competition systems as in the control system, whereas *C.*

analis eventually disappeared from the competition system. In the small beans, *C. phaseoli* could not establish itself in the competition system and, hence, *C. analis* dominated. Experimental results of interspecific competition support the predicted correspondence between bean size and competition type.

CONTEST AND SCRAMBLE COMPETITION WITHIN SPECIES

We explored intraspicy contest and scramble competitions between the two strains of *C. maculatus*. In Figure 1(c) we show a similar contrast between the contest and scramble strains of *C. maculatus* to that in the interspecie competition systems. This contrast, however, may disappear due to hybridization between the two competition types in the intraspecific competition systems.

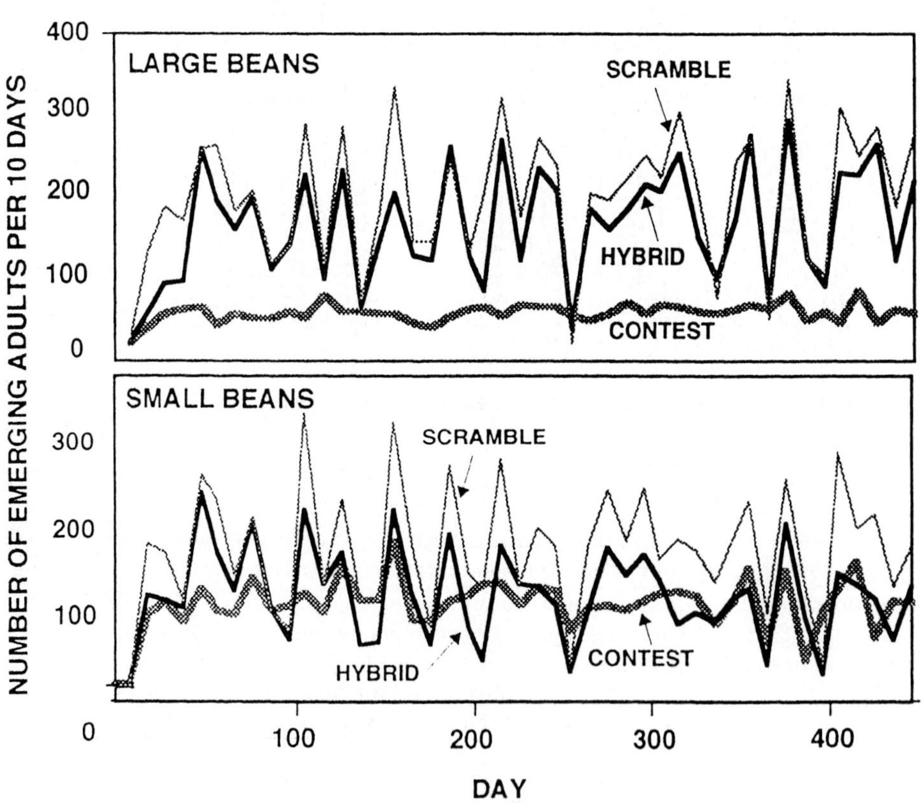

FIGURE 5 Multiple-generation competition between *C. maculatus* strains.

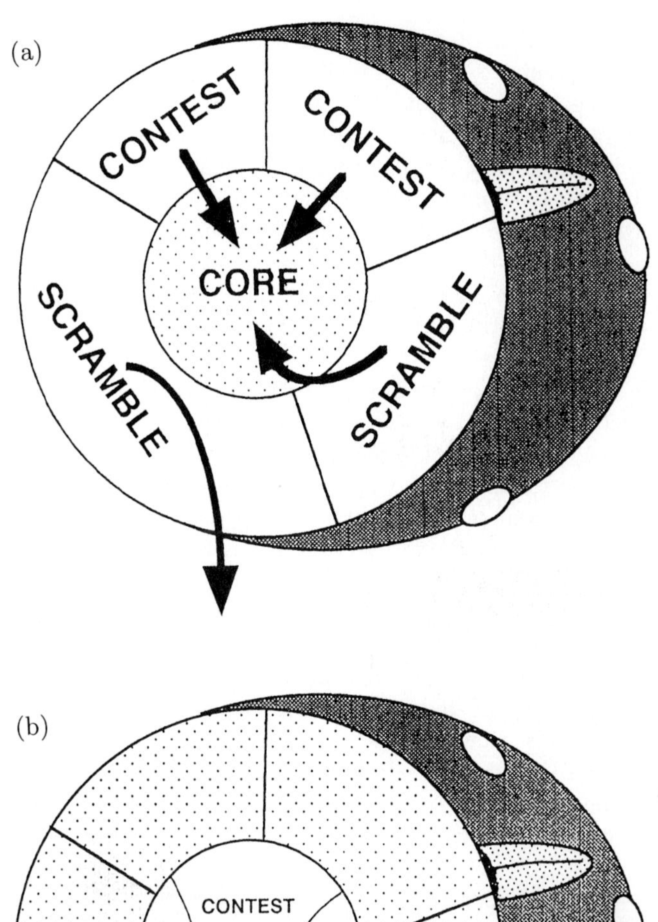

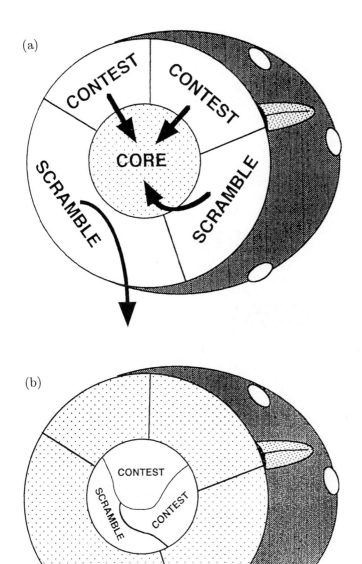

FIGURE 6
Competition regime inside a bean.

We conducted three types of reciprocal hybridization between the two strains (parental crossing, F1 crossing, and back-crossing of F1) to investigate whether the competition types were genetically determined. Each pair was released in a Petri dish with 40 large mung beans. All beans with two hatched eggs were isolated in the wells of plastic cell-culture boxes and the following adult emergence was recorded. Whether an egg was contest or scramble type could be judged only by the number of adults emerging from a bean. If each female laid eggs randomly on beans, and if egg hatchability is 100%, only mortality during larval stages was caused by competition. In such a situation, we would expect single emergence from a bean if both or one of the two eggs were contest type. If both eggs were scramble type, two adults could emerge from a bean. From the emergence pattern from the beans with two eggs, we estimated the proportion of contest-type eggs laid by a female. The plots of the proportion of contest-type eggs laid by each crossing pair almost lay on a straight line (see Figure 4), which suggests that the additive genetic mechanism controls the competition types.

We performed multiple-generation competition experiments in the same way as in the interspecific competition. After hybridization, there was no morphological distinction between the two types. Thus, we compared population dynamics of hybrid lines with those of control lines that consisted of only contest or only scramble strain. The population dynamics of the hybrid lines converged to those of contest and scramble strains in the small and the large beans, respectively (Figure 5). The prediction of the relationship between bean size and competition type was supported also in the intraspecific competition systems.

To investigate genetic changes during competition, on the 220th day we also tried back-crossing of the hybrid lines to the scramble strain The proportion of contest-type eggs laid by females of the hybrid lines also converged to that of the scramble type in the large beans. However, the proportion of the contest type remained at the level of females of F1 hybrid (see Figure 4) in the small beans. Thus, there was no convergence to the contest type in the small beans at the genetic level, suggesting the discrepancy between genetic and population levels.

THE ARTIFICIAL WORLD IN SILICO

Contest and scramble competition can be considered as two different strategies at the phenotypic level.[28] Several theoretical works have tried to connect those phenotypic strategies directly to the underlying genetic mechanisms (e.g., Thomas,[32,33] Hofbuer and Sigmund,[5] Lomnicki,[17] and Smith[27]). However, contest and scramble competitions should be considered the phenomena at the population level and we should go down the hierarchy to the individual and the genetic levels to understand their mechanisms. The individual level may be further split into behavioral and physiological levels.

Quantitative genetic approaches have been used to explain interacting life history characteristics at the genetic level in several selection regimes (e.g., Lande,[13,14] and Lande and Arnold[15]). However, those models rely on unrealistic simplifications which have been criticized from a theoretical point of view (e.g., Crespi and

Bookstein[1]). Moreover, the verification of quantitative genetic theories often requires tremendous amounts of experimental work which are hardly practical.[26] Large gaps between different levels emphasis the need for more powerful approaches.

The GA is now familiar as a specific remedy for practical optimization problems in engineering. However, it was originally intended to abstract and rigorously explain the adaptive processes of natural systems.[7,4] This compound of simple rules and operators neatly abstracts the genetic mechanisms of real organisms in nature. We tried to equip an artificial system with a GA to simulate evolution of the competition types (higher level phenomena) in terms of life history characteristics (lower level phenomena) controlled by genetic mechanisms.[8,9]

DECOMPOSITION OF CONTEST AND SCRAMBLE COMPETITION INTO LIFE HISTORY PARAMETERS

Understanding the phenomena at the population level in terms of the genetic level requires choosing the proper genetically controlled life history parameters. Several short-term laboratory experiments showed that the following three larval parameters affected the competition types: interference behaviors, developmental periods, and centripetal tendency inside a bean.[36,37,38] The interference behavior consists of just biting and killing the opponents rather than cannibalism. The centripetal tendency represents the larval propensity to dig into the core part of a bean. If this tendency is high, larvae should encounter each other eventually and may have to fight at the core of a bean. On the other hand, they could avoid meeting each other if this tendency is low. Smith[27] also discussed the importance of this parameter in his game theoretical model.

In addition to larval characteristics, the egg distribution pattern by adult females among beans is also important because it determines the initial larval density inside a bean.[19,28,35] In total, four (three larval and one adult) parameters were coded into bit strings, each string acting as an individual in our artificial world.

GENETIC CODING AND MATING

We adopted a diploid GA system with two sexes to keep reality high. We used two different sets of coding for inter- and intraspecific competition systems, respectively (see Figure 7). In the interspecific competition, the first bit of a string determined whether an individual had the ability to execute interference behavior against the opponents. The individual bit and killed the opponents inside a bean if the first string bit was 1. If the bit was 0, the individual shared the available resource with the opponents. Mating was allowed only between those whose pairs of the first bits were the same.

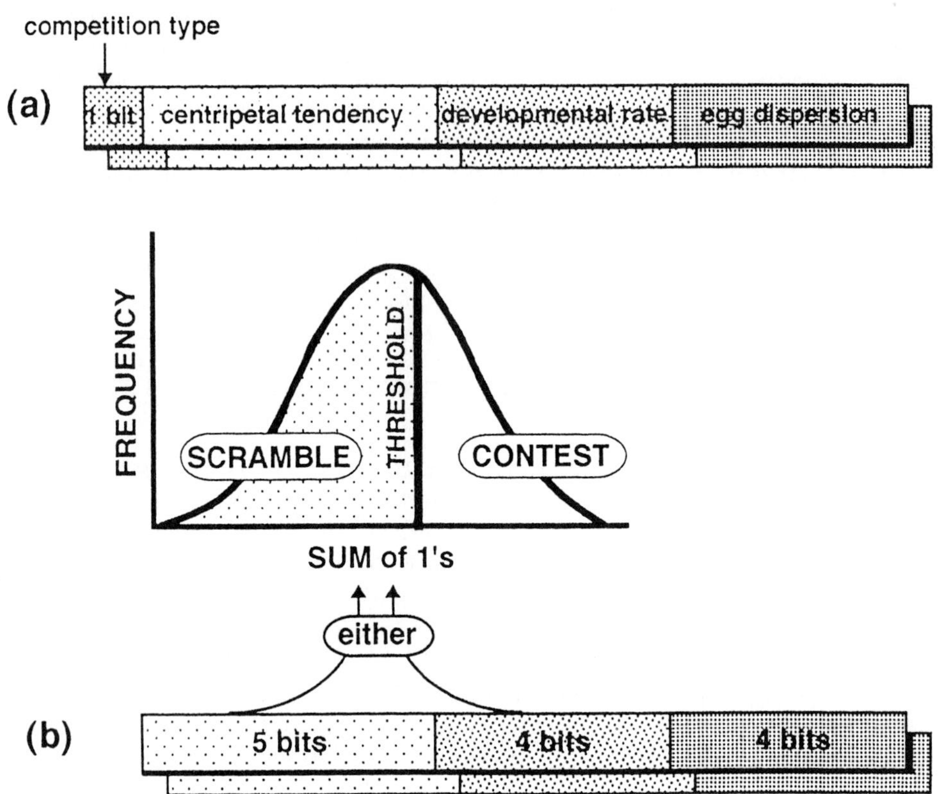

FIGURE 7 Genetic implementation of three life history parameters.

The lower stream of strings coded the larval centripetal tendency (5 bits), the larval developmental rate (4 bits), and the egg distribution pattern by adults (4 bits). Each parameter value was calculated by summing 1's in each parameter region for a pair of strings. This genetic coding simulated additive genetic mechanisms well. Larval centripetal tendency ranged from 0% to 100%. Larval developmental rate (expressed in terms of resource consumption rate per unit time) ranged from 1.0 mg to 5.0 mg. Egg dispersion was expressed by the number of eggs per bean (integer) which ranged from 0 to 4 per bean.

In the intraspecific competition, most of the genetic coding was the same as that in the interspecific competition except that there was no first bits that determined interference behavior. Contest and scramble competition are dichotomous phenomena, and it is difficult to express them in terms of additive genetic mechanisms. We overcame this difficulty by using a threshold model[25] which could translate all-or-none phenomena into an underlying binomial distribution that represented additive genetic mechanisms.

For the underlying function, we used either larval centripetal tendency or developmental period. In one version, if centripetal tendency was greater than 50%, the individual was assumed to have a potential to execute interference behavior. In another version, if consumption rate was greater than 2 mg, the individual was assumed to be able to bite and kill the opponent(s) inside a bean. Thus, the three larval parameters and one adult parameter were embedded in a 13-bit string in the intraspecific competition systems. Such genetic coding allowed us to predict which life history characteristic controlled by genetics, thereby allowing the competition types at the population level. Contrary to the interspecific competition systems, any pair could mate and reproduce irrespective of differentiation of potential interference behavior in the intraspecific competition.

We applied conventional GA operators, such as one-point crossing over and point mutation. We adopted diploid GA systems; therefore, the crossing over occurred at meiosis. We set the mutation rate at .001 for all simulations. Our system consisted of males and females with primary sex ratio of 1:1. Sex and other parameters described below were recorded outside the GA strings.

COMPETITION REGIME IN SILICO

There was no explicit fitness function in our competition systems. The fitness landscape was a dynamic result during evolution of competition types. The artificial world *in silico* consisted of BEANS, each of which consisted of two parts; core and periphery (see Figure 6). Both the large (70 mg) and the small (35 mg) BEANS had the same amount of core (30 mg) but differed in the amount of periphery. The artificial BEAN WEEVILS had two developmental phases: adult and larval. The larval phase was further divided into two periods: the developmental period at the periphery and that at the core of the BEANS. Although these binary simplifications may be against the spirit of the ALife approach, this simplification reproduced successfully the density-dependent competition results between contest and scramble species.[35]

We used the same initial simulation conditions that we did in the real experimental systems. The BEAN WEEVILS lived in virtual Petri dishes containing 5 g of either small (about 140) or large (about 70) BEANS. Initially we introduced five pairs of each competition type into the system. We renewed 5 g of BEANS in discrete fashion, creating a non-overlapping generation system. Each generation consisted of three periods: the larval period in the periphery of BEANS, that in the core, and the adult oviposition period outside of the BEANS.

We introduced two asymmetrical characteristics between contest type and scramble type to reflect our observations of laboratory experimental systems.[37,40] First we set the minimal amount of resource required to complete development to be larger for contest type (20 mg) than that for scramble type (10 mg), although the maximum requirement was the same (30 mg) for both types. This asymmetry encourages different resource-sharing patterns between the two competition types (see Figure 1). Next we introduced a developmental constraint in the fighting ability of the contest type. The contest-type individuals could execute its interference behavior only when it obtained more than half of the maximum weight requirement for an adult (i.e., 15 mg).

Adult BEAN WEEVILS laid eggs on BEANS, and their larvae hatched immediately and entered into the peripheral portion of the beans (see Figure 8). In the peripheral region, the interference behavior was not allowed so there was only exploitative competition in which competitors got the resource according to their developmental rate. After the peripheral region was exhausted, larvae emerged out of the bean (1) if they obtained the maximum requirement of resource for adults or (2) if they got more than the minimum requirement, abandoned the BEAN, and went into the core region. The second decision was simulated by generating

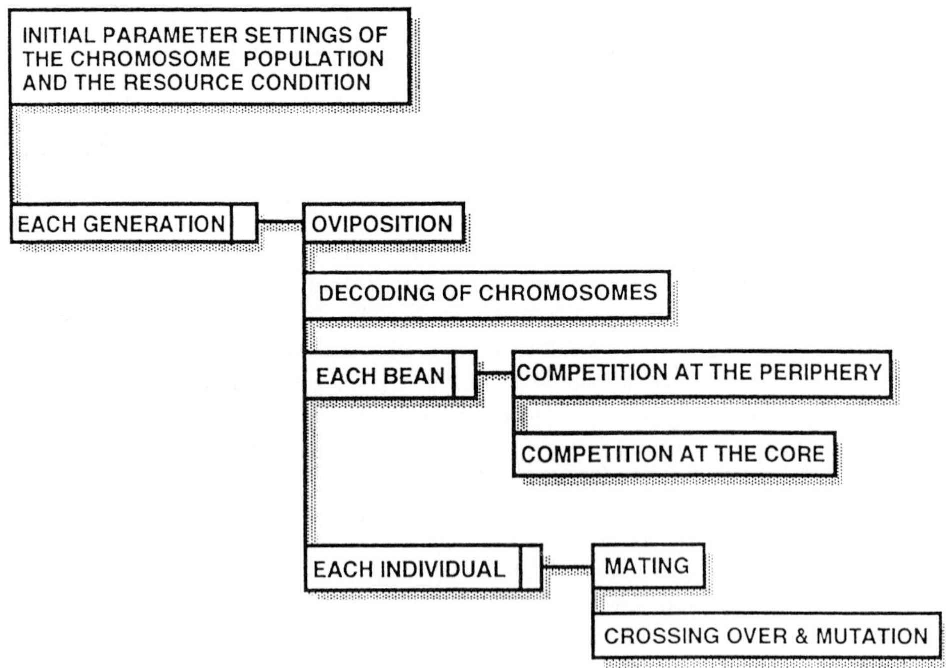

FIGURE 8 PAD of the simulation.

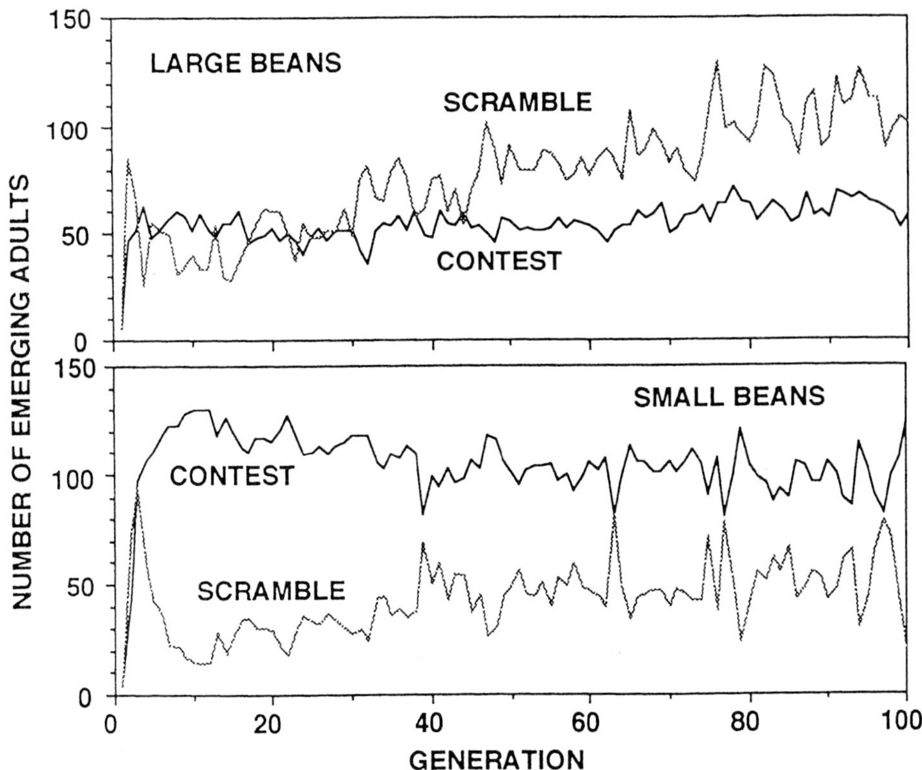

FIGURE 9 Multiple-generation competition without genetic interactions.

random numbers ranging from 0 to 1 and comparing the number with the centripetal tendency encoded in each pair of GA strings. If the random number was less than the centripetal tendency, the individual entered the core region.

At the core of a BEAN, if there was no individual who could execute interference behavior, all surviving competitors shared the resource at the core according to their developmental rates (exploitative competition). If there was only one contest-type individual that matured enough to execute interference behavior, it killed all the opponents and monopolized the core region. If there were more than one contest-type individual, only one of them survived and dominated the core resource.

Adult BEAN WEEVILS that emerged from BEANS had a 1:1 sex ratio. They mated immediately after emergence and exchanged genetic materials as described above. Females laid eggs on newly supplied BEANS according to their egg-distributing propensity encoded in their GA strings. Fecundity of each female was proportional to its body size multiplied by a factor of 2.

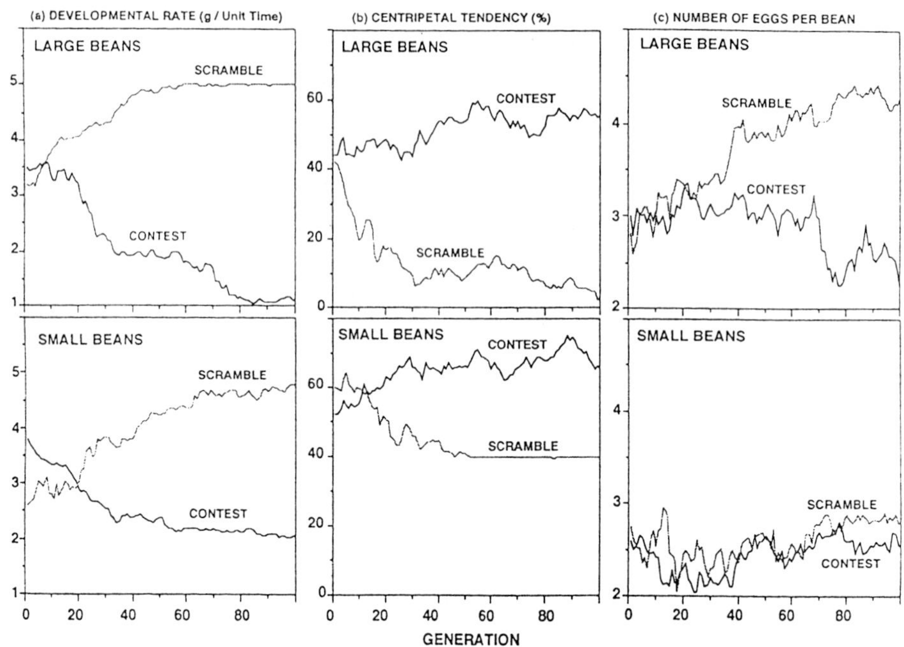

FIGURE 10 Divergence of life history parameters in interspecific competition.

We continued resource renewals for 100 generations in both inter- and intraspecific competitions. This length corresponded to about 3000 days for real experimental systems. In Figure 8 we show the whole process of the virtual multiple-generation competition.

RESULTS

In the interspecific competition, trends in population dynamics observed in real laboratory systems were almost fully reproduced. The contest BEAN WEEVILS overcame the scramble ones in the small BEANS whereas the reverse was true in the large BEANS (see Figure 9), although there was no extinction of inferior species during the 100 generations in either bean-size condition. In Figure 10 we show more surprising results of evolution at the individual level. The contest individuals increased their larval centripetal tendency but decreased their developmental rate in the course of competition, irrespective of bean size. On the contrary, the scramble BEAN WEEVILS fed more at periphery (i.e., emerging earlier) than in later generations (i.e., decreasing centripetal tendency). They also increased their developmental rate in later generations. In the large BEANS, contest-type individuals dispersed their eggs more in later generations whereas scramble ones tended to clump their eggs. These tendencies corresponded well to those observed in the real world.[34,36]

In the intraspecific competition where crossing between the two competition types was allowed, the tendency observed in the laboratory experiments was reproduced in terms of population dynamics when we adopted the centripetal tendency as the underlying variable of the threshold trait (see Figure 11). The proportion of individuals adopting each strategy in the hybrid line also supported the predicted relationships that the contest type was favored in small BEANS but the reverse

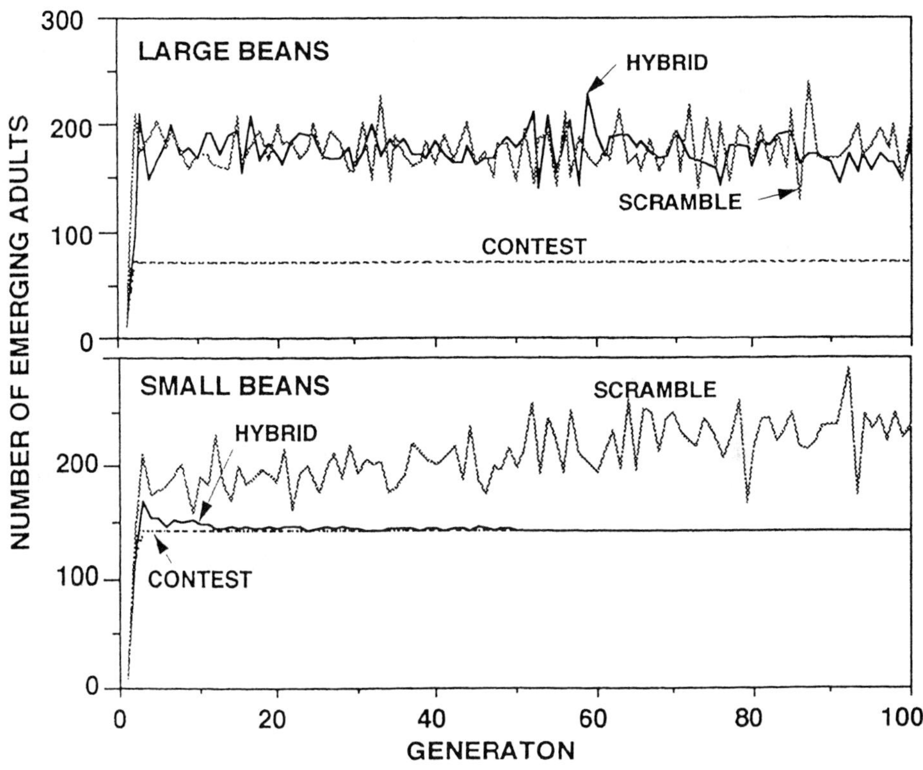

FIGURE 11 Multiple-generation competition with genetic interactions when larval centripetal tendency is assumed to be the underlying variable of the threshold trait.

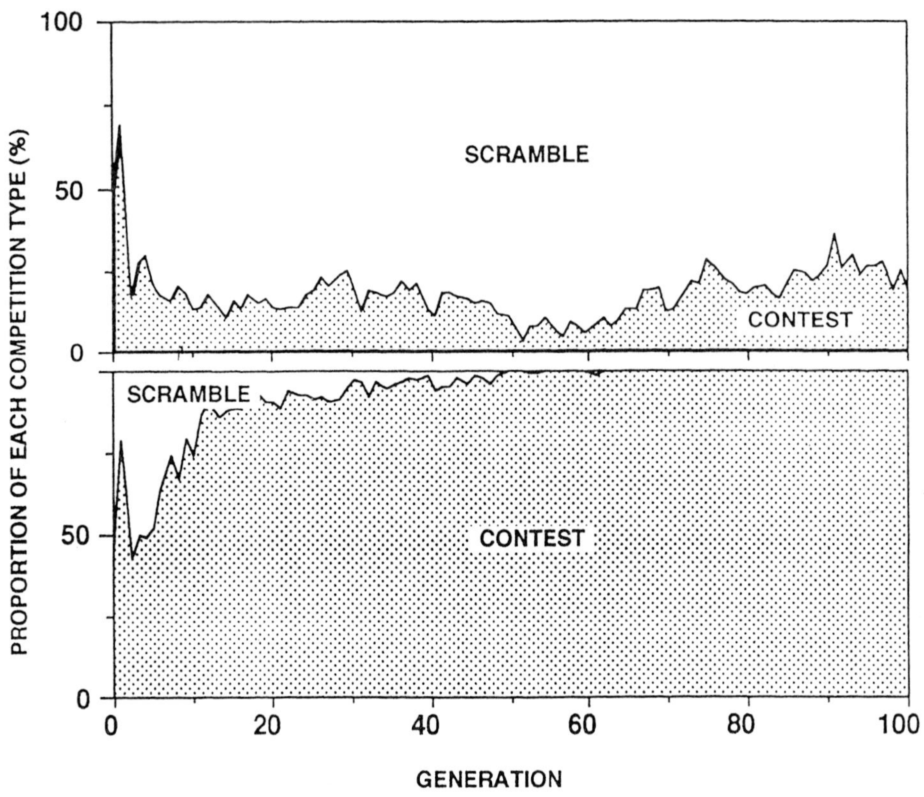

FIGURE 12 Changing patterns of proportions of the two strategies.

was true in the large BEANS (see Figure 12). The case was not true when we used the developmental rate as the underlying variable. No convergence of competition types was obtained in either bean size. These results suggest that the competition type of bean weevils was determined by the larval centripetal tendency that was under the control of the additive genetic mechanisms.

AFTER THE ROUND TRIP TO THE THREE WORLDS

In the three worlds described above, we have investigated the evolution of contest and scramble competition in bean weevils. The real world proposed a hypothesis that contest and scramble competition were selected for in small and large beans, respectively. The "bottle experiment" supported this hypothesis in terms of population dynamics. Nevertheless, it also showed a discrepancy between phenomena at the population level and the genetic level. In the ALife world, we incorporated behavioral and physiological characteristics to bridge the genetic and population levels. The GA played an important role in the emergence of the predicted phenomena during the experiments *in silico*.

We reimported the GA to investigate genetic interactions in a real biological system. The GA provides two benefits. First, it requires no sophisticated mathematical knowledge, such as diffusion equations and comprehensive matrix algebra. Those mathematical techniques are the main obstacles for students learning conventional population genetics. The GA is a compound of strikingly simple local rules. All we have to do is just to decide the genetic coding and to let the system run. Several GA packages are now available (e.g., GAucsd). They were developed originally for optimization problems in engineering. It is not difficult to modify them for the investigation of biological evolutionary problems. One can also consult Goldberg's[4] excellent book that provides simple GA source codes in Pascal.

The second good reason to use the GA for biological problems is its degree of freedom. The number of loci and chromosomes are at our discretion. One can also incorporate original genetic operations, such as "real" INVERSION and TEMPLATES,[4] as well as conventional point mutation and crossing over. Genetic coding could be further elaborated by changing degrees of epistasis and pleiotropy.[11] This high flexibility in the GA has its origin in the confusion of the phenomena at the molecular level and those at the chromosomal level. However, this confusion may bring about a breakthrough that cannot be achieved by the conventional population genetics approach. Thus, the GA can add a touch of ALife to our competition system.

The choice of life history parameters and the adopted competition scenario inside a bean had great influence on the evolutionary pattern in our simulation system. The parameters and the scenario were the product of our five-year dialogue with the laboratory systems. In this sense, we used our bottle experiment as a Socratic method to extract phenomena at the individual level. Our artificial world *in silico* further extends the Socratic method to the genetic level. To answer the questions posed by the artificial Socrates, we have to feedback obtained insights into laboratory experiments as well as into additional computer simulations.

Our ALife approach started from real biological systems and may be categorized as a weak ALife.[16] Our bottle experiment should be taken as a wet ware ALife at the individual level. Several macro-level ALife systems have been already proposed (e.g., MIRROR by Hogeweg and RAM by Taylor in *Artificial Life II*). While these examples refer directly to natural systems without operational wet ware, our systems have a wet ware interface between real and ALife worlds. Shifting our emphasis to computer simulations even enables us to use a strong ALife. We have the freedom to create a pile of any blocks of life history parameters we choose. We can adopt any genetic coding. We can change environments, such as bean sizes and generations. Thus, we can go back and forth between weak and strong ALife by changing our emphasis on either bottle experiments or computer simulations.

It should be mentioned that our ALife model did not answer fully the problem of the discrepancy between population and genetic levels observed in our laboratory experiments. In laboratory experiments with hybridization, the scramble type seemed to be genetically fixed in the large beans after 220 days. In our ALife systems, this phase corresponded to about 10 generations and neither competition type had fixed yet (see Figure 13). However, the fixation of competition type occurred in the small beans rather than in the large beans in the following generations. Further analysis *in silico* is needed to simulate hybridization and genetic coding of life history parameters. Strong ALife should be a good probe for the investigation of underlying mechanisms of phenomena at multiple levels. This study suggests that the GA can be used as the first step for the genetic analysis of complex ecological and evolutionary systems. Simulations proposed in this study have faces of both "strong" and "weak" ALife approaches. Such a "moderate" ALife approach surely should prove useful to the many biologists suffering from the difficulty of studying the inherent complexity of biological systems.

ACKNOWLEDGMENTS

We thank Dr. C. Langton and Prof. C. Taylor who encouraged us to contribute this paper to the ALife III proceedings and gave us fruitful comments to earlier manuscripts. We also thank the members of ALife Seminar at GLOCOM, especially Dr. T. Sakura, for giving valuable comments on our work. This work was supported in part by grants from the Ministry of Education, Japan (01540541, 03269101, and 04264102).

REFERENCES

1. Crespi, B. J., and F. L. Bookstein. "A Path-Analytic Model for the Measurement of Selection on Morphology." *Evolution* **43** (1989): 18–28.
2. Falconer, D. S. "Introduction to Quantitative Genetics." *Pitman, Bath* (1985): 340.
3. Gause, G. F. "Ecology of Populations." *Quart. Rev. Biol.* **7** (1932): 27–48.
4. Goldberg, D. E. *Genetic Algorithms.* Reading, MA: Addison-Wesley, 1989.
5. Hofbauer, J., and K. Sigmund. *The Theory of Evolution and Dynamical Systems.* Cambridge, MA: Cambridge University Press, 1988.

6. Hogeweg, P. "Mirror Beyond Mirror: Puddles of Life." In *Artificial Life*, edited by C. G. Langton. Santa Fe Institute Studies in the Sciences of Complexity, Proc. Vol. VI, 297–316. Redwood City, CA: Addison Wesley, 1989.
7. Holland, J. H. *Adaptation in Natural and Artificial Systems*. Ann Arbor: University of Michigan Press, 1975.
8. Hoshino, T., M. Ichinose, and Y. Toquenaga. "Learning and Evolution of Ethological Strategy by Genetic Algorithm." In *Genetic Algorithm*, edited by H. Kitano. Sangyo-Tosho (in Japanese), 1993.
9. Ichinose, M. "Study of Adaptation of Bean Weevil by Using the Genetic Algorithm." Master Thesis in Sciences and Engineering, University of Tsukuba (in Japanese), 1992.
10. Janzen, D. H. "Seed-Eaters Versus Seed Size, Number, Toxicity and Dispersal." *Evolution* **23(1)** (1969): 1–27.
11. Kauffman, S. "Towards a General Theory of Adaptive Walks on Rugged Landscapes." *J. Theor. Biol.* **128** (1987): 11–45.
12. Kiritani, K. "Ecology and Establishment as Pests in Bruchidae." *Shin-Kontyu* **9** (1956): 7–11 (in Japanese).
13. Lande, R. "Quantitative Genetic Analysis of Multivariate Evolution, Applied to Brain: Body Size Allometry." *Evolution* **33** (1979): 402–416.
14. Lande, R. "A Quantitative Genetic Theory of Life History Evolution." *Ecology* **63** (1982): 607–615.
15. Lande, R., and S. J. Arnold. "The Measurement of Selection on Correlated Characters." *Evolution* **37** (1983): 1210–1226.
16. Langton, C. G., ed. *Artificial Life*. Santa Fe Institute Studies in the Sciences of Complexity, Proc. Vol. VI. Reading, MA: Addison-Wesley, 1988.
17. Łomnicki, A. *Population Ecology of Individuals*. Princeton: Princeton Univer-University Press, 1988.
18. Merz, D. B., and D. E. McCauley. "The Domain of Laboratory Ecology." *Conceptual Issues in Ecology*, edited by E. Saarinen. Dordrecht: D. Reidel, 1982.
19. Mitchell, R. "The Evolution of Oviposition Tactics in the Bean Weevil *Callosobruchus maculatus* (F.)." *Ecology* **56** (1975): 696–702.
20. Nicholson, A. J. "An Outline of the Dynamics of Animal Populations." *Aust. J. Zool.* **2** (1954): 9–65.
21. O'Neill, R. V., D. L. DeAngelis, J. B. Waide, and T. F. H. Allen. *A Hierarchical Concept of Ecosystems*. Princeton, NJ: Princeton University Press, 1986.
22. Park, T. "Experimental Studies of Insect Populations." *Am. Nat.* **71** (1937): 21–33.
23. Pearl, R., and S. L. Parker. "Experimental Studies on the Duration of Life. IV. Data on the Influence of Density of Population on Duration of Life in Drosophila." *Am. Nat.* **56** (1922): 312–321.
24. Peters, R. H. *A Critique for Ecology*. Cambridge: Cambridge University Press, 1991.

25. Roff, D. A. "The Evolution of Wing Dimorphism in Insects." *Evolution* **40** (1986): 1009–1020.
26. Shaw, R. G. "The Comparison of Quantitative Genetic Parameters Between Populations." *Evol.* **45** (1991): 143–151.
27. Smith, R. H. "Adaptation of Callosobruchus Species to Competition." In *Bruchids and Legumes: Economics, Ecology and Coevolution*, edited by K. Fujii et al., 351–360. Dordrecht: Kluwer, 1990.
28. Smith, R. H., and M. Lesselles. "Oviposition, Ovicide, and Larval Competition in Granivorous Insects." In *Behavioural Ecology: Ecological Consequences of Adaptive Behavior*, edited by R. M. Sibly and R. H. Smith, 423–448 . Oxford: Blackwell, 1985.
29. Southgate, B. J. "Biology of the Bruchidae." *Ann. Rev. Entomol.* **24** (1979): 449–473.
30. Taylor, C. E., D. R. Jefferson, S. R. Turner, and S. R. Goldman. "RAM: Artificial Life for the Exploration of Complex Biological Systems." In *Artificial Life*, edited by C. G. Langton. Santa Fe Institute Studies in the Sciences of Complexity, Proc. Vol. VI, 275–296. Redwood City, CA: Addison Wesley,1989.
31. Thanthianga, C., and R. Mitchell. "Vibration Mediates Prudent Resource Exploitation by Competing Larvae of the Bruchid Bean Weevil, Callosobruchus Maculatus." *Entomol. Exp. Appl.* **44(1)** (1987): 15–21.
32. Thomas, B. "Genetical ESS-Models. I. Concepts and Basic Model." *Theor. Pop. Biol.* **28** (1985): 18–32.
33. Thomas, B. "Genetical ESS-Models. II. Multi-Strategy Models and Multiple Alleles." *Theor. Pop. Biol.* **28** (1985): 33–49.
34. Toquenaga, Y. "The Mechanisms of Contest and Scramble Competition in Bruchid Species." In *Bruchids and Legumes: Economics, Ecology and Coevolution*, edited by K. Fujii et al., 341–350. Dordrecht: Kluwer, 1990.
35. Toquenaga, Y. "A Mechanistic Model of Contest and Scramble Competition Observed in Callosobruchus Species." Abstracts of Fukuoka Symposium of Theoretical Ecology, vol. 46, 1990.
36. Toquenaga, Y., and K. Fujii. "Contest and Scramble Competition Between Two Bruchid Species, *Callosobruchus analis* and *C. phaseoli* (Coleoptera: Bruchidae). I. Larval Competition Curves and Interference Mechanisms." *Res. Popul. Ecol.* **32** (1990): 349–364.
37. Toquenaga, Y., and K. Fujii. "Contest and Scramble Competitions Between Two Bruchid Species, *Callosobruchus analis* and *C. phaseoli* (Coleoptera: Bruchidae). II. Larval Competition Experiment." *Res. Popul. Ecol.* **31** (1991): 129–139.
38. Toquenaga, Y., and K. Fujii. "Contest and Scramble Competitions Between Two Bruchid Species, *Callosobruchus analis* and *C. phaseoli* (Coleoptera: Bruchidae). III. Multiple-Generation Competition Experiment." *Res. Popul. Ecol.* **33** (1991): 187–197.

39. Toquenaga, Y., and K. Fujii. "Contest and Scramble Competitions in *Callosobruchus maculatus* (Coleoptera: Bruchidae). I. Larval Competition Curves and Resource Sharing Patterns." *Res. Popul. Ecol.* **33** (1991): 199–211.
40. Toquenaga, Y., and K. Fujii. "Contest and Scramble Competitions in *Callosobruchus maculatus* (Coleoptera: Bruchidae). II. Larval Competition and Interference Mechanisms." *Res. Popul. Ecol.* **35** (1993): 57–68.
41. Utida, S. "Studies on Experimental Population of the Azuki Bean Weevil, *Callosobruchus chinensis* (L.). I. The Effect of Population Density on the Progeny Populations." *Mem. Coll. Agr.* **48** (1941): 1–30.

J. Stephen Lansing and James N. Kremer
Department of Anthropology, University of Southern California, University Park, Los Angeles, California, 90089-0032; e-mail: lansing@quark.usc.edu and jnkremer@mizar.esc.edu.

Emergent Properties of Balinese Water Temple Networks: Coadaptation on a Rugged Fitness Landscape

One might think that few subjects could be further removed from artificial life than the structure of age-old ritual ties between agricultural temples on the Indonesian island of Bali. Yet a major theme in ALife research has been the process of self-organization, more specifically the dynamics of coadaptation on a rugged fitness landscape.[10,14] This article presents an analysis of such a process occurring at a human scale, over a time period measurable in decades.

The subject is a simulation model (the "Bali Model") of the role of networks of "water temples" in the ecology of Balinese wet rice terraces. The model assumes that the domesticated ecology of Balinese rice terraces is the product of a stochastic evolutionary process, in which each generation of farmers seeks to maximize harvest yields by reducing losses due to water stress and pest damage. The Balinese believe that irrigation water is the gift of a goddess, and decisions about water scheduling are made by the farmers at local water temples. Each temple is not an autonomous unit, but part of a network of temples that allows groups of farmers to synchronize cropping patterns and control the cycle of wet and dry phases in the terrace ecosystem. The relationship between these temple networks and the productivity of the terrace ecosystem was the focus of our research. Simulation modeling allowed us to observe the emergence of artificial water temple networks through a process of coadaptation to a rugged (and changing) fitness landscape, and to compare these artificial networks with real water temples.

Our most recent model (described here) was inspired by the work of Stuart Kauffman and his collaborators, who have created mathematical models of natural selection in which they are able to tune the ruggedness of the adaptive landscape and observe the results (see *Artificial Life II*). Kauffman was particularly interested in the process of coevolution: what happens if the fitness for a given locus (n) in sequence space depends not only on itself, but on (k) other sites? If everything depends on everything else ($n = k$), then the system is not able to respond well to perturbations and is "frozen" into place. Conversely, if $k = 0$ (the fitness of each variant depends on itself alone), then the system behaves rather chaotically. Tuning k changes the structure of the fitness landscape.

In the Bali model of coadaptation, local farmer's associations called subaks try out various cropping patterns, each coded with a number, all of which are variants of some general plane like "2 crops of high-yielding rice and a vegetable crop." Different crop codes merely scramble the dates or planting sequences. Each of these variants is tested for each subak and given a fitness or yeild value. But we're not really interested in which one is best (there is no analogue to the "wild type" for a crop code). Instead, the testing process is used to model selection, which yields—not a best or fittest crop code—but rather a best or fittest pattern of coordination (synchronized cropping) among the subaks. The "fitness" of each crop code for each subak depends on what the other subaks do, since this affects the flow of water and growth of rice and pests. Any starting pattern of crop codes (i.e., any collection of "mutants") subject to selection will undergo a process of change, leading to higher average fitness values when the best structure of coordination emerges. We're searching for something like Kauffman's k values. But unlike Kauffman's studies of the emergence of complex structures in the immune system, in the Bali model k varies by location because of the ecological dynamics of river flows and rice and pest biology. In other words, there is no single optimal k value for all subaks. Nonetheless, regardless of the initial conditions (as long as they are biologically plausible), the same *structures* of social coordination emerge spontaneously through a process of self-organization set in motion by the selection process.

For over a thousand years, generations of Balinese farmers have gradually transformed the landscape of their island, clearing forests, digging irrigation canals and terracing hillsides to enable themselves and their descendants to grow irrigated rice. Paralleling the physical system of terraces and irrigation works, the Balinese have also constructed intricate networks of shrines and temples dedicated to agricultural deities. Ecological modelling shows that water temple networks can have macroscopic effects on the topography of the adaptive landscape, and may be representative of a class of *complex adaptive systems* that have evolved to manage agroecosystems.

INTRODUCTION

In 1984, Eric Alden Smith published a devastating critique of the uses of systems ecology and simulation modeling in anthropology.[20] While this chapter is in part a defense of these methods, we do not take issue with any of Smith's conclusions. Instead, we hope to demonstrate that systems models can serve a different heuristic purpose than the naive functionalist, energy-maximization or group-selection models skillfully demolished by Smith. In particular, we hope to show that simulation models are uniquely appropriate for addressing the issues of adaptation and determinism in the development of complex social systems like the water temples of Bali. But before we turn to the uses of simulation models, it may be useful to sketch out how our approach differs from those criticized by Smith.

In this chapter we will try to show how the techniques of ecological simulation modeling can help to illuminate this historical process. The major difference between the modeling process that we will describe, and the sorts of models used by biologists, is that here our interest focuses on the effects of human agency in reshaping ecosystems. History only happened once, of course, but with simulation models we can "rewind the tapes" and investigate the consequences of changing social and ecological parameters. This approach can help avoid one of the most common pitfalls in materialist approaches to social theory: the assumption that whatever social institutions happen to exist at a particular place and time are the deterministic results of environmental circumstances. For while the evolution of productive systems like irrigation networks is undoubtedly shaped by material constraints, it does not follow that such constraints mandate a specific set of cultural and ecological responses.

The need for an approach that will enable us to analyze the effects of interactions between social and ecological variables over time is the reason for our disagreement with Eric Smith. Smith objects to simulation models on the grounds that they appear to be teleological, depicting ecosystems as self-regulating or functionally integrated.[20] In preference to systems ecology, which tries to study whole ecosystems, Smith urges us to adopt the perspective of evolutionary ecology, which focuses on calculations of the "fitness value" or payoff of specific behavioral strategies to individual social actors. But successful applications of evolutionary ecology in anthropology have been confined to studies of hunting and gathering societies, where arguably we are concerned with "external" rather than "humanized" nature. In this chapter, our major goal is to comprehend the emergence of cooperative behavior among Balinese farmers. Here, the "fitness value" or payoff of different farming strategies changes as a result of complex interactions between irrigation networks and the domesticated ecology of the rice terraces. To foreshadow the most interesting results, a spontaneous process of self-organization occurred when we allowed water temples to react to changing environmental conditions over time in a simulation model. Artificial cooperative networks appeared that bore a very close resemblance to actual temple networks in the study area. As these networks formed, average harvest yields rose to a new plateau. Subsequently, irrigation systems organized into artificial temple networks were able to withstand ecological perturbations (such as pest outbreaks or drought) much better than in otherwise-identical models that lacked temple networks. It appears, then, that networks of water temples may have a definite structure, which leads to higher sustained productivity than would be the case if they were randomly ordered. Further, these structures can emerge without conscious planning, through a stochastic process of coadaptation. Thus water temple networks may represent a hitherto-unnoticed type of social organization: a self-organizing managerial system, shaped by a process of coadaption on a *rugged fitness landscape* (see Palmer,[18] and also Kauffman and Johnsen[10]).

We begin this chapter with a brief analysis of the ecological role of water temples, and a description of a simulation model we developed to explore the ecological role of water temples along two rivers. Tests of the model's predictions against two years of historical data added empirical support to the theoretical argument in *Priests and Programmers*[16] that water temples optimize rice harvests. In the second half of the chapter, we turn to the question of how the water temple networks manage to find optimal or near-optimal scales of coordination in water management. The entire ecological simulation model described in the first part of the chapter becomes a single time step in a nonlinear model designed to explore the relative importance of trial and error versus conscious design in the evolution of temple networks. Finally, we return to the methodological issues raised by Smith.

THE ECOLOGICAL ROLE OF WATER TEMPLES

In Bali, rice is grown in paddy fields fed by irrigation systems dependent on rainfall (there are no storage dams in the rivers). Rainfall varies by season and elevation, and in combination with groundwater inflow determines river flow. Traditional Balinese irrigation systems begin with a weir in a river, which shunts all or part of the flow into a tunnel that emerges some distance downstream, at a lower elevation, where the water is routed through a system of canals and aqueducts to the summit of a terraced hillside. Thus the flow of water into each farmer's fields depends on the seasonal flow of water in the rivers and streams, which in turn depends on rainfall and groundwater flow.

The role of water in the rice paddy ecosystem goes far beyond providing water to the roots of the rice plants. By controlling the flow of water into the terraced fields, farmers are able to create pulses in several important biochemical cycles. The cycle of wet and dry phases alters soil pH; induces a cycle of aerobic and anaerobic conditions in the soil that determines the activity of microorganisms; circulates mineral nutrients; fosters the growth of nitrogen-fixing cyanobacteria; excludes weeds; stabilizes soil temperature; and over the long term governs the formation of a plough pan that prevents nutrients from being leached into the subsoil. On a larger scale, the flooding and draining of blocks of terraces also has important effects on pest populations. If farmers with adjacent fields can synchronize their cropping patterns to create a uniform fallow period over a sufficiently large area, rice pests are temporarily deprived of their habitat and pest populations can be sharply reduced. How large an area must be fallow, and for how long, depends on the species characteristics of the rice pests. However, if too many farmers follow an identical cropping pattern in an effort to control pests, they will experience peak irrigation demand at the same time, and there might not be enough water for all, especially because the distance between weirs on Balinese rivers is usually only a few kilometers. Water sharing and pest control are thus opposing constraints, and the optimal scale for the coordination of cropping patterns depends on local conditions.

Paralleling the physical system of terraces and irrigation works, the Balinese have also constructed intricate networks of shrines and temples dedicated to agricultural deities and the Goddess of Waters. These temples play an instrumental role in the productive process by providing farmers with a structure to coordinate cropping patterns and the phases of agricultural labor. An analysis of the "ritual technology" that makes this possible is beyond the scope of this article (see Lansing[15,16]), but the ecological effects of temple networks can be clearly seen in the relationship between neighboring irrigation systems in the upper reaches of the Petanu river in southern Bali (see Figure 1).

As the map indicates, the Bayad weir provides water for a hundred hectares of rice terraces organized as a single subak, or farmer's association. A few kilometers downstream from the Bayad weir, the Manuaba weir provides water for 350 hectares of terraces, organized into ten subaks. The water temple hierarchy at Bayad consists of a weir shrine (*pura ulun empelan*) and a "Head of the Ricefields" temple (*pura ulun swi*) situated above the terraces. The larger Manuaba system also begins with a shrine at the weir, but includes two Ulun Swi temples, one for each major block of terraces. The congregations of both Ulun Swi temples also belong to a larger Masceti temple that is symbolically identified with the entire Manuaba irrigation system. Representatives of all ten subaks meet once a year at the Masceti temple to decide on a cropping pattern. Subsequently, the ritual calender carried out at the Ulun Swi temples provides a template for the phases of agricultural labor.

For example, all ten subaks belonging to the Masceti temple planted IR 64, a high yielding Green Revolution rice, in mid-September 1988 and harvested an average of 6.5 tons/hectare in mid-December. Subsequently, they planted kruing (another high-yielding rice) in early February and harvested 6 tons/ha in May. In June, they all planted vegetables and harvested approximately 2 tons/ha in August. During this time period, the flow of irrigation water into the Manuaba weir was as shown in Figure 2. This cropping pattern synchronized harvests for all ten subaks, encompassing 350 hectares of rice terraces, thereby possibly helping to keep down pest populations. Pest infestations for this period were reported to be minimal: less than one percent damage to the crops, primarily from brown planthoppers. This compares to pest losses of up to 50% of the crop in the late 1970s, when each subak planted rice continuously and cropping patterns were very disorganized (see Lansing[16]). However, the average flow of approximately 3 liters/second/hectare is less than the recommended average flow of 5 liters/sec/hectare, and suggests that the crops may have experienced some water stress. Certainly there was never any excess water. In that light, it is interesting to note that the Bayad subak upstream followed exactly the same cropping pattern as Manuaba, except that they began two weeks earlier. In general, irrigation demand is highest at the beginning of a new planting cycle, because the dry fields must become saturated. By starting two weeks after their upstream neighbors, the Manuaba subaks could help avoid water shortages at the time when irrigation demand peaks.

There would thus appear to be good ecological reasons for the Manuaba subaks to coordinate their cropping patterns with their upstream neighbors. The Bayad subak might also find that it is in their interest to coordinate their fallow periods

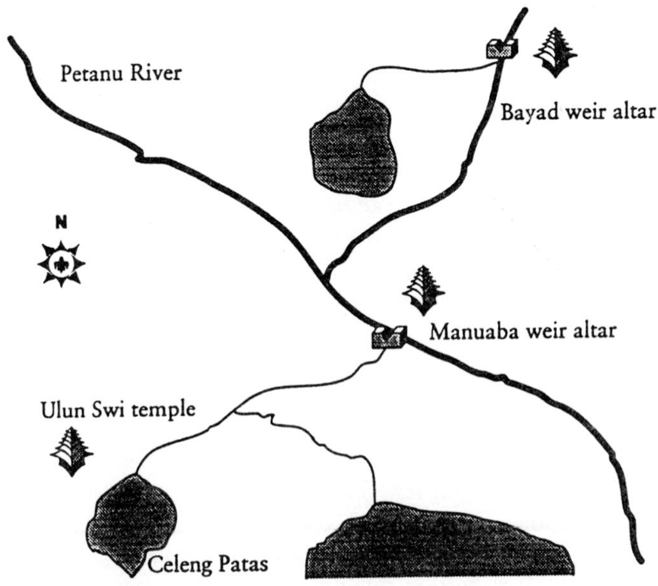

FIGURE 1 Bayad and Manuaba irrigation systems.[16]

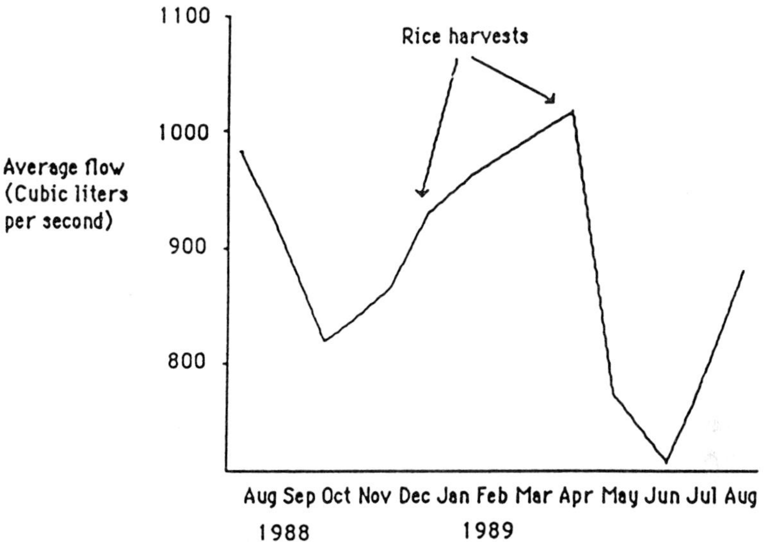

FIGURE 2 Flow at the Manuaba weir (1988-1989).

with Manuaba, so as to keep down pest populations. As it happens, the Manuaba subaks regularly send a delegation to the Bayad weir to request holy water, and the interdependency of the two irrigation systems is given symbolic expression by ritual ties between the deities of the two weirs.

A SIMULATION MODEL OF TWO RIVERS

In the case we have just considered, the water temples play their part by helping the subaks to balance two opposing constraints: water sharing and pest control. The effect of these constraints varies by location. In 1988, we built a simulation model to explore the effects of synchronized cropping patterns along two entire rivers, the Oos and the Petanu.[12,13] The model allowed us to simulate the effects of coordination by water temples under varying ecological conditions, and also to simulate other possible levels of coordination. At one extreme, all subaks follow exactly the same cropping pattern; at the other, each subak sets its own unique cropping pattern. The actual water temple scale of coordination lies between these extremes.

The watershed of the Oos and Petanu rivers includes approximately 6,136 hectares of irrigated rice terraces (see Figure 3). Based on topographical maps, we divided the Oos-Petanu watershed into 12 subsections specifying the catchment basins for each weir for which hydrological data was available. For each of the 172 subaks located in these basins, we specified the name, area, basin in which it resides, the weir from which it receives irrigation water, and the weir to which any excess is returned. We also defined the real spatial mosiac connecting these subaks. Given this geographical setting, the program simulates the rainfall, river flow, irrigation demand, rice growth stage, and pest levels for all watersheds and all subaks. At the appropriate times, the harvest is adjusted for cumulative water stress and pest damage, yields are tallied, and the next crop cycle is initiated (see Kremer[11]).

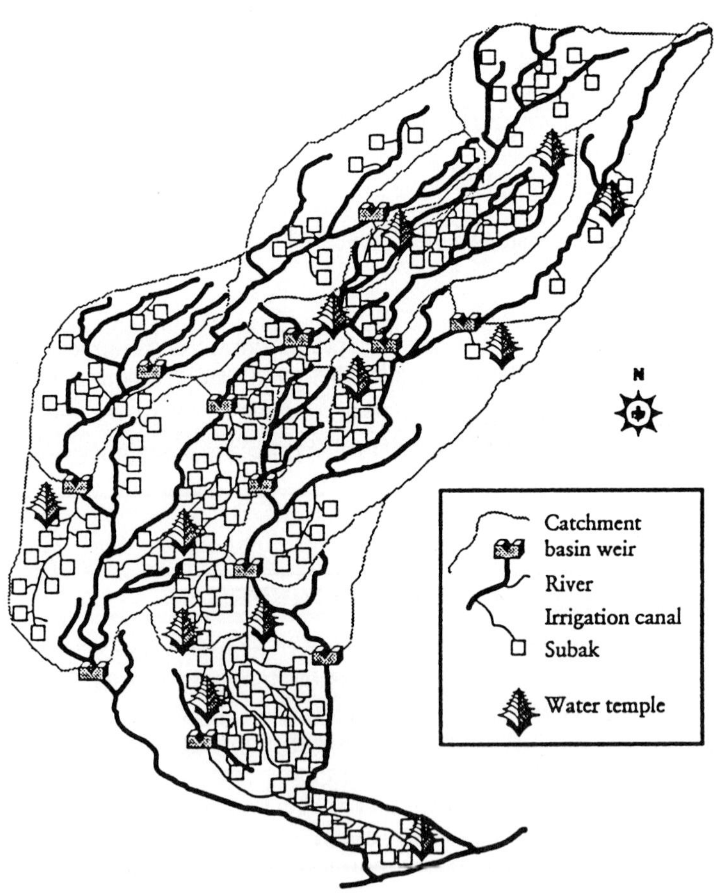

FIGURE 3 The Oos and Petanu rivers in south-central Bali (not to scale).[16]

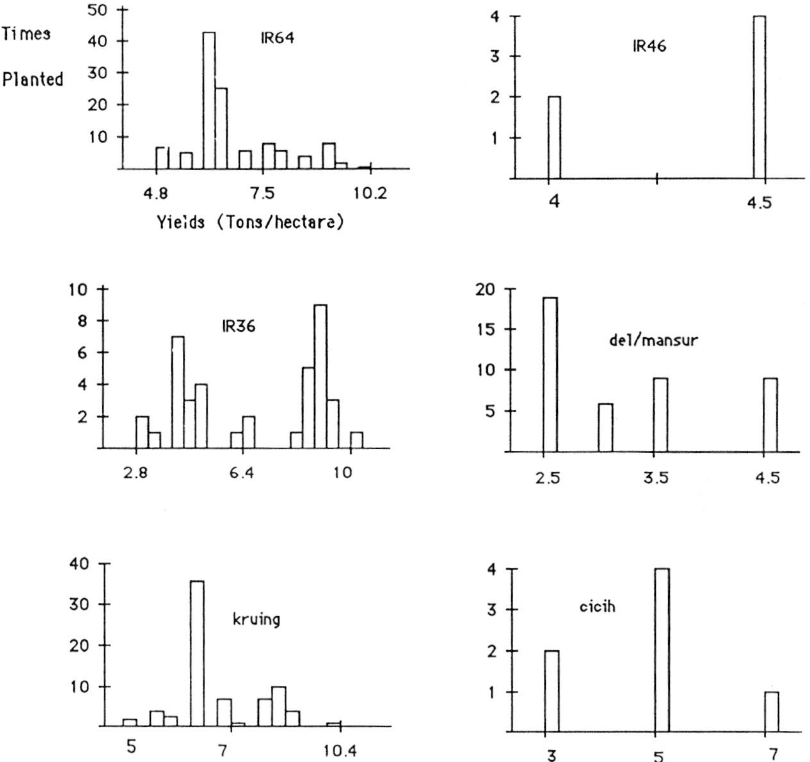

FIGURE 4 Distribution of rice harvest yeilds (tons/hectare) for six varieties planted in 1988-1989. In the graph in the upper left corner, about 45 subaks planted IR 64, a high-yielding rice variety, and harvested about six tons per hectare. However, some subaks harvested as much as ten tons per hectare. Del, mansur, and cicih are native Balinese rice varieties, which seldom yield more than five tons/hectare. Note that there is a wide distribution of yields for all varieties, both high-yielding and native.

In order to test the predictions of this model, in 1989 Lansing began to work with a team of Balinese students to gather real data on rainfall, irrigation flow, crop yields, water stress, and pest damage. As Figure 4 shows, there was a great deal of variation in actual harvest yields reported by the subaks (a point we will return to later on). Subsequently, we loaded the model with rainfall records based on real monthly averages; assigned each subak a cropping pattern that approximated its real cropping pattern, and compared the results with the actual distribution of yields. (See Figure 5.) The model's predictions improve for the second harvest, after the simulation of pest damage and water stress has been allowed to continue longer.

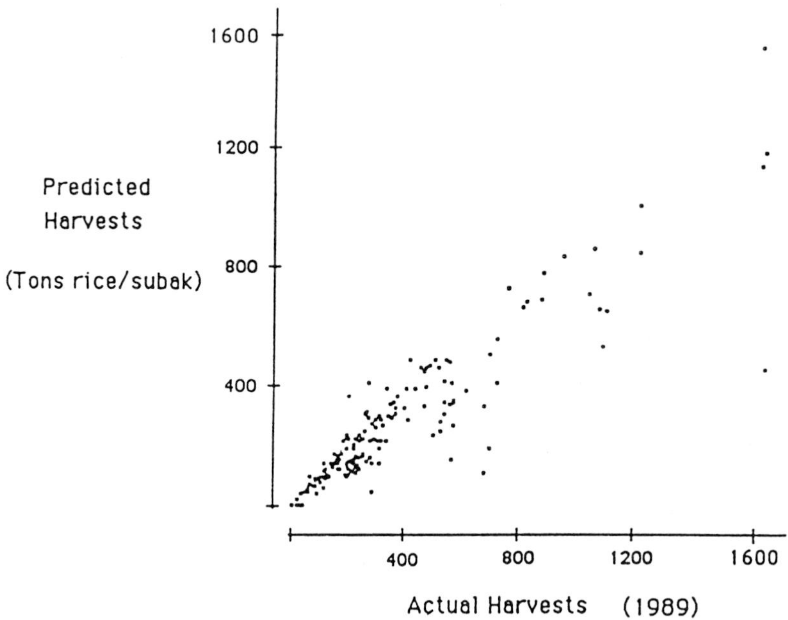

FIGURE 5 Comparison of model predictions with acutal rice harvests by subak (1989). Pearson's product-moment correlation $r = 0.90$.[13]

Considering the simplicity of the model, yields per hectare were also well correlated, with $r = 0.5$. To eliminate the possibility that model results were simply not very responsive to variations in cropping plans, we ran additional simulations in which we disrupted the local coordination implicit in the planting schedules followed by the subaks in 1989. When planting dates were randomized but the actual crops planted remained the same, the correlation for the second crop in 1989 dropped from 0.50 to 0.01.[13]

Comparison of the effects of different scales of coordination by the subaks for many simulation runs showed that the scale of coordination which most closely resembles the actual pattern of water temple control achieves the highest rice yields by optimizing the tradeoff between water sharing and pest control, as shown in Figure 6.

On the far left in Figure 6, when the cropping pattern is set by each individual subak, there is high pest damage. On the far right, a single cropping pattern for the whole watershed reduces pest damage but maximizes water stress. The highest

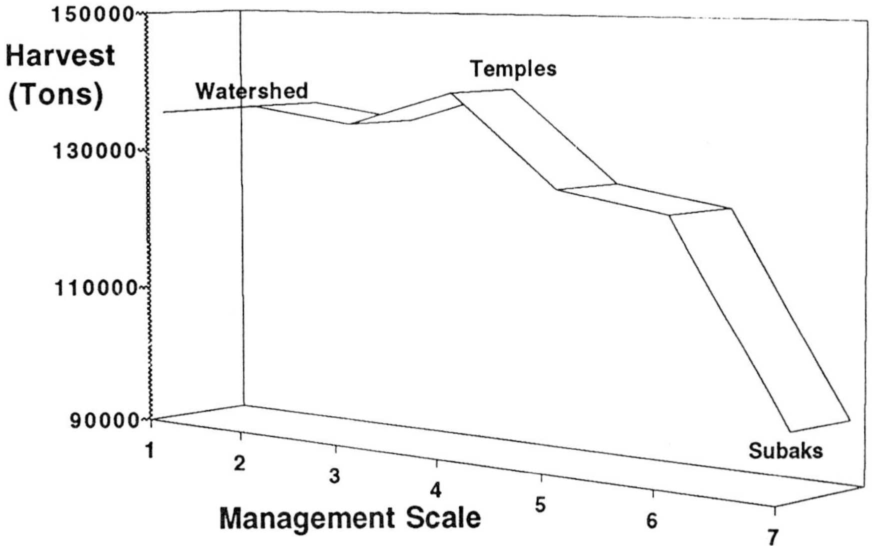

FIGURE 6 Effects of seven different scales of social coordination on rice yields, pest damage, and water stress.

peak is achieved by the scale of coordination which most closely approximates the temple scale of coordination. These results suggest two initial conclusions:

1. Most of the observed variation in harvest yields is due to reductions in yields caused by water stress or pest damage (see Figure 5).
2. The scale of social coordination in the management of irrigation has important effects on both of these variables (see Figure 6).[13]

MODELING ADAPTATION ON A RUGGED FITNESS LANDSCAPE

These results encouraged us to shift our attention from the mechanisms at work at the level of individual water temples, to the possible existence of system-level properties of temple networks. Lansing's approach to this question borrows from the theory of "fitness landscapes" in biology (see Wright,[22] Palmer,[18] and Kauffman and Johnsen[10]). As we have just seen, it is possible to calculate theoretical optima for rice production, and then compare these results to the actual system of temple coordination along each river. In biology, the idea of a fitness landscape is based on the notion that the fitness of an organism (species, population,etc.) depends not only on its own intrinsic characteristics ("genotype"), but also on its interaction with a local environment. The term "landscape" comes from visualizing a geographical landscape of fitness "peaks," where each peak represents an adaptive solution to a problem of optimization. Figure 6 may therefore be viewed as an example of a fitness landscape, in which the highest fitness peak is achieved by the temple scale of coordination. Note, however, that in biological models optimization occurs through "blind" natural selection, whereas here we propose a different mechanism: the deliberate efforts of farmers to cooperate in setting cropping patterns so as to maximize their harvests.

But it was not yet clear from our earlier analysis how the temple networks manage to find an optimal scale of coordination in the management of irrigation. Are differences in the structure of temple networks from one river to the next the result of deliberate planning by farmers, priests, or royal engineers? Or are they the product of trial and error adjustments by generations of farmers? Could the water temple network of River A do as well managing River B, or is each temple network a uniquely optimal solution to the specific ecological conditions of a particular region? And finally, how do temple networks respond to changing conditions?

A different type of simulation analysis provides a way to investigate such questions, which relate to the dynamic behavior of temple networks. Kremer's model (described above) became a single time step in a program developed by Lansing to explore the process of adaptation on a changing fitness landscape. Imagine that the water temple system does not exist, but that all the known ecological conditions remain unchanged along both rivers. As a new year begins, each of the 172 subaks in the model begins to plant rice or vegetables. At the end of the year, harvest yields are calculated for each subak. Subsequently, each subak checks to see whether any of its neighbors got higher yields. If so, the target subak copies the cropping pattern of its (best) neighbor. The model then simulates another year of growth, tabulates yields, and continues to run until each subak has reached its local optimum. What will happen?

In Figure 7 we show the results of such a simulation for the traditional Balinese cropping pattern, kerta masa. This cropping pattern begins with a long-maturing rice variety (*del* or *mansur*), followed by a fallow period, and a faster-maturing second rice crop (*cicih*). In this simulation, all subaks follow this cropping pattern. However, starting dates (when to begin planting the first crop) are randomly assigned to each subak.

After the first run, the average yield (tons rice/hectare/year) for the subaks was slightly more than five tons. Each subak then compared its yield with those of its four closest neighbors. Eighty six subaks discovered that one of their neighbors had a higher yield, and copied their neighbor's cropping pattern. The next year, average yields went up dramatically, and 94 subaks changed their cropping patterns. After eight years, average yields peaked, and all but 20 subaks stopped changing their cropping patterns. One hundred and fifty two subaks had reached a local optimum, in which their yields would not improve if they adopted a neighbor's cropping pattern. The remaining 20 subaks keep swapping cropping patterns with their neighbors indefinitely, as first one and then another obtains a slightly better yield.

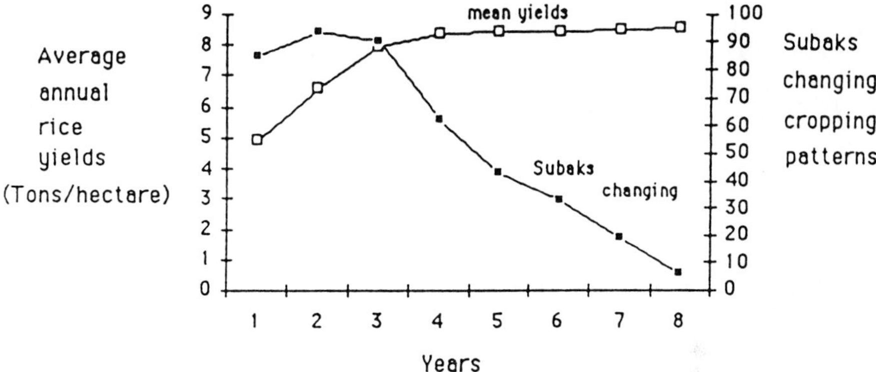

FIGURE 7 Increase in average rice yields as artificial temple networks appear.

FIGURE 8 First run of a model of coadaptation. Each symbol indicates a different cropping pattern. Note that they are randomly distributed among the subaks shown in this map of the Oos-Petanu watersheds. The average harvest was 4.9 tons of rice per hectare.

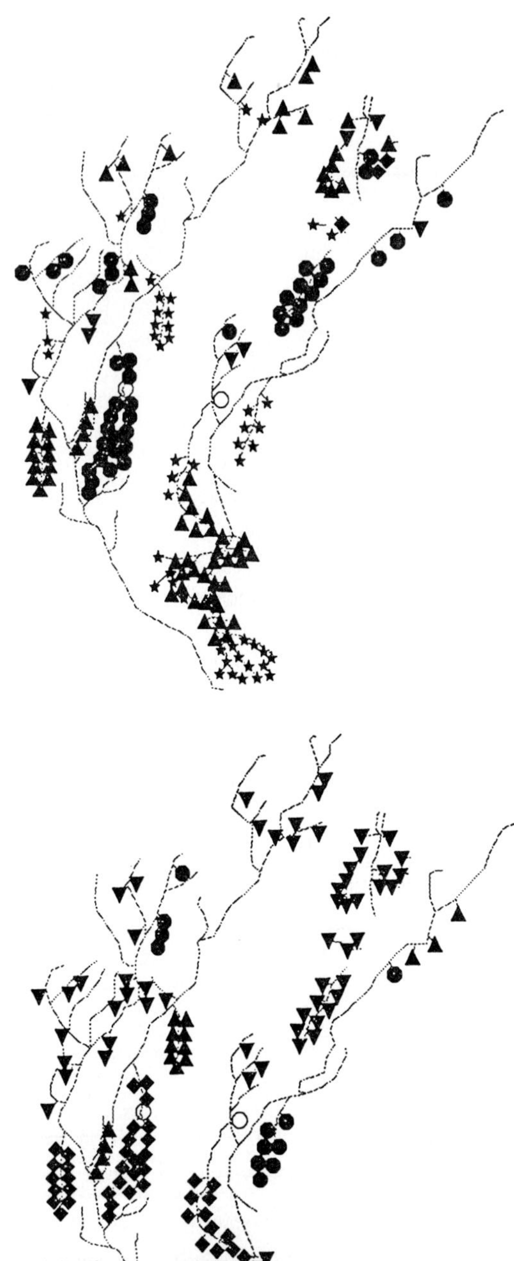

FIGURE 9 Last run of the model. The average harvest rose to 8.57 tons/hectare. Note that the pattern of synchronized cropping patterns (indicated by identical symbols) closely resembles the pattern created by temples, shown in Figure 10.

FIGURE 10 The distribution of synchronized cropping patterns in the traditional system of water temple networks.

In Figure 8 we map the distribution of cropping patterns for subaks in the first run; in Figure 9 for the last run of the simulation, and in Figure 10 the pattern of coordinated cropping patterns achieved by the water temple network. The resemblance between the last run of the hill-climbing program (Figure 9), and the temple system (Figure 10) will be evident to the reader. Competition to achieve maximal yields led to the formation of cooperative units with synchronized cropping patterns, that bear a very close resemblance to actual water temple networks.

Should we therefore conclude that patches of coordination resembling water temples will spontaneously develop as subaks seek to optimize their harvest yields? It is worthwhile to note that the number of possible distributions of cropping patterns that could occur in this model is astronomically large, and the chances that a temple like system of coordination would occur by chance are correspondingly small. But perhaps the temple networks are uniquely fitted to the traditional cropping pattern as a result of centuries of trial and error by the subaks? To test this possibility, many more simulations were conducted in order to vary not only the cropping patterns (crop varieties and start dates), but also the ecological parameters (rainfall, evapotranspiration, pest growth rates, dispersal rates, and damage coefficients). The same pattern occured in all cases: after 8–35 years, a complex structure of coordinated cropping patterns emerged, which bore a remarkable similarity to the actual pattern of water temple coordination along these rivers.

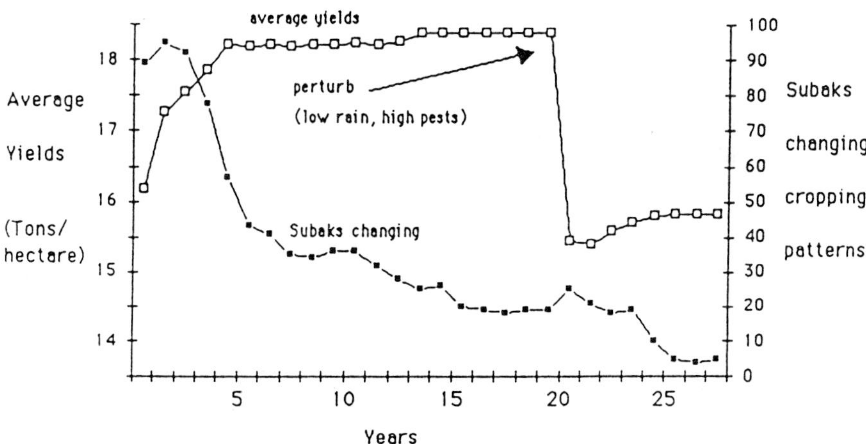

FIGURE 11 The effects of perturbation on yields. The initial cropping patterns included two crops of high yielding rice, and a vegetable crop, with randomized start dates. Rainfall was reduced to 80% normal and pest coefficients were increased in the twenty-first year.

Finally, consider the outcome of an experiment in which the network is perturbed: In Figure 11 we show the results of a simulation in which each subak grew two crops of high-yielding rice, as well as a vegetable crop. Using normal values for all ecological variables, average yields improved from 16 to 18.5 tons/hectare/year after twenty years, and stabilized with 20 subaks still changing their cropping patterns. In the 21st year a plague of pests was visited on the region by increasing the pest growth, dispersal and damage rates. Simultaneously, rainfall was decreased to 80% normal. Yields fell to 15.3 immediately but recovered to 15.8 within 7 years. Yet when the conditions of low rain/high pests occur from the very beginning of an otherwise identical simulation, it took twice as long to reach the same optimal yield value.

DISCUSSION

Before pursuing the implications of these results, it is worthwhile to linger for a moment over the mathematical logic that produced them. In the sequences of simulation runs described above, each subak alters ecological conditions for its neighbors when it varies its cropping pattern. A coadaptive process begins as each subak responds to these changing conditions by seeking to optimize rice yields. The actions of the subaks affect conditions for their neighbors, and the fitness landscape changes for most subaks with each run of the model. In other words, the actions of the subaks influence ecological variables such as irrigation flows, which in turn affect future decisions by other subaks. For example, in Figure 12 we show differences in water shortages at one of the twelve major irrigation systems in the model (Klutug) for a typical simulation. Major water shortages in March and April, which led to reduced yields for some subaks in the early runs, have disappeared by the 54th run. A similar pattern always occurs at the other twelve weirs: water shortages are gradually reduced. Note that this coadaptive process does not find the ideal cropping pattern for all; rather it finds locally optimal scales of coordination in the synchronization of cropping patterns. Let us call these artificial water temples, to distinguish them from real water temples. The interesting result here is that the same phenomenon occurs every time, regardless of the initial distribution of cropping patterns, or ecological parameters such as flow rates or pest biology. Within 8 to 35 years, depending on ecological conditions, the subaks spontaneously self-organize into a network of artificial water temples where all subaks are at or near a local optimum.

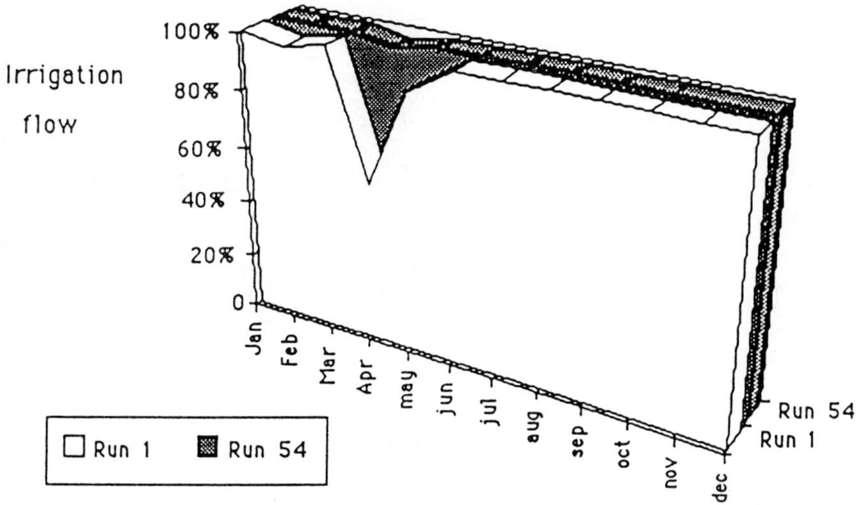

FIGURE 12 Water shortages at the Klutug dam.

Once this structure appears, the entire network displays an interesting emergent property: the ability to recover from external perturbations (such as low rainfall or high pest levels), as shown in Figure 11. Such disturbances initially generate a cascade of changes that propagate through one or more clusters of subaks, but soon lead to a new equilibrium. The ability of the system as a whole to react to changes is thus a property of the network itself, rather than the accidental consequence of the actions of individual subaks or temples. Put another way, the ability of each local group of farmers to reclimb the shifting peaks of the fitness landscape depends not only on their own initiative, but on the ability of the system as a whole to respond to changes.

Recent work in the mathematical theory of optimization on rugged fitness landscapes sheds an interesting light on this process, suggesting that these results are not artifacts of our computer program, but predictable outcomes from a process of coadaptation on a rugged fitness landscape. Both of the properties we have noted—the maximization of sustained yields, and the enhanced ability of the entire network to cope with perturbations—appear to occur for a wide range of coevolving complex systems. Artificial water temple networks fulfill the formal definition of a complex adaptive system:

1. It consists of a network of interacting agents (processes, elements);
2. It exhibits a dynamic, aggregate behavior that emerges from the individual activities of the agents; and
3. Its aggregate behavior can be described without a detailed knowledge of the behavior of the individual agents.

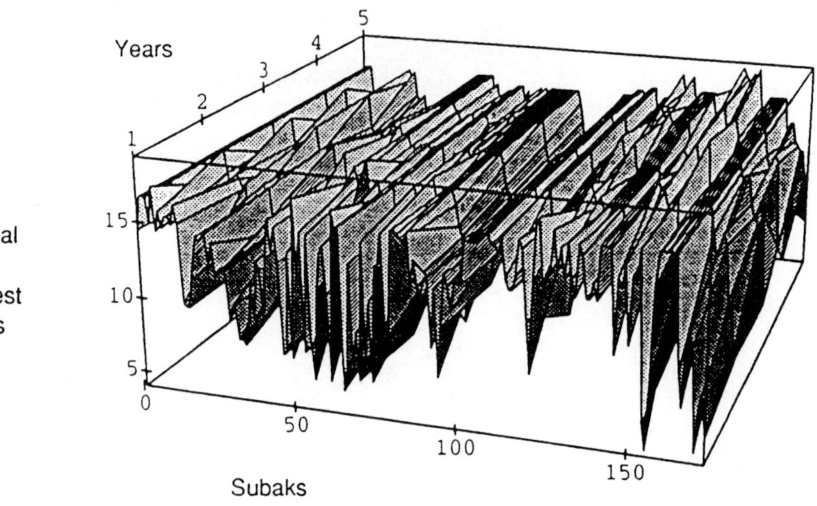

FIGURE 13 The first five years of a hill-climbing simulation of the Oos-Petanu watersheds. Note that yields change every year for most subaks, in a process of coadaptation.

TABLE 1 Increase in mean fitness (f_i) after temple networks appear

Cropping patterns:	Traditional	High Yielding	Low rain, high pests
First run	4.9	15.91	13.67
Last run	8.57	18.08	7.66

An agent in such a system is adaptive if it satisfies an additional pair of criteria: the actions of the agent in its environment can be assigned a value (performance, payoff, fitness or the like); and the agent behaves so as to increase this value over time. A complex adaptive system, then, is a complex system containing adaptive agents, networked so that the environment of each adaptive agent includes others in the system (see Holland and Miller[8]).

If each agent (in this case, each subak) always acts independently, the system as a whole behaves chaotically. Alternatively, if each agent is linked to all the others, the system is stable but only massive perturbations can cause alterations in behavior. Metaphorically, the system is frozen into place. Complex behaviors occur when these frozen components begin to *melt*, and different-sized islands of synchronized agents emerge. This transitional state between order and chaos is the narrow zone of complex or periodic behavior.[14,21] Recent work in the theory of complex systems suggests that coevolution on a rugged fitness landscape may drive many kinds of networks towards this zone or class of behaviors. The biologist Stuart Kauffman recently observed that, as if by an invisible hand, coevolving complex entities may mutually attain the poised boundary between order and chaos.[9] Here, mean sustained payoff, or fitness, or profit, is optimized.

By now we have come a long way from our original point of departure, and it may be helpful to say a few concluding words about the relationship between the models described in this paper, and evolutionary ecology. One major difference is the level of analysis: evolutionary ecology focuses on determining the fitness value or payoff of specific strategies for individuals, whereas from a systems perspective we are interested in whether it is possible to predict changes in these values over time by analyzing the interaction of social and ecological variables. Our model tracks the behavior of 172 subaks along two rivers, each of which sets an annual cropping pattern (rice, vegetables, fallow interludes). In the model, for $i = 1$ to N cropping patterns: x_i = frequency of cropping pattern i; f_i = fitness payoff (yield, expressed as tons rice/hectare/year), and f_{opt} = a local optimum fitness value.

In the first year, we calculate the fitness payoff (yield) for each cropping pattern, f_i. In succeeding years, x_i varies as its fitness value changes with respect to the average fitness ($\overline{f}$). For each cropping pattern tested:

$$\dot{x}_i = f_i - \overline{f}$$

where

$$\overline{f} = \sum_{i-1}^{N} \frac{f_i}{N}.$$

As time goes on, and patches of coordinated cropping patterns appear which improve the balance between water sharing and pest control, the mean fitness level of the entire system increases:

$$\overline{f} \to f_{\text{opt}}$$

A three-dimensional illustration of this process is shown below in Figure 13. Ultimately, regardless of the initial choice of cropping patterns, the mean value of f_i increases to a local optimum after artificial water temple networks come into existence. For example, compare the average yields for the initial and final runs with three different cropping patterns in Table 1.

While evolutionary ecology also begins with calculations of fi, it makes different assumptions about how these values are found. The key issue is the process of selection: as Eric Smith observes, "in evolutionary ecology adaptation via natural selection in a finite environment is the primary causal force."[20] Natural selection is thus obliged to try out most of the possible variants (see Eigen[4]). In our model, we could mimic this process by randomly changing cropping patterns for each subak every year and selecting good ones as they appear, a procedure that would take a very long time, even on a computer. Instead, we try to model the effects of deliberate selection, in this case the annual meetings of the subaks, where the farmers discuss the outcomes of last year's cropping patterns in their vicinity and try to pick a good one for the next cycle. In mathematical terms, they are searching the peaks of their local fitness landscape. This process of selection is not blind, but on the contrary highly efficient.

A second key distinction is the change in the fitness landscape itself brought about by this search strategy. As we have seen, the emergence of artificial temple networks increases the height of the fitness peaks, an effect which occurs regardless of our initial assumptions about the physical and biological systems. Our model demonstrates that cultural systems like the water temple networks can have macroscopic effects on the topography of the adaptive landscape. But such effects are not apparent from within the horizons of evolutionary ecology, since they are properties of systems rather than individuals.

CONCLUSION

The analytic techniques described in this chapter have enabled us to shift the level of analysis progressively from the individual farmer and field to the subak and water temple, and ultimately to the historical development of temple networks. We have shown that the structure of water temple networks could have developed through a process of spontaneous self-organization, rather than deliberate planning by royal engineers or other planners. This idea could be subjected to a more rigorous test, since our methodology is capable of predicting the response of temple networks to historical changes. The resulting patterns of spatial organization could be compared with the actual sequence of development as revealed by archaeological investigations.

We have also shown that the emergence of temple networks leads to higher average harvest yields, and improvement in sustainability (the ability to cope efficiently with ecological perturbations). Since these effects occur in our model as consequences of a process of coadaptation, they are probably not unique. Balinese water temples may be representative of a class of complex adaptive systems that have evolved to manage agroecosystems. As Roy Rappaport[19] argued a generation ago, it is likely that ritual often plays an important role in such "traditional" systems of resource management.

It also follows from our analysis that that the wide range of harvest yields reported by the subaks for 1988-1989 (see Figure 4) is likely to be the signature of suboptimal conditions, in which the subaks are prevented from rediscovering an optimal distribution of cropping patterns. In other words, under present conditions new crops and cropping patterns are tried out each year, in an effectively random process. Some subaks do well, others poorly, and the network as a whole remains in a state of continuous perturbation, corresponding to the early runs in our simulation model. As we have argued elsewhere, these conditions are after-effects of the Green Revolution in Bali: the replacement of native Balinese rice varieties with high yielding imported varieties, coupled with new management plans based on the assumption that production will be optimized if each farm or subak is an autonomous productive unit (see Lansing[16]). Our results suggest to the contrary, that self-organizing temple networks are intrinsically capable of a better job of water management than either autonomous subaks or centralized hierarchical control (see Figures 8, 9, and 10). But as long as research on agroecosystems remains focused on the behavior of individuals, the productive role of "traditional" systems of resource management like the water temples will remain invisible. Ironically, recent plans for the improvement of Balinese irrigation systems by international development agencies foresee an end to the productive role of water temples as an almost inevitable result of technical progress.[2]

ACKNOWLEDGMENTS

The research described in this paper was supported by grants from the National Science Foundation (BNS 8705400 and BNS 9005919). The simulation model of the Oos and Petanu rivers was tested with data collected by a team of Balinese undergraduate students from Udayana University: I Gde Suarja, Dewa Gde Adi Parwata, Ni Made Sri Tutik Andayani, and I Made Cakranegara. Many friends and colleagues have contributed to this research, especially Lene Crosby, Charles Taylor, Kevin Arrigo, Gary Seaman, Liane Gabora, John Miller, Chris Langton, Walter Fontana, Mike Simmons, Tyde Richards, Thierry Bardini, David Jefferson, Stuart Plattner, Jill Schroeder, and David Rudner; and in Bali the Jero Gde Mekalihan, Wayan Pageh, Dr. Nyoman Sutawan, Ida Pedanda Sidemen, Dr. Gusti Ngurah Bagus, Guru Nengah Tekah, Ir. Cokorde Raka , Ir. Jelantik Sushila, and Dr. Andrew Toth.

REFERENCES

1. Arrigo, K. R. "A Simulated Antarctic Fast-Ice Ecosystem." Ph.D. Dissertation, University of Southern California, 1991.
2. Asian Development Bank. "Bali Irrigation Project in Indonesia." Project Performance Audit Report PE-241 L 352-INO. Asian Development Bank, Post Evaluation Office, Manila, Philippines, 1988

3. Braudel, F. *On History.* Translated by S. Matthews. Chicago, IL: University of Chicago Press, 1980 [1969].
4. Eigen, M., and R. Winkler-Oswatitsch *Steps Towards Life: A Perspective on Evolution.* English translation by Paul Wooly. Oxford: Oxford University Press, 1992.
5. Giddens, A. *A Contemporary Critique of Historical Materialism.* Berkeley, CA: University of California Press, 1981.
6. Gould, S. J. *Wonderful Life: The Burgess Shale and the Nature of History.* New York: W.W. Norton, 1990.
7. Habermas, J. *Knowledge and Human Interests.* Boston: Beacon Press, 1971.
8. Holland, J. H., and J. H. Miller. "Artificial Adaptive Agents in Economic Theory." *Am. Econ. Assoc. Papers & Proc.*, **81(2)** (1991).
9. Kauffman, S. "The Sciences of Complexity and the Origins of Order." *Philos. Sci.*, in press.
10. Kauffman, S., and S. Johnsen. "Co-Evolution to the Edge of Chaos: Coupled Fitness Landscapes, Poised States, and Co-adaptive Avalanches." In *Artificial Life II*, edited by C. G. Langton, C. Taylor, J. Farmer, and S. Rasmussen, 325–370. Santa Fe Institute Studies in the Sciences of Complexity, Proc. Vol. X. Redwood City, CA: Addison-Wesley, 1991.
11. Kremer, J. N. "Technical Report on the Ecological Simulation Model." In *Priests and Programmers: Technologies of Power in the Engineered Landscape of Bali,* edited by J. S. Lansing, 153–158. Princeton, NJ: Princeton University Press, 1991.
12. Kremer, J. N., and J. S. Lansing. "Modeling Water Temples and Rice Irrigation in Bali: A Lesson in Socioecological Communication." In *Maximum Power: The Application of the Ideas of Howard Odum to Ecology,* edited by C. A. S. Hall. Department of Environment and Engineering, University of Colorado. Boulder, CO: University of Colorado Press, 1992.
13. Kremer, J. N., and J. S. Lansing. "Landscape Coordination by Temples Optimizes Rice in Bali (Indonesia)." Unpublished manuscript, 1992.
14. Langton C. G. "Life at the Edge of Chaos." In *Artificial Life II,* edited by C.G. Langton, C. Taylor, J. Farmer, and S. Rasmussen, 41–92. Santa Fe Institute Studies in the Sciences of Complexity, Proc. Vol. X. Redwood City, CA: Addison-Wesley, 1991.
15. Lansing, J. S. "Balinese Water Temples and the Management of Irrigation." *Am. Anthrop.* **89(2)** (1987): 326–341.
16. Lansing, J. S. *Priests and Programmers: Technologies of Power in the Engineered Landscape of Bali.* Princeton, NJ: Princeton University Press, 1991.
17. Marx, Karl. *Economic and Philosophical Manuscripts of 1844.* Moscow: Foreign Languages Publishing House, 1961.
18. Palmer, R. "Optimization on Rugged Landscapes." In *Molecular Evolution on Rugged Fitness Landscapes,* edited by A. S. Perelson and S. Kauffman, 3–26. Santa Fe Institute Studies in the Sciences of Complexity, Proc. Vol. IX. Redwood City, CA: Addison-Wesley, 1989.

19. Rappaport, R. A. "The Sacred in Human Evolution." *Rev. Ecol. & Sys.* **2** (1971): 23–44.
20. Smith, E. A. "Anthropology, Evolutionary Theory and the Explanatory Limits of the Ecosystem Concept." In *The Ecosystem Concept in Anthropology,* edited by E. Moran, 51–87. AAAS Selected Symposium 1992. Boulder, CO: Westview Press, 1992.
21. Weisbuch, G. *Complex System Dynamics: An Introduction to Automata Networks.* Santa Fe Institute Studies in the Sciences of Complexity, Lecture Proceedings Vol II. Redwood City, CA: Addison-Wesley, 1990.
22. Wright, S. "The Roles of Mutation, Inbreeding, Crossbreeding, and Selection in Evolution." In *Proceedings of the Sixth International Congress on Genetics*, Vol. 1, (1932): 356–366.

John R. Koza
Computer Science Department, Margaret Jacks Hall, Stanford University, Stanford, CA, 94305-2140, e-mail: koza@cs.stanford.edu

Artificial Life: Spontaneous Emergence of Self-Replicating and Evolutionary Self-Improving Computer Programs

This chapter reports on the spontaneous emergence of computer programs exhibiting the ability to asexually reproduce, to reproduce by combining parts from two parents, and to improve their performance through evolution from a primordial ooze of primitive computational elements. The computational elements are a computationally complete set and compositions of them are capable of agglomerating themselves to one another.

INTRODUCTION

John von Neumann (Burks[7]), E. F. Codd,[9] Thatcher,[52] John Devore (Devore & Hightower[12]), Christopher Langton,[35] Thomas Ray,[44,45,46] Skipper,[48] Rasmussen et al.,[42,43] and others have designed various self-reproducing automata and computer programs. Each of these self-replicating entities required considerable human ingenuity to design. None of these self-reproducing entities were small enough or simple enough that one could expect to conduct an experiment in any reasonable amount of time in which the entity might spontaneously emerge by means of a blind generative search. The issue of the amount of computer time required to witness spontaneous emergence is important because it sheds some light on the origins of self-reproducing entities and, as a practical matter, because it determines whether it is possible to conduct computer experiments concerning the phenomenon of spontaneous emergence (either now or in the foreseeable future). In addition, although several of the above self-reproducing entities designed by humans are computationally universal, none of them are programmed to exhibit evolutionarily self-improving behavior nor do they have the ability to learn. In contrast, living organisms exhibit both self-improving behavior and the ability to learn.

As Farmer and Belin state,

> "Discovering how to make...self-reproducing patterns more robust so that they can evolve to increasingly more complex states is probably the central problem in the study of artificial life."[18]

Moreover, all of the above self-reproducing entities are very brittle in that they do not correctly function if there is almost any perturbation from the structure that their human programmer specified. In contrast, living organisms exhibit considerable ability to continue to function under perturbation.

This chapter addresses the question as to whether self-replicating computer programs can spontaneously emerge from a primordial ooze of primitive computational elements by means of a blind random generative process within a reasonable amount of computer time. And, if so, can such spontaneously emergent self-replicating programs exhibit evolutionarily self-improving behavior and learning ability? In this chapter, we answer both questions affirmatively.

OVERVIEW

In nature, the evolutionary process occurs when the following four conditions are satisfied:

- An entity must have the ability to reproduce itself.
- There must be a population of such self-reproducing entities.
- There must be some variety among the population.
- There must be some difference in ability to survive in the environment associated with the variety.

In a population of entities, the presence of some variability that is associated with some difference in the survival rate is almost inevitable. As Charles Darwin observed in *On the Origin of Species by Means of Natural Selection* (1859),

"I think it would be a most extraordinary fact if no variation ever had occurred useful to each being's own welfare.... But if variations useful to any organic being do occur, assuredly individuals thus characterised will have the best chance of being preserved in the struggle for life; and from the strong principle of inheritance they will tend to produce offspring similarly characterised. This principle of preservation, I have called, for the sake of brevity, Natural Selection."

Thus, the first of the above four conditions (the ability to self-replicate) is, in practice, the crucial one for starting the evolutionary process. Once the entity is self-reproducing, evolution is the inevitable result of naturally occurring variations that are correlated with different rates of survival and reproduction. Of course, evolution is not purposive in the sense that it seeks to reach some predefined goal. Instead, the different rates of survival and reproduction of individuals in their environment produce, over time, changes in the genetic makeup of the population.[8,11,50]

This chapter presents an example of a population of computer programs learning to perform a simple task wherein the individuals in the population are active (as they are in nature) and where the fitness measure is implicit (as it is in nature). The essential ingredient for the individuals to be active is that they have the ability to reproduce themselves (either alone or in conjunction with a mate). Specifically, this chapter explores the organizational complexity required for a computational structure to engage in three levels of behavior: asexual reproduction (i.e., self-replicability), sexual reproduction (wherein parts of two parents genetically recombine to produce a new offspring which inherit traits from both parents), and evolutionarily self-improving behavior (i.e., learning).

If an individual has the ability to perform or partially perform some task beyond self-reproduction and if performance or partial performance of that task increases its probability of survival and reproduction (or increases its number of offspring), then the processes of natural selection and evolution exert selective pressure in favor of that individual. Thus, populations of self-reproducing individual computer programs are able to learn to perform tasks.

BACKGROUND ON SELF-REPRODUCING AUTOMATA AND PROGRAMS

In his 1949 lectures at the University of Illinois, John von Neumann explored the abstract question of what level of organizational complexity is required for self-replication to occur. von Neumann did not finish or publish his work on this subject prior to his death, but Arthur Burks, a colleague of von Neumann, extensively edited and completed many of von Neumann's manuscripts on this subject. Burks[6] states that von Neumann in his 1949 lectures inquired as to

> "What kind of logical organization is sufficient for an automaton to reproduce itself? This question is not precise and admits to trivial versions as well as interesting ones. von Neumann...was not trying to simulate the self-reproduction of a natural system at the level of genetics and biochemistry. He wished to abstract from the natural self-reproduction problem its logical form."

By abstracting out the chemical, biological, and mechanical details of the molecular structures that successfully engage in self-replication, von Neumann was able to highlight the essential requirements for the self-replicability of a structure. In particular, von Neumann demonstrated the possibility of self-replicability of a computational structure by actually designing a self-reproducing automaton consisting of a two-dimensional cellular arrangement containing a large number of individual 29-state automata cells.[5,6,7] The next state of each 29-state automaton was a function of its own current state and the state of its four neighbors (N, E, W, and S) in the two-dimensional cellular space.

von Neumann designed his self-reproducing automaton to perform the functions of a universal Turing machine and showed how to embed both a computation-universal automaton and its associated input tape into the two-dimensional cellular space. His computation-universal automaton was *a fortiori* construction universal in the sense that it was capable of reading the input tape (composed of cells of the cellular space), interpreting the data on the tape, and using a constructing arm to construct the configuration described on the tape in an unoccupied part of the cellular space. His automaton was also capable of backspacing the tape, constructing a copy of the tape, attaching the copy to the configuration just constructed, signaling to the configuration that the construction process had finished, and retracting the constructing arm. By putting a description of the constructing automaton itself on the input tape, von Neumann created a self-reproducing automaton.[1,26,27,41,49,52] Although never implemented, von Neumann's self-reproducing automaton was very large (involving many millions of cells).

von Neumann's self-reproducing automaton treats the information on the input tape to the Turing machine in two distinct ways.[35,36] First, the information on the input tape is interpreted as active instructions to be executed by the constructor in order to cause the construction of a particular configuration in an unoccupied part of the cellular space. Second, the information on the input tape is interpreted as passive data which is to be copied, in an uninterpreted way, to become the tape of the new machine. In other words, the information on the tape is interpreted as both a program and as data at different stages of the overall process.

Although von Neumann's 1949 lectures predated the discovery and elucidation by Watson and Crick in 1953 of the self-replicating structure and genetic role of deoxyribonucleic acid (DNA), the dual role of the information found in von Neumann's cellular automaton has a direct analogy in nature. In particular, the process of actively constructing a configuration based on the information contained on the input tape is analogous to the translation of the nucleotide bases into a chain of amino acids constituting a protein. And, the process of passively copying the input tape is analogous to the transcription of the nucleotide bases of DNA to messenger ribonucleic acid (mRNA) or to a complementary strand of DNA.[2,47] As Rasmussen et al. observes,[42] in chemical reactions involving molecules, entities interact in such a way that the entity can serve as both the entity executing a reaction (the program) and the entity being acted upon by the reaction (the data).

E. F. Codd[9] reduced the number of states required for a computationally universal, self-reproducing automaton from 29 to only eight. Codd's self-reproducing automaton occupied about 100,000,000 cells.[10,12]. In the early 1970s, John Devore's self-reproducing automaton simplified Codd's automaton and occupied only about 94,794 cells.[12] A computer simulation of Devore's automaton can be seen on videotape.[21] Thatcher[52] designed another self-reproducing automaton.

Langton[35] observed that the capability of computational universality found in the self-reproducing automata of von Neumann, Codd, Devore, and Thatcher is not known to be present in any self-replicating molecules (which may be the building blocks of the earliest and simplest forms of life) or in the biological structure of any known living entity. Consequently, Langton achieved a substantial reduction in size by abandoning computation universality. In particular, Langton designed a self-reproducing cellular automaton that occupied an area of only 100 cells in its cellular space. Like Codd's and Devore's automata, each cell in Langton's automaton had only eight states. Like the designs of von Neumann, Devore, and Codd, each cell in Langton's self-reproducing automaton communicated only with its four neighbors. However, unlike the other self-reproducing automata, the coded description of Langton's self-reproducing automaton was not on a static tape, but instead endlessly circulated in a manner reminiscent of the delay-line storage devices used in early computers. A computer simulation of Langton's automaton can be seen on videotape.[40]

All of the above self-reproducing cellular automata take advantage of the particular transition function applied to all cells in the cellular space (just as molecular self-replication takes advantage of the particular physical laws governing the molecules involved). If the particular transition function is too powerful, the issue of self-reproduction becomes trivial. Langton's design, like those of von Neumann, Codd, Devore, and Thatcher, avoided triviality by requiring that the responsibility for construction of the copy be actively directed by instructions residing primarily in the configuration itself. In addition, Langton's design shared an important characteristic with the other designs and with DNA, namely that the stored information has dual roles in that it is treated both as instructions to be actively interpreted and executed and as data to be passively copied into the new automaton. In addition, triviality is avoided because the crucial copying event copies only one primitive element at a time.

Ray[44,45] wrote a clever 80-line self-reproducing computer program in assembly code and demonstrated how his program could evolve over time as a consequence of mutation. A visualization of this work appears as a videotape in Ray.[46] Ray wrote his program in a special assembly language (Tierra) for a special virtual machine. Ray's virtual machine was intentionally imperfect and introduced random mutations to his original hand-written 80-line program (called the "ancestor"). Ray observed the emergence, over a period of hundreds of millions of time steps, of an impressive variety of different entities (some self-reproducing, some parasitically self-reproducing, some symbiotically self-reproducing, and some not self-reproducing) and a dazzling array of biological phenomena including parasitism, defenses against parasitism, hyperparasitism, symbiosis, and social parasitism.

In Ray's Tierra system, certain assembly code instructions can search memory backwards or forwards from their location for the first occurrence of a sequence of consecutive NOP (No-Operation) instructions having a specified sequence of bits in their one-bit data fields. The bit pattern used in such searches is called a *template*. As in nature, complementary matching is performed with the template. In writing his self-reproducing program, Ray put a certain identifying four-bit template at the beginning of his program and another identifying four-bit template at its end. Ray's program is capable of performing a self-examination to locate these two templates. His program is then able to calculate its own size by subtracting (using registers of his computer) the memory addresses of the two templates discovered by this self-examining search. Ray then wrote a loop in which each instruction between the now-located beginning and now-located end of his program was moved (copied), one instruction at a time, from its location in memory into an unoccupied area of memory. Thus, Ray's program is able to make an exact copy of itself. A special assembly code instruction MAL ("Memory Allocation") causes the operating system to allocate an unoccupied area of memory to an offspring and a special cell division instruction activates that new area as an independent new entity. Ray's program uses data in the same dual way used by DNA and the designs of von Neumann, Codd, Thatcher, Devore, and Langton in that it first actively executes its 80 assembly code instructions as instructions at one point and it passively copies its 80 instructions as data at another point. However, in contrast to DNA and the other designs, Ray's program does not use a separate coded description of the entity to be reproduced; instead, it uses the program itself as its own description. Laing[33,34] discusses such "automata introspection" where there is no separate description of the entity being reproduced.

An important design feature of Ray's Tierra system is that it is never necessary to refer to a numerical memory address (either absolute or relative) in the operand of an instruction. Ray achieves this in two ways.

First, he avoids numerical memory addresses entirely as the operands of his assembly code instructions by designing his virtual machine with only four registers, a flag bit, a program counter, and one small stack. Thus, both the operation code of each assembly code instruction and its associated operand (if any) fits into only five bits. Some of Ray's operation codes occupy three of the five bits while others occupy four bits. For the operations whose operation code occupies three bits, the remaining two bits designate a particular one of four registers to be used in conjunction with the operation. For the operations whose operation code occupies four bits, the remaining bit is a data bit.

Second, Ray's biologically motivated scheme for template matching eliminates the need to refer to a location in memory by means of a numerical memory address. Specifically, a template consists of consecutive NOP instructions, each with a one-bit data field. A searching template is deemed to have found a match in memory if it finds consecutive NOP instructions whose data bits are complementary to the data bits of the searching template. The search is limited to a specified small bounded area of memory. Thus, it is possible to transfer control (i.e., jump) to a particular location in memory without ever referring to a specific numerical address by searching memory for the nearest occurrence of a matching (i.e., complementary) template. Similarly, a particular location in memory (such as the beginning and end of the program) may be located without referring to its specific numerical memory address in the operand of an assembly code instruction.

Holland's ECHO system[24,25] for exploring evolution in a miniature world of co-evolving creatures employs a similar template-matching scheme. In Holland's system, a search template (consisting of all or part of a binary string) is deemed to match another binary string if the bits match (starting from, say, the left) for the number of positions they have in common.

Skipper[48] has presented a model of evolution incorporating several of the above features, including template matching.

Self-reproducing computer programs have been written in FORTRAN, C, LISP, PASCAL, and many other languages.[3,4] Computer viruses and worms are self-reproducing programs that operate within various computers.[14,15,16,51] Programs entered into the Core Wars tournaments typically have self-replicating features.[13,15,42,43]

In nature, carbon-based life forms exploit energy available from the environment (primarily the sun) to organize matter available from the environment in order to survive, reproduce, and evolve. This exploitative and organizational process operates within the constraints of the rules of physical interaction governing the matter involved, notably the rules of physical interaction of atoms of carbon, hydrogen, oxygen, nitrogen, and other elements found in organic molecules.

The field of artificial life contemplates alternatives to the single carbon-based paradigm for life found on earth. One such alternative involves computational structures (such as programs within the digital environment of a computer) that exploit computer time to organize memory in order to survive, reproduce, and improve themselves.[37,38,39,40] This exploitative and organizational process operates within the constraints of the rules of interaction governing the milieu of computational structures. In this analogy, computer time corresponds to energy available from the environment and computer memory corresponds to matter available from the environment.

COMMON FEATURES OF SELF-REPRODUCING AUTOMATA AND PROGRAMS

All of the foregoing self-reproducing automata and programs have the crucial common feature of creating a copy of a given item at one critical point in their operation. In each instance, they create the copy by exploiting the ability of the milieu in which they are operating to change the state of some part of the computer memory (in the manner permitted by the rules of interaction of the milieu) to a specified desired new state, which is, in each instance, a copy of something. That is, there is an assumption that there is sufficient free matter and free energy available in the milieu to create a copy of a specified thing.

In the self-reproducing cellular automata of von Neumann, Codd, Devore, and Thatcher, the copy is created by interpreting the coded description on the tape as a sequence of instructions which cause the construction, cell by cell, of the new machine in an unoccupied part of the cellular space. The copying actually occurs at the moment when the state of some previously quiescent cell immediately adjacent to the tip of the constructing arm is changed to the desired new state. This crucial state transition can occur because the rules of the milieu permit it and because there is sufficient free matter and free energy available in the milieu to permit the creation of a copy of the desired thing. The free matter arises from the plasticity of the state of each cell in the cellular space. That is, every cell is free to change to any desired new state in accordance with the underlying physics (i.e., the state transition rules of the cellular space). The force that causes the change of state corresponds to the free energy of the system (i.e., computer time). Free energy must be supplied to the system to enable it to change states. The copying is done without affecting or depleting the original coded description in any way because of this availability of free matter and free energy. The fact that the automaton is copied at a different time than the coded description on the tape emphasizes the dual use of the coded description on the tape.

In Langton's self-reproducing cellular automaton, the coded description circulates endlessly in a channel of cells. For most points along the channel, there is only one cell that is poised (either immediately ahead or around a corner) to receive whatever was emitted by its immediately adjacent predecessor in the channel. However, at one particular point in the channel, there are two automata that are poised to receive whatever is emitted by the immediately adjacent predecessor. At this fan-out point in the channel, one copy of the coded description continues circulating around the original channel while another copy of the original description goes off toward an unoccupied part of the cellular space. The copying actually occurs at the moment when the states of the two previously quiescent cells adjacent to the cell at the fan-out point are changed to the specified new state. This crucial state transition can occur because the rules of the milieu permit it and because there is sufficient free matter (i.e., states of the cellular space) and free energy (i.e., the force that causes the change of state) available in the milieu to create a copy of a specified thing (i.e., the copying is done without affecting or depleting the original coded description).

In Ray's self-reproducing program, the 80-instruction program serves as its own description and the MOVE-INDIRECT instruction inside an iterative loop make a copy of what the program finds from a self-examination. Inside the iterative loop, each assembly code instruction of the program is copied, one by one, from its location in memory into an unoccupied area of memory. Copying occurs when the state of a memory location in an unoccupied (or inactive) area of memory is changed to the specified new state by the MOVE-INDIRECT instruction. This crucial state transition can occur because the rules of the milieu permit it and because there is sufficient free matter and free energy available in the milieu to create a copy of a specified thing without affecting or depleting that which is copied.

In each instance, the copying of the entire entity is done cell by cell over many time steps.

Fontana[19,20] and Rasmussen et al.[42,43] have addressed similar issues.

In nature, an individual strand of DNA is able to replicate itself because there is a plentiful supply of nucleotide bases available in the milieu to bind (in a complementary manner) to the bases of both of the existing strands of DNA when the DNA unwinds. Similarly, messenger RNA (mRNA) can be translated by a ribosome into a string of amino acids (a protein) by means of the genetic code because there is a plentiful supply of *transfer RNA* (tRNA) available in the milieu to bind (in a complementary manner) to the codons of the messenger RNA.

ORGANIZATIONAL COMPLEXITY OF EXISTING SELF-REPRODUCING AUTOMATA AND PROGRAMS

Each of the above self-reproducing computer programs was conceived and written by a human programmer using an impressive amount of ingenuity. None is small and none is simple.

von Neumann recognized that a certain "minimum number of parts" was a necessary precondition for self-reproduction. (See also Dyson.[17]) von Neumann observed that a machine tool that stamps out parts is an example of a synthesizer that is more complex than that which it synthesizes. The machine tool is

> "an organization which synthesizes something [that is] necessarily more complicated...than the organization it synthesizes," so that "complication, or [re]productive potentiality in an organization, is degenerative." [7]

von Neumann also recognized that living organisms, unlike the machine tool, can produce things as complicated as themselves (by means of reproduction) and can produce things more complicated than themselves (by means of evolution). von Neumann concluded

> "there is a minimum number of parts below which complication is degenerative, in the sense that if one automaton makes another, the second is less complex than the first, but above which it is possible for an automaton to construct other automata of equal or higher complexity."[7]

It would appear that the "minimum number of parts" to which von Neumann referred must be exceedingly large.[23] Certainly von Neumann's, Codd's, Thatcher's, and Devore's self-reproducing automata each had a very large number of parts. If the minimum number of parts for a self-reproducing entity is large, that entity is going to be very rare in the space of possible entities of which it is an instance. Therefore, the probability of finding that entity in a random search of that space of entities is very small. Accordingly, the probability of spontaneous emergence of that entity is very small.

The self-reproducing programs written by von Neumann, Codd, Thatcher, Devore, Langton, and Ray are each just one program from an enormous space of possible computer programs. It would not be possible to find any of these programs in any reasonable amount of time by means of any kind of blind random search of the space of programs.

We can get a rough idea of the probability of finding one of their programs in a blind random search if we make four simplifying assumptions. First, suppose that we ignore the milieu itself in making this rough calculation (e.g., the rules of the cellular space or the rules governing the computer and operating system executing the assembly code instructions). Second, suppose we limit the search to only entities whose size is no larger than the precise size that the human programmer involved actually used. In other words, since Langton's design occupies 100 cells in a cellular space, we consider only entities with 100 or fewer cells. Third, suppose we limit the search to spaces that do not contain any features beyond the minimum set of features actually used by the programmer involved. Fourth, suppose that we ignore all symmetries and equivalences in functionality (i.e., we assume that there is only one such solution).

von Neumann's 29-state self-reproducing automaton has many millions of cells. An apparently grossly erroneous underestimate of 200,000 cells is cited in Kemeny.[27] Using this underestimate, the probability of discovering this particular automaton in a blind random generative search is still one in $29^{200,000} \approx 10^{292,480}$.

Codd's eight-state self-reproducing automaton has about 10^8 cells.[12] Using the above simplifying assumptions, the probability of discovering this particular automaton in a blind random generative search is about one in $8^{10,000,000} \approx 10^{9,000,000}$. Thatcher's self-reproducing automaton is also very large.

Devore's version of Codd's program fits into a rectangle of 259 cells by 366 cells (i.e., 94,794 cells). The probability of randomly generating this particular automaton is about one in $8^{94,794} \approx 10^{85,315}$.

Langton's program is much smaller and is known to occupy 100 cells. Therefore, the probability of randomly generating this automaton is one in $8^{100} \approx 10^{91}$.

Ray's hand-written program consists of 80 lines of 5-bit assembly code instructions. Therefore, the probability of randomly generating this automaton is one in $32^{80} \approx 10^{120}$.

Interestingly, Ray observed the evolution, over hundreds of millions of time steps, of a self-reproducing program consisting of only 22 assembly code instructions in the Tierra language. The probability of randomly generating this evolutionarily-derived self-reproducing program is one in $32^{22} \approx 10^{33}$.

The rough calculations above suggest that the probability of creating a self-reproducing computer program at random must be exceedingly small. That is, the chance is remote that a self-reproducing computer entity can spontaneously emerge merely by trying random compositions of the available ingredients (i.e., states of an automaton or assembly code instructions). The probability of randomly generating Langton's automaton and the probability of randomly generating the evolutionarily-derived 22-line program are the least remote. However, even these probabilities (i.e., 10^{91} and 10^{33}, respectively) are far too remote to permit spontaneous emergence with existing available computer resources or any foreseeable computer resources or other resources for conducting simulations and experiments. For example, even if it were possible to test a billion (10^9) points per second and if a blind random search had been running since the beginning of the universe (i.e., about 15 billion years), it would be possible to have searched only about 10^{27} points.

However, once spontaneous emergence of a self-replicating entity occurs in a particular milieu, the situation changes. As von Neumann recognized, once created, self-reproducing structures can not only reproduce, but multiply. He noted,

> "living organisms are very complicated aggregations of elementary parts, and by any reasonable theory of probability or thermodynamics, highly improbable...[however] if by any peculiar accident there should ever be one of them, from there on the rules of probability do not apply, and there will be many of them, at least if the milieu is reasonable."

von Neumann further noted,

> "produc[ing] other organisms like themselves...is [the] normal function [of living organisms]. [T]hey wouldn't exist if they didn't do this,and it's plausible that this is the reason why they abound in the world."[7]

Darwin recognized that once a population of different, self-replicating structures has been created, the ones that are fitter in grappling with the problems posed by the environment tend to survive and reproduce at a higher rate. This natural selection causes the evolution of structures that are improvements over previous structures.

In nature, individuals in the population that do not respond effectively to a particular combination of environmental conditions suffer a diminished probability of survival to the age of reproduction. That is, the environment communicates a negative message (in the extreme, immediate death) to unfit entities. If entities in the population have a finite lifetime (which is the case in this chapter), this means that the potential offspring of entities that do not respond effectively to the environment do not become a part of the population for the next generation.

The size, complexity, and intricacy of von Neumann's, Codd's, Thatcher's, Devore's, Langton's, and Ray's self-reproducing entities suggests that the organizational complexity required for self-replicability is very high, and, therefore, the possibility of spontaneous emergence of self-replicability (much less sexual reproduction and evolutionary self-improving behavior) is very low.

In the remainder of this chapter, we show that the probability of spontaneous emergence of self-reproducing and self-improving computer programs is very much more likely than the foregoing discussion would suggest. In fact, we will demonstrate experimentally that spontaneous emergence of self-reproducing and self-improving computer programs is possible, in an appropriate milieu, with a fast workstation.

The design of this experiment is severely constrained by considerations of the available computer resources. Accordingly, certain compromises must be made to improve the probability of spontaneous emergence. Each compromise is accompanied by an explanation of the approach that could be pursued if there were no such considerations of the available computer resources. However, it will be seen that each of the rare events in which we are interested would nevertheless have occurred within a reasonably small amount of time (albeit more time than we have available on our computer), even if the compromises had not been made.

The set of basic computational operations that we chose are not the most elementary, and they do not purport to accurately model the chemical, biological, or mechanical details of the actual organic molecules that carry out asexual reproduction, sexual reproduction, and evolutionary self-improving behavior in carbon-based life forms. However, these basic computational operations perform functions that bear some resemblance to the functions performed by organic molecules in nature, and they perform these functions in a way that bears some resemblance to the way organic molecules perform in nature.

Spontaneous emergence of computational structures capable of complex behavior might occur in a sea of randomly created computer program fragments that interact in a random way. Such a sea and the turbulent intermixing of entities in this sea of entities has some degree of biological plausibility since biological macromolecules randomly move and randomly come into contact in their cellular milieu.

One of the basic operations permits the incorporation of one entire computer program into another. That is, larger entities can grow by means of an agglomeration of smaller entities. This agglomerative operation is a highly simplified form of recombination (crossover). Another basic operation permits the emission of new entities into the sea.

The approach used here bears some relation to both the genetic algorithm[22,25] and genetic programming[28,29,30] as illustrated on videotape.[32] In genetic methods, the calculation of the fitness measure is usually an explicit calculation and the individuals in the population are passive. The fitness of each individual is measured by means of some explicit calculation and the controller of the genetic process applies various genetic operations to the passive individuals on the basis of that computed fitness. The controller exists outside of the individuals. The user of a genetic method may think of the fitness of a living organism as a numerical quantity reflecting the probability of reproduction weighted by the number of offspring produced (the fecundity); however, this numerical fitness measure is not known to the individual or explicitly used by the individual. Instead, this fitness value is used by the outside controller to execute the genetic process.

In contrast, in nature, fitness is implicit and the individuals are active. There is no explicit numerical calculation of fitness. Individuals act independently without centralized control or direction. If the individual is successful in grappling with its environment, it may survive to the age of reproduction and reproduce. Moreover, the individuals in the population are active in that each individual has the capacity for reproduction (either alone or in conjunction with a mate). The fact that they reproduce means that they are fit, and their fecundity indicates the degree of their fitness.

PRIMITIVE FUNCTIONS AND TERMINALS FROM WHICH THE COMPUTATIONAL ENTITIES ARE COMPOSED

Our focus here will be on a primordial ooze of primitive computational entities composed of functions from a function set $\mathcal{F}$ and of terminals from a terminal set $\mathcal{T}$. The terminal set consists of three terminals and is

$$\mathcal{T} = \{\text{D0, D1, D2}\}.$$

The three terminals D0, D1, and D2 can be thought of as sensors of the environment. They are variables that can evaluate to either NIL (i.e., false) or something other than NIL (i.e., T = true).

The function set contains eight functions and is

$$\mathcal{F} = \{\text{NOT, SOR, SAND, ADD, SST, DWT, LR, SR}\},$$

taking one, two, two, one, one, two, zero, and zero arguments, respectively.

The first three of the functions in $\mathcal{F}$ (NOT, SOR, and SAND) together constitute a computationally complete set of Boolean functions. They have no side effects.

The function not takes one argument and performs an ordinary Boolean negation.

The function SOR ("Strict OR") takes two arguments and performs a Boolean disjunction in a way that combines the behavior of the OR macro from Common LISP and the PROGN special form from Common LISP. In particular, SOR begins by unconditionally evaluating both of its arguments in the style of a PROGN function (thereby executing the side effects, if any, of both arguments) and then returns the result of evaluating the first of its two arguments that evaluates to something other than NIL. If both arguments evaluate to NIL, SOR returns NIL. In contrast, once the ordinary OR in Common LISP finds an argument that evaluates to something other than NIL, it does not evaluate any remaining argument(s) and therefore does not execute any side effects of those remaining argument(s).

The function SAND ("Strict AND") takes two arguments and performs a Boolean conjunction in a "strict" manner similar to SOR.

Each of the other five functions (ADD, SST, DWT, LR, and SR) have return values; however, as we will see, their real functionality lies in their side effects. The ADD ("agglomerate") and SST ("Search Sea for Template") functions act as identify functions insofar as their return values are concerned (i.e., they return the result of evaluating their one argument). The DWT ("Do ≪WORK≫ until ≪TRAILING-TEMPLATE≫ matches") function returns the result of evaluating its second argument (i.e., it acts as an identity function on its second argument); its return value does not depend at all on its first argument. The return value of the LR ("Load REGISTER") function is T and the return value of the SR ("Store REGISTER") function is NIL.

Three global variables are used in conjunction with the SST, DWT, LR and SR functions: MATCHED-TREE, OPEN-SPACE, and REGISTER. Each is initially NIL when evaluation of a given program begins.

There is a sea (population) of computer programs each composed of ingredients from $\mathcal{F}$ and $\mathcal{T}$. As a program circulates in the sea of computer programs, both the ADD function and the SST function search the ever-changing sea for a program whose top part matches the top part of the argument subtree of the ADD or the SST, much as a freely-moving biological macromolecule circulates in its milieu by means of random diffusion and seeks other molecules with a receptor site that matches (in a complementary way) its own receptor site. If the search is successful, the ADD function and the SST function then perform their specified side effects.

Only the top part of the argument subtree of the ADD function and SST function is used in the search; that part is called the *template.* This method of template matching has a degree of biological plausibility and is an extension into the space of program trees from the space of the linear bit strings in Holland's ECHO templates and the space of linear strings of assembly code instructions in Ray's Tierra templates. (See also Skipper.[48])

Two points from the argument subtree of the ADD function and the SST function are used in the comparison with the other programs in the sea. The template consists of the top point of the argument subtree of the ADD or SST and the point immediately below the top point on the leftmost branch of the argument subtree. If the argument subtree of an ADD or SST contains only one point (i.e., is insufficient), the search fails and no side effect is produced.

If there is a side-effecting ADD or SST in the template of an ADD or SST, the ADD or SST in the template is not executed. Instead, the ADD or SST being executed merely uses the ADD or SST in the template symbolically in trying to make a match. The points in the argument subtree of an ADD or SST other than the two points constituting the template are not relevant to the template matching process in any way. In particular, if there is another ADD or SST located inside the argument subtree, but beneath the template, they do not initiate any searching.

The comparison is made only to the top part (the site) of the program being searched (consisting of the root of the program tree being searched and the point immediately below the root on the leftmost branch).

The sea (called a *Turing gas* by Fontana[19,20]) is turbulent in the sense that the immediate neighbor of a given program is constantly changing. The search of the sea initiated by an ADD or SST begins with the program itself and then proceeds to the other programs in the sea. A template is never allowed to match to itself (i.e., to the argument subtree of the very ADD or SST currently being executed). The search of the sea continues until 100% of the programs in the sea have been searched.

When the SST function ("Search Sea for Template") searches the sea of programs, and the SST function finds a match, the global variable MATCHED-TREE is bound to the entire matching program.

If the template of the ADD function of the program being executed successfully matches the top part of a program in the sea, the ADD function has the side effect of substituting the entire matching program into the program being executed. Specifically, the entire matching program replaces the ADD being executed and the entire argument subtree of the ADD (i.e., not just the template of the ADD). In other words, the entire matching program is permanently incorporated into the physical structure of the program being executed, so the physical size of the program being executed usually increases as a consequence of the ADD function. The inserted program is not executed at the time of its agglomeration. However, since it has now become part of a larger program, its parts may be executed on the next occasion when the program as a whole is executed. The execution of the ADD function is finished as soon as a single match and substitution occurs. If an ADD function cannot find a match, the search fails and the ADD has no side effect.

The ADD function is the basic tool for creating structures capable of more complex behavior. The ADD function operates by agglomerating free-floating material from the sea. In the initial generation of the process, this agglomerative process merely combines programs that were initially in the sea; however, in later generations, this process combines programs that are themselves the products of earlier combinations. This agglomerative process thereby creates, over time, large structures that contain a hierarchy of smaller structures. In nature, when the active receptor sites of two freely moving biological molecules match (in a complementary way), a chemical reaction takes place to which both entities contribute. The specific reaction generally would not have taken place had they not come together. Thus, there is some resemblance between the operation of the ADD function and the biochemical reactions that occur in nature. As will be seen, the physical growth by means of agglomeration yields a growth in functionality from generation to generation.

The number of ADD or SST functions in the program being executed generally increases as execution of ADDs incorporate programs into the program being executed. We call these new occurrences of the ADD or SST functions *acquired* functions.

The function LR ("Load REGISTER") works in conjunction with the global variables MATCHED-TREE and REGISTER. This function views the program (i.e., parse tree, LISP S-expression) in MATCHED-TREE as an unparenthesized sequence of symbols. For example, the program (i.e., parse tree)

(SAND (SOR (NOT D1) D0) D2)

is viewed as the sequence

{SAND SOR NOT D1 D0 D2}.

When each function has a specific number of arguments (as is the case here with the function set $\mathcal{F}$), an unparenthesized sequence of symbols is unambiguously equivalent to the original program (i.e., parse tree, LISP S-expression). This view is similar to that employed in the FORTH programming language, where functions are applied to a specified number of arguments, but where parentheses are not used or needed. The function LR removes the first element (i.e., function or terminal) from MATCHED-TREE (viewed as an unparenthesized sequence of symbols) and moves it into the REGISTER. For example, if LR were to act once on the above program, the unparenthesized sequence associated with MATCHED-TREE would be reduced to

{SOR NOT D1 D0 D2}

and the REGISTER would contain SAND.

The function SR ("Store REGISTER") works in conjunction with the global variables REGISTER and OPEN-SPACE. The function SR takes the current contents of REGISTER (which is the name of a single function or terminal) and places it in OPEN-SPACE. REGISTER is not changed by the SR. Thus, an LR and an SR together can move a primitive element of a program from MATCHED-TREE to OPEN-SPACE.

The two-argument iterative operator DWT iteratively evaluates its first argument, ≪WORK≫, until its entire second argument, ≪TRAILING-TEMPLATE≫, matches the contents of MATCHED-TREE. The DWT operator is similar to the iterative looping operator found in many programming languages in that it performs certain work until its termination predicate is satisfied. The DWT operator is terminated by means of a biologically-motivated template-matching scheme.

If MATCHED-TREE contains a program because of the action of a previous SST ("Search Sea for Template"), a DWT ("Do ≪WORK≫ until ≪TRAILING-TEMPLATE≫ matches") that happens to contain both an LR ("Load REGISTER") and an SR ("Store REGISTER") in that order in its first argument subtree will iteratively load the REGISTER with the individual primitive elements of the program in MATCHED-TREE and successively store them into OPEN-SPACE one at a time. This iterative process will continue until the ≪TRAILING-TEMPLATE≫ of the DWT matches the current part of MATCHED-TREE.

If DWT fails to find a subtree that matches ≪TRAILING-TEMPLATE≫ while it is traversing MATCHED-TREE, the contents of OPEN-SPACE are discarded. If the tree placed into the OPEN-SPACE is not syntactically valid, the contents of OPEN-SPACE disintegrate, i.e., are discarded.

When the dwt operator terminates, the string of symbols in OPEN-SPACE is once more should be viewed as the equivalent parse tree.

Because of limited computer time, it is a practical necessity that deeply nested loops be avoided. Therefore, a DWT contained in the first argument of a DWT is not executed. If DWT fails to terminate for any reason within a reasonable number of steps or the symbols placed into OPEN-SPACE by the side effects of the ≪WORK≫ do not form a valid program, the contents of OPEN-SPACE are discarded for that DWT.

If there is an occurrence of a DWT, SST, or ADD function in the second argument subtree of a DWT, the inner function is never used to perform a search; the outer function merely uses it in trying to make a match.

SPONTANEOUS EMERGENCE

We start by creating a sea (population) of 12,500,000 computer programs consisting of independently created random compositions of the eight functions from the function set $\mathcal{F}$ and the three terminals from the terminal set $\mathcal{T}$ described above. Figure 1 shows one of the programs that was actually generated by this random generative process of 12,500,000 programs. The program is shown as a rooted,

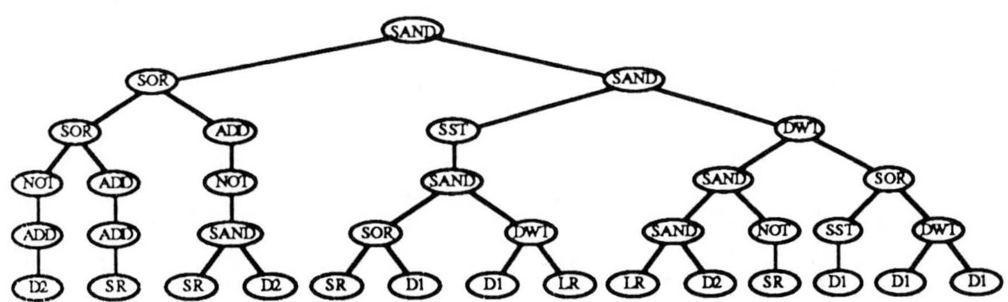

FIGURE 1 Spontaneously emergent 36-point program.

point-labeled tree with ordered branches. In the tree representation of a program, the external points of the tree are associated with terminals (i.e., inputs to the program) or functions taking no arguments (i.e., LR or SR here) and the internal points are associated with functions taking one or more arguments. There are as many lines emanating downward from each internal point as arguments taken by the function. The particular program shown has 36 points (i.e., functions and terminals).

This 36-point program tree can be presented in "prefix" form as used in the Common LISP programming language:

(SAND (SOR (SOR (NOT (ADD D2)) (ADD (ADD (SR))))
(*AND* (*NOT* (*SAND* (*SR*) *D*2))))
(SAND (*SST* (*SAND* (*SOR* (*SR*) (*D*1) (*DWTD*1(*LR*))))
(DWT **(SAND (SAND (LR) D2) (NOT(SR)))**
(SOR (SST D1) (DWT D1 D1)))))).

When this 36-point program is executed with one of the eight combinations of the three terminals D0, D1, and D2 as its inputs, it produces a certain Boolean value as its output. Each of the three logical functions (NOT, SOR, and SAND) return a specified Boolean output based on their one or two arguments; each of the five side-effecting functions (ADD, SST, LR, SR, and DWT) return T, return NIL, or act as the identity function. When this 36-point program is executed with all eight combinations of the three terminals D0, D1, and smcaps d2 as its inputs, it is discovered to perform the three-argument Boolean function shown in Table 1. We say that this program performs three-argument Boolean rule 204 because 11001100_2 is 204_{10}.

TABLE 1 Truth table for the spontaneously emergent 36-point program.

	D2	D1	D0	Rule 204
0	NIL	NIL	NIL	NIL
1	NIL	NIL	T	NIL
2	NIL	T	NIL	T
3	NIL	T	T	T
4	T	NIL	NIL	NIL
5	T	NIL	T	NIL
6	T	T	NIL	T
7	T	T	T	T

As we will shortly see, this spontaneously emergent 36-point program is a self-replicator. We use the term self-replicator to mean that a program reproduces itself exactly, reproduces itself approximately, reproduces itself mutually as part of a group of two or more programs, or reproduces itself in any of the three forgoing ways with changes due to agglomeration of new material.

The italicized portion of this 36-point program contains an executable ADD function. This ADD function has an argument consisting of four points of which the functions NOT and SAND constitute its template. Each time this program is evaluated, the ADD function searches the sea for a program with the function NOT at its root and SAND as the point below the root in the leftmost branch. The search starts with the program itself. The only match found within this program is with the template of the argument of the very ADD function being executed. Since this kind of match is disallowed, the result is that no match is found within this program. The search therefore continues to the neighbor of this program in the sea. A match may occur between the template of this ADD (i.e., the NOT and SAND) and some neighboring program in the sea having a NOT at its root and a SAND immediately below its root (i.e., having a site consisting of NOT and SAND). If and when this happens, the add function in the first program will perform the side effect of substituting the entire second program into the first program. The five (of the 36) points of the first program consisting of the ADD and the four points below the ADD in the argument subtree of the ADD will be replaced by the entire second program.

Note that the SR function within the subtree argument to this ADD is not executable. Note the second of the two ADDs in the first underlined portion of the program is not executable because its template is too small. Note that the first of these two ADDs is executable, but its search will generally be fruitless since any potential matching program having an ADD at its root would have metamorphosed itself into something else.

The underlined and italicized portion of this program contains an SST subtree that searches the sea with a template of SAND and SOR. The SST function ("Search Sea for Template") searches the sea of programs for a program whose top part (i.e., site) matches the top part (i.e., template) of the argument subtree of the SST function. When an SST function finds a match, the global variable MATCHED-TREE is bound to the entire matching program. Note that the SR and DWT functions within the subtree argument to this SST are not executable.

The first argument to the executable two-argument DWT function on line 4 is in boldface, while the second argument is underlined in boldface. The first argument to this DWT executes both the LR ("Load REGISTER") and the SR ("Store REGISTER") functions, thereby causing the copying, one piece at a time, of the contents of MATCHED-TREE to the REGISTER. The six-item second argument to this DWT is the ≪TRAILING-TEMPLATE≫. Since this DWT function appears at the far right of the program, a match between the trailing template and the contents of MATCHED-TREE occurs precisely on completion of the copying of the entire contents of MATCHED-TREE.

This 36-point program is not the only self-replicator that we found. In fact, we found 604 self-replicators in this random search of 12,500,000 programs, so that the probability of finding a self-replicator from a blind random search of programs composed of terminals from the terminal set $\mathcal{T}$ and functions from the function set $\mathcal{F}$ is about one in 20,695.

The discovery of this 36-point program and the 603 other self-replicators in the 12,500,000 programs verifies that spontaneous emergence of self-reproducing computer programs is possible.

If we had unlimited computer resources, we could proceed using all 12,500,000 programs, however, proceeding with the 12,500,000 programs would be prohibitive with the available computer resources. Moreover, proceeding with the 604 self-replicators would also be prohibitive since they contain many excessively time-consuming and memory-consuming features. Many of the randomly created self-replicators contain multiple occurrences of the ADD function that cause the permanent incorporation of various matching programs into the program being executed, thereby increasing the size of the program at an exponential rate. Moreover, in practice, this growth in program size coincides with multiple occurrences of the DWT function. The combined effect of multiple occurrences of both the ADD and DWT functions in the initial generation is that the available computer memory is almost immediately exhausted. Therefore, we decided to screen the 604 self-replicators so that only less prolific individuals are included in our sea of programs. If we had unlimited computer resources, we would not need to do this screening.

As initial screening criteria, we require that each program be capable of asexual reproduction, have exactly one executable ADD function (i.e., one with a non-trivial template), and that the executable ADD function appears before (i.e., to the left of) exactly one executable DWT function. It was necessary to require that the one executable occurrence of the ADD function appear to the left of the one executable occurrence of the DWT function so that the ADD function would always be executed prior to the DWT function. For purposes of this screening, a function is considered executable provided it is not in the argument subtree of an SST function or an ADD function. The reason is that, if there were only one DWT function in a program and if it were executed before execution of the ADD function, the emission produced by the DWT would necessarily always occur before the substitution caused by the ADD. Consequently, the substitution would never be permanently incorporated into a program for the next generation, and it would be impossible to evolve an improvement requiring more than one generation to evolve. Even with this screening of the initial population, the exponential growth of the programs caused by the ADD function and the large number of emissions produced by the DWT functions kept the entire process on the edge of collapse. After applying these initial screening criteria, there are only eight self-replicators satisfying these initial screening criteria, so that the probability of finding a screened self-replicator with the function set $\mathcal{F}$ is about one in 1,562,500.

The following five additional criteria make a self-replicator even more suitable for our purposes here. First, there is exactly one executable SST function. Second, the template of the one executable ADD and SST function must not refer to a program that would never be present in a population of self-replicators. Third, the ADD, SST, and DWT functions must appear in that order when the LISP S-expression is evaluated in the usual way. Fourth, the executable ADD and SST functions must not be part of the first argument to the DWT function. Fifth, the first argument to the DWT function must have one executable LR and one executable SR function appearing in that order.

If we impose these additional five screening criteria, then there are only two fully screened self-replicators in 12,500,000, so that the probability of finding a fully screened self-replicator is about one in 6,250,000. The 36-point self-replicator shown above is one of these two.

Finding 500 fully screened self-replicators by means of blind random search would require processing about 3,125,000,000 individuals (i.e., 500 times 6,250,000). Moreover, since numerous separate runs will be required in this chapter, many different initial random populations of 500 fully screened self-replicators will be necessary. Even if a single pool of, say, 3,000 replicators were used (with a different random drawing of 500 individuals being made for each particular run), 18,750,000,000 individuals would still have to be processed to yield the desired pool of 3,000. Both of these approaches are totally prohibitive.

The spontaneous emergence of 604 self-replicators, the eight screened self-replicators, and two fully screened self-replicators out of 12,500,000 randomly generated individuals already established that the spontaneous emergence of self-reproducing computer programs is possible. Finding additional self-replicators is merely a matter of computer time. Since the goal now is to perform experiments involving the emergence of evolutionary self-improving behavior, we decided to synthesize a pool of 3,000 self-replicators modeled after the fully screened self-replicators that were actually discovered. In synthesizing the pool of 3,000, we made certain that it contained a wide variety of different sizes and shapes representative of the sizes and shapes of the self-replicators that were actually discovered. No duplicate checking was done.

EVOLUTION OF SELF-IMPROVING BEHAVIOR

Actual biological entities interact with their environment. Each program in the sea is an entity which must grapple with its environment. The determination of whether an individual in the population has learned a task is made by evaluating how well it performs the task for all possible situations that it may encounter in the environment.

The domain of Boolean functions provides a relatively simple domain in which learning can take place. We represent these situations by means of combinations of the environmental sensors (i.e., inputs to the program). Specifically, the environment consists of all $2^3 = 8$ possible combinations of the three Boolean variables D2, D1, and D0. Each program in the population is evaluated once for each of the eight possible combinations of the environmental sensors D2, D1, and D0. If the program performs correctly (i.e., emulates the behavior of the target Boolean function to be learned) for a particular combination of environmental sensors, its emission, if any, will be admitted into the OPEN-SPACE. If the emission then survives random culling (i.e., decimation unrelated to its performance), the emitted entity will become part of the sea for the next generation. For example, if an entity emulates a particular target Boolean function for six of the eight possible combinations of the environmental sensors, it has six opportunities to produce emissions. In contrast, if an entity emulates the target Boolean function for three of the eight possible combinations, it will have only half as many opportunities to emit items. If the individual entity happens to be a self-replicator, it will emit a copy of itself into the OPEN-SPACE if it successfully performs the unknown task for a particular combination of environmental conditions. A self-replicator that performs the unknown task for six combinations of environmental conditions will have twice the chance of becoming part of the sea for the next generation than a self-replicator that correctly performs the task for only three combinations.

The programs that are in the sea in one generation of the population do not automatically get into the next generation. They have a limited life (i.e., one generational time-step). The only way to get into the new generation is by means of a SR function that places symbols of a program into the OPEN-SPACE. In a self-replicator, there is at least one DWT function with appropriate invocations of LR and SR that together copy the program into the OPEN-SPACE. In addition, getting into the next generation requires correctly emulating the performance of a target three-argument Boolean function for one or more combinations of the environmental sensors. In other words, an entity survives only if it correctly grapples with the problems posed by its environment. Fitter programs successfully emulate the target function for more combinations of the environmental sensors and therefore are more likely to appear in the next generation.

Note that there is no explicit numerical calculation of fitness here. Instead, the entities in the sea are independently acting and self-replicating entities. Each entity is presented with the entire spectrum of possible combinations of environmental sensors. If an entity successfully performs the task for a combination of environmental conditions and the entity is also capable of self-reproduction, it has a chance of being admitted into the next generation. Effectiveness in grappling with the unknown environment (i.e., correct performance of the unknown task) is required for an emission to have a chance to be admitted to the next generation. Since the entities have a finite lifetime (one generational time step), the failure of an entity to respond effectively to a combination of environmental conditions means that its emission does not become a part of the population for the next generation. Thus, the requirements for continued viability in the population are both the ability to self-replicate (exactly, approximately, mutually, or with changes due to agglomeration) and the ability to effectively grapple with the environment. In the context of these self-replicating entities in this sea, natural selection operates on individuals in the population that are more or less successful in grappling with their environment.

Because the sea is constantly changing, the neighbor of a given program changes as each combination of the environmental sensors is presented to each program for testing.

If we had unlimited computer resources, we would allow 100% of the emissions to move from the OPEN-SPACE into the sea for any combination of environmental sensors for which the program emulates the target function. However, since our computer resources are limited and we intend to present each computer program in the sea with all combinations of its environmental sensors, we need to randomly cull these emissions. In particular, only 30% of the emissions coming from programs that correctly emulate the behavior of the target function to be learned for a particular combination of environmental sensors are actually admitted to the sea from. This culling is conducted using a uniform random probability distribution and is not related to performance, so an entity that performs better will have a greater chance of having its emissions end up in the sea of the next generation. Because the template size is 2, and because the function set has eight functions, and because the terminal set has three terminals, there are 40 (i.e., 5×8) possible templates. Since the neighbor of a given program changes as each different combination of environmental sensors is presented to a program, it is desirable that the sea contain at least as many possible matches as there are combinations of its environmental sensors (i.e., eight) for the typical template, and preferably more. This suggests a population somewhat larger than 320. In fact, we settled on a population size of 500.

If the population is to consist of about 500 individuals capable of emission and if a randomly created computer program emulates the performance of a target three-argument Boolean function for about half of the eight possible combinations of the three environmental sensors, then there will be about 2,000 emissions as a result of testing 500 individuals as to their ability to grapple with their environment. If the programs in the sea improve so that after a few generations the typical program emulates the performance of a target three-argument Boolean function for about six of the eight possible combinations of environmental sensors, then, the number of emissions rises to about 3,000. When these emissions are randomly culled so as to leave only 30% of their original number on each generation, about 900 remain.

In addition, programs generally acquire additional occurrences of the DWT function when an ADD function makes a substitution, thereby causing the number of occurrences of DWT functions to increase rapidly. These occurrences of acquired functions are identified with the suffix -ACQ herein. The templates of these acquired DWT functions are rarely helpful or relevant to self-replication. If we had unlimited computer resources, we could retain all of these acquired DWT functions.

In practice, we heavily cull the emissions produced by these acquired DWT functions so that only 0.01% of the emissions from programs that correctly emulate the behavior of the target function are admitted to the sea.

The population is monitored with the goal of maintaining it at 500. The culling percentages selected for this problem are intended to initially yield somewhat more than 500 emissions. If the number of emissions after this culling is still above 500, the remaining emissions are randomly culled again so that exactly 500 emissions are placed into the sea. Occasionally, the sea contains fewer than 500 programs. If we had unlimited computer resources, there would be no need to carefully control the size of the population in this way.

The ADD function bears some resemblance to the crossover (recombination) operation of genetic programming. The two operations are similar in that they permanently incorporate a program tree from one program into another program. They differ in four ways. First, the ADD function here produces only one offspring. Second, the sea changes as each new combination of environmental sensors is considered, so the matching program is generally not constant between such combinations. Third, there is no random selection of a crossover point within the program; the entire matching program is always inserted into the program being executed. Fourth, the incorporation of genetic material into the program is initiated from inside the program itself. The program being executed is active, and its code causes the incorporation of genetic material into itself and causes the reproduction of the program into the next generation.

Computer memory as well as computer time was limited. The initial sea containing the desired 500 fully screened self-replicators would have come from a run containing about 3,124,999,500 programs that were not fully screened self-replicators. Note that programs that are not even self-replicators may nevertheless contain numerous occurrences of the DWT function and may produce numerous emissions. Since we did not have sufficient memory to handle these 3,124,999,500 other programs, much less their emissions, and we did not have sufficient computer time to execute 6,249,999 useless programs for each potentially interesting program, we started each run with 500 fully screened individuals.

In implementing the ADD function, the COPY-TREE function is used to make a copy of both the matching program and the program being executed. Only a copy of the matching program is permanently incorporated into a copy of the program being executed, thus permitting the unchanged original program to be exposed to all possible combinations of its environmental sensors within the current generation and permitting the matching program to interact more than once with the program being executed on the current generation. If we had unlimited computer resources, the sea could be sufficiently large and rich to permit a matching program to be devoured by the program into which it is incorporated. In that event, essentially the same process would occur progressively over many generations, thereby producing some extraordinarily large programs.

Even with both types of culling mentioned above and the full screening of generation 0, our experience is that the growth in the physical size of the programs and the growth in emissions is such that the process can be run for only about six generations. Therefore, it is necessary to find a task that can be learned in fewer than about six generations with an initial population of size 500, but which is not so simple that it is likely to be produced at random.

Among three-argument Boolean functions generated at random using the function set {AND, OR, NAND, NOR}, Boolean rule 235 lies roughly in the middle of all three-argument Boolean functions in difficulty of discovery by genetic programming.[30] Rule 235 is equivalent to

(OR (EQV D2 D1) D0),

and is a moderately difficult rule to learn because it contains the even-2-parity (i.e., not-exclusive-or) function EQV.

RESULTS OF ONE RUN

We proceed using Boolean function 235 as the environment for the programs in the sea and a pool of 3,000 synthesized programs patterned after the actually discovered, fully screened self-replicators. In one run, one of the individuals in generation 0 of our sea of programs was the following 28-point program, which performs the three-argument Boolean function 250:

(SOR (NOT (SAND (NOT D0) (ADD (NOT (SOR D0 D2)))))

(SAND *(SST (SOR (NOT D2) (ADD D2)))*

(DWT (SOR (LR) (SR))

(SOR (SAND D0 D2) (SAND D0 D0)))))).

The ADD function and its entire argument subtree are underlined in the program above; the ADD has a template of NOT and SOR. The SST function has a template of SOR and NOT. The SST subtree is shown in italics. The DWT function, its first argument (SOR (LR) (SR)), and its second ≪TRAILING-TEMPLATE≫ argument consisting of (SOR (SAND D0 D2) (SAND D0 D0)) are shown in boldface. This 28-point program is a self-replicator, because its SST function has a template consisting of SOR and NOT, thereby matching its own root and causing MATCHED-TREE to be bound to itself. The ≪TRAILING-TEMPLATE≫ of the DWT function matches itself, thereby causing the copying of the entire program into the sea (subject to culling) for the next generation for those combinations of environmental sensors that emulate rule 235.

In Figure 2 we graphically depict this 28-point program.

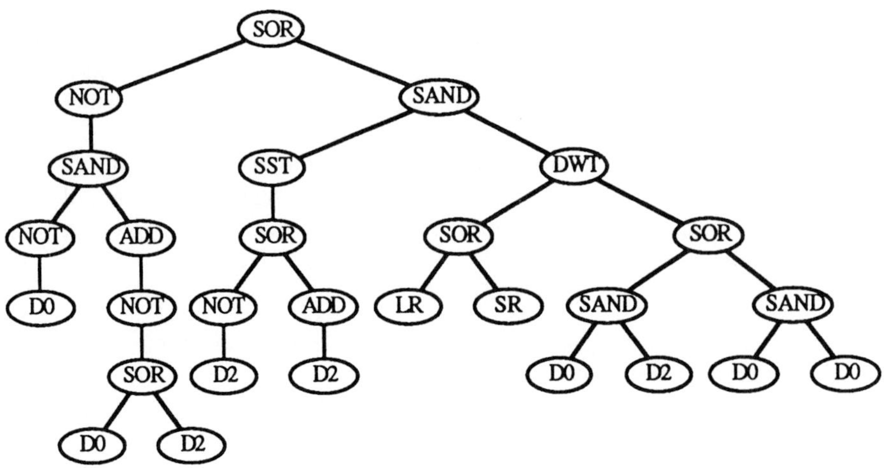

FIGURE 2 28-point self-replicator performing rule 250 from generation 0.

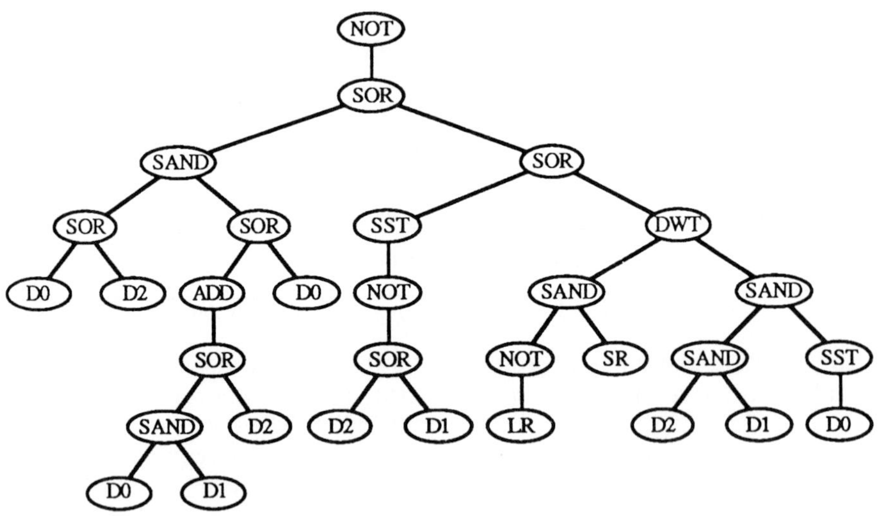

FIGURE 3 31-point self-replicator performing rule 004 from generation 0.

Note that a human programmer could write a much smaller and simpler self-replicator composed of the functions from the function set $\mathcal{F}$ and terminals from the terminal set $\mathcal{T}$. However, the random compositions that spontaneously emerge in a blind random search of this space of compositions are unlikely to be minimal in size.

The first neighbor in the sea in generation 0 that matched the template of NOT and SOR of the ADD function was the following 31-point program, which performs the three-argument Boolean rule 004:

> (**NOT** (**SOR** (SAND (SOR D0 D2) (SOR (ADD (SOR (SAND D0 D1) D2)) D0)) (SOR (SST (NOT (SOR D2 D1))) (DWT (SAND (NOT (LR)) (SR)) (SAND (SAND D2 D1) (SST D0)))))).

This 31-point program is also a self-replicator whose root is NOT and whose point immediately below and to the left of the root is SOR, as shown in boldface above.

In Figure 3 we graphically depict this 31-point program.

The side effect of the execution of the ADD function in the 28-point individual is that the following 54-point individual, which performs the three-argument rule 251, is created:

> (SOR (NOT (SAND (NOT D0) **(NOT (SOR (SAND (SOR D0 D2) (SOR (ADD (SOR (SAND D0 D1) D2)) D0)) (SOR (SST (NOT (SOR D2 D1))) (DWT-ACQ (SAND (NOT (LR)) (SR)) (SAND (SAND D2 D1) (SST D0)))))))**) (SAND (SST (SOR (NOT D2) (ADD D2))) (DWT (SOR (LR) (SR)) (SOR (SAND D0 D2) (SAND D0 D0))))).

The 31-point subtree in boldface (with a NOT at its root and with SOR immediately below and to the left of the root) was inserted by the ADD function from the 28-point individual. The 28-point individual lost 5 points and gained 31 points, thus producing the 54-point individual. Note that this new 54-point individual has an acquired DWT-ACQ function that came from the 31-point individual.

In this run, this 54-point individual eventually enters the sea and becomes part of generation 1.

In Figure 4 we graphically depict this 54-point program.

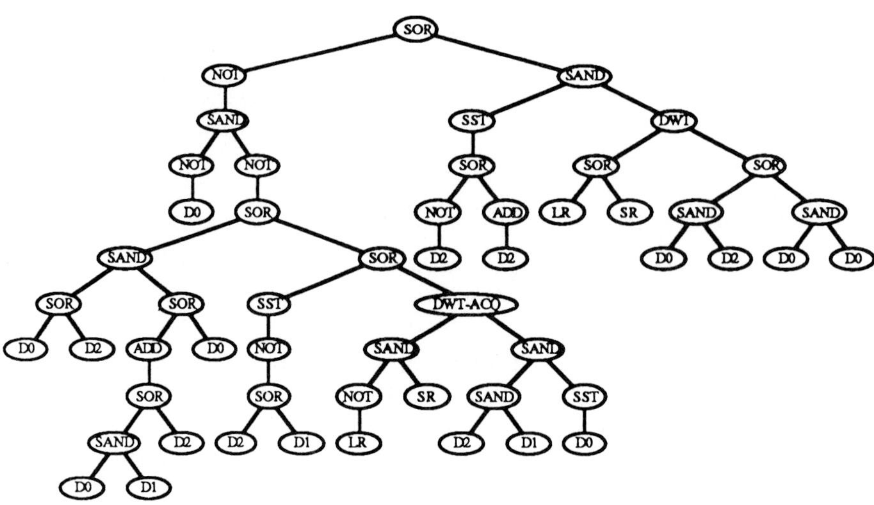

FIGURE 4 54-point self-replicator performing rule 251 from generation 1 with sor at its root and with NOT immediately below and to the left of the root.

Another of the individuals in generation 0 of our sea of programs was the following 41-point program, which performs the three-argument Boolean function 238:

> (SOR (SAND (NOT **(ADD (SOR D0 (SAND D2 D0))**)) (SAND (NOT (SOR (SAND D0 D1) D2)) (SST (SOR (SAND D1 D2) D0)))) (SOR D1 (DWT (SOR (SAND (NOT D0) (NOT (LR))) (SAND D0 (SR))) (SOR (SOR D0 D1) (SAND D2 D0))))).

The ADD function and its argument subtree in this program is shown in boldface. Its template consists of SOR and D0. As it happens, there is no program in the sea that matches this template. However, because this individual is a self-replicator that emulates rule 235 for some combinations of the environmental sensors, this 41-point individual is copied (subject to culling) into the sea of generation 1.

In Figure 5 we graphically depict this 41-point program.

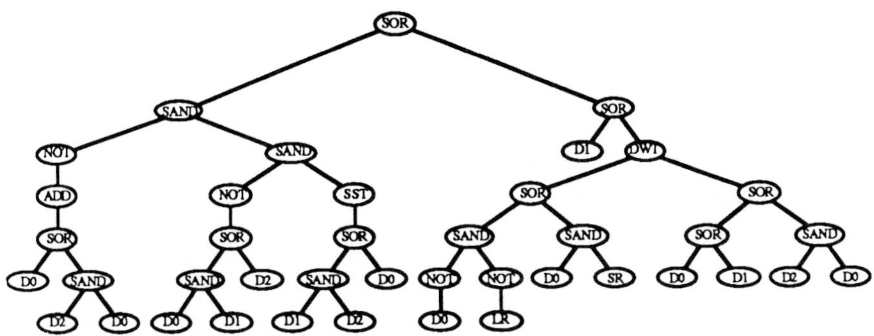

FIGURE 5 41-point self-replicator performing rule 238 in generation 0.

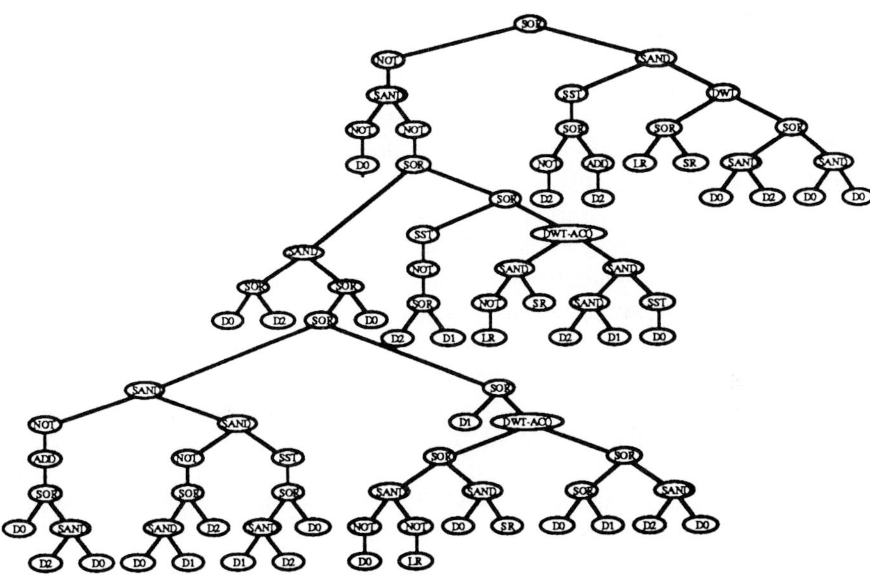

FIGURE 6 89-point self-replicator performing rule 235 in generation 2.

The above 54-point individual contains an ADD function with a template of SOR and SAND. When it was evaluated during generation 1, it encountered the 41-point individual (which has SOR at its root and SAND immediately below and to the left of the root) as a neighbor. Therefore, the ADD template of the above 54-point program matched the 41-point neighbor. The side effect of the match is the creation of an 89-point individual (i.e., 54 points plus 41 points minus 6 points for the subtree argument of the ADD function being executed). This 89-point program performs rule 235:

(SOR (NOT (SAND (NOT D0) (NOT (SOR (SAND (SOR D0 D2) (SOR **(SOR (SAND (NOT (ADD (SOR D0 (SAND D2 D0)))) (SAND (NOT (SOR (SAND D0 D1) D2)) (SST (SOR (SAND D1 D2) D0)))) (SOR D1 (DWT (SOR (SAND (NOT D0) (NOT (LR))) (SAND D0 (SR))) (SOR (SOR D0 D1) (SAND D2 D0)))))** D0)) (SOR (SST (NOT (SOR D2 D1))) (DWT (SAND (NOT (LR)) (SOR (NOT D2) (ADD D2))) (SR)) (SAND (SAND D2 D1) (SST D0)))))))) (SAND (SST (DWT (SOR (LR) (SR)) (SOR (SAND D0 D2) (SAND D0 D0))))).

The 41-point subtree that was inserted by the ADD function is shown in boldface in the above program.

This 89-point individual performing three-argument Boolean rule 235 eventually enters the sea for generation 2.

In Figure 6 we graphically depict this 89-point program.

Figure 7 is a genealogical tree showing the three grandparents from generation 0, the two parents from generation 1, and the final 89-point individual performing rule 235 in generation 2 from Figure 9. The 28-point grandparent performing rule 250 comes from Figure 2. The 31-point grandparent performing rule 004 comes from Figure 3. The 54-point parent performing rule 251 comes from Figure 4.

In Figure 8 we show, by generation, the progressive improvement for the run in which the 89-point self-replicator performing rule 235 was found in generation 2. In this figure, standardized fitness is the zero-based (i.e., zero is best) measure of fitness based on the number of errors (out of eight) made by a particular program. In particular, it shows the average standardized fitness of the population as a whole and the value of the standardized fitness for the best-of-generation individual and the worst-of-generation individual.

In Figure 9 we show, by generation, the average structural complexity of the population as a whole and the values of structural complexity for the best-of-generation individual. Because of the way that the ADD function incorporates entire individuals in the population within existing individuals, structural complexity rises very rapidly.

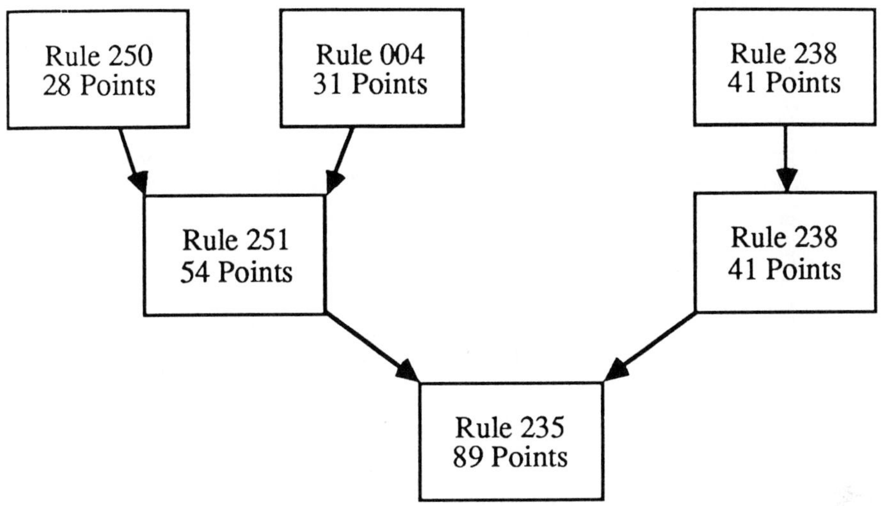

FIGURE 7 Genealogical tree for the 89-point self-replicator performing rule 235 in generation 2.

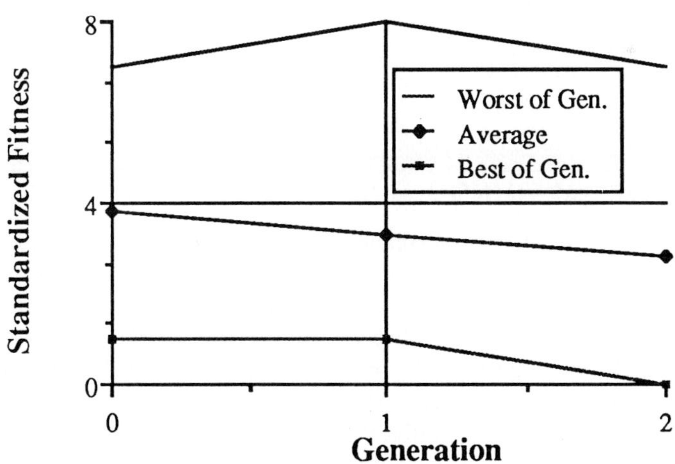

FIGURE 8 Fitness curves.

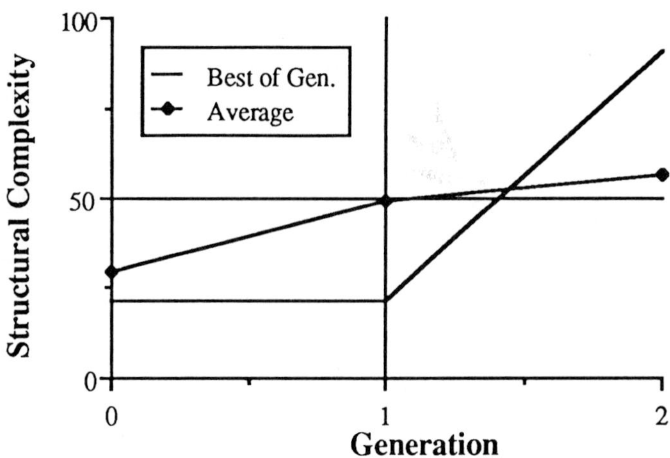

FIGURE 9 Structural complexity curves.

In Figure 10 we show the hits histograms for generations 0, 1, and 2 of this run. One can see that the population as a whole has improved over even the small number of generations shown. The arrow points to the two individuals scoring 8 hits on generation 2. In this figure, "hits" is the fitness measure based on the number of correct responses made by a program to the combinations of environmental inputs it experiences.

RESULTS OF 242 RUNS

In Figure 11 we present the performance curves showing, by generation, the cumulative probability of success $P(M, i)$ and the number of individuals that must be processed $I(M, i, z)$ to yield, with 99% probability, at least one program producing the correct value of rule 235 for all eight combinations of environmental sensors. The graph is based on 242 runs. The population was 500 and the maximum number of generations to be run was 2. The rising curve in Figure 11 shows, by generation, the experimentally observed cumulative probability of success, $P(M, i)$, of solving the problem by generation i (i.e., finding at least one program in the population which produces the correct value for all combinations of environmental sensors). The falling curve shows, by generation, the number of individuals that must be processed, $I(M, i, z)$, to yield, with $z = 99\%$ probability, a solution to the problem by generation i. $I(M, i, z)$ is derived from the experimentally observed values of $P(M, i)$ and is the product of the population size M, the generation number i, and the number of independent runs $R(z)$ necessary to yield a solution to the problem with probability z by generation i. The number of runs $R(z)$ is, in turn, given by

$$R(z) = \left\lceil \frac{\log(1 - z)}{\log(1 - P(M, i))} \right\rceil,$$

where the brackets indicate the ceiling function or rounding up to the next highest integer. The cumulative probability of success, $P(M, i)$, is 2.9% by generation 1 and 4.5% by generation 2. The numbers in the oval indicate that if this problem is run through to generation 2, processing a total of 148,500 (i.e., 500 $\times$ 3 generations $\times$ 99 runs) individuals is sufficient to yield a solution to this problem with 99% probability.

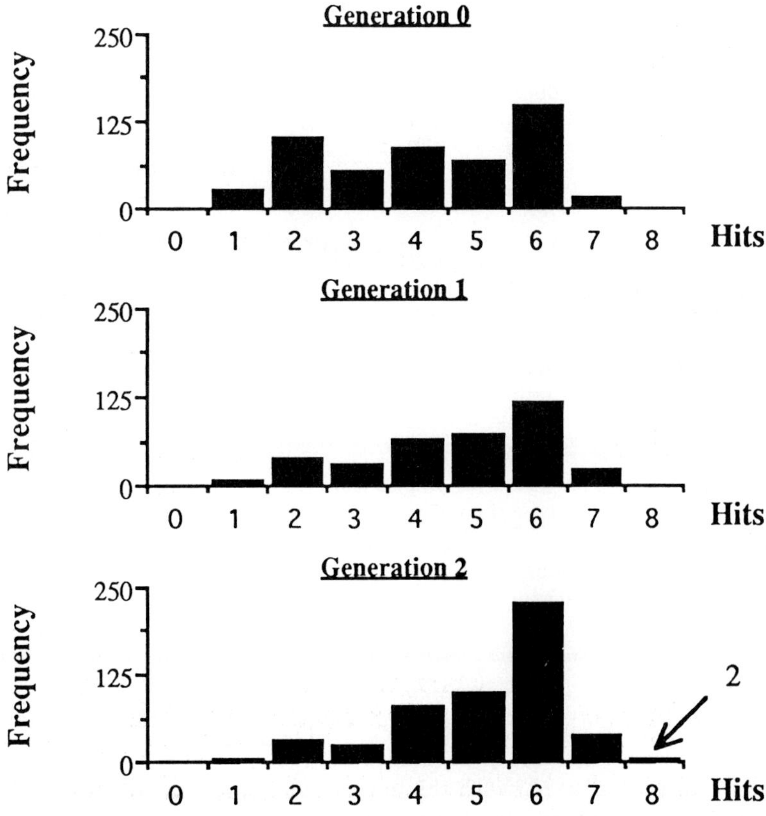

FIGURE 10 Hits histograms for generations 0, 1, and 2.

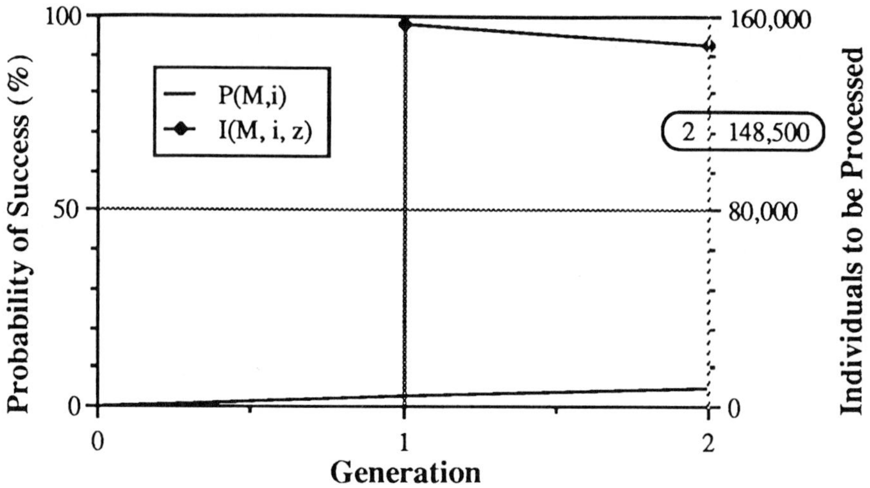

FIGURE 11 Performance curves show that 148,500 individuals must be processed to yield a solution to the problem of learning rule 235.

TABLE 2 Summary.

Probability of a program performing rule 235	1:33,177
Number of individuals that must be processed to find a program performing rule 235 with 99% probability	158,251
Probability of a self-replicator	1:20,695
Probability of a screened self-replicator	1:1,562,500
Probability of a fully screened self-replicator	1:6,250,000
500 times probability of a fully screened self-replicator	3.125×10^9
Number of self-replicators that must be processed to evolve a self-replicator performing rule 235 with 99% probability	148,500

When we tested rule 235 with the function set $\mathcal{F}$, we found that it has a probability of about one in 33,177 of being generated at random. This means that the processing of 158,251 individuals is sufficient to yield a random program realizing rule 235 with 99% probability. Thus, for a population size of 500, we would be very unlikely to encounter a solution to rule 235 merely as a consequence of blind random search on any one run. Thus, the selection of rule 235 as the "environment" was a reasonable one. The 148,500 individuals required to be processed here is less than the 158,251 individuals that must be processed to find an program performing rule 23 at random. It is interesting that rule 235 can be learned with this agglomerative ADD function.

SUMMARY

In summary, in spite of a number of simplifications and compromises necessitated by the limited available computer resources, we have demonstrated the spontaneous emergence of self-replicating computer programs, sexual reproduction, and evolutionary self-improving behavior among randomly created computer programs using a computationally complete set of logical functions.

In Table 2 we summarize the key features of this experiment.

The above experiment is well within the range of currently available computational resources. The fact that spontaneous emergence can occur with a probability of the order of 10^{-6} to 10^{-9} with a function set such as $\mathcal{F}$ suggests that many fruitful experiments on spontaneous emergence and evolutionary self-improvement can be conducted at this time.

ACKNOWLEDGMENTS

James P. Rice of the Knowledge Systems Laboratory at Stanford University made numerous contributions in connection with the computer programming of the above. Simon Handley made helpful comments on the above.

REFERENCES

1. Arbib, Michael A. "Simple Self-Reproducing Universal Automata." *Infor. & Control* **9** (1966): 177–189.
2. Bagley, Richard J. and J. Doyne Farmer "Spontaneous Emergence of a Metabolism." In *Artificial Life II*, edited by C. G. Langton, C. Taylor, J. Doyne Farmer, and S. Rasmussen, 93–140. Santa Fe Institute Studies in the Sciences of Complexity, Proc. Vol. X. Redwood City, CA: Addison-Wesley, 1991.
3. Bratley, Paul, and Jean Millo "Computer Recreations: Self-Reproducing Programs." *Software Practice and Experience* **2** (1972): 397–400.

4. Burger, John, David Brill, and Filip Machi "Self-Reproducing Programs." *Byte* **5** (1980): 72–74.
5. Burks, Arthur W. *Theory of Self Reproducing Automata* Urbana, IL: University of Illinois Press, 1966.
6. Burks, Arthur W. *Essays on Cellular Automata* Urbana, IL: University of Illinois Press, 1970.
7. Burks, Arthur W. "von Neumann's Self-Reproducing Automata." In *Papers of John von Neumann on Computing and Computer Theory,* edited by Wm. Aspray and A. Burks, 491–552. Cambridge, MA: MIT Press, 1987.
8. Buss, Leo W. *The Evolution of Individuality.* Princeton, NJ: Princeton University Press, 1987.
9. Codd, Edgar. F. *Cellular Automata.* New York: Academic Press, 1968.
10. Codd, Edgar F., Private communication, June 12, 1992.
11. Dawkins, Richard. *The Blind Watchmaker.* New York: W. W. Norton, 1987.
12. Devore, John, and Ronald Hightower *The Devore Variation of the Codd Self-Replicating Computer.* Presented at the Third Workshop on Artificial Life, Santa Fe, New Mexico. Draft: November 30, 1992.
13. Dewdney, A. K. "In a Game Called Core War Hostile Programs Engage in a Battle of Bits" *Sci. Am.* **250** (1984): 14–22.
14. Dewdney, A. K. "A Core War Bestiary of Viruses, Worms and Other Threats to Computer Memories." *Sci. Am.* **252** (1985): 14–23.
15. Dewdney, A. K. "A Program Called MICE Nibbles its Way to Victory at the First Core War Tournament" *Scientific American* **256** (1987): 14–20.
16. Dewdney, A. K. "Of Worms, Viruses, and Core War." *Sci. Am.* **260** (1989): 110–113.
17. Dyson, Freeman. *Origins of Life.* Cambridge, MA: Cambridge University Press, 1985.
18. Farmer, J. Doyne and Alletta d'A. Belin "Artificial life: The Coming Evolution." In *Artificial Life II,* edited by C. G. Langton et al., 818–838.[38]
19. Fontana, Walter. "Functional Self-Organization in Complex Systems." In *1990 Lectures in Complex Systems,* edited by L. Nadel and D. L. Stein, 407–426. Redwood City, CA: Addison-Wesley, 1991.
20. Fontana, Walter. "Algorithmic Chemistry" In *Artificial Life II,* edited by C. G. Langton et al, 159–209.[38]
21. Hightower. In *Artificial Life II* video prooceedings, edited by C. G. Langton.[39]
22. Holland, John H. *Adaptation in Natural and Artificial Systems.* Ann Arbor, MI: University of Michigan Press, 1975.
23. Holland, John H. "Studies of the Spontaneous Emergence of Self-Replicating Systems Using Cellular Automata and Formal Grammars." In *Automata, Languages, Development,* edited by A. Lindenmayer and G. Rozenberg,384-404. Amsterdam: North-Holland, 1976.
24. Holland, John H. "ECHO: Explorations of Evolution in a Miniature World," Presented at the Second Workshop on Artificial Life in Santa Fe, New Mexico, February 1990.

25. Holland, John H. *Second Edition of Adaptation in Natural and Artificial Systems.* Cambridge, MA: MIT Press, 1992.
26. Kampis, George *Self-Modifying Systems in Biology and Cognitive Science.* Oxford, UK: Pergamon Press, 1991.
27. Kemeny, John G. "Man Viewed as a Machine." *Sci. Am.* **192(4)** (1955): 58–67.
28. Koza, John R. "Hierarchical Genetic Algorithms Operating on Populations of Computer Programs." In *Proceedings of the 11th International Joint Conference on Artificial Intelligence* Vol. I, 768–774. San Mateo, CA: Morgan Kaufmann, 1989.
29. Koza, John R. "Genetic Evolution and Co-Evolution of Computer Programs." In *Artificial Life II*, edited by C. G. Langton et al., 603–629.[38]
30. Koza, John R. *Genetic Programming: On the Programming of Computers by Means of Natural Selection.* Cambridge, MA: MIT Press, 1992.
31. Koza, John R., and James P. Rice "A Genetic Approach to Artificial Intelligence." In *Artificial Life II*, edited by C. G. Langton. Santa Fe Institute Studies in the Sciences of Complexity, Video Proc. Redwood City, CA: Addison-Wesley, 1991.
32. Koza, John R., and James P. Rice *Genetic Programming: The Movie.* Cambridge, MA: MIT Press, 1992.
33. Laing, Richard "Automaton Introspection." *Computer & System Sci.* **13** (1976): 172–183.
34. Laing, Richard "Automaton Models of Reproduction by Self-Inspection." *Theor. Bio.* **66** (1977): 437–456.
35. Langton, Christopher G. "Self-Reproduction in Cellular Automata." In *Cellular Automata*, edited by J. Doyne Farmer, T. Toffoli, and S. Wolfram, 135–144. Proceedings of an Interdisciplinary Workshop, Los Alamos, NM, March 7-11, 1983. Amsterdam: North-Holland Physics Publishing, 1983. Also in *Physica D* **10** (1984): 135–144.
36. Langton, Christopher G. "Studying Artificial Life with Cellular Automata." In *Evolution, Games, and Learning,* edited by J. Doyne Farmer, A. Lapedes, N. Packard, and B. Wendroff, 120–149. Amsterdam: North-Holland, 1986. Also in *Physica D* **22D** (1986): 120–149.
37. Langton, Christopher G., ed. *Artificial Life.* Santa Fe Institute Studies in the Sciences of Complexity, Proc. Vol. VI. Redwood City, CA: Addison-Wesley, 1989.
38. Langton, Christopher G., Charles Taylor, J. Doyne Farmer, and Steen Rasmussen, eds. *Artificial Life II.* SFI Studies in the Sciences of Complexity, Proc. Vol. X. Redwood City, CA: Addison-Wesley, 1991.
39. Langton, Christopher G., ed. *Artificial Life II.* SFI Studies in the Sciences of Complexity, Video Proceedings. Redwood City, CA: Addison-Wesley 1991.
40. Langton, Christopher G. "Self-Reproducing Loops and Virtual Ants." In *Artificial Life II* video prooceedings, edited by C. G. Langton.[39]

41. Myhill, John "The Abstract Theory of Self-Reproduction." In *Essays on Cellular Automata*, by W. A. Burks, 206–218. Urbana, IL: University of Illinois Press, 1970.
42. Rasmussen, S., C. Knudsen, R. Feldberg, and M. Hindsholm, "The Coreworld: Emergence and Evolution of Cooperative Structures in a Computational Chemistry." In *Emergent Computation: Self-Organizing, Collective, and Cooperative Computing Networks*, edited by S. Forrest, 111–134. Cambridge, MA: MIT Press, 1990. Also in *Physica D* **42D** (1990): 111–134.
43. Rasmussen, S., C. Knudsen, and R. Feldberg, "Dynamics of Programmable Matter." In *Artificial Life II*, edited by C. G. Langton et al., 211–254.[38]
44. Ray, Thomas S. "An Approach to the Synthesis of Life." In *Artificial Life II*, edited by C. G. Langton et al., 371–408.[38]
45. Ray, Thomas S. "Is it Alive or is it GA?" In *Proceedings of the Fourth International Conference on Genetic Algorithms*, edited by R. Belew and L. Booker, 527–534. San Mateo, CA: Morgan Kaufmann, 1991.
46. Ray, Thomas S. "Population Dynamics of Digital Organisms." In *Artificial Life II* video proceedings, edited by C. G. Langton.[39]
47. Schuster, P. "Evolution of Self-Replicating Molecules—A Comparison of Various Models for Selection." In *Dynamical Systems and Cellular Automata* edited by J. Demongeot, E. Goles and M. Tchuente, [need page numbers]. London, UK: Academic Press, 1985.
48. Skipper, Jakob. "The Complete Zoo Evolution in a Box." In *Toward a Practice of Autonomous Systems: Proceedings of the first European Conference on Artificial Life,* edited by F. J. Varela and P. Bourgine, 355–364. Cambridge, MA: MIT Press 1992.
49. Smith, Alvy Ray "Simple Non-Trivial Self-Reproducing Machines." In *Artificial Life II*, edited by C. G. Langton, et al., 709–725.[38]
50. Smith, John Maynard *Evolutionary Genetics.* Oxford, UK: Oxford University Press. 1989.
51. Spafford, Eugene H. "Computer Viruses—A Form of Artificial Life?" In *Artificial Life II*, edited by C. G. Langton et al., 727–745.[38]
52. Thatcher, J. W. "Self-Describing Turing Machines and Self-Reproducing Cellular Automata." In *Essays on Cellular Automata*, edited by A. W. Burks, 132–186. Urbana, IL: University of Illinois Press, 1970.

Larry Yaeger
Apple Computer, Inc., One Infinite Loop, MS 301-4H, Cupertino, CA 95014;
e-mail: larryy@apple.com

Computational Genetics, Physiology, Metabolism, Neural Systems, Learning, Vision, and Behavior or PolyWorld: Life in a New Context

1. INTRODUCTION

The study of living systems has taken many forms, from research into fundamental physical processes to ethological studies of animal behavior on a global scale. Traditionally these investigations have focused exclusively on "real" biological systems existing in our world's ecological system. Only recently have investigations of living systems begun to occur in "artificial" systems in computers and robotic hardware.

The potential benefits of an enhanced understanding of living systems are tremendous. Some are of a grand scale and are intuitively obvious, such as improvements in our ability to manage our own real ecosystems, the development of true machine intelligence, and the possibility of understanding our own mental and physiological processes. Some are of a more prosaic scale, but more accessible thereby, and perhaps of more immediate utility, such as simple learning systems, robust pattern classifiers, general-purpose optimization schemes, robotic controllers, and evolvable software algorithms. The technological issues of the study of Artificial Life (ALife) were well laid out by Langton[27] in the proceedings of the first ALife workshop; the societal and philosophical implications of ALife are well presented by Farmer and Belin[16] in the proceedings of the second ALife workshop.

This paper discusses a computer model of living organisms and the ecology in which they exist called PolyWorld. PolyWorld attempts to bring together all the principle components of real living systems into a single artificial (manmade) living system. PolyWorld brings together biologically motivated genetics, simple simulated physiologies and metabolisms, Hebbian learning in arbitrary neural network architectures, a visual perceptive mechanism, and a suite of primitive behaviors in artificial organisms grounded in an ecology just complex enough to foster speciation and interspecies competition. Predation, mimicry, sexual reproduction, and even communication are all supported in a straightforward fashion. The resulting survival strategies, both individual and group, are purely emergent, as are the functionalities embodied in their neural network "brains." Complex behaviors resulting from the simulated neural activity are unpredictable, and change as natural selection acts over multiple generations.

In many ways, PolyWorld may be thought of as a sort of electronic primordial soup experiment, in the vein of Urey and Miller's[33] classic experiment, only commencing at a much higher level of organization. While one could claim that Urey and Miller really just threw a bunch of ingredients in a pot and watched to see what happened, the reason these men made a contribution to science rather than *ratatouille* is that they put the *right* ingredients in the *right* pot...and watched to see what happened. Here we start with software-coded genetics and various simple nerve cells (light-sensitive, motor, and unspecified neuronal) as the ingredients, and place them in a competitive ecological crucible which subjects them to an internally consistent physics and the process of natural selection. And we watch to see what happens.

Due especially to its biological verisimilitude, PolyWorld may serve as a tool for investigating issues relevant to evolutionary biology, behavioral ecology, ethology, and neurophysiology. The original motivations for its design and implementation, however, were three-fold: (1) to determine if it is possible to evoke complex ethological-level survival strategies and behaviors as emergent phenomena (without their being programmed in), (2) to create *artificial* life that is as close as possible to *real* life, by combining as many critical components of *real* life as possible in an *artificial* system, and (3) to begin exploring Artificial Life as a path toward Artificial Intelligence, utilizing the same key elements that led to natural intelligence: the *evolution* of *nervous systems* in an *ecology.*

This paper will discuss the design principles employed in PolyWorld, the "species" that have evolved in various simulations, and the group and individual behaviors observed in these simulations.

2. BACKGROUND

This work owes much in terms of inspiration to the work of Walter,[46,47,48] Braitenberg,[3] Dawkins,[9,10,11] Holland,[21] Linsker,[29,30,31] and Pearson.[37]

Walter's early work with simple electronic "turtle" nervous systems and Braitenberg's "vehicles" suggested the whole *bottom-up*, synthesis-before-analysis approach, along with the viability of making somewhat arbitrary connections between simple sensory inputs and simple motor controls. PolyWorld (PW) diverges from these works by encapsulating its synthetic organisms in a simulated world, by employing neural systems and learning rules from the world of computational neurophysiology, and by supporting a range of interactions between organisms. And though a number of other researchers (Travers,[45] Wharton and Koball,[49] and even a commercial product from Bascom software, the author of which is not known) have built simple Braitenberg vehicle simulators (or actual physical models in the case of Wharton and Koball), these typically concentrated on a wiring-diagram user interface, and implemented vehicles through only level number 2 (of 14). PW, on the other hand, takes note of the fact that by as early as Vehicle 4, Braitenberg invoked a form of natural selection, and supports the *evolution* of its organisms' "wiring diagrams," rather than having them specified by hand. The neural systems of PW also utilize Hebbian learning during the lifetime of an individual, which is undoubtedly purposefully similar to Braitenberg's "mnemotrix wire."

Dawkins' writings communicate both the beauty and the effectiveness of evolutionary dynamics. In personal communications, he has also brought out key issues in speciation, such as the isolation of populations and the reduced viability of divergent species interbreeding, that have become important elements of this simulator.

The artificial neural systems employed in PW are based on Hebbian learning, and a novel approach to network architecture specification. Besides the obvious importance of Hebb's[18] research and speculations, their instantiation in the work by Linsker and Pearson has guided the selection of these particular techniques for use in PW. Linsker's work demonstrated that Hebbian learning, as employed in PW, can and will self-organize important types of neural response patterns observed in early visual systems of real organisms. Pearson, working with Edelman, utilized a variant of Hebbian learning and successfully demonstrated important principles of neuronal and synaptic self-organization—cooperation and competition (for representing their observed inputs)—that again correspond well to phenomena observed in real living systems. PolyWorld takes this unsupervised learning technique and embeds it in arbitrary, evolving neural architectures, and then confronts the simulated neural system with survival tasks in a simulated ecology.

In the last couple of decades, a number of researchers have developed computational ecologies targeted at various scientific issues. Conrad[7,8] and Packard[35] have built systems to explore fundamental principles of evolutionary dynamics. Jefferson et al.[23] and Collins and Jefferson[6] have constructed systems dealing with evolutionary mechanisms, behavioral goals, and learning architectures (Finite State

Automata vs. Neural Networks). Taylor et al.[42] developed a system to investigate the relationship between individual behavior and population dynamics. Ackley and Littman[1] built such a simulator to demonstrate a novel mechanism by which evolution can guide learning. Todd and Miller[32,43,44] have explored evolutionary selection for different learning algorithms in organisms with simple vision systems and an innate sense of "smell" that functions with varying degrees of accuracy. Hillis[19,20] has used simple computational ecologies to evolve "ramps," and exchange-sort algorithms. Core Wars[13,14,15] is a nonevolving ecology of code fragments, and Rasmussen's VENUS[38] is an evolving system based largely on Core Wars. Ray[39] has also developed a computational ecology, Tierra, based on evolving code fragments. And Koza[26] has developed a system for evolving LISP functions that he terms "Genetic Programming." PolyWorld, in its original conception, was targeted principally at the evolution of neural architectures for systems faced with complex behavioral tasks; however, its biologically motivated behavioral and reproductive strategies, and the evolutionary mechanisms employed, also make it suitable for use in behavioral ecology and evolutionary biology. The extent of PW's fidelity to biological systems, together with its unique use of a naturalistic visual perceptive system to ground its inhabitants in their environment, distinguish it significantly from previous ecological simulators.

Holland's ECHO system explicitly models a form of predation, involving "offense" and "defense" genes that determine the outcome of violent encounters. Holland notes that in his system, this form of predation was essential to the evolution of complex genomes. Though not as crucial to PW's genetic complexity, predation was also designed into PW from the beginning. In PW, genes also affect the outcome of violent encounters between organisms, but more indirectly through their "physiological" characteristics (strength and size). There is also a behavioral component to the outcome of these encounters in PW, namely the degree of "volition" associated with the "fighting" behavior (the activation level of a predefined "fight" neuron), that differs from ECHO's handling of predation.

Belew et al.[2] give an excellent overview of recent work in the area of evolving neural networks. Reviewed briefly there, and presented in detail in their own paper, Harp et al.[17] have developed a scheme for evolving neural architectures that has an element of ontogenetic development. Their approach involves a set of synaptic projection radii between neuronal "areas." PW's scheme for evolving architectures relies on the specification of connection densities and topological distortion of connections between neuronal groups. These architectural criteria are represented in the genome, and then expressed as an organism's neural architecture at "birth." This technique, though perhaps not quite as *developmental* as Harp's approach, or the non-neural yet very biologically motivated cellular growth work of de Boer et al.,[12] has the strengths of being much more developmental (and representationally compact) than a simple encoding of synaptic efficacy in the genes, and being computationally very efficient. It attempts to capture the statistical results of development, without the necessity of modeling the developmental process itself.

Chalmers[4] has experimented with evolving supervised neural network learning algorithms, successfully evolving the classic "delta rule" for a linear, single-layer perceptron, and speculated on applying this "genetic connectionism" approach to other architectures and learning algorithms. He also varies the diversity of his learning tasks, and demonstrates a correlation between this diversity and the generality of the evolved learning algorithm, similar to the correlation observed between amount of training data and generalization in supervised, "Back-Prop" neural networks. Though the evolution of unsupervised learning algorithms is an area of special interest to the author, the current version of PW has the classic "Hebb rule" built in. Neural architectures are, however, evolved in PW. Interestingly, by permitting free movement in a simulated environment, PW effectively can generate an unlimited amount of diverse input for the neural mechanisms employed by its denizens.

Nolfi et al.[34] and Parisi et al.[36] have explored evolving the connection strengths in small, fixed-architecture feed-forward neural networks controlling simple movement strategies in organisms evolved to seek food. The organisms are directly provided with angle and distance to food items, and are alone in their environment. Nolfi and Parisi et al. also introduce a "self-supervised" learning technique, using the traditional back-propagation-of-error algorithm, and demonstrate an improvement in *evolved* foraging efficiency associated with a *learned* ability to predict the sensory consequences of motor activity. PW employs an unsupervised learning algorithm and arbitrary neural architectures, with a more biologically motivated vision mechanism, as well as a competitive ecology.

For the purpose of computer graphics animation, Renault et al.[41] have experimented with visual systems for controlling computer-generated characters. Their system goes beyond visual processing, however, to include unique object identification and distances to objects as part of the input to the character control programs. These control programs are rule-based and completely hand crafted, specifically to provide obstacle avoidance. In contrast, PW uses only the pixel colors associated with visual processing, and provides these as input to the nonrule-based neural systems of evolving organisms, without specifying the meaning or use of this information.

Cliff[5] has implemented a neural visual system for a simulated fly, and states that it is only by a grounding perceptive mechanism such as vision that neural models can be made sense of. For the purposes of his simulation, the model fly is attached to a rotating, but otherwise unmoving test-stand similar to real experimental setups. Organisms in PW use vision as their primary sense mechanism, but are free to explore their environment, and must do so effectively—using their vision to guide a suite of primitive behaviors—in order to survive and reproduce.

One of the first decisions necessary when commencing an investigation into artificial living systems is that of scale: At what level of detail is it desirable to specify the parameters and underlying models of the simulation, and at what level does one wish to observe the resultant behaviors? The study of real living systems has spanned many physical and temporal scales: from molecular-level biochemical processes that take place in nanoseconds, through cellular-level neural processes with time scales of a few milliseconds, to global evolutionary processes occurring over geological time scales. Given current constraints on compute power, it is simply not feasible to begin computation with subatomic physics and expect to observe ethological behaviors. Since ecology-level dynamics were the desired output level of the system being designed, it was clear that behavior models for PW's individual organisms could not be *too* complex. However, a desire to avoid rule-based behavior specification led to a decision to model the organisms' behaviors at the neuronal level. Since even natural evolutionary forces are constrained by their previous successes, the real world has filled up with organisms exhibiting a wide range of variations on assemblages of neuronal cells (in addition to other cell types, of course). Modeling PW's organisms at this level permits us to sidestep millions of years of evolution, while still taking advantage of its results to date.

3. OVERVIEW

PolyWorld is an ecological simulator of a simple flat world, possibly divided up by a few impassable barriers, and inhabited by a variety of organisms and freely growing "food." The inhabiting organisms use vision as input to a neural network brain that employs Hebbian learning at its synapses. The outputs of this brain fully determine the organisms' behaviors. These organisms and all other visible constituents of the world are represented by simple polygonal shapes. Vision is provided by rendering an image of the world from each organism's point of view, and using the resulting pixel map as input to the organism's brain, as if it were light falling on a retina.

A small number of an organism's neurons are predetermined to activate a suite of possible primitive behaviors, including eating, mating, fighting, moving forward, turning, controlling their field of view, and controlling the brightness of a few of the polygons on their bodies. Organisms expend energy with each action, including neural activity. They must replenish this energy in order to survive. They may do so by eating the food that grows around the environment. When an organism dies, its carcass turns into food. Because one of the possible primitive behaviors is *fighting*, organisms can potentially damage other organisms. So they may also replenish their energies by killing and eating each other. Predation is thus modeled quite naturally.

The organisms' simulated physiologies and metabolic rates are determined from an underlying genome, as are their neural architectures. When two spatially overlapping organisms *both* express their *mating* behavior, reproduction occurs by taking the genetic material from the two haploid individuals, subjecting it to crossover and mutation, and then expressing the new genome as a child organism.

One way to look at this artificial world is as a somewhat complex energy balancing problem. The fittest organism will be the one that best learns to replenish its energies by eating, and to pass on its genes by mating. The particular patterns of activity that a successful organism engages in—the methods by which it sustains and reproduces itself—will be optimal for some particular fitness landscape. But since that fitness landscape depends upon the behavior of the world's other inhabitants, it must, per force, be a dynamic landscape. Since there is considerable variation in the placement and behavior of food and other organisms in the world, that fitness landscape is also fundamentally stochastic. Indeed, if the "fittest organism in the world" fails to find a suitable mate in order to pass on the important bits of its genetic material, then those genes will be lost. . .possibly for all time. Accordingly, every world has the potential to be quite different from every other world.

Throughout this paper, the term *created* is applied to organisms spontaneously generated by the system (like the initial seed population), while *born* is used to refer to organisms resulting from the mating behaviors of the organisms. Populations of organisms that have evolved a set of behaviors that allow them to replenish their numbers through *births*, with no further *creations* (after some point in time), are said to exhibit a Successful Behavior Strategy (SBS), or simply to be *successful.* Once an SBS has emerged, there is *no fitness function except survival.* Until an SBS has emerged, PW is run in a sort of "on-line Genetic Algorithm (GA)" mode (also known as a steady-state GA), with an *ad hoc* fitness function. During this stage, a minimum number of organisms may be guaranteed to populate the world. If the number of deaths causes the number of organisms extant in the world to drop below this minimum, either another *random* organism may be *created* by the system, or the offspring of two organisms from a table of the N fittest may be *created* or, rarely, the best organism ever may be returned to the world unchanged (known as an *elitist* strategy in traditional GAs). This *ad hoc* fitness function rewards organisms for eating, mating, living their full life span, dying with reserve energies, and simply moving. Each reward category is normalized by the maximum possible reward in each category, and has a specifiable scale factor to permit easy tuning of the fitness function. Some simulation runs acquire an SBS in the first seed population and never require this on-line GA stage. Others never acquire an SBS and are considered unsuccessful simulations.

Current high-end simulations typically involve over 300 organisms, with approximately 200 neurons each, and require about 13 seconds per time step on a Silicon Graphics Iris 4D/240-GTX. With an average life span of about 500 time steps, and a time-to-first-offspring of about 100 time steps, this means that 500 generations can be run at this complexity in about one week. More modest simulations with around 100 comparable organisms require about 4 seconds per frame, and take a day or two for the same task. And at the low complexity end, simple demonstration worlds can be run in "real time," at a few frames per second, and allow a more interactive experience for learning the system. Color Plate 17 shows a sample view of the PolyWorld environment.

4. GENETICS

An organism's genes completely encode both its "physiology" and its neural architecture. Table 1 lists the full complement of genes present in the organisms of PolyWorld.

All genes are eight bits in length, and may be Gray-coded or binary-coded. All but the ID gene are used to provide eight bits of precision between a specifiable minimum and maximum value for the corresponding attribute. For example, if the minimum possible size is minSize, and the maximum possible size is maxSize, and the value of the size gene (scaled by 255 to lie between 0.0 and 1.0) is valSizeGene, then the size of the organism with this gene will be:

$$\text{size} = \text{minSize} + \text{valSizeGene} * (\text{maxSize} - \text{minSize})$$

These extrema values, along with a variety of other controlling parameters for the simulation, are contained in a "worldfile" that is read by the simulator at startup.

The first eight genes control the organism's simulated physiology. Its size and strength affect both the rate at which it expends energy and the outcome of "fights" with other organisms. In addition, its size is related directly to the maximum energy that it can store internally. The next gene, maximum speed, also affects its "metabolic" rate.

TABLE 1 List of genes in organisms of PolyWorld.

size
strength
maximum speed
ID (green coloration)
mutation rate
life span
fraction of energy to offspring
number of neurons devoted to red component of vision
number of neurons devoted to green component of vision
number of neurons devoted to blue component of vision
number of internal neuronal groups
number of excitatory neurons in each internal neuronal group
number of inhibitory neurons in each internal neuronal group
initial bias of neurons in each noninput neuronal group
bias learning rate for each noninput neuronal group
connection density between all pairs of neuronal groups and neuron types
topological distortion between all pairs of neuronal groups and neuron types
learning rate between all pairs of neuronal groups and neuron types

The ID gene's only function is to provide the green component of the organism's coloration at display time. Since organisms can actually see each other, this could, in principle, support mimicry. For example, a completely passive species could evolve to display the green coloration of a very aggressive species if it were of selective advantage. It might also be possible to attract potential mates by displaying the green coloration of food, though this might be of limited survival value. (In practice, however, neither of these somewhat sophisticated evolutionary responses has yet been observed.)

Mutation rate, the number of crossover points used during reproduction, and maximum life span were placed in the genes in order to permit a kind of meta-level genetics, and in recognition of the fact that these parameters were themselves evolved in natural systems. They are, however, typically constrained to operate within "reasonable" limits; 0.01 to 0.1 for mutation rate, 2 to 8 for number of crossover points, and a few hundred to a few thousand "time steps" for life span.

The final *physiology* gene controls the fraction of an organism's remaining energy that it will donate to its offspring upon birth. The offspring's total available energy at birth is the sum of these contributions from the two parents. Accordingly, at least one aspect of sexual reproduction may be captured by PW's evolutionary "biology": it is entirely possible for two interbreeding subspecies to be almost identical genetically, differing only in the amount of personal energy devoted to the reproductive process. PW has not yet been instrumented to observe for this phenomenon.

The remaining genes are used to define the organism's neural architecture. These control parameters will be discussed in the section on neurons and learning. It should be noted here, however, that one of the motivations for this method of specifying the neural architecture was to reduce the number of genes necessary to specify the neural system. Early versions of PW used a simpler, fully recurrent neural architecture and maintained a complete matrix of synaptic efficacies between all pairs of neurons in the genes. For 200 (N_N) neurons, this older model required 40,000 (N_N^2) genes. The current scheme supports evolving neural architectures that are fully specified by $12N_G^2 + 232N_G + 1026$, where N_G is the number of *internal* neuronal "groups" or clusters (*output* group sizes are fixed to 1, and *input* groups do not need biases, bias learning rates, or incoming synaptic connections). Thus, for 4 internal groups, with up to 32 neurons per group, plus up to 16 neurons per vision group (of which there are 3, one for each color component: red, green, blue), plus 2 other input groups (one neuron per group), plus the standard 7 output groups (one neuron each), a network of up to 185 neurons can be fully specified by just 2,146 genes. The large constants in this equation (232 and 1026) are due to the fixed set of input and output groups, and, especially, the desire to maintain each output neuron as a distinct group. Though the number of specifications are significantly reduced from a full crossbar matrix, this number still heavily outweighs the number of genes devoted to physiology. To permit a more robust exploration of the space of possible physiologies, then, one crossover during genetic reproduction is always forced to occur somewhere within the set of physiology genes. Note that since the minimum number of crossover points is typically set to 2, crossover will also be employed at some point(s) in the neurophysiology genes.

An organism's genome is allocated and interpreted such that space is available for the maximum possible number of neuronal groups. That is, one of the parameters specified per pair of groups in a network with 3 groups out of a maximum of 5 groups would be accessed as:

$$1,1 \quad 1,2 \quad 1,3 \;--\; 2,1 \quad 2,2 \quad 2,3 \;--\; 3,1 \quad 3,2 \quad 3,3 \quad --$$

where the entries marked "–" serve simply as place holders. This is as opposed to an access scheme looking like:

$$1,1 \quad 1,2 \quad 1,3 \qquad 2,1 \quad 2,2 \quad 2,3 \qquad 3,1 \quad 3,2 \quad 3,3$$

where the entries are contiguous. The reason for this is to permit a smoother evolution of these neural architectures. The addition of a fourth group would leave the old connections intact in the first representation, but not in the second. It is even possible for a useful subcomponent of the architecture to ride along dormant in the genes to be expressed at a later time.

Though learning is supported in the neural network model employed in PW, only the architecture and some initial values are encoded in the genes; hence, evolution in PW is purely Darwinian, not Lamarckian.

As in most GAs, when an organism is *created* from scratch, the bits in its genes are first zeroed, and then turned on with a certain bit probability. Unlike most GAs, it is possible to specify a range of legal bit probabilities, rather than always using 0.5. The bit probability for an individual organism is then randomly selected from this range and used to initialize the organism's bit-string genome. So the probability of a bit being on in a particular organism will depend on the value randomly selected from the specified range, while the probability of a bit being on in the population as a whole will just be the mean of the specified range (0.5 if the range is 0.0 to 1.0). This permits a wider variance in local, organism-specific bit probabilities in early populations, rather than depending entirely on mutation and crossover to so shuffle the bits. Whether this is of any real value should be tested in a simpler GA system, and may be problem-specific in any event. Here it was felt that both the older fully recurrent neural network architecture and the later evolving neural architectures were more likely to have behaviorally/evolutionarily useful solutions with lower bit densities; this provided a mechanism for so biasing the initial seed population without ruling out selection towards the unexpected end of the spectrum.

There is an optional "miscegenation function" (so dubbed by Richard Dawkins), that may be used to probabilistically influence the likelihood of genetically dissimilar organisms producing viable offspring; the greater the dissimilarity, the lower the probability of their successfully reproducing. This function is not typically invoked until after a (specifiable) "significant" number of births without an intervening creation in order to allow the early stages of the simulation to explore as many genetic recombinations as possible. It can also be turned off entirely.

5. PHYSIOLOGY AND METABOLISM

As discussed above, the simulated physiology of PolyWorld's organisms is determined by their genes. The *size* of the organism directly affects the maximum amount of energy that the organism can store. If an organism's size is allowed to range between minSize and maxSize, and its energy capacity ranges between minECap and maxECap, then a given organism's actual energy capacity, ECap is given as:

$$\text{ECap} = \text{minECap} + (\text{size} - \text{minSize}) \times (\text{maxECap} - \text{minECap})(\text{maxSize} - \text{minSize}).$$

Similar linear relations are used to determine the influence of an organism's size on the rate at which it expends energy during forward or turning movement (relative to a specifiable maximum-size penalty), and the size advantage it will have during a fight with another organism (relative to a specifiable maximum-size advantage).

An organism's *strength* also affects both its energy expenditure and its advantage in a fight. Strength directly scales the total energy used in a given time step, and thus usually ranges around 1.0 (typically 0.5 to 2.0). An attacker's strength also scales the effect on the victim's energy loss (fighting is discussed in more detail below in the section on behavior).

The energy expended by an organism's neural processing is determined linearly from the number of neurons and the number of synapses it has. A maximum number of neurons and synapses is determined from the control parameters for the entire world, then each individual's neural energy expenditure is computed relative to these maxima. Globally applied "neuron-to-energy" and "synapse-to-energy" conversion factors then multiply these scaled neuron and synapse counts to determine the actual energy expended per time step.

There are similar *behavior*-to-energy conversion factors for each of the primitive behaviors (eating, mating, fighting, moving, turning, focusing, and lighting). The total energy expended in a time step is then the activation (0. to 1.) of the corresponding output/behavior neuron multiplied by that behavior's energy-conversion factor, summed over all behaviors, plus the neural energy expenditure, plus a specifiable fixed energy drain, with this sum finally scaled by the organism's strength.

As should be evident, there are clear energy conservation benefits to being small and weak, yet there are clear predatory advantages to being large and strong. Size also permits an overall greater capacity to store energy, thus making energy available for additional behavioral activity, including reproduction. The interplay between these opposing advantages is intended to produce niches in the fitness landscape, which may change over time. There are similar opposing pressures between energy expenditure and visual acuity on the number of input neurons devoted to vision.

There are two classes of energy storage in each organism: health energy and food-value energy. Both are replenished by eating food. Both are depleted by neural activity and by engaging in the various behaviors. But when an organism is attacked, only its health energy is depleted by the attack. If this health energy reaches zero, the organism dies. When an organism dies, it is converted into a piece of food containing an amount of energy equal to the organism's food-value energy. This separation of health energy from food-value energy makes the predator-prey interactions quite natural; i.e., it is possible for an organism to be killed by having its health energy driven to zero, while still maintaining a relatively high food value for the attacker.

An organism's food-value energy will always be greater than or equal to its health energy, yet both classes of energy have the same maximum capacity. Accordingly, an organism may continue to eat to replenish its health energy after its food-value energy has reached capacity. It is the health energy that is provided as input to the neural network (see next section), and that is used to determine the amount of energy to be transferred to offspring.

Purely for the purposes of display, an organism's length and width are scaled by (the square root of) its maximum speed, length being multiplied, width being divided. Thus faster individuals will appear longer and sleeker, while slower individuals will appear shorter and bulkier. Since an organism's visual acuity is subject to evolutionary pressures, it is conceivable that an organism might emerge that was able to ascertain another organism's maximum speed purely from its shape, if there was a great enough advantage to the acquisition of this information.

6. NEURAL SYSTEMS AND LEARNING

The *inputs* to an organism's neural network "brain" are its "vision," the current normalized level of its internal health-energy store, and a random value. The *outputs* are the suite of seven possible primitive behaviors (eating, mating, fighting, moving, turning, focusing, and lighting). The *internal* neurons and all of the synaptic connections have *no* prespecified functionality; their utility is determined entirely by genetics and natural selection.

The form of an organism's brain, or neural system, is fully characterized by a set of parameters that are encoded in its genes. Referring back to Table 1, notice that the number of neurons devoted to each color component of vision is specified separately, permitting a specialization for more resolution in the most effective color, should this be of selective advantage. These numbers typically range between 1 and 16 neurons per color.

Next is a parameter that specifies the number of internal neuronal groups or clusters. This typically ranges from 1 to 5. In addition, there are 5 input groups (red vision, green vision, blue vision, energy level, and random), plus 7 output groups (the behaviors listed above).

Each neural group may have distinct populations of excitatory (e-) and inhibitory (i-) neurons. The number of e- and i-neurons are specified on a per-group basis, and typically range between 1 and 16 neurons of each type. Synaptic connections from e-neurons are always excitatory (ranging from 0.0 to a specifiable maximum efficacy). Synaptic connections from i-neurons are always inhibitory (ranging from −1.e–10 to the negative of the maximum efficacy).

Though the bias on each of the noninput neurons varies during the simulation, the initial values for these biases and their learning rates are specified on a per-group basis, for each of the noninput neural groups. Biases are updated by a Hebbian learning rule, as if it were a synaptic connection to a neuron that was always fully activated but, unlike other synapses in this network, the bias may change sign. Biases typically range from −1.0 to 1.0, and bias learning rates typically range from 0.0 to 0.2.

The remaining parameters—connection density (CD), topological distortion (TD), and learning rate (LR)—are all specified for each pair of neuronal groups and neuron types. That is, separate values for each of these parameters are specified for the excitatory-to-excitatory (e–e), excitatory-to-inhibitory (e–i), inhibitory-to-inhibitory (i–i), and inhibitory-to-excitatory (i–e) synaptic connections between group i and group j, for each pair of groups i and j.

Connection density, as the name suggests, is used to determine the extent of the connectivity between neuronal groups. The number of e–e synapses between group i and group j is given by the nearest integer to $\mathrm{CD}_{e-e}(i,j) * \mathrm{N}_e(i) * \mathrm{N}_e(j)$, where $\mathrm{CD}_{e-e}(i,j)$ is the e–e CD from group j to group i, $\mathrm{N}_e(i)$ is the number of e-neurons in group i, and $\mathrm{N}_e(j)$ is the number of e-neurons in group j. Similar expressions hold for the other types of connections between all pairs of groups. CD can range from 0.0 to 1.0.

Topological distortion is used to determine the degree of disorder in the mapping of synaptic connections from one group to the next. That is, for a TD of 0.0, synapses are mapped to perfectly contiguous stretches of neurons in the adjacent layer; for a TD of 1.0, synapses are mapped in a completely random fashion between adjacent layers. Thus retinatopic maps such as those observed in natural organisms can be enforced (or not) at the architectural level (as well as resulting from the learning process). TD typically ranges from 0.0 to 1.0.

Learning rate controls the Hebbian learning process at the synapses between each pair of neuronal groups. This permits natural selection to favor hardwired, "instinctive" connections for some neural pathways, while supporting learning in other pathways. LR typically ranges from 0.0 to 0.2.

This method of specifying the neural architecture is fairly general and is not biased for any particular neural organization. Possibly, one might expect to evolve a preponderance of inhibitory connections, especially locally, if the simulated neural architectures evolve to match real neural systems; yet the possibility exists for establishing local excitatory connections (such as are found in CA3 in the hippocampus). The technique does not, however, explicitly model architectures whose characteristics are heavily based upon spatial organization (such as the parallel fibers originating from the granule cells in the cerebellum). A straightforward extension to the current method, that allowed unique specifications of the same parameters along multiple spatial dimensions, could account for such organizational schemes. However, with the limited compute resources currently being applied to PW simulations, and thus the limited number of neurons permitted in each brain, it was not deemed worthwhile to further decompose the groups into these spatial subcategories.

When an organism's brain is "grown" from its underlying genome, the synaptic efficacies are randomly distributed between specifiable minimum and maximum values. The brain is then exposed to a sequence of mock visual inputs consisting of random noise, for a specifiable number of cycles. This is all prior to birth. In this fashion, it is unnecessary to store *any* synaptic efficacies in the genes. This approach was inspired by Linsker's simulations of visual cortex, which gave rise to on-center-off-surround cells, orientation-selective cells, and so on, when exposed only to noise. The crucial aspects of the networks in this case are their architecture—layered receptive fields in Linsker's case, evolved arbitrary topology in PW—and the learning rule—Hebbian learning in both cases.

It was debated whether to update all the organisms' brains *synchronously* or not. That is, whether each organism's neural network should be allowed to make a complete neural activation and synaptic learning pass with each time step. Even though it was desired to penalize organisms that evolved additional neurons and synapses, synchronous updating was ultimately selected, primarily because the corresponding structures in nature are executed in parallel, and penalties based on their serial implementation would be excessive. The penalty is more properly derived from the additional energy use associated with these additional neural structures.

At each time step, the input neurons are set to the appropriate values, corresponding to the organism's visual field, its current health-energy level, and a random number. New neuronal activations are computed by the simple formulae:

$$x_i = \sum a_j^t s_{ij}^t$$

$$a_i^{t+1} = \frac{1}{(1 + e^{-\alpha x^i})}$$

where a_j^t is the neuronal activation of neuron j at time t (the beginning of this time step), s_{ij}^t is the synaptic efficacy from neuron j to neuron i at time t, a_i^{t+1} is the neuronal activation of neuron i at time $t+1$ (the end of this time step), and α is a specifiable logistic slope.

The synaptic efficacies are then updated according to a Hebb rule, as:

$$s_{ij}^{t+1} = s_{ij}^t + \eta_{kl}^c (a_i^{t+1} - 0.5)(a_j^t - 0.5)$$

where s_{ij}^{t+1} is the synaptic efficacy from neuron j to neuron i at time $t+1$, and η_{kl}^c is the learning rate for connections of type c (e–e, e–i, i–i, or i–e) from group l to group k. An optional multiplicative decay term may also be applied to the synaptic efficacy.

This simple "summing and squashing" neuron and Hebbian update rule are certainly coarse abstractions of the complexities observed in real neural systems. Credence is lent to these particular abstractions by the previously quoted simulation work of Linsker, Pearson, and others, and by Linsker's and others' information-theoretic analytical work on such systems, which suggest that they may capture the information-processing attributes of real neural systems, if not their precise method of action. These neuronal and learning models were selected for use in PW based on these results and the models' computational tractability.

During the course of a simulation, neural and synaptic activities may be monitored for a number of organisms in the world (the top five "fittest," according to the *ad hoc* fitness function discussed earlier, even if it is not being used to create new organisms). A few examples are shown in Color Plate 18.

Early simulations with PW had much simpler, fully recurrent neural architectures. Though not particularly representative of real biological neural architectures, acceptable behavior strategies were evolved, and some of the results being presented are from organisms using these early networks.

7. VISION

The color vision supplied as input to the organism is first rendered at the minimum window size permitted on the Iris + 1 (because even-sized buffers can be accessed faster), or 22 × 22 pixels. The pixel row just above vertical center is then properly anti-aliased into whatever number of visual neurons an organism has. Even though organisms and the environment of PW are three-dimensional, the organisms' vision consists of just this one-dimensional strip of pixels, rather than the complete pixel map. Since the organisms are confined to motion on a ground plane, it was felt that the benefit derived from computational efficiency outweighed the small loss of information resulting from this restriction.

As indicated in Color Plate 18 the number of neurons devoted to each of the color components is evolved independently (though they are adjacent on the genome, and so may tend to crossover together).

As was discussed in the Neurons and Learning section, an organism's vision is shown in the display of the brain internals that may be invoked interactively for some of the "fittest" individuals. In addition, the full 22 × 22 pixel map for each of the organisms is usually displayed at the top of the screen. This is mostly for a "reality check"—visual reassurance that the organisms are seeing what they would be expected to see, and may be disabled for a slight speed gain.

The vertical field of view of the organisms is fixed at 10°, since they only see a strip of pixels just above the center of the image. Their horizontal field of view, however, is under their own "volitional," neural control. That is, the activation of the *focusing* neuron is mapped between a minimum and maximum field of view (typically 20° to 120°). In principle, this might permit some depth of field determinations based on cyclic focusing operations, though it is highly doubtful that anything so sophisticated could emerge in the limited neural systems employed by the organisms so far.

This type of direct perception of the environment should answer one of cognitive psychology's most frequently sounded complaints against traditional AI: The organisms of PW are "grounded" in their environment by their sense of vision.

8. BEHAVIOR

A suite of primitive behaviors is made available to all organisms in PolyWorld, namely:

- eating
- mating
- fighting
- moving
- turning
- focusing
- lighting

All of these behaviors are expressed by raising the activation level of a prespecified neuron in the brain. Given computational constraints, it was felt that a minimum number of cycles should be devoted to motor activity, hence this simple one-neuron-one-behavior mapping. The first three behaviors, eating, mating, and fighting, all have some associated threshold that must be exceeded before the activity is initiated. Energy is expended by each of the behaviors, including eating. The energy expenditure rates are controllable by scale factors (see the Physiology and Metabolism section) in the "worldfile" (see The New Context section).

Eating is an organism's method for replenishing depleted energy stores. In order to eat, an organism's position must cause it to overlap a piece of food. The amount of energy consumed is proportional to the activation of the *eating* neuron, once that activation exceeds a specifiable threshold.

Mating is an organism's method for reproducing. In order to reproduce, an organism's position must cause it to overlap another organism, and *both* organisms must express their mating behavior in excess of a specifiable threshold. The outcome of the reproductive attempt may be affected by the miscegenation function (see the Genetics section), or by the maximum number of organisms permitted in the world (see The New Context section). The organism's "desire" to mate (the activation level of its *mating* neuron) is mapped onto its blue color component for display purposes; this coloration is visible to other organisms as well as to human observers.

Fighting is an organism's method for attacking another organism. In order to successfully attack the other organism, the attacker's position must cause it to overlap the attackee. Only *one* organism need express its fighting behavior to successfully attack another. The energy that is depleted from the prey is a function of the volitional degree of the attack (the activation of the predator's *fight* neuron), the predator's current health-energy level, the predator's strength, and the predator's size. The product of these contributing factors from the predator is scaled by a global attack-to-energy conversion factor to make the final determination of the amount of energy actually depleted from the prey. If both organisms are expressing their fight behavior, the same computation is carried out reversing the roles of predator and prey. Note that while this system of fighting/predation will permit every extreme of interaction, including the equivalent of an ant attacking an elephant (or the equally pointless act of an elephant attacking an ant for food), such actions will be of an evolutionarily useful, survivability-enhancing value comparable to those same ridiculous examples of real-world behavior. In addition, since the level of expression of this behavior is under the volitional control of the organisms' nervous systems, a full spectrum from complete pacificity to uninterrupted fighting is possible (and exhibited in the "dervishes" species discussed in the Results section below). Each organism's desire to fight is mapped onto its red color component for display purposes; this coloration is visible to other organisms as well as to human observers.

Moving refers to an organism's forward motion. Unless an organism encounters a barrier, or the edge of the world, it will move forward by an amount proportional to the activation of its *moving* neuron.

Turning refers to a change in an organism's orientation on the ground plane (yaw). An organism will turn about its vertical axis by an amount proportional to the activation of its *turning* neuron.

Focusing refers to an organism's control over its horizontal field of view. As discussed in the Vision section, the activation of an organism's *focusing* neuron will be linearly mapped onto a range of possible angles to provide its horizontal field of view. This makes it possible for an organism to use its vision to survey most of the world in front of it or to focus closely on smaller regions of the world.

Lighting refers to an organism's control over the brightness of a cap of several polygons on the front face of its "body." The activation of an organism's *lighting* neuron is linearly mapped onto the full 0 to 255 brightness range in all color components of these front polygons. Accordingly, a simple form of visual communication is possible, in principle, for the organisms inhabiting PW. (No evidence of their use of this form of communication has yet been found nor sought to date, though evidence of the organisms' use of vision for controlling locomotion has been observed.)

9. THE NEW CONTEXT

The "world" of PolyWorld is a flat ground plane, possibly divided up by a few impassable barriers, filled with randomly grown pieces of food, and inhabited by the organisms previously described.

The number of organisms in the world is controllable by several means. First, a maximum number of organisms is specifiable, in order to keep the problem computationally tractable. Second, a minimum number of organisms is specifiable to keep the world populated during the early on-line GA stage (see the Genetics section). Finally, an initial number of organisms is specifiable to determine how many individuals to seed the world with at the start of the simulation.

Food is grown at a specifiable rate up to a specifiable maximum number of *grown* food items. The number of food items may be guaranteed to be kept between a specifiable minimum and maximum food count. Subject to this maximum, food is also generated as the result of an organism's death. The amount of energy in a piece of food that is grown is randomly determined between a specifiable minimum and maximum food energy. The amount of energy in a piece of food resulting from the death of an organism is that organism's food-value energy (see the Physiology and Metabolism section) at death, or a specifiable minimum food energy at death.

An arbitrary number of barriers may be placed in the world, which inhibit movement of the organisms. These can serve to partially or completely isolate populations of organisms, and as such can contribute significantly to speciation (genetic diversity). For reasons of computational efficiency, they are typically placed parallel to the z-(depth) axis, though this is not strictly necessary.

It is possible to manage the minimum, maximum, and initial numbers of organisms and food items, along with the *ad hoc* fitness statistics, simultaneously for a number of different independent "domains." These domains *must* be aligned parallel to the z-(depth) axis, and typically, though not necessarily, coincide with the divisions imposed on the world by the barriers. This permits the simultaneous "culturing" of completely independent populations when barriers extend the full length of the world, or limits the spread of genes between domains to those resulting from actual movement of organisms when the barriers are arranged so as to leave gaps for organisms to travel through. If the domain fitness statistics were not kept separately, then genes from one domain could migrate to another domain by virtue of their global fitness during the start-up on-line GA phase.

It is possible to set a flag such that the edges of the world act as barriers (the usual), wrap around, or are not there at all. In this last case, PW's ground plane acts much like Braitenberg's *table top*, with organisms that move past the edge of the world dying instantly.

Various monitoring and graphing tools exist to assist in following the progress of a simulation and in developing an understanding of the evolutionary and neural dynamics at work. As was mentioned earlier (in the section on Neural Systems and Learning), a display of the internal workings of any of the five "fittest" organisms may be called up at any time. In addition, a small window that maintains an overhead view of the world will automatically track that same organism upon request. This overhead window may also be zoomed in and out to follow the organism more closely.

Also available are graphic displays of the time histories of certain quantities of interest, including: (1) population sizes (overall and per domain), (2) the past maximum, current maximum, and current average values of the *ad hoc* fitness function, (3) the ratio of the number of organisms "born" (by mating) to the sum of the number of organisms born and created, and (4) the ratio of the difference of food energy *in* and food energy *out* to the sum of these two values. These last two items in particular are important gauges of the course of the simulation. Item (3) will start at 0.0 and asymptote to 1.0 for successful simulations, in which at least one species has emerged with an SBS; it will peak well below 1.0 for unsuccessful simulations. Item (4) ranges from −1.0 to 1.0, and should asymptote to 0.0, for a world where energy is conserved. Three values are actually plotted for item (4): (a) the total food energy, including the initial seeding of the world, which starts at 1.0 and should asymptote to 0.0; (b) the average food-energy, excluding the initial seeding of the world, which starts at 0.0, and rapidly becomes negative, but should also asymptote to 0.0; and (c) the current food energy on a time-step by time-step basis, which fluctuates rapidly, but should cluster around the average food energy.

One additional display can graphically present the results of an analysis of the genetic variability in the population. All pairs of organisms are examined to determine the magnitude of the Hamming distance between them in gene space, normalized by the maximum possible genetic distance between two organisms. These normalized distances are divided into as many distinct histogram bins as there are vertical pixels in the graph, and pixel brightness is used to indicate how many pairs of organisms fell into each bin. A single column of pixels thus shows the distribution of "genetic separation" for the entire population at a single point in time. A new column of pixels is added each time the genetic makeup of the population changes (each birth or death). The result is a complete time history of genetic variability in the world. This approach's strength is that it is able to show such a complete temporal evolution of population-wide genetic variability at a glance. Its weakness is that by reducing genetic differences to a single number, one cannot tell the difference between many genes that are only slightly different and a few genes that are very different; nor is it possible, of course, to tell *which* genes differ.

All of the simulation control parameters and display options are defined in a single "worldfile" that is read at the start of the simulation. In addition, some of the display options can be invoked interactively at run time.

There is not space to go into many details of the code itself. However, it may be worth noting that it consists of about 15,000 raw (not compiled) lines of C++, and is entirely object oriented, *except* for a single routine devoted to handling the organism-organism and organism-food interactions (for reasons of computational efficiency). The organisms, food, and barriers are maintained in doubly linked lists sorted on a single dimension (x). This simple data structure has minimal maintenance overhead, yet rules out most nonintersections very well, and permits a sorting algorithm to be used that capitalizes on the expected frame-to-frame coherency of organism positions. It runs on a Silicon Graphics Iris (to take advantage of its hardware renderer for all the vision processing), and uses a set of object-oriented C++ graphics routines (included in the line count above) that wrap around the standard Iris graphics library.

The tools currently available in PW for tracking population histories, fitnesses, genetic makeup, and so on, are primarily graphical and exist largely for the purpose of monitoring the progress of the simulation. They exist more to help create intuitions than to support detailed quantitative analysis. The only exception to this rule is one particular measure of population-wide genetic diversity (the normalized Hamming distance between all pairs of organisms, discussed above), which can be recorded to a file on demand. The current version of PW does not purport to provide a universal set of tools for all the possible evolutionary biology and behavioral ecology experiments that one might wish to perform with it; it is doubtful whether all such possible tools could even be imagined in advance. This is one of the principle reasons that PW's full source code has been made available, in the hopes that anyone wishing to apply the PW simulation environment to a particular study could add their own data-gathering and data analysis tools. And, hopefully, the object-oriented style used for programming PW will make it relatively easy to add a new graphical analysis tool or to put in hooks to capture the precise data required for a particular study. The author has provided his e-mail address and is willing to help interested researchers navigate in the code to help facilitate their research needs. If time permits, coding help may also be available from the author, and tools added for other studies will be integrated into the baseline code.

10. RESULTS: SPECIATION AND COMPLEX EMERGENT BEHAVIORS

Despite the variability inherent in different worlds, certain recurring "species" have occurred in a number of the simulations run to date. By "species" is meant: groups of organisms carrying out a common individual behavior that results in distinctive group behaviors. Since the selection of these behaviors are derived from the activity of their neural network brains, and the success of these behaviors is partially a function of their physiologies, both of which are in turn based on the genome of the organism, the behavioral differences may generally be traced to the organism's genetic code. Hence, these behavioral differences are representative of different genetic species. No effort has been made to date to quantify or to uniquely identify the genetic differences between these species, due mostly to constraints on the author's time (PW is still very much a work in progress). Examining potentially subtle differences in thousands of genes (or tens of thousands in the older fully recurrent nervous systems) will require some well-designed and well-implemented graphical analysis tools (though some thought has gone into just what such a tool might look like, no code has been written). For now, these ethological-level behaviors (one of the original motivations for building PW, recall) may be the best way to begin developing some understandings and intuitions about the evolutionary dynamics possible in such a system.

A simulation is considered "successful" if and only if some number of species emerge with a Successful Behavior Strategy (SBS); these populations are capable of sustaining their numbers through their mating behaviors, and thus organism *creations* cease. The observational reports below only refer to such "successful" simulations. Accordingly, there are typically many tens to many thousands of examples of any particular behavior, especially in the simpler, more homogeneous species, where all members of the species exhibit approximately the same behavior, and where that species has usually occurred in multiple simulations.

The first of these species has been referred to as the "frenetic joggers." In an early simulation without barriers, without a miscegenation function, and with borders that wrap around (essentially forming a torus), a population emerged that basically just ran straight ahead at full speed, always wanting to mate and always wanting to eat. That particular world happened to be benign enough, in that it turned out they would run into pieces of food or each other often enough to sustain themselves and to reproduce. It was an adequate, if not particularly interesting solution for that world. And without the miscegenation function or any physical isolation due to barriers, whatever diversity was present in the early world population was quickly redistributed and blended into a single species that completely dominated the world for as long as the simulation was run.

The second recurring species has been referred to as the "indolent cannibals." These organisms "solve" the world energy and reproduction problem by turning the world into an almost zero-dimensional point. That is, they never travel very far from either their parents or their offspring. These organisms mate with each other, fight with each other, kill each other, and eat each other when they die. They were most prevalent in simulations run before the parents were required to transfer their own energies to the offspring; the organisms of these worlds were exploiting an essentially free energy source. With proper energy balancing, this behavior was reduced to only an occasional flare-up near corners of the world, where some organisms with limited motor skills naturally end up congregating, sometimes for quite extended periods of time. It turns out that the primary evolutionary benefit associated with this behavior was the ready availability of mates, rather than the "cannibalistic" food supply. This was determined by completely eliminating the food normally left behind by an organism's death, yet still observing the emergence of such species. Large colonies of these indolent cannibals look from above like a continuous (nongridded) version of Conway's game of LIFE.

The third recurring species has been referred to as the "edge runners." These organisms take the next step up from the cannibals, and essentially reduce their world to an approximately one-dimensional curve. They mostly just run around and around the edge of the world (system constraints prevent them from falling off in most simulations). This turns out to be a fairly good strategy, since, if enough other organisms are doing it, then some will have died along the path, ensuring adequate supplies of food. And mates are easily found by simply running a little faster or a little slower, running in the opposite direction, or simply stopping at some point and waiting for other runners to arrive (all of which behaviors have been observed). A form of this behavior persists even when barriers block access to the rest of the world; organisms still sometimes congregate along any edges, including the barriers. It has been suggested[22] that this may be a form of *behavioral isolation*, permitting this species to retain its genetic identity to the exclusion of other species.

Another species recently emerged as the first evolutionarily stable solution to a Braitenberg-style, "table top" world—one with deadly, drop-off edges. These "dervishes" evolved a simple rapid-turning strategy that kept them away from the dangerous edges of the world, and yet explored enough of the world to bring them into contact with food and each other. While this basic behavioral strategy persisted for many hundreds of generations, the dervish populations continued to explore optimum degrees of predation, in a sort of continuous prisoner's dilemma over optimum degrees of cooperation. Indeed, largely uniform behaviors (as indicated by the organisms' coloration) in whole subspecies were suggestive of a Tit-For-Tat strategy; even though variations in predation were being introduced into the communities from their peripheries (or through mutation), whole populations soon tended to adopt the same basic behavior patterns. Waves of varying levels of expression of this fighting behavior could be observed sweeping through several distinct populations, with the greatest variation in behaviors clearly seen at the boundaries between these populations.

The most interesting species and individuals are not so easily classified. In some worlds, individuals' behaviors have been so varied as to preclude any obvious classification into distinct species. In other worlds there appear to be multiple distinct species, with no single species obviously dominating. It is especially in these simulations that a number of complex, emergent behaviors have been observed, including:

1. responding to visual stimuli by speeding up,
2. responding to an attack by running away (speeding up),
3. responding to an attack by fighting back,
4. grazing (slowing upon encountering each food patch),
5. expressing an attraction to food (seeking out and circling food), and
6. following other organisms.

The first item is important in that it implies that conditions have been found that will cause evolution to select for the use of the organisms' vision systems. All four of the earlier, simpler species' behaviors would be appropriate even if these vision systems did not exist. Yet PW was built on the assumption that vision would be a powerful, useful sense mechanism that evolution could not fail to employ. Even a simple speeding-up in response to visual stimulation could result in reaching food or a potential mate more effectively, and this was the first observed visual response to emerge.

The second and third items both represent reasonable responses to attack by a predator. Fleeing may reduce the effect of the attack, and fighting back is an energy-efficient use of the organism's own ability to fight (as opposed to expressing the fight behavior continuously which would expend unnecessary energy).

Strategies four and five represent efficient feeding strategies. As simple a survival skill as grazing might seem—to simply notice when one's internal energy is going up, and cease moving until it stops going up—it was not observed in early simulations. It is still not a wide-spread phenomenon, though a few instances have now been observed. Only the most recent simulation, as of this writing, has given rise to a population of organisms that seem to be able to actively seek out food and "orbit" it while eating; such "foraging" is clearly a valuable survival trait. These organisms appear to be drawn to the food as if there were a magnet or some point attractor located in the food and controlling the organisms' behavior, though no such mechanism exists in PW. Their attraction to the food is purely a result of selection forces acting on the neural architecture connecting their vision systems to their motor systems.

The final, "following" strategy has also emerged only in this most recent simulation. Clearly of value, whether for seeking a prey or a mate, this represents the most complex coupling of the vision sense mechanism to the organisms' motor controls yet observed. Small "swarms" of organisms, and one example of a few organisms "chasing" each other were even suggestive of simple "flocking" behaviors. The "swarm" appears to be a fairly stable formation, persisting for as long as it could be observed in the particular simulation. Having observed the driving forces in the simpler species in PW, this can be reasonably well understood: By staying within the swarm, reproductive partners are readily found (as with the "cannibals"), yet the swarm drifts, allowing its members to find new food sources as the old ones are depleted (more like the "dervishes"). To achieve this limited "flocking" or "swarming" behavior, an attraction to other bright (or moving—the neural architectures of these organisms have not been analyzed, and most definitely should be) objects is all that is required. As with the foraging behavior, this attraction is mediated exclusively through the action of natural selection on the organisms' nervous systems.

Except for the last two behaviors, all of the species and behaviors discussed in this section have been observed in multiple, distinct simulations. And these final two behaviors are exhibited by tens or hundreds of organisms in the single most recent simulation in which they occurred.

All of these behaviors, being inherently temporal phenomena, require some sort of temporal medium for display. Short video clips of most of the above species and behaviors should be available in a companion videotape released by the publisher of this book.

The various species and individual behaviors discussed in this section are necessarily snapshots of a moving target. The PW code has undergone almost continuous development throughout the period of time covered by the simulations discussed above. In particular, and probably significantly, the most complex "foraging" and "following" behaviors were never observed in organisms based on the older, fully recurrent networks. It will not come as a surprise to anyone with even the most limited knowledge of neurophysiology that the architecture of an organism's nervous system is vitally important. The arbitrary neural architectures of the most recent versions of PW appear to offer significant advantages for exploring the space of "useful" nervous systems. And though it would be difficult for anyone, the author included, to reproduce precisely many of the early simulation results (though snapshots do exist of the code in various stages of development along with the "worldfiles" that generated all of the organisms discussed here), it should be quite straightforward for anyone to reproduce the more significant and interesting later results, since both the source code for PW and the worldfiles that define the starting conditions for these simulations have been made freely available (see the Future Directions section below). (Many hours of videotapes and sometimes detailed, sometimes sparse, scientific journals also exist for most of these simulations.)

11. DISCUSSION

Real benefits have already begun to accrue from the studies of artificial neural systems. Meanwhile, the study of artificial evolution—genetic algorithms—is yielding insights into problems of optimization and into the dynamics of natural selection. One form of the study of Artificial Life is the perhaps obvious combination of these two fields of research. Adding computer graphics visualization techniques yields the basic substrate of PolyWorld.

One of the primary goals set out for PW has already been met: the evolution of complex emergent behaviors from only the simple suite of primitive behaviors built into the organisms of PW, their sense mechanisms, and the action of natural selection on their neural systems. These recognizable behavioral strategies from real living organisms, such as "fleeing," "fighting back," "grazing," "foraging," "following," and "flocking," are purely emergent in the PW environment. And built as they are from simple, known primitive behaviors, in response to simple, understandable ecological pressures, they may be able to remove a little bit of the mystery, if not the wonder, at the evolution of such behaviors in natural organisms.

The simple but effective strategies evolved by organisms in the earlier, simpler simulations may be valuable as sort of "null hypotheses" about certain forms of animal behavior. In particular, aggregation and wall-following amongst these simple organisms occurs without need for elaborate behavioral strategies. It is sufficient that corners and walls obstruct the simple trajectories of limited motor skills. Yet if enough organisms occupy these locational niches, it becomes a behavioral niche as well, by providing readily available mates, and an easily achieved form of behavioral isolation.

It is, perhaps, easier to contemplate and understand these behaviors in the simulated organisms of PW than it is in natural organisms, precisely because they are simulated. The blessing and the curse of Artificial Life is that it is much more difficult for humans to anthropomorphize (zoomorphize? biomorphize?) these organisms in a machine than it is natural organisms. This frees us from prejudices and preconceptions when observing and analyzing the behaviors of artificial organisms, yet the most highly motivated of ALife researchers is going to find it difficult to look at an artificial organism and declare it unequivocally alive.

As more and more sophisticated computational models of living systems are developed, it will be only natural to ask whether they are in fact really alive. To answer this, however, requires a resolution to probably the greatest unanswered question posed and addressed by the study of Artificial Life: "What is life?" Farmer and Belin[16] offer an analogous question for consideration: "If we voyage to another planet, how will we know whether or not life is present?" One might also ask: If we "voyage" to an artificial world, how will we know whether or not life is present? In a tentative first step towards trying to answer such a question, Farmer and Belin offer a set of "properties that we associate with life." Here is a brief analysis of how well the organisms of PolyWorld meet these criteria:

- "*Life is a pattern in spacetime*, rather than a specific material object."
  By this, Farmer and Belin mean to point out that even a specific living organism is really the process (that persists), rather than the nutrients or specific chemical constituents in which that process is embedded (which typically do *not* persist). Inherent in this observation is that the process may very well be taking place in something quite different than the traditional hydro-carbon chains associated with biological life (BLife). PW organisms are indeed patterns in a computer, rather than any traditional substrate. And as with BLife organisms, in order to persist they extract nutrients from the environment which, while necessary for life, are not the *process* of life. So PW organisms neither extend nor violate this first condition.
- "*Self-reproduction.*"
  PW organisms certainly reproduce within the context of their world. The initiation of their act of reproduction is, akin to higher level BLife organisms, a result of the processing of a moderately complex nervous system—a "volitional," behavioral action. The method of their reproduction is dependent on the software genetics built into PW—the pseudo-physics and psuedo-biology of PW. This pseudo-biology is not itself an emergent property, but an assumed capability. Whether this, therefore, represents *self*-reproduction or not may be arguable.
  Some might argue that the step-by-step process of reproduction itself must be wholly contained within the simulated physics of an artificial world, perhaps in

the manner of cellular automata models, in order to qualify as *self*-reproduction, and that an assumed capability and corresponding software mechanism for reproduction is too high a level of abstraction to support *real* artificial life (though one might counterargue that the assumed, software physics of such cellular automata systems are just as *ad hoc* a construction as these software genetics, just as unreal, and that even if a model of real molecular-level chemical interactions were used as the substrate, the physics underlying those interactions were *ad hoc*, and so on *ad infinitum*, until we reach levels of subatomic physics that cannot be modeled, since they are not known).

On the other hand, all ALife systems for the foreseeable future *must* make certain assumptions, and select levels of abstraction—levels of organization and complexity—at which they will develop their models. As Chris Langton puts it, they all must write an I.O.U. at some level of detail. The genetic code of PW's organisms is, by design, in the system's software architecture, and the mechanism for combining—crossing over and mutating—those genetic codes, and then interpreting them as a new organism is also embedded in the software that defines PW's physics and biology. It is not obvious that organisms that evolve to exploit these software reproduction mechanisms are any less *alive* than organisms that continue to evolve characteristics and behaviors which, though represented there, are otherwise unrelated to the biochemical mechanisms for the reproduction of DNA. (A longer-necked giraffe is selected for not because of the way its genes are copied, but because its longer neck offers a behavioral advantage; a more intelligent ape descendent is selected for that intelligence, not for the method of its genetic coding. Though a nonevolving genetic code does significantly limit the open-endedness of a simulation, only for organisms sufficiently primitive that the method of storing and reproducing their genetic code is still being explored and optimized are these characteristics an undeniably essential aspect.) Nor is it obvious that a PW organism's dependence on these software reproductive mechanisms is any less lifelike than the dependence that an organism in Ray's Tierra has on a built-in copy instruction, or the dependence that a software virus has on a computer system's built-in duplication functions, or a biological virus's dependence on existing mechanisms for copying its host's DNA.

Finally, a simulator may also be thought of as both the world context it creates *and* the underlying software. Even though the software was hand-crafted, rather than evolved, in a very real sense it represents the fundamental physics and, possibly, the low-level biological mechanisms of the simulated ecology. If a simulated organism is thought of as the combination of its simulated physiology, its simulated neurophysiology, and its real software, then even the high-level, abstracted form of reproduction employed in PW might be thought of as *self*-reproduction. Whether this violates the spirit of Farmer and Belin's criterion or not is unclear.

- "*Information storage of a self-representation.*"

PW organisms use an analog of the same storage mechanism Farmer and Belin mention for natural organisms: their genetic representation. Here again, that representation is stored and interpreted within PW's software context, not directly within the resulting simulated world. The representation itself cannot evolve within the current PW framework (though the author has considered a number of alternative schemes for supporting this), and is not itself an emergent property. These limitations do not violate Farmer and Belin's stated criterion, but do indicate some constraints imposed on the system by its chosen level of abstraction.

- "*A metabolism.*"
  A PW organism's metabolism effectively converts food found in the environment into the energy it needs to carry out its internal processes and behavioral activities, just as is the case in natural organisms. In PW's current biochemistry, there is only one type of nutrient required to sustain life and permit reproduction of its organisms (though this need not be the case in later versions of PW). In contrast to Holland's ECHO, for example, the single food type in PW is converted directly to an energy that is available to all of an organism's behavioral systems, rather than accumulating in a reservoir of components until all the constituent elements are available to support reproduction; given the existence of only a single nutrient type in PW, there is little point in treating it otherwise. But as with ECHO, sufficient quantities of this nutrient must be available in order to support reproduction. And in PW, the energy derived from this nutrient is then expended through every action, as well as the neural processing, of the organism; and the effects of one of those actions—fighting—is directly scaled by this available energy. So in some ways the metabolism of PW's organisms is simpler than that in ECHO, but in some ways it is more complex. In any event, the metabolism in PW organisms is certainly *much* simpler than the metabolism in biological organisms, but if the basic functionality is the same, does the complexity of the underlying process matter?
- "*Functional interactions with the environment.*"
  PW organisms certainly do interact with their environment. Besides eating food and expending the resultant available energy, they interact extensively with other organisms in the environment. In fact, as in BLife, the other organisms comprise the most important element of that environment. The more sophisticated organisms in PW respond behaviorally to changes in the environment, and such responses are purely under the control of the organism.
- "*Interdependence of parts.*"
  Following Farmer and Belin's reasoning, PW organisms can and would die were they somehow separated from their internal energy store. And severing an organism's brain in two would not produce two organisms with behavior anything like the original. As was discussed in the Results section above, the most sophisticated behaviors in PW only emerged once the arbitrary neural architectures were implemented. The particular clusters of neurons in an organism's brain and the precise pattern of connections between those clusters define that organism's behaviors; altering those clusters and their connections produces a

different (and not necessarily viable) organism. Stepping outside the bounds of the simulation, they would also "die" if their various procedures and data were destroyed or isolated. In either case, half an organism is no longer that organism, if it is any organism at all.

- "*Stability under perturbations.*"
  PW organisms can survive small changes to their environment. Indeed, whole species have reemerged in entirely different simulations. Again stepping outside the simulation, whole species have emerged with and without any of a variety of errors in the code.
- "*The ability to evolve.*"
  PW organisms clearly can and do evolve. There are undoubtedly limits to their evolution; e.g., they could not possibly evolve a sense of smell without programmer intervention. However, all natural organisms we know of have limits to their evolutionary capabilities: It is highly unlikely that humans could evolve a steel appendage; if Einstein is correct, it is absolutely impossible for them to evolve a method of personal locomotion that would exceed the speed of light. All organisms, natural or artificial, are bound by the physics of their universe. Similar to the question about metabolism, does the complexity of the underlying physics matter?

So, with the above caveats, questions, arguments and counterarguments, it would appear that the organisms of PolyWorld come surprisingly close to fulfilling Farmer and Belin's set of criteria; indeed, they *may* do so entirely. Farmer and Belin would not argue with the conclusion that we need to refine further our constraints on the definition of life. It is unknown which side they might argue of the knowingly contentious statement that we may already need to welcome a new *genus* to the world.

This issue of just what really defines life—what really is and is not alive—will continue to be both a driving force and a thorn in the side of the field of ALife. Some, including Chris Langton,[28] the father of the field, would argue that perhaps all of Farmer and Belin's criteria must themselves be self-organized. It might also be argued that none of the above criteria are especially necessary or appropriate: Consider a hypothetical, Turing-certified, artificially intelligent computer. It might not be able to reproduce itself. It might not have a complete self-representation. Its metabolism would be no more complicated than that of a PW organism, "eating" nothing but electricity. It may not evolve. Yet if you could discuss Tolstoy and Terminator with it, if it understood the concepts of *noblesse oblige* and the gentleman farmer, and if it, perhaps, shared some of your musical tastes...could you fail to consider it *alive*? But, on the other hand, perhaps this argument unnecessarily conflates *intelligence* and *life*. Or, more damningly, perhaps, like Searle's famous Chinese Room argument, it postulates an impossibility to make its (therefore invalid) point. Indeed, intelligence may only be achievable through a process of evolution—through life, natural or artificial. (Even if one could copy every nuance of a human being's nervous system into a computer, thus sidestepping computational evolution, the thing being copied is itself the product of an evolutionary process.)

Ray[40] has suggested that perhaps all of ALife research is really just modeling *aspects* of life, rather than actually creating "capital-L" *Life*, and each effort, to a large degree, echoes the interests and biases of the individual researcher. Langton's Loops met his criterion at the time—self-reproduction, even though their reproduction is that of a precise crystalline form which cannot evolve. Ray's Tierra organisms met his criteria, too—self-reproduction and evolution, though their behaviors are limited exclusively to the reproductive process. PolyWorld's organisms met their creator's criteria—reproduction, evolution, and ethological-level behaviors, though the intelligence of these organisms is limited, at best, to that found on the lowest rungs of BLife organisms. Perhaps extensions to PolyWorld, or the next researcher's ALife environment, will successfully evolve more intelligent, more obviously *alive* organisms. Or perhaps Langton's Swarm work—an eco-simulator based on a multiplane CA—will reproduce the higher-level behaviors of PolyWorld, but build them on the lower-level physics of CAs. Curiously, manmade life seems almost easy to model, almost impossible to accept as truly alive.

Ultimately, the resolution to this question of real life in artificial organisms may have to be based on a consensus, as with Turing's famous test for artificial intelligence. Perhaps in this case, however, the consensus of a knowledgeable and informed jury is needed, rather than that of Turing's unspecified, presumably average group of individuals. As with the debate about the "aliveness" of natural viruses being properly resident with biologists, the question of "aliveness" in artificial organisms is probably best argued by a combination of computer-aware biologists and biology-aware computer scientists.

Still, researchers in this field seem to have a feeling that if you have a box of stuff, there really ought to be a way to tell if something is alive in there. The best approach may yet be an information-theoretic one, harkening back to the suggestions of von Neumann and Schrödinger that the crucial, defining aspect of living organisms is that they are information processors—information-rich islands in a sea of background information. If, say, the ability to predict the chemical composition or electrical charge or some other measure of state, in either the current location over time, or in adjacent locations at the same time, were measured—at some scale—in a bounded volume, then perhaps we could use such information measures to make an assessment of the amount, or degree, of life contained therein. Or perhaps if such a determination were made at a *variety* of scales within the contained region, living things might stand out clearly from the nonliving background by the manner in which their information content scales. But for now, no such quantitative measures (or even particularly cogent theories) exist, so intelligent conjecture, argument, and opinion will have to suffice.

12. FUTURE DIRECTIONS

The various species and behaviors that have emerged in the different simulations suggest that PW may be a rich enough simulation environment to pursue further evolutionary studies. In particular, a way of "benchmarking" PW—the way one compares the results of a computational fluid dynamics code to known analytical solutions for flow over a flat plate or a cylinder, or measured flows over an airfoil in a windtunnel—may be possible in the form of optimal foraging strategies as studied in the field of behavioral ecology. A simple, canonical foraging experiment has been defined and analyzed, and some preliminary simulations run in PW. Agreement or disagreement with the analytical model should be examined and understood.

The neural architectures that provide the most useful survival strategies should be analyzed and understood. It would also be fairly straightforward to encode an entire range of learning algorithms in the genes of the organisms in PW, and attempt to evolve the most effective learning algorithm, rather than assuming it to be Hebbian. (Some consideration has even been given to the possibility of having the fundamental genetic representation of information—the genetic code—evolve.) At least it might be worthwhile implementing cluster-to-cluster initial connection strengths, initial connection strength variances, and maximum connection strengths, to begin to hint at distinct cell types. Or it may be more worthwhile to jump directly to a more sophisticated cell model, capable of capturing the actual temporal dynamics of spike trains rather than average firing rates.

Though the statistical approach to the specification of neural architectures currently employed in PW can, to a certain extent, finesse the need for an ontogenetic, developmental process, this is felt to be one of the most potentially valuable directions for future ALife work. A richer, more biologically motivated developmental process might provide as significant an improvement in the process of searching the space of evolutionarily useful neural architectures as the current scheme did over the fixed, fully recurrent networks. An ontogenetic process for the organisms' neurophysiologies (and physiologies) might serve both to smooth the fitness landscape and, occasionally, to introduce useful cliffs in that landscape, as it is conjectured to do in natural evolving systems. And the converse may also be true; PW may be a very effective testbed for alternative ontogenetic theories and algorithmic models of development.

More environmental interactions should be supported, including the ability for the organisms to pick up, carry, and drop pieces of food, and perhaps even pieces of barrier material. This should yield useful reasons for organisms to cooperate, other than simply to reproduce.

The simple metabolisms dictated by the current use of a single type of nutrient could readily be replaced with a more complex biochemistry. Multiple nutrients distributed amongst multiple food types would make for a considerably richer environment. (Even the potentially substantial impact of ordered, rather than random, food growth has yet to be explored.) It might even be feasible to incorporate a simple set of internally consistent chemical reactions, with appropriate energetics, catalytic relationships, and so on.

Though not discussed in the earlier parts of the paper, the gross energetics of the system have been observed to be crucial to the evolution of successful survival strategies. Mirroring the differences between energy-rich tropical zones and energy-starved polar zones in our one known, natural ecosystem, artificial life flourishes in energy-rich simulations, and languishes in energy-starved simulations. Perhaps someday it may be possible to make useful predictions about viable ranges of energy flux for natural systems from artificial ecologies like PW.

A quantitative assessment of the degree to which the isolation of populations affects speciation may be possible with PW. Some tentative first steps have already been taken in this direction, though questions remain about the most appropriate comparisons to make and the appropriate times at which to make these comparisons. This coupled with the problems associated with assuring the emergence of an SBS in every population, and the simple magnitude of processing time required to perform the simulations has delayed a complete series of experiments of this nature.

There are thousands of other interesting experiments that one might perform with this system, including: monitoring brain size in otherwise stable populations, such as the "dervishes" (are smaller and smaller nervous systems actually being selected for?); monitoring the frequency and magnitude of attacks on other organisms as a function of their genetic (dis)similarity; monitoring the amount of energy given to offspring in a single species (is there any indication of an asymmetric split into different relative contributions?); hand-tailoring a good neural architecture or two and seeding the world with these engineered organisms); providing multiple internal, neural time cycles per external, action time cycle; and evolving three completely independent domains of organisms, with barriers in place, and then removing the barriers to observe the interspecies dynamics. It may even be possible to model the entire population of Orca whales that frequent the waters around Vancouver, and look for an evolutionary split into pods that travel little and eat essentially stationary food sources versus pods that travel widely and feed on fish, a very mobile food source. And on and on. In hopes that others may find PolyWorld to be a useful tool for exploring these kinds of questions, it has been made available via ftp from ftp.apple.com (130.43.2.3) in /pub/polyworld. Complete source code and some sample "worldfiles" are provided.

In a more fanciful, and perhaps more visionary vein, it is hoped that, someday, one of the organisms in PolyWorld that demonstrates all the survival behaviors observed to date, plus a few others, could be transferred from its original environment to, say, a maze world, and become the subject of some classical conditioning experiments. Klopf's[24,25] success at demonstrating over 15 classical conditioning phenomena in a single neuron using differential Hebbian learning (he called it "drive-reinforcement" learning), strongly suggests that such phenomena should be demonstrable in PolyWorld's organisms.

And then, of course, there is simply "more, bigger, and longer": more organisms, with bigger neural systems, evolving longer. As a *gedanken* experiment, consider just how *much* "more, bigger, and longer" might be useful: The current 3×10^2 organisms, 3×10^2 neurons, and 10^3 generations (approximately) could be expanded to 10^6 organisms, neurons, and generations, through an increase in compute power of about 10^{10}. (Though this sounds like a tremendous increase to ask for, consider that the current simulation is running on a single, scalar workstation processor; not a vectorized, massively parallel processor, then extend today's trends in compute power, and this ceases to be such a daunting request. Also, the compute power needed to model quite complex organisms may be significantly less than this due to the greatly reduced motor and autonomic nervous systems that would be required by artificial organisms.) By coincidence, it turns out that this is a fairly reasonable amount of compute power with which to consider modeling a complete human brain—basically devoting one of today's fast computers to every neuron—but no one understands how to actually construct such an artificial brain. However, this same amount of compute power might be used to evolve the equivalent of several new species of computational lab rats every week...and *this* is how: by combining evolution, neural systems, and ecological dynamics.

If there is any question about why one would wish to pursue these research directions, it is always possible to point to the benefits to be derived in the evolutionary, ecological, biological, ethological, and even computer science fields. But it may also turn out to be the only "right" way to approach machine intelligence.

One view of intelligence is as an evolved, adaptive response to a variable environment, that due to historical constraints and opportunism on the part of nature happens to be based upon neuronal cells. One might further recognize that intelligence is actually a near-continuum—a spectrum from the simplest organisms to the most complex—rather than some singular event, unique to human beings. Then, by utilizing both the method—Natural Selection—and the tools—assemblies of neuronal cells—used in the creation of natural intelligence, PolyWorld is an attempt to take the appropriate first steps toward modeling, understanding, and reproducing the phenomenon of intelligence.

For while one of the grand goals of science is certainly the development of a functioning human level (or greater) intelligence in the computer, it would be an only slightly less grand achievement to evolve a computational *Aplysia* that was fully knowable—fully instrumentable and, ultimately, fully understandable. And perhaps it is only through such an evolutionary approach that it will be possible to provide the important milestones and benchmarks—sea slug, rat, simian,...—that will let us know we are on the right scientific path toward that grander goal.

ACKNOWLEDGMENTS

The author would like to thank Alan Kay and Ann Marion of Apple's Vivarium Program for their support and encouragement of this admittedly exotic research. He would also like to thank Richard Dawkins for early encouragement and suggestions; Steve Nowlan, Terry Sejnowski, and Francis Crick for their interest and kind words; and Chris Langton, Luc Steels, and Tom Ray for some wonderfully thought provoking feedback and discussions. Finally and emphatically, he would like to thank his wife, Levi Thomas, for her support, understanding, and humor (especially for christening PolyWorld's organisms "C-monkeys").

REFERENCES

1. Ackley, D., and M. Littman. "Interactions Between Learning and Evolution." In *Artificial Life II*, edited by C. Langton, C. Taylor, J. Farmer, and S. Rasmussen. Santa Fe Institute Studies in the Sciences of Complexity, Proc. Vol. X. Redwood City, CA: Addison-Wesley, 1992.
2. Belew, R. K., J. McInerney, and N. N. Schraudolph. "Evolving Networks: Using the Genetic Algorithm with Connectionist Learning." In *Artificial Life II*, edited by C. Langton, C. Taylor, J. Farmer, and S. Rasmussen. Santa Fe Institute Studies in the Sciences of Complexity, Proc. Vol. X. Redwood City, CA: Addison-Wesley, 1992.
3. Braitenberg, V. *Vehicles: Experiments in Synthetic Psychology.* Cambridge, MA: A Bradford Book, MIT Press, 1984.
4. Chalmers, D. "The Evolution of Learning: An Experiment in Genetic Connectionism." In *Connectionist Models, Proceedings of the 1990 Summer School*, edited by D. S. Touretzky, J. L. Elman, T. J. Sejnowski, and G. E. Hinton. San Mateo, CA: Morgan Kaufmann, 1991.
5. Cliff, D. "The Computational Hoverfly; a Study in Computational Neuroethology." In *From Animals to Animats, Proceedings of the First International Conference on Simulation of Adaptive Behavior*, edited by J.-A. Meyer and S. Wilson. Cambridge and London: MIT Press, 1991.
6. Collins, R. J., and D. R. Jefferson. "AntFarm: Towards Simulated Evolution." In *Artificial Life II*, edited by C. Langton, C. Taylor, J. Farmer, and S. Rasmussen. Santa Fe Institute Studies in the Sciences of Complexity, Proc. Vol. X. Redwood City, CA: Addison-Wesley, 1992.
7. Conrad, M., and M. Strizich. "EVOLVE II: A Computer Model of an Evolving Ecosystem." *Biosystems* **17** (1985): 245–258.
8. Conrad, M. "Computer Test Beds for Evolutionary Theory." Oral Presentation at Artificial Life I Conference, 1987.

9. Dawkins, R. *The Selfish Gene.* Oxford: Oxford University Press, 1976.
10. Dawkins, R. *The Extended Phenotype: The Gene as a Unit of Selection.* Oxford: Oxford University Press, 1983.
11. Dawkins, R. *The Blind Watchmaker.* New York: W. W. Norton, 1986.
12. de Boer, M. J. M., F. D. Fracchia, and P. Prusinkiewicz. "Analysis and Simulation of the Development of Cellular Layers." In *Artificial Life II*, edited by C. Langton, C. Taylor, J. Farmer, and S. Rasmussen. Santa Fe Institute Studies in the Sciences of Complexity, Proc. Vol. X. Redwood City, CA: Addison-Wesley, 1992.
13. Dewdney, A. K. "Computer Recreations: In the Game Called Core War Hostile Programs Engage in a Battle of Bits." *Sci. Am.* **250(5)** (1984): 14–22.
14. Dewdney, A. K. "Computer Recreations: A Core War Bestiary of Viruses, Worms, and Other Threats to Computer Memories." *Sci. Am.* **252(3)** (1985): 14–23.
15. Dewdney, A. K. "A Program Called MICE Nibbles Its Way to Victory at the First Core War Tournament." *Sci. Am.* **256(1)** (1987): 14–20.
16. Farmer, J. D., and A. d'A. Belin. "Artificial Life: The Coming Evolution." In *Artificial Life II*, edited by C. Langton, C. Taylor, J. Farmer, and S. Rasmussen. Santa Fe Institute Studies in the Sciences of Complexity, Proc. Vol. X. Redwood City, CA: Addison-Wesley, 1992.
17. Harp, S., T. Samad, and A. Guha. "Towards the Genetic Synthesis of Neural Networks." In *Proceedings Third International Conference on Genetic Algorithms*, edited by J. D. Schaffer. San Mateo, CA: Morgan Kaufmann, 1990.
18. Hebb, D. O. *The Organization of Behavior.* New York: John Wiley, 1949.
19. Hillis, D. "Simulated Evolution and the Red Queen Hypothesis." Oral Presentation at Artificial Life II Conference, 1990.
20. Hillis, D. "Co-Evolving Parasites Improve Simulated Evolution as an Optimization Procedure." In *Artificial Life II*, edited by C. Langton, C. Taylor, J. Farmer, and S. Rasmussen. Santa Fe Institute Studies in the Sciences of Complexity, Proc. Vol. X. Redwood City, CA: Addison-Wesley, 1992.
21. Holland, J. "Echo: Explorations of Evolution in a Miniature World." Oral Presentation at Artificial Life II Conference, 1990.
22. Hugie, D. Personal Communication, Behavioral Ecology Department, Simon Fraser University, Vancouver, B.C., Canada, 1992.
23. Jefferson, D., R. Collins, C. Cooper, M. Dyer, M. Flowers, R. Korf, C. Taylor, and A. Wang. "Evolution as a Theme in Artificial Life: The Genesys/Tracker System." In *Artificial Life II*, edited by C. Langton, C. Taylor, J. Farmer, and S. Rasmussen. Santa Fe Institute Studies in the Sciences of Complexity, Proc. Vol. X. Redwood City, CA: Addison-Wesley, 1992.
24. Klopf, A. H. "A Drive-Reinforcement Model of Single Neuron Function: An Alternative to the Hebbian Neuronal Model." In *Neural Networks for Computer*, edited by J. S. Denker. AIP Conference Proceedings 151, New York: American Institute of Physics, 1986.

25. Klopf, A. H. "A Neuronal Model of Classical Conditioning." AFWAL-TR-87-1139, Air Force Wright Aeronautical Laboratories, October, 1987.
26. Koza, J. R. "Genetic Evolution and Co-Evolution of Computer Programs." In *Artificial Life II*, edited by C. Langton, C. Taylor, J. Farmer, and S. Rasmussen. Santa Fe Institute Studies in the Sciences of Complexity, Proc. Vol. X. Redwood City, CA: Addison-Wesley, 1992.
27. Langton, C. G., ed. *Artificial Life.* Santa Fe Institute Studies in the Sciences of Complexity, Proc. Vol. VI. Redwood City, CA: Addison-Wesley, 1989.
28. Langton, C. G. Personal Communication, 1993.
29. Linsker, R. "Towards an Organizing Principle for a Layered Perceptual Network." In *Neural Information Processing Systems*, edited by D. Z. Anderson. New York: American Institute of Physics, 1988.
30. Linsker, R. "Self-Organization in a Perceptual Network." *Computer* **21(3)** (1988): 105–117.
31. Linsker, R. "An Application of the Principle of Maximum Information Preservation to Linear Systems." In *Advances in Neural Information Processing Systems 1*, edited by D. S. Touretzky. San Mateo, CA: Morgan Kaufmann, 1989.
32. Miller, G. F., and P. M. Todd. "Exploring Adaptive Agency I: Theory and Methods for Simulating the Evolution of Learning." In *Connectionist Models, Proceedings of the 1990 Summer School*, edited by D. S. Touretzky, J. L. Elman, T. J. Sejnowski, and G. E. Hinton. San Mateo, CA: Morgan Kaufmann, 1991.
33. Miller, S. M., and L. E. Orgel. *The Origins of Life.* Englewood Cliffs, NJ: Prentice-Hall, 1974.
34. Nolfi, S., J. L. Elman, and D. Parisi. "Learning and Evolution in Neural Networks." CRL Technical Report 9019, Center for Research in Language, University of California, San Diego, La Jolla, CA, 1990.
35. Packard, N. "Intrinsic Adaptation in a Simple Model for Evolution." In *Artificial Life*, edited by C. Langton. Santa Fe Institute Studies in the Sciences of Complexity, Proc. Vol. VI. Redwood City, CA: Addison-Wesley, 1989.
36. Parisi, D., S. Nolfi, and F. Cecconi. "Learning, Behavior, and Evolution." Technical Report PCIA-91-14, Department of Cognitive Processes and Artificial Intelligence, Institute of Psychology, C.N.R. - Rome, June 1991. (To appear in *Proceedings of ECAL-91—First European Conference on Artificial Life*, December 1991, Paris.)
37. Pearson, J. "Competitive/Cooperative Behavior of Neuronal Groups in Brain Function." Oral Presentation at Artificial Life I Conference, 1987. (And in *Neural Darwinism: The Theory of Neuronal Group Selection*, edited by G. M. Edelman. New York: Basic Books, 1987.)
38. Rasmussen, S., C. Knudsen, R. Feldberg, and M. Hindsholm. "The Coreworld: Emergence and Evolution of Cooperative Structures in a Computational Chemistry." In *Emergent Computation*, edited by Stephanie Forrest. A Special Volume. *Physica D* **42** (1990): 1–3.

39. Ray, T. S. "An Approach to the Synthesis of Life." In *Artificial Life II*, edited by C. Langton, C. Taylor, J. Farmer, and S. Rasmussen. Santa Fe Institute Studies in the Sciences of Complexity, Proc. Vol. X. Redwood City, CA: Addison-Wesley, 1992.
40. Ray, T. S. Personal Communication, 1993.
41. Renault, O., N. M. Thalmann, and D. Thalmann. "A Vision-Based Approach to Behavioural Animation." *J. Visual. & Comp. Animation* **1** (1990): 18–21.
42. Taylor, C. E., D. R. Jefferson, S. R. Turner, and S. R. Goldman. "RAM: Artificial Life for the Exploration of Complex Biological Systems." In *Artificial Life*, edited by C. Langton. Santa Fe Institute Studies in the Sciences of Complexity, Proc. Vol. VI. Redwood City, CA: Addison-Wesley, 1989.
43. Todd, P. M., and G. F. Miller. "A General Framework for the Evolution of Adaptive Simulated Creatures." Oral Presentation at Artificial Life I Conference, 1987.
44. Todd, P. M., and G. F. Miller. "Exploring Adaptive Agency II: Simulating the Evolution of Associative Learning." In *From Animals to Animats, Proceedings of the First International Conference on Simulation of Adaptive Behavior*, edited by J.-A. Meyer and S. Wilson. Cambridge and London: MIT Press, 1991.
45. Travers, M. "Animal Construction Kits." In *Artificial Life*, edited by C. Langton. Santa Fe Institute Studies in the Sciences of Complexity, Proc. Vol. VI. Redwood City, CA: Addison-Wesley, 1989.
46. Walter, W. G. *The Living Brain.* New York: W. W. Norton, 1963.
47. Walter, W. G. "An Imitation of Life." *Sci. Am.* **182(5)** (1950): 42–45.
48. Walter, W. G. "A Machine that Learns." *Sci. Am.* **185(2)** (1951): 60–63.
49. Wharton, J., and B. Koball. "A Test Vehicle for Braitenberg Control Algorithms" Oral Presentation at Artificial Life I Conference, 1987.

Inman Harvey
School of Cognitive and Computing Sciences, University of Sussex, Brighton BN1 9QH, England; e-mail: inmanh@cogs.susx.ac.uk

Evolutionary Robotics and SAGA: The Case for Hill Crawling and Tournament Selection

This paper will look at an evolutionary approach to robotics; partly at pragmatic issues, but primarily at theoretical issues associated with the evolutionary algorithms that are appropriate. Genetic Algorithms (GAs) are not suitable in their usual form for the evolution of cognitive structures, which must be done in an incremental fashion. SAGA—Species Adaptation Genetic Algorithms—is a conceptual framework for extending GAs to variable-length genotypes, where evolution allows a species of individuals to evolve from simple to more complex.

In the context of species evolution the metaphor of hill crawling as opposed to hill climbing is introduced, and appropriate mutation rates discussed. On both pragmatic and theoretical grounds, it will be suggested that there are good reasons for using Tournament Selection in evolutionary robotics.

1. WHY EVOLUTIONARY ROBOTICS?

Subsumption-style cognitive architecture for robots[4,5] in theory analyzes independent behaviors of a robot, and "wires them in" largely independently from sensor input to motor output. Later "wired in" behaviors interact with earlier ones either through the environment or by suppression or inhibition mechanisms.

As the number of layers or modules within such an architecture increases, the number of potential interactions increases much faster. The foresight needed to design by hand will soon be outstripped by the exponentially increasing complexity.[21] As with so many Artificial Intelligence (AI) problems, progress beyond relatively toy domains becomes infeasible. So the possibility of automatic evolution of the cognitive architecture without explicit design becomes very attractive. Natural evolution is the existence proof for the viability of this approach, given appropriate resources.

Genetic Algorithms (GAs) are a form of search technique, modeled on Darwinian evolution, primarily used for function optimization.[15] An evolutionary approach to robotics necessarily means an incremental approach, and yet this is something that standard GAs cannot handle; in fact, standard GAs, though borrowing ideas from natural evolution, are themselves of no use for *applied* evolution. Below, I shall introduce SAGA, a framework that extends GAs for applied evolution. This will demonstrate that necessarily, in incremental evolution of, for example, a robotic cognitive system, the population will be genetically converged; in other words, the cognitive structures of all the robots will be fairly similar, and the genotypes will be positioned around some hill in a fitness/genotype landscape.

From this it will follow that the evolutionary search process will involve hill crawling as much as hill climbing in the fitness landscape; this has implications for mutation rates and for the selection mechanism for the evolutionary algorithm. This, in turn, leads to theoretical reasons why tournament selection is appropriate; there are, in any case, *practical* reasons why it might be appropriate for evolutionary robotics.

2. RELATED WORK

Evolutionary approaches, often using variants of GAs, to Artificial Life in simulations have been widespread.[25,24] Evolutionary robotics was proposed for philosophical reasons by Cariani.[7] It is only recently that serious proposals have been made to apply evolutionary approaches to real-world robots.[6,21,29] Earlier, a student of Brooks discussed some of the issues involved, with reference to subsumption architectures.[33] De Garis[10] proposed using GAs for building behavioral modules for artificial nervous systems, or artificial embryology. Beer[2] used GAs to synthesize a walking behavior for a six-legged agent. In a more traditional robotics context, mention is made of an evolutionary approach by Barhen et al.[1]

Recently the Japanese government research laboratories, ATR in Kyoto, have set up a well-funded research group for Evolutionary Robotics in their Evolutionary Systems department. Similar work is pursued at the research laboratory ETL in Japan, and there is interest from Japanese industry; Mitsubishi sponsored a symposium on Evolutionary Robotics in March 1993. At the Simulation of Adaptive Behavior 1992 conference[9] in Hawaii, a group of papers closely related to this field were presented. Since 1992 the Evolutionary Robotics Group at Sussex has been artificially evolving control systems for mobile robots—coevolved with sensor attributes—for visual navigation tasks. This work started with simulations of a real physical robot, and is now using a specialized piece of hardware allowing real vision to be used in a robot that can have a succession of control systems rapidly and automatically evaluated in sequence.[17,18] The control systems evolved are noisy dynamic recurrent networks.[16] We agree with Beer[3] in his advocacy of a dynamical systems perspective on autonomous agents, and are generally sympathetic to the *enactive* approach to cognition[32]; in this paper these concerns will not be elaborated on.[1]

3. SIMULATION VERSUS REALITY

Any evolutionary technique is going to need a large number of trials of robots, and practical constraints mean that these should be done on simulated robots if this is viable. Traditionally, and for good reason, those who have built real robots have tended to scorn simulations as implicitly assuming that all the really hard real-world problems have been solved. To quote Brooks[5]:

> First, there is no notion of the uncertainty that the real world presents.... Second, there is a tendency to not only postulate sensors which return perfect information (e.g., the cell ahead *contains food*—no real perception system can do such a thing) but there is a real danger of confusing the global world view and the robot's view of the world....

It is standard practice for a commercial pilot to convert to flying a new model of a plane by training in a flight simulator, so that the first real flight in the new plane is carrying passengers. Two things should be noted: the conversion is from one plane type to another similar one, and commercial flight simulators are so complex that, although cheaper than flying the real thing, the cost is not of a totally different order.

[1]Papers from the Sussex Evolutionary Robotics Group may be obtained from the present author, or by anonymous ftp from 192.33.16.70; files pub/reports/csrp/csrp???.ps.Z, for ??? = 219,220,221,222,223,256,264,265,267,278.

If robot simulations are being used to save money, then they are likely to be the equivalent of a flight simulator on a workstation rather than a commercial airline simulator. The usefulness of robot simulations may be compared with the usefulness of practising on a personal flight simulator program for a pilot learning for the first time to fly a real light aircraft. There can be no substitute for experience of the real thing, though the simulations can be of benefit.

The simulator is limited by the programmer's knowledge of relevant factors to be included, and if, for instance, no account of the effects of wind shear is put into a flight simulator, the first encounter with this in the real world will be hazardous. So any benefits brought to an evolutionary approach by using simulations will inevitably have to be paid for by putting major effort into the realism of the simulator. Trials of evolved architectures on real robots will have to be carried out at frequent intervals for the dual purpose of validating the fitnesses *and* providing feedback for improvement of the simulator. At Sussex we have found that doing simulations of vision, using ray-tracing, has been so computationally expensive that we have built hardware to allow automation of multiple evaluations with real vision[9]; it is faster.

4. SAGA AND A GRADUAL INCREASE IN COMPLEXITY

Some hints from natural evolution have been used by the GA community to produce effective search techniques for complex multidimensional search spaces. But this use of GAs for function optimization is problem-solving in what is, although enormous, a predefined space of possibilities of known size—this size being a maximum of a^l when genotypes are of length l with a possible alleles at each position. But the most impressive feature of natural evolution is how, over aeons, organisms have evolved from simple organisms to ever more complex ones, with an associated increase in genotype lengths. This aspect of evolution has been completely ignored in the standard GA literature. GAs have been adapted to problem solving, and the problem-solving metaphor or frame of mind is, I believe, much of the time inappropriate for considering both natural evolution and potential robot evolution.

The theoretical underpinning for GAs, Holland's Schema Theorem,[20,15] is no longer valid when the genotypes within a population vary in length. Some GA systems have used variable lengths, e.g., Smith's LS-1 classifiers[31] and Koza's genetic programming[23]; but the analyses offered in these two examples do not satisfactorily extend the notion of a schema such that schemata are preserved by the genetic operators.

The conceptual framework of SAGA was introduced in 1991 in order to try to understand the dynamics of a GA when genotype lengths are allowed to increase.[18] Working with a finite population, a standard GA often starts with a random distribution that spans the whole search space; the genetic operators, particularly

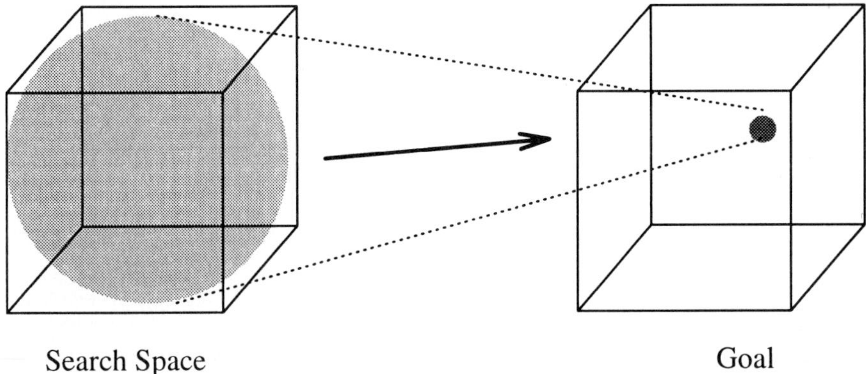

FIGURE 1 The evolution of a standard GA in a fixed-dimensional search space; the population initially spans the whole space, and in the end focuses on the optimum.

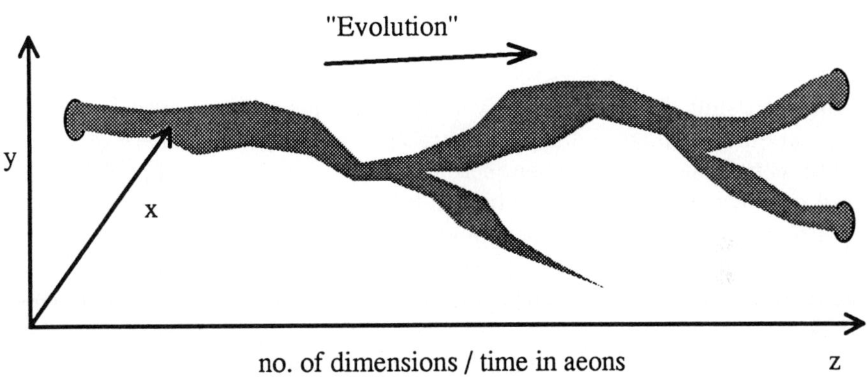

FIGURE 2 The progress of the always compact course of a species; the horizontal axis indicates both time and the (loosely correlated) number of dimensions of the current search space. The other axes represent the current dimensions in genotype space. The possibility of splitting into separate species, and of extinction, are indicated in the sketch, although not here discussed.

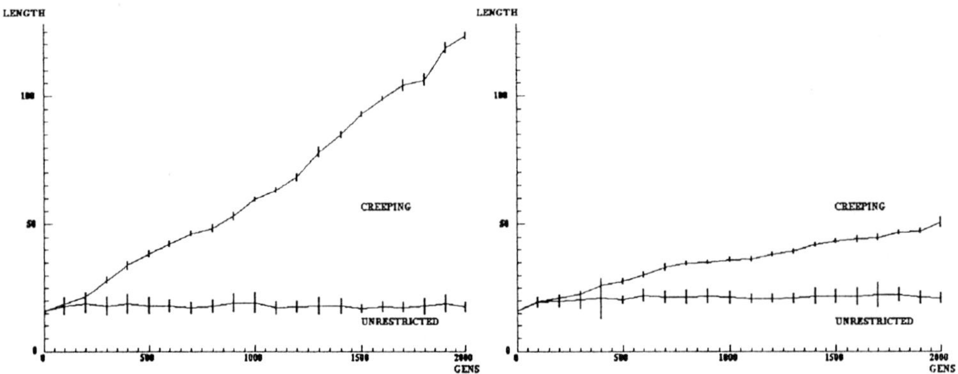

FIGURE 3 Average genotype lengths against generations; vertical bars show standard deviations. Effects of "creeping" and "unrestricted" increased-length genetic operators on a population with the same fitness conditions, epistasis $K = 2$. Left graph: linkage with neighbouring genes. Right graph: random linkage.

recombination, shift the population over successive generations until hopefully it converges around some optimum (see Figure 1). If genotype lengths are going to be allowed to increase indefinitely, then there is no finite search space of predetermined size, and this picture can no longer be valid. In a paper by Harvey,[18] it is shown, using concepts of epistasis and fitness landscapes drawn from theoretical biology,[22] that progress through such a genotype space will only be feasible through relatively gradual increases in genotype length. A general trend towards an increase in length is associated with the evolution of a *species* rather than global search. The word *species* I use to refer to a fit population of relative genotypic homogeneity.[2]

In contrast to the goal-seeking metaphor of Figure 1, a journey through SAGA space can be characterized in the form of Figure 2. The conclusion of Harvey's paper,[18] that only gradual increases in genotype length are likely to be viable, means that the finite resources of the population in searching around its current focus should be concentrated on just such gradual increases. The analysis given was supplemented by experimentation using an NK model[22] which gave the confirmatory results shown in Figure 3. This conclusion accords well with, for instance, Brooks' approach of "wiring in" new behaviors one at a time, and waiting until current behaviors are thoroughly debugged before "wiring in" the next.

[2]This is only indirectly related to a biological definition of the word. However, it follows from my definition that crosses between members of the same species have a good chance of being another fit member of the same species, whereas crosses between different species will almost certainly be unfit.

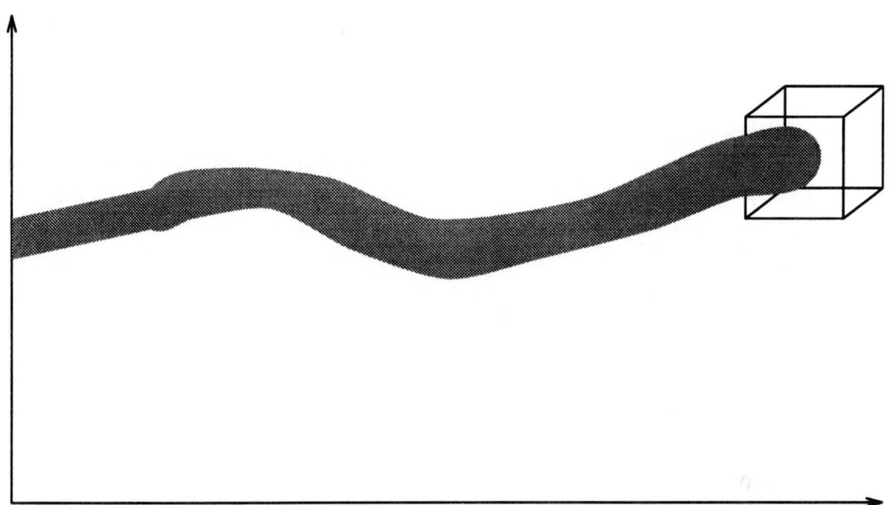

FIGURE 4 As a species evolves through SAGA space, the search for higher fitness only takes place in a very local search space around the current focus of the species.

In general, the "problem solving," or "goal seeking" metaphor for evolution (Figure 1) is misleading. Within a SAGA space, however, it can still be useful to use this metaphor in the restricted sense of searching around the current focus of a species for neighboring regions that are fitter or, in the case of neutral drift, not less fit (see Figure 4). Such a search takes place through application of genetic operators such as mutation or change length.

A change-length genetic operator, when subject to the restrictions of only allowing small changes in length in any single application, can be translated into a form equivalent to mutation by sleight-of-hand. In the case of an increase by one gene in a binary genotype with possible alleles 0 and 1, this gene can be considered instead to have 3 possible values 0, 1, and $-$, where the latter value is equivalent to "absent"; the new appearance of this gene can be considered as a mutation from $-$ to either 0 or 1. If more than one gene appears in one reproduction event, then this would be equivalent to a simultaneous set of mutations, which for low mutation rates is highly unlikely.

For this reason in the next sections of this paper we will consider mutation only, though bearing in mind that through this sleight-of-hand it will be possible to extend some of the conclusions to an increased-length operator which works at very low rates. In addition we will start by assuming that there is no recombination; i.e., reproduction is asexual. Having sketched out the important factors using this assumption, the impact of recombination on this sketch will be discussed.

5. SPECIES HILL CRAWLING

Usually in GAs, premature convergence of a population is something to be avoided. In SAGA, we are continually working with a converged population and are interested in encouraging search around the local focus while being careful not to lose the gains that were made in achieving the current *status quo*. In the absence of any mutation (or change length), genetic operator selection will concentrate the population at the current best. The smallest amount of mutation will hill-climb this current best to a local optimum. As mutation rates increase, the population will spread out around this local optimum, searching the neighborhood but, if mutation rates become too high, then the population will disperse completely, and the search will become random with the previous hilltop lost.

The problem is that of *Muller's ratchet.*[26] Call the genotype that represents the very peak of the hill the master sequence (or the "wild type"). As a converged population, the other members will be quite close in Hamming distance to this master sequence and, hence, far more mutations will increase this Hamming distance than will decrease it. The only force opposed to this pressure is that of selection

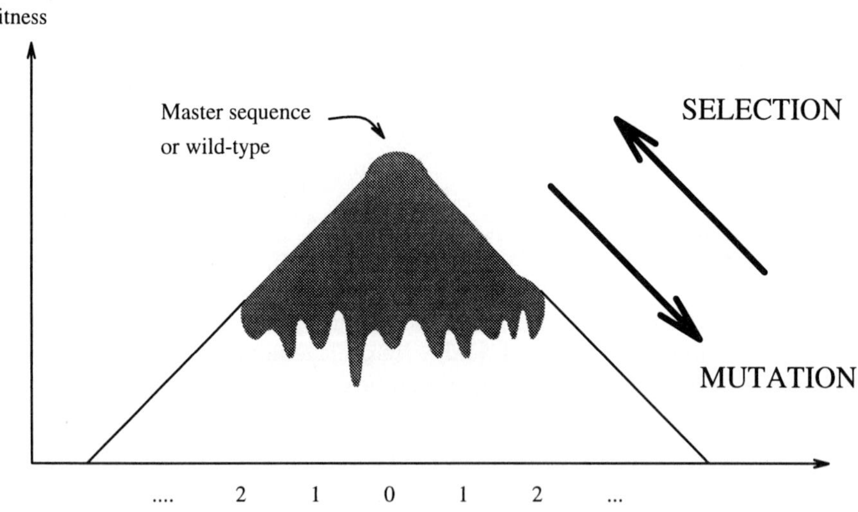

FIGURE 5 The opposing forces of mutation and selection on a population centered around a local optimum, where the Hamming distance from the master sequence is directly related to fitness ranking.

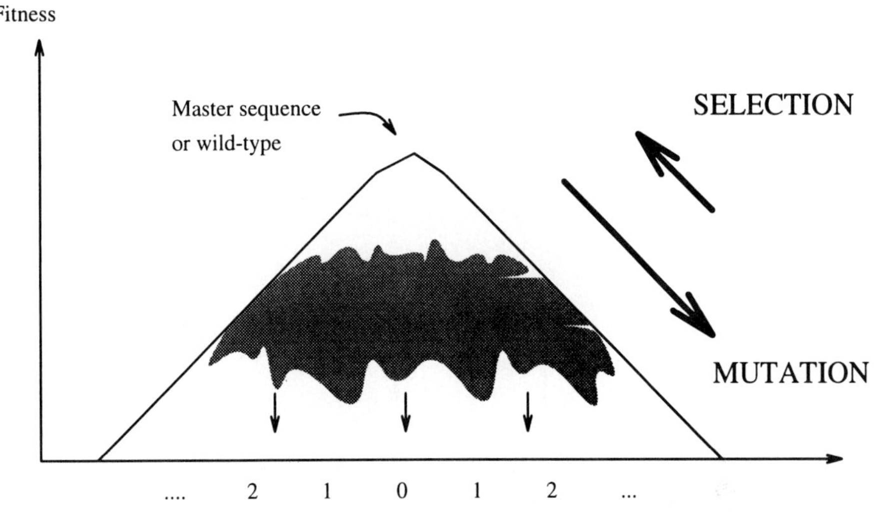

FIGURE 6 When mutation outweighs selection so that the fittest rank can be lost, Muller's ratchet inexorably drives the population down the hill.

preferentially reproducing the master sequence and its nearest neighbors in sufficiently large numbers to allow an occasional copy of the master with fortuitously no mutation; the other possibility, of fortuitous back-mutation from a near neighbor to the master sequence, is so small as to be usually negligible.

In Figure 5 we sketch the effects of mutation on a population centered around a local optimum. The vertical axis represents fitness or selective values. The horizontal axis indicates distance in sequence space from the master sequence at the top of the hill. The shape of the hill indicates the assumption that, within this local neighborhood, an increase in the number of mutations is monotonically related to a decrease in fitness. In Figure 6 we demonstrate the effect of Muller's ratchet when mutation is high enough to cause loss of information. In Figure 7 we sketch the effects when mutation is high enough (without bringing Muller's ratchet into play) for some elements of the population to crawl down the hill far enough to reach a ridge of high selective values. As discussed by Eigen,[13] this results under selection in a significant proportion of the population working their way along this ridge, and making possible the reaching of outliers further in Hamming distance in that particular direction from the master sequence.

If any such outliers reach a second hill that climbs away from the ridge, then parts of the population can climb this hill. Depending on the difference in fitness and the spread of the population, it will either move *en masse* to the new hill as a better local optimum, or share itself across both of them.

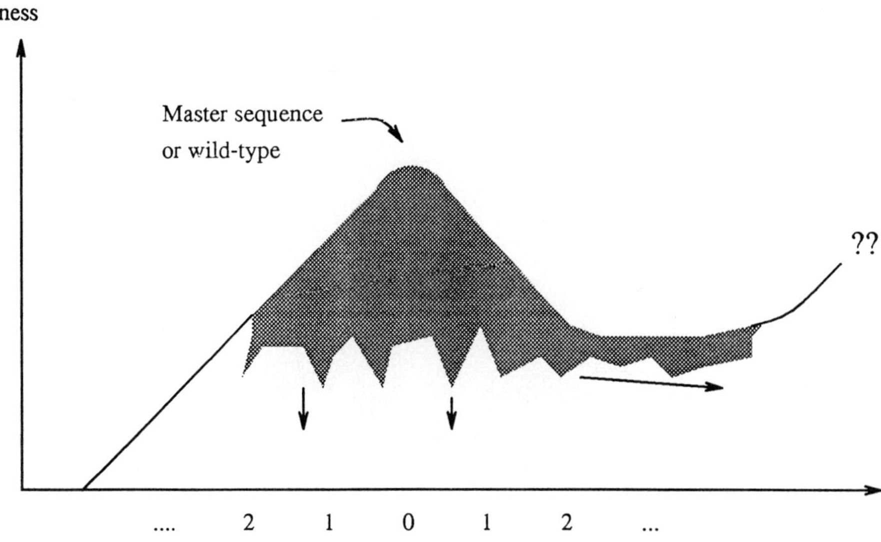

FIGURE 7 If the population can crawl down the hill far enough to reach a ridge of relatively high fitness, it will spread along it, potentially reaching new hills.

So in a SAGA setup of evolution of a converged species, we want to encourage through the genetic operators such hill crawling downward towards ridges to new hills, subject to the constraint that we do not want to lose track of the current hill. To quote Eigen et al.[13]:

> In conventional natural selection theory, advantageous mutations drove the evolutionary process. The neutral theory introduced selectively neutral mutants, in addition to the advantageous ones, which contribute to evolution through random drift. The concept of quasi-species shows that much weight is attributed to those slightly deleterious mutants that are situated along high ridges in the value landscape. They guide populations toward the peaks of high selective values.

6. SAGA AND MUTATION RATES

Although progress of a species through a fitness landscape is not discussed in the standard GA literature, in theoretical biology there is relevant work in the related field of molecular quasi-species.[12,13] In particular, analysis of "the error catastrophe" shows that, subject to certain conditions, there is a maximum rate of mutation which allows a quasi-species of molecules to stay localized around its current optimum. This critical maximum rate balances selective forces which tend to increase the number of the fittest members of the population against the forces of mutation which tend, more often than not, to drag offspring down in fitness and away from any local optimum. But a zero mutation rate allows for no further local search beyond the current species and, other things being equal, increased mutation rates will increase the rate of evolution. Hence, if mutation rates can be adjusted, it would be a good idea to use a rate close to but less than any critical rate that causes the species to fall apart. A further possibility, in the spirit of simulated annealing, is to temporarily allow the rate to go *slightly above* the critical rate—to allow exploration—and then to cut it back again to consolidate any gains thus made.

For an infinite population, in the particular context of molecular evolution, Eigen and Schuster[12] show that these forces just balance for a mutation rate

$$m = \frac{\ln \sigma}{l}$$

where l is the genotype length and σ is the *superiority* parameter of the master sequence—the factor by which selection of the master sequence exceeds the average selection of the rest of the population. The diagrams they show for the very sharp cutoff at the critical rate refer to a fitness landscape with a single "needle" peak for the master sequence, all the rest of the population taken to be equally (un-)fit; where the hill slopes more gently from the master sequence, the cutoff is less abrupt (see Figure 14). For typical values of σ between 2 and 20, the upper limit of mutation before a quasi-species "loses its grip" on the current hill would be between $0.7/l$ and $3/l$. When the population is of finite large size, yet small enough for stochastic effects of genetic drift to start having an effect, the same overall picture holds except for a reduction in this critical mutation rate (the "error threshold").[28] Expressed in terms of the single-digit accuracy of replication $q = 1 - m$, then the critical value of q for a population of size N is related to that for an infinite population by

$$q_N = q_\infty \left(1 + \frac{2\sqrt{\sigma - 1}}{l\sqrt{N}} + \dots\right).$$

Nowak and Schuster suggest that the approximations made in deriving this equation mean that it should only be relied on for large populations of significantly more than 100. Nevertheless, the presence of l, genotype length, in the denominator suggests that for genotypes of length order 100, and populations of size order 100, the error threshold will be extremely close to that for an infinite population.

Since it is the natural logarithm of the *superiority* parameter σ which enters into the equation for m, variations in this of an order of magnitude do not affect the error threshold as significantly as variations in genotype length. In conventional GAs, choice of mutation rates tends to be a low figure, typically 0.01 or 0.001 per bit as a background operator, that is decided upon without regard to the genotype length. This despite suggestions from experimentation by Schaffer[30] that optimal rates $m_{\text{opt}} = \alpha/(N^{0.9318} l^{0.4535})$, for some constant α; by Hesser[19] that earlier higher values should decrease exponentially towards $m_{\text{opt}} = \alpha'/(N\sqrt{l})$; and by DeJong[11] quoted by Hesser[19] that recommend $m_{\text{opt}} = 1/l$.

But these rates, and also the error threshold given by Eigen and Schuster, are based on particular assumptions about the selective forces on the population. The use of tournament selection provides a significantly different range of selective forces to a population, which means the above analyses cannot be relied on. However, in all simulations of hill crawling where different mutation rates are tried, a typical U-shaped curve is found, giving the shortest time needed to reach another hill at around the mutation rates suggested by theory.

In Figure 13, an artificial fitness space was created with two fitness hills. The fitnesses of any point in the space was given by (the negative of) the minimum of the Hamming distances to each hill. At the beginning the population was entirely concentrated at the peak of one hill, and then was allowed to drift under various mutation rates, with tournament selection, until the other hill was first reached by a single member. Obviously with zero mutation it would never be found; with excessively large mutation, random search would result, also taking excessively long times. The typical U-shaped curve is shown in Figure 14. In this case the difference between with and without recombination was only marginally in favour of the former.

Recently in the GA community there has been some discussion of the surprising success (in some circumstances) of what has come to be called *Naive Evolution*, i.e., mutation only, contrary to normal GA folklore which emphasizes the significance of crossover. It would be interesting to check on those circumstances where it has been found useful, and see whether the population is in fact converged, with hill crawling being the motive force for progress. The optimal mutation rates that are appropriate when hill crawling is feasible have been obscured in the GA literature by the usual practice of quoting mutation rates per bit or per symbol, rather than per genotype. It is the optimal mutation rates per genotype that can be found within a band that is nearly invariant over all genotype lengths.

7. TOURNAMENT SELECTION FOR PRACTICAL REASONS

Realistic simulations take time to run, and it will be necessary to do a large number in parallel. As each simulation is complex, parallel machines with SIMD are of no use; for instance, an individual workstation per simulation would be appropriate. In almost all networks of workstations there is a vast unused computational capacity that can be used effectively by running background processes. It then becomes attractive to use an evolutionary algorithm which allows asynchronous processing of the individuals.

Standard GAs tend to evaluate the whole current population, to select from these, and to apply genetic operators to produce the next generation. A steady-state algorithm such as GENITOR[34] replaces an individual at a time rather than a generation at a time. But since it always replaces the current worst member of the population, it requires global communication of statistics about the whole population before carrying out such a replacement. In a network of processes running asynchronously, with the possibility of individual machines being down for periods of time, this negates some of the benefits of parallelism.

Tournament selection operates by taking two (or sometimes more) members of the population chosen at random, and choosing the best of this tournament to contribute genetic material to a new individual. There are a variety of ways to choose which old individual should sacrifice its place for the new. In a tournament of size two, a copy of the winner, after application of genetic operators such as mutation or crossover (recombination), could replace the loser, or replace a randomly chosen member of the population. Sexual recombination can take place between the winners of two different tournaments.

The practical advantage of this procedure when using a network of workstations is that it can be truly asynchronous and decentralized. Communication between machines is largely limited to occasional passing of genotype strings. If one, or a whole group of machines, slows due to loading or even is down for a time, the algorithm can carry on regardless on the remaining machines. It even becomes feasible to use several different networks only occasionally communicating with each other by electronic mail. Most tournaments would be "local" within one network, as it is only necessary for there to be one transmission of genetic material per "generation" (i.e., set of tournaments such that the number of participants equals the population size) between two otherwise isolated genetic pools for them to stay together. It thus becomes possible to use enormous amounts of computing power that is currently little used on present facilities.

For robotics applications it should be noted that the use of tournament selection reduces the evaluation of the robots to a very simple question: of two given robots, which is (probably) the "best"? If the robots are tested (in reality or in simulation) on a series of tasks of increasing complexity in a noisy environment, then the evaluation will become something like: which of the two got further before stopping? The power of evolution allows complexity to be built up through a succession of such trivial "questions and answers," each containing at the very maximum 1 bit of information. As with any selection mechanism that is equivalent to ranking, it is not necessary to have an evaluation function that returns a scalar value, which may simplify matters greatly. However, the extent to which noise in tournament selection—occasionally selecting the less fit by mistake—affects the stability of a species will be considered below.

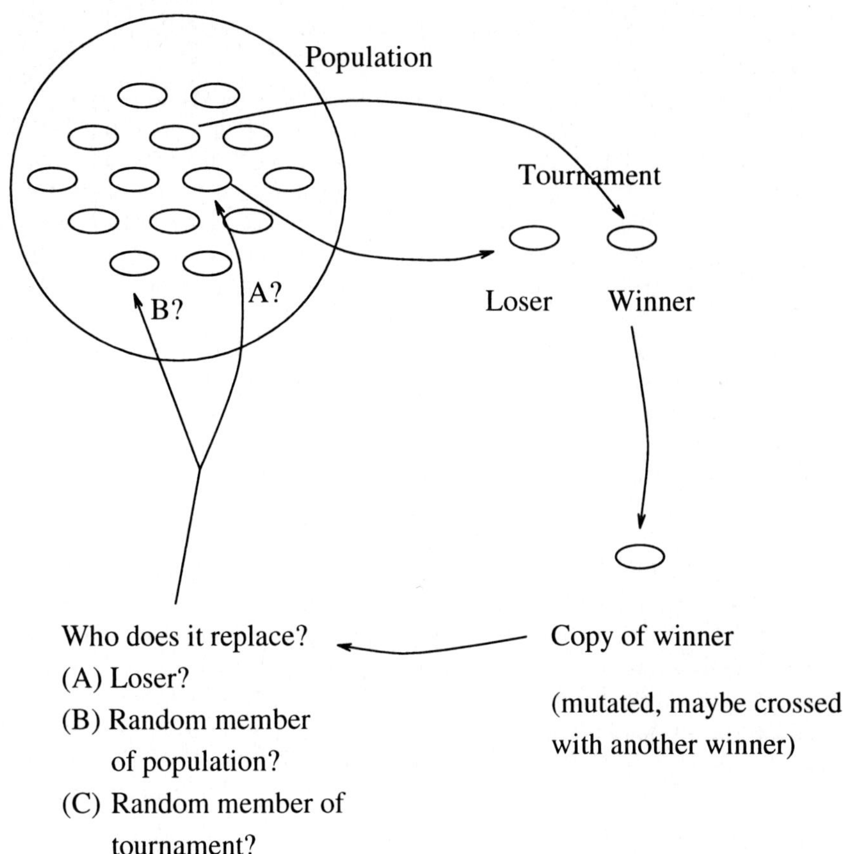

FIGURE 8 Tournament Selection. Possible choices for who dies to make way for the copy of the winner.

In the game of *Twenty Questions* each reply can provide a maximum of 1 bit of information, hence at the end potentially discriminating between 2^{20} choices; but answers to inappropriate questions will provide much less than 1 bit, and frequently zero information. In an evolutionary approach to robotics it is the task of the genetic operators such as mutation, change length, and recombination to generate new test cases that "ask new questions" which tend to be appropriate. The following sections look at the problem from a theoretical perspective, and demonstrate that tournament selection can aid the genetic operators in this task.

8. TOURNAMENT SELECTION FOR THEORETICAL REASONS

Tournament selection relies on selecting randomly from the population a small number of contestants for a tournament, and taking the winner for further genetic processing to contribute to the future population. We will only consider a tournament of size 2, but within this constraint there is still a wide range of possibilities.

Tournament selection in GAs is generally used on a generational basis; i.e., the current generation is held fixed, and an appropriate number of tournaments are held to build up a genetic pool for mating or mutation, after which this becomes the new generation. It can be shown[14] that this is equivalent to a ranking scheme,[3] in which the highest rank on average contributes two members to the genetic pool, the middle rank one member, and the bottom rank none. In Eigen and Schuster's terms, the superiority $\sigma = 2$. Such a method loses the advantage of asynchronous parallel computing mentioned earlier.

A steady-state method, which can be done asynchronously, involves replacing the offspring derived from the winner after each tournament. In the case of a draw, the winner is chosen at random. The interesting question is who dies so as to make way for the offspring, and some possibilities are (see Figure 8):

1. A randomly chosen member of the population.
2. A randomly chosen member of the tournament.
3. The loser of the tournament.
4. A subtle variation—do *not* reproduce from the winner, but remove the loser of the tournament and replace it by an offspring of a randomly chosen member of the population.

The effects of these methods will be described in terms of a notional generation, when a set of tournaments such that the number of participants equals the population size have been run. The first one is similar in effect to the original generational basis, except that the superiority of the first rank is approximately e rather than 2—based on the fact that $(1 + 1/N)^N$ tends towards e for large N. This superiority only holds true while each member of the population is in general separately ranked, and ceases being valid as soon as a significant number of the tournaments are draws—in other words, Eigen and Schuster's analysis no longer becomes directly applicable as their selection mechanism is radically different from tournament selection.

[3] The advantages of a ranking scheme are discussed by Whitley.[34]

Scenario 1: Exactly m bits flipped, chosen at random. Genotype is so long that chance of favourable back-mutation towards master-sequence is negligible.

Scenario 2: Exactly m bits flipped, chosen at random. Finite length genotype with significant possibility of back-mutation.

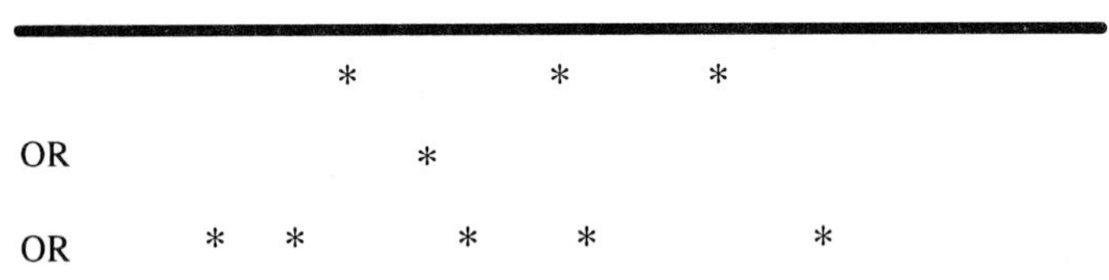

Scenario 3: Each bit is flipped with probability of m/(genotype length). I.e. on average m mutations on whole genotype.

FIGURE 9 Possible ways to apply a given mutation rate.

The second method has the same effect as the first—a randomly chosen member of the tournament will, in the long run, be just as "averagely fit" as a random member of the whole population. Since in half of these tournaments the winner will be replacing itself, then N such tournaments are equivalent, as far as σ is concerned, to $N/2$ tournaments using method 3. Hence a full notional generation with method 3 yields a σ value of e^2. Surprisingly, method 4 gives the same value of σ as methods 1 and 2.

Method 3 will be the one discussed here, and the effect of mutation rates on a converged species (with a binary genotype) will be assessed for three different scenarios (see Figure 9):

- A mutation rate of μ bits flipped per genotype, where μ is a small integer and the genotype length is so long that the possibility of back-mutations towards the current master sequence can be ignored.
- A similar mutation rate of μ bits per genotype on a genotype of length l, with the possibility of back-mutations.

- An average mutation rate of μ bits per genotype, calculated independently at the rate of μ/l at each locus.

8.1 LONG GENOTYPES

In the first scenario, we can classify each member of the population by the Hamming distance from the master sequence. This will increase by μ at each replication, giving possible distances of 0, μ, $2\mu, \ldots$, so without loss of generality we need only consider $\mu = 1$. In the context of hill crawling, our interest is in how the population distributes itself within different mutant classes of size r_i, whose members have Hamming distance i from the master sequence. Given a tournament between members of distance i and j, for $i < j$ the winner i will remain in the population, and j will be replaced by a mutant of distance $i + 1$. A tournament draw between members of the same distance i results in a winner i and the loser replaced by $i+1$.

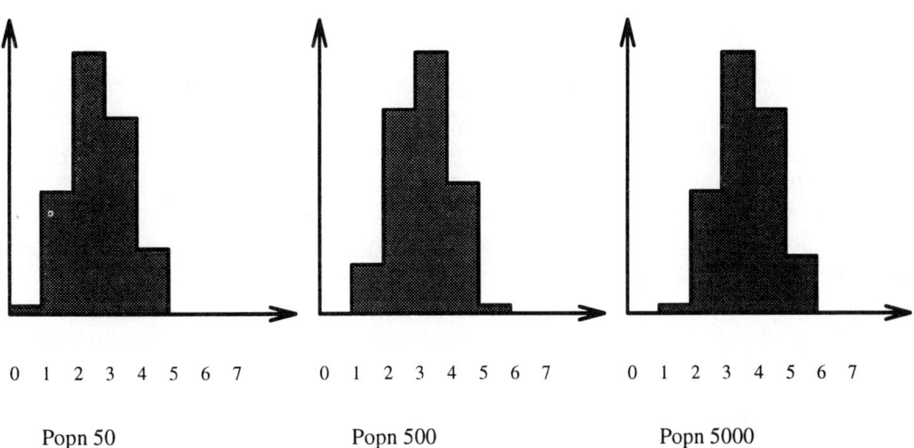

FIGURE 10 Numbers in each class, ranked by the Hamming distance from the master sequence. Effectively infinitely long genotype, in that there are no back mutations, but exactly one mutation in genotype at replication of the winner of tournaments. Results shown are after simulations of a number of tournaments equal to 4000 times population size.

For maximum hill crawling without losing the master sequence (of distance 0) from the population, the long-term fate of this master sequence should be considered. It can be seen that all tournaments between a 0 and a 0 result in the loss of one 0 to the population, and there is no other way in which 0s can be gained. If all tournaments are constrained to be between two different individuals, then r_0 will soon reduce to one member which will thereafter survive alone for ever. This member, the "wild type," will win all its tournaments and continually replenish the flow of mutants down the hill away from it. Histograms of results from populations of various sizes run in a computer simulation are shown in Figure 10. An equation that allows one to iteratively derive the expected size of each class is derived in Appendix A.

If the same individual can be chosen twice for a tournament, resulting in the replacement of itself by a mutated copy, then the wild type will eventually be lost through just such an incident, and Muller's ratchet will start to operate. However, if tournaments are between different individuals, then the wild type will never be lost, whatever the size of μ. We thus have a selection mechanism which can move the bulk of the population crawling down the hill as much as is desired, without ever encountering the error threshold by Eigen and Schuster. The banding into multiples of μ can be broken up by alternating between two different integers for μ.

BUT CAREFUL .... There is a dangerous potential flaw in this. We are relying on the choice of winner of a tournament being 100% reliable, and in the context of evolutionary robotics, as discussed earlier, this may very easily not be the case. If the reliability of choice is $p < 1$, then sooner or later the wild type will be lost and Muller's ratchet will start. A possible counter to this will be to only mutate the replica with probability $q < 1$, and otherwise leave it unchanged. In Appendix B it is shown that this will save the situation in an infinite population for values of $q < (2p-1)/p$, this upper bound being independent of the value of μ. For example, if $p = 0.9$, we should have $q < 0.888\ldots$. In the case of a finite population, q should be reduced further to allow for the stochastic effects of genetic drift. The analysis of this case is not attempted here.

8.2 SHORTER GENOTYPES

Experiments run in simulation with populations of various sizes with genotypes of length 100, and with a mutation rate of exactly one bit flipped at replication, demonstrate what happens when the possibility of back-mutation is no longer negligible. The histograms in Figure 11 demonstrate a similar shape to those in the previous case, except that much more of the population stays close to the wild type. It can be seen that in the case of unreliable choice discussed above, this effect will supplement that of any given value of q in countering the loss of the wild type.

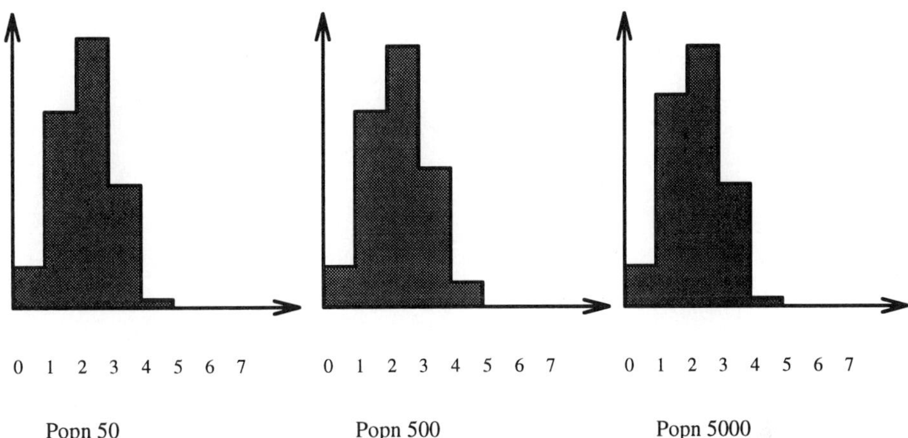

FIGURE 11 Similar simulations to those shown in previous figure, except that genotypes are of length 100. Exactly one mutation per genotype at replication, which means a significant number of back mutations towards the master sequence.

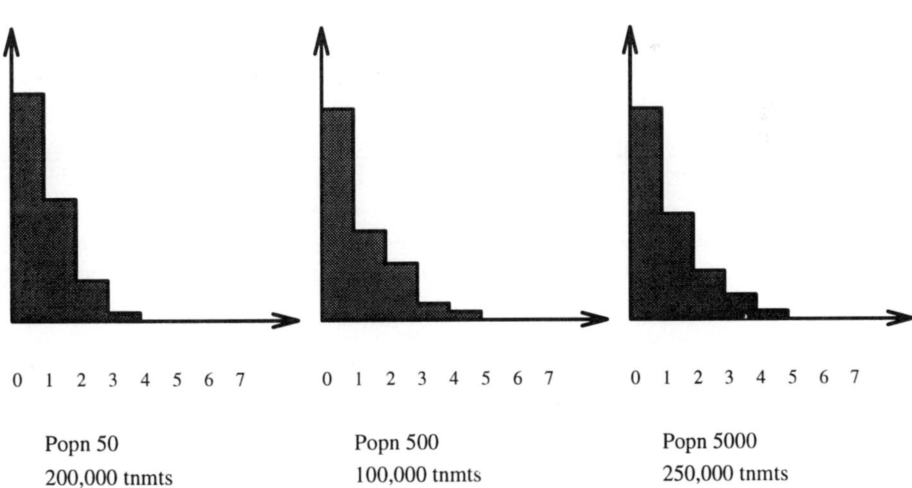

FIGURE 12 Similar conditions to previous figure, except that there is a 1/100 chance of a mutation at each bit of the genotype, which is of length 100.

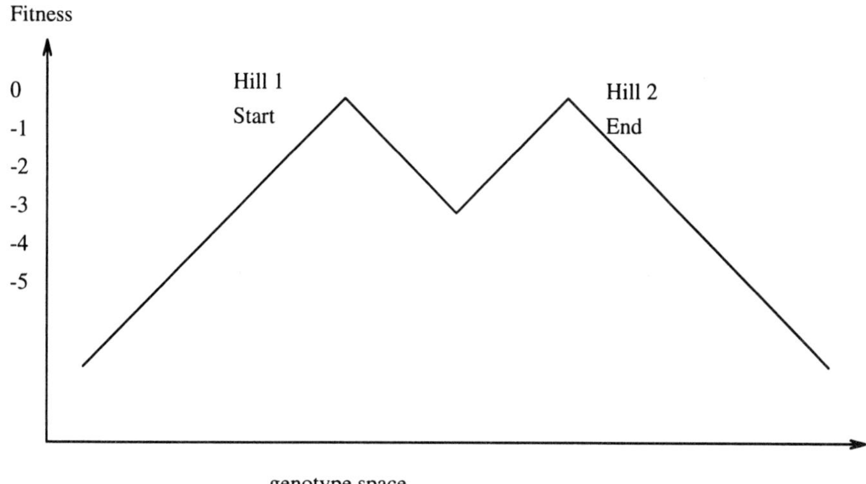

FIGURE 13 Two genotypes are designated as start and goal, each of them with maximum fitness (of 0). The fitness of any other genotype is defined as (the negative of) the minimum of the Hamming distances to start and to goal. This gives a fitness landscape of two hills as shown. The entire population starts off at the peak of the "start hill."

8.3 MUTATIONS ASSESSED INDEPENDENTLY AT EACH LOCUS

When the same experiment is run on genotypes of length 100—but in this case instead of exactly one bit flipped per genotype, there is a $1/100$ chance of flipping at each locus on the genotype—the results, shown in Figure 12, are startlingly different. The reason for this is that although the expected number of mutations per genotype is 1, this is made up from a probability of about $1/e$ of no mutations, about $1/0.99e$ of one mutation, about $1/1.98e$ of 2 mutations, and so on. The significant probability of there being no mutation has a similar effect to that given by the deliberate introduction of a probability $1 - q$ of no mutation, discussed above in the long genotype scenario and in Appendix B. Appendix C shows that the proportion of the population expected to be the wild type is $2/(1 + e^{\mu})$, which when $\mu = 1$ gives a value of about 0.538. When there is a probability p of making a mistake in a tournament, it is also shown in this appendix that in an infinite population the wild type will not be lost if $p > e^{\mu}/(1 + e^{\mu})$, which for $\mu = 1$ gives a minimum value for p of 0.731. To use this equation in the other direction, if it is known that $p > 0.9$, then the maximum value of μ would be $\ln 9 \approx 2.19$. These figures would be worse in a finite population due to stochastic effects.

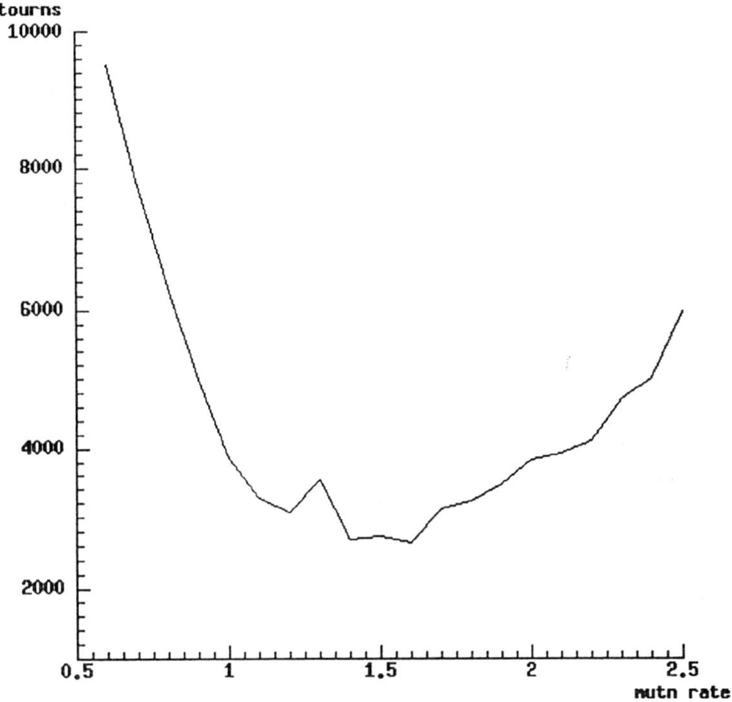

FIGURE 14 Number of tournaments for a population centered at one "hilltop" to have a first member reach a nearby hilltop. Rate specified is the average number of mutations per genotype. Recombination has been used.

9. SAGA AND RECOMBINATION

It has been suggested above that the application of a change-length genetic operator at the very low rates required in SAGA can be treated in a similar fashion to low mutation rates, although the sleight-of-hand used is equivalent to increasing the number of possible alleles at the relevant loci. In the context of adjusting genetic operators so as to be able to influence hill crawling without losing the current wild type, the introduction of recombination makes a major impact.

One virtue of recombination within a species is that when two different favorable but improbable mutations take place within two different members of a population, then sexual replication can produce at a stroke an individual combining both; whereas asexual replication would require both improbable events to occur within the same single line of descent. It is this virtue that is stressed in standard GAs, yet here we will concentrate on another virtue—the other side of the same coin—which is that recombination functions can act as a form of repair mechanism protecting against Muller's ratchet.[26]

With tournament selection, candidates for recombination would be the winners of two separate tournaments, and the two offspring, after crossover and mutation, can replace the two losers. In general, the crossover will produce one offspring closer to the wild type than the average of the two parents' distances, and another offspring further away than this average; after which mutation adds its toll. This constitutes a force producing a restorative flow towards the wild type, allowing larger mutation rates without loss of the current local optimum. Simulations confirm this.

There are practical computational problems in dealing with recombination with the variable-length genotypes that are necessary in evolutionary robotics; given a crossover point in one parent genotype, where should the crossover point in the other parent genotype be? A discussion of this problem, and an algorithm which provides an efficient technique, are presented by Harvey.[17]

10. ELITISM IN NOISY FINITE POPULATIONS

We have been seeking ways of avoiding the loss of the wild type, while promoting appropriate exploration. In GAs, the policy of retaining the best, unchanged, for inclusion in the next generation is known as *elitism*, and we have seen that tournament selection gives you elitism for free, when the tournaments are 100% reliable. In the presence of noise, we can counter the operation of Muller's ratchet by some of the measures mentioned above which guarantee preservation of the wild type in nonzero proportions in infinite populations. For finite populations of a practical size, stochastic effects are significant, and the counter measures are less effective.

11. CONCLUSIONS

An evolutionary approach to robot design, working from simple towards more complex cognitive architectures, implies species evolution within the SAGA conceptual framework. This requires a very different analysis from standard GAs, and the abandonment of the goal-seeking metaphor associated with them. A new metaphor of hill crawling of a converged species has been introduced, and this needs an analysis of the conflicting forces of exploitation and exploration—which here means efficiently searching down the current hillside and along high value ridges in the fitness landscape while being careful not to lose track of the current hilltop.

Whereas theoretical biologists are trying to analyze the selection mechanisms that they believe exist in the natural world, in simulated evolution we can choose our own selection mechanism. Arguments have been presented that tournament selection can be used for hill crawling, with significantly higher mutation rates than are used in conventional GAs; higher mutation rates enable a faster rate of evolution. It has also been argued that with the complex simulations that would be needed in evolutionary robotics, requiring serious computing power for each individual being evaluated, tournament selection allows a practical evolutionary setup to be highly distributed over an asynchronous network or networks of machines with minimal intercommunication. In addition, tournaments reduce the selection process to a succession of binary decisions as to which of two individuals is the better, avoiding scaling problems with any evaluation function.

Analytical results have been shown for the effects of tournament selection in the case of infinite populations, with and without reliable tournament decisions. Stochastic effects of genetic drift in small populations alter these results. Results from simulations with finite populations under different conditions have been shown, and practical ways to overcome Muller's ratchet have been suggested.

APPENDIX A: RANK SIZES WITH LONG GENOTYPES

Consider a population of size N, with binary genotypes, and classify each individual by Hamming distance from the current master sequence or wild type, which without loss of generality can be taken to be a genotype of all 0s. It is assumed that fitness monotonically decreases with this distance and, without loss of generality, the negative fitness can be the number of 1s in the genotype. At each replication there is a single mutation, and the genotypes are so long that the chance of a back-mutation is negligible; i.e., all mutations are deleterious.

The number of individuals with i 1s is defined as r_i. The loss from and gain to this class should balance at equilibrium.

Some individual of rank i is lost in a tournament if there is a tournament between a rank i and and a rank $< i$, with probability

$$2\frac{r_i}{N}\sum_{j=0}^{i-1}\frac{r_j}{(N-1)},$$

and also for one between two different members of rank i, probability

$$\frac{r_i}{N}\frac{(r_i-1)}{(N-1)}.$$

Some individual of rank i is gained in a tournament between rank $(i-1)$ and a rank $> i$, with probability

$$2\frac{r_{i-1}}{N}\sum_{j=i+1}^{N}\frac{r_j}{(N-1)},$$

and also for one between two different members of rank $(i-1)$, probability

$$\frac{r_{i-1}}{N}\frac{(r_{i-1}-1)}{(N-1)}.$$

Setting gains equal to losses and eliminating $N(N-1)$, we have

$$2r_i\sum_{j=0}^{i-1} r_j + r_i(r_i-1) = 2r_{i-1}\sum_{j=i+1}^{N} r_j + r_{i-1}(r_{i-1}-1).$$

This gives a value for r_i based on values for lower i's. Since we know that $r_0 = 1$, we have

$$2r_1 + r_1(r_1-1) = 2(N-1-r_1).$$

This quadratic equation yields a positive solution of

$$r_1 = \frac{1}{2}(-3+\sqrt{8N+1}).$$

For $N = 100$ this gives $r_1 \approx 12.65$. Successive values for higher i can be found by iteratively solving successive quadratics.

APPENDIX B: NOISY DECISIONS IN AN INFINITE POPULATION

Consider a similar situation to that in Appendix A, but where there is a probability p of correctly deciding a tournament and, hence, $(1-p)$ of making a mistake. We will be considering the long-term possibility of losing all members of rank 0, and will assume a population of infinite size so as to ignore stochastic effects of genetic drift. Let the proportion of the population in rank 0, the master sequence, be a. To keep up the value of a, we will impose a probability q of mutating on replication and, hence, $(1-q)$ of there being no mutation.

A wild type of rank 0 will be lost from the population when rank 0 meets rank 0 and there *is* a mutation, probability a^2q; and also when rank 0 meets rank ≥ 1 and the wrong one wins, probability $2a(1-a)(1-p)$.

A rank 0 will be gained when a rank 0 meets a rank ≥ 1, the right one wins, and there is *no* mutation, probability $2a(1-a)p(1-q)$.

Setting gains equal to losses and dividing by a, we have

$$aq + 2(1-a)(1-p) = 2(1-a)p(1-q),$$

$$a(4p+q-2pq-2) = 2(2p-pq-1).$$

We can assume that $p > 0.5$, say, $p = 0.5 + s$ for positive s. The factor on the l.h.s. of the equation, multiplying a, then becomes $(2+4s+q-q-2sq-2)$, which reduces to $2s(2-q)$. We know that s is positive and $(2-q)$ is positive, so the condition for a to be positive is that the r.h.s. of the above equation is also positive.
Hence $2(2p-pq-1) > 0$, which gives $q < (2p-1)/p$ as the condition for the proportion a of rank 0 to remain positive.

APPENDIX C: MUTATIONS ASSESSED INDEPENDENTLY AT EACH LOCUS

Consider an population of N binary genotypes each of length l. N is assumed to be large enough to avoid stochastic genetic drift, and l is assumed large enough for approximations to be made below.

If there are on average μ mutations per genotype, this is a probability of μ/l at each locus. So the chance of there being no mutation at all l loci on a genotype is $(1-\mu/l)^l$. For small μ and large l this is close to $e^{-\mu}$.

Let the proportion of rank 0 in the population be a. Then a rank 0 will be lost to the population when rank 0 meets rank 0 and there *is* a mutation, giving a probability of

$$a^2\left(1-e^{-\mu}\right);$$

a rank 0 will be gained when a rank 0 meets a rank ≥ 1 and there is *no* mutation, a probability of

$$2a(1-a)e^{-\mu}.$$

Setting gains equal to losses, and multiplying through, we have

$$a(e^{\mu}-1)=2(1-a).$$

Hence the proportion of the population expected to be the wild type at equilibrium is:

$$a=\frac{2}{1+e^{\mu}}.$$

Let us now reconsider this scenario with noise added, when the winner of a tournament is selected with probability p. A rank 0 will still be lost to the population when rank 0 meets rank 0 and there *is* a mutation, giving a probability of

$$a^2\left(1-e^{-\mu}\right);$$

but a rank 0 will *also* be lost when rank 0 meets rank ≥ 1, and noise makes it lose; the probability is

$$2(1-p)a(1-a).$$

A rank 0 will be gained when a rank 0 meets a rank ≥ 1, wins, and there is *no* mutation, a probability of

$$2pa(1-a)e^{-\mu}.$$

Setting gains equal to losses, and multiplying through, we have

$$a(e^{\mu}-1+2e^{\mu}(1-p)(1-a))=2p(1-a).$$

Hence

$$a(e^{\mu}(2p-1)-1)=2p-2(1-p)e^{\mu}.$$

The contents of the bracket on the l.h.s. are always positive, so that the condition for $a > 0$ is that the r.h.s. is positive,

$$p > (1 - p)e^{\mu}.$$

Hence, for preservation of the wild type, we need

$$p > \frac{e^{\mu}}{1 + e^{\mu}}.$$

The proportion of the population expected to be the wild type at equilibrium is:

$$a = \frac{2}{1 + e^{\mu}}.$$

REFERENCES

1. Barhen, J., W. B. Dress, and C. C. Jorgensen. "Applications of Concurrent Neuromorphic Algorithms for Autonomous Robots." In *Neural Computers*, edited by R. Eckmiller and C. von der Malsburg, 321–333. Berlin: Springer-Verlag, 1989.
2. Beer, R. D., and J. C. Gallagher. "Evolving Dynamic Neural Networks for Adaptive Behavior." Technical Report CES-91-17, Case Western Reserve University, Cleveland, Ohio, 1991.
3. Beer, R. D. "A Dynamical Systems Perspective on Autonomous Agents." Technical Report CES-92-11, Case Western Reserve University, Cleveland, Ohio, 1992.
4. Brooks, R. A. "Achieving Artificial Intelligence Through Building Robots." A.I. Memo 899, MIT A.I. Lab, May 1986.
5. Brooks, R. A. "Intelligence Without Representation." *Art. Int.* **47** (1991): 139–159.
6. Brooks, R. A. "Artificial Life and Real Robots." In *Toward a Practice of Autonomous Systems: Proceedings of the First European Conference on Artificial Life*, edited by F. J. Varela and P. Bourgine, 3–10. Cambridge, MA: MIT Press/Bradford Books, 1992.
7. Cariani, P. "On the Design of Devices with Emergent Semantic Functions." Ph.D. Thesis, State University of New York at Binghamton, 1989.
8. Cliff, D. T., I. Harvey, and P. Husbands. "Incremental Evolution of Neural Network Architectures for Adaptive Behaviour." In *Proceedings of the First European Symposium on Artificial Neural Networks, ESANN'93*, edited by M. Verseylen, 39–44. Brussels: D facto, 1993.

9. Cliff, D. T., P. Husbands, and I. Harvey. "Evolving Visually Guided Robots." In *From Animals to Animats 2, Proceedings of the Second International Conference on Simulation of Adaptive Behaviour (SAB92)*, edited by J.-A. Meyer, H. Roitblat, and S. Wilson, 374–383. Cambridge, MA: MIT Press Bradford Books, 1993.
10. de Garis, Hugo "The Genetic Programming of Steerable Behaviors in Gen-Nets." In *Toward a Practice of Autonomous Systems: Proceedings of the First European Conference on Artificial Life*, edited by F. J. Varela and P. Bourgine, 272–281. Cambridge, MA: MIT Press/Bradford Books, 1992.
11. De Jong, K. "Analysis of the Behavior of a Class of Genetic Adaptive Systems." Ph.D. Thesis, University of Michigan, 1975.
12. Eigen, M., and P. Schuster. *The Hypercycle: A Principle of Natural Self-Organization.* Berlin: Springer-Verlag, 1979.
13. Eigen, M., J. McCaskill, and P. Schuster. "Molecular Quasispecies." *J. Phys. Chem.* **92** (1988): 6881–6891.
14. Goldberg, D. E., and K. Deb. "A Comparative Analysis of Selection Schemes Used in Genetic Algorithms." Technical Report TCGA-90007, TCGA, The University of Alabama, 1990.
15. Goldberg, David E. *Genetic Algorithms in Search, Optimization and Machine Learning.* Reading, MA: Addison-Wesley, 1989.
16. Harvey, I., P. Husbands, and D. T. Cliff. "Issues in Evolutionary Robotics." In *From Animals to Animats 2: Proceedings of the Second International Conference on Simulation of Adaptive Behaviour (SAB92)*, edited by J.-A. Meyer, H. Roitblat, and S. Wilson, 364–373. Cambridge, MA: MIT Press Bradford Books, 1993.
17. Harvey, Inman "The SAGA Cross: The Mechanics of Crossover for Variable-Length Genetic Algorithms." In *Parallel Problem Solving from Nature, 2*, edited by R. Männer and B. Manderick, 269–278. North-Holland, 1992.
18. Harvey, Inman "Species Adaptation Genetic Algorithms: The Basis for a Continuing SAGA." In *Toward a Practice of Autonomous Systems: Proceedings of the First European Conference on Artificial Life*, edited by F. J. Varela and P. Bourgine, 346–354. Cambridge, MA: MIT Press/Bradford Books, 1992.
19. Hesser, J., and R. Männer. "Towards an Optimal Mutation Probability for Genetic Algorithms." In *Parallel Problem Solving from Nature*, edited by H.-P. Schwefel and R. Männer. Lecture Notes in Computer Science, Vol. 496. Berlin: Springer-Verlag, 1991.
20. Holland, John *Adaptation in Natural and Artificial Systems.* Ann Arbor, MI: University of Michigan Press, 1975.
21. Husbands, P., and I. Harvey. "Evolution Versus Design: Controlling Autonomous Robots." In *Integrating Perception, Planning and Action, Proceedings of 3rd Annual Conference on Artificial Intelligence, Simulation and Planning*, 139–146. IEEE Press, 1992.

22. Kauffman, Stuart "Adaptation on Rugged Fitness Landscapes." In *Lectures in the Sciences of Complexity*, edited by Daniel L. Stein. Santa Fe Institute Studies in the Sciences of Complexity, Lect. Vol. I, 527–618. Reading, MA: Addison-Wesley 1989.
23. Koza, John R. "Genetic Programming: A Paradigm for Genetically Breeding Populations of Computer Programs to Solve Problems." Technical Report STAN-CS-90-1314, Department of Computer Science, Stanford University, 1990.
24. Langton, C. G., J. D. Farmer, S. Rasmussen, and C. Taylor, eds. *Artificial Life II.* Santa Fe Institute Studies in the Sciences of Complexity, Proc. Vol. XI. Reading, MA: Addison Wesley, 1991.
25. Langton, C. G., ed. *Artificial Life.* Santa Fe Institute Studies in the Sciences of Complexity, Proc. Vol. VI. Reading, MA: Addison Wesley, 1989.
26. Maynard Smith, John *The Evolution of Sex.* Cambridge: Cambridge University Press, 1978.
27. Meyer, J.-A., H. Roitblat, and S. Wilson, eds. *Proceedings of the Second International Conference on Simulation of Adaptive Behaviour (SAB92).* Cambridge, MA: MIT Press Bradford Books, 1993.
28. Nowak, Martin, and Peter Schuster. "Error Thresholds of Replication in Finite Populations, Mutation Frequencies and the Onset of Muller's Ratchet." *J. Theor. Biol.* **137** (1989): 375–395.
29. PRANCE. "Perceptive Robots: Autonomous Navigation and Cooperation Through Evolution." Unpublished research proposal, by the PRANCE consortium: University of Sussex, Cap Gemini Innovation, École Normale Supérieure (Paris), and Université Libre de Bruxelles, 1991.
30. Schaffer, J. D., R. A. Caruana, L. J. Eshelman, and R. Das. "A Study of Control Parameters Affecting Online Performance of Genetic Algorithms for Function Optimization." In *Proceedings of the 3rd ICGA*, edited by J. D. Schaffer. San Mateo, CA: Morgan Kaufmann, 1989.
31. Smith, Stephen F. *A Learning System Based on Genetic Adaptive Algorithms.* Ph.D. Thesis, Department of Computer Science, University of Pittsburgh, 1980.
32. Varela, F., E. Thompson, and E. Rosch. *The Embodied Mind.* Cambridge: MIT Press, 1991.
33. Viola, P. "Mobile Robot Evolution." Bachelors Thesis, Massachusettes Institute of Technology, 1988.
34. Whitley, D. "The Genitor Algorithm and Selection Pressure: Why Rank-Based Allocation of Reproductive Trials is Best." In *Proceedings of the 3rd ICGA*, edited by J. D. Schaffer. San Mateo, CA: Morgan Kaufmann, 1989.

Craig W. Reynolds
Electronic Arts, 1450 Fashion Island Boulevard, San Mateo, CA 94044; e-mail: creynolds@ea.com and cwr@red.com

An Evolved, Vision-Based Model of Obstacle Avoidance Behavior

Obstacle avoidance behavior can emerge from evolution under selection pressure from an appropriate fitness measure, using a simple computational model of visual perception and locomotion. The Genetic Programming paradigm is used to model evolution. Both the structure of the visual sensor array, and the mapping from sensor data to motor action is determined by an evolved control program. The agent's forward motion is constant and innate, its steering rate is limited. Collisions can be avoided only by effective steering. Fitness is based on the number of simulation steps the agent can run before colliding with an obstacle.

INTRODUCTION

In the work described here, a behavioral controller for a two-dimensional *critter* (artificial animal, autonomous agent, robot, or what-have-you) is obtained through simulated evolution. A stimulus-response controller for a *vision-based obstacle avoidance* task can spontaneously emerge under selection pressure from a fitness measure based on the task. The evolutionary process starts with primitive computational elements that provide simulated perception and motor control, as well as arithmetic and logical connections between them. The fitness of an individual controller is determined by installing it in a simulated critter in a simulated world and by judging the performance it achieves there. The critter moves forward at a constant rate and the controller must "steer" to avoid collisions with obstacles in its simulated environment. Collisions are considered fatal. A critter's fitness is based on how long it can run before hitting an obstacle. Over time, under the patient guidance of fitness testing, the evolutionary process constructs an increasingly effective mapping from perception to motor control which allows the critters to become increasingly effective at avoiding collisions.

This work is not intended to be a realistic model of the evolution of vision-based obstacle avoidance in natural animals. Instead it provides an abstract example of how this behavior can arise in an evolutionary process. It provides a computational model for examining theories about how specific selection pressures and various environmental factors affect the evolution of obstacle avoidance strategies.

While admittedly unconnected to natural evolution of corresponding behaviors in real animals, this work aspires to a certain level of plausibility by the *closed* nature of its simulated world. The controller's action is fully determined by the information it obtains about the simulated world through its simulated perceptions. The fitness of a controller is fully determined by the performance of its simulated behavior in the simulated world. The critter's behavior is *grounded* in its perception of the world, and its perception directly reflects the consequences of its behavior.

Maja Mataric has observed that "obstacle avoidance is what robots spend most of their time doing."[26] It is only after these seemingly simple behaviors are correctly handled that the presumably more complex behaviors can be built on top of them. That obstacle avoidance has such an important and central role may seem counterintuitive. We humans pay scant attention to our own obstacle avoidance because it happens on such a low, almost subconscious, level. To become (painfully!) aware of how much time we spend engaged in active vision-based obstacle avoidance, we need only close our eyes while walking through an unfamiliar place to see how many obstacles we "discover."

One approach to evolving a sensor-based obstacle avoidance strategy is to postulate a certain fixed sensor array, specifying the number of sensors and their geometric configuration. Then the problem becomes the evolution of a control structure that maps those sensor channels into motor action. This approach would be appropriate if, for example, we were dealing with a pre-existing robot vehicle and wanted a controller for it. In contrast, natural animals *coevolve* their visual systems and motor control systems over the millennia; neither is fixed in advance. One goal of this project was to include this kind of coevolution in the model. Part of the motivation was a bias for a "natural" as opposed to "robotic" model. Another motivation was a concern that dictating sensor density or placement would indirectly dictate which obstacle avoidance strategy would emerge. The hope was that the unifying concept of simulated Darwinian evolution could simultaneously solve both the sensor placement problem and the obstacle avoidance problem. The results presented here will demonstrate that such coevolution can occur.

PREVIOUS WORK

Because the work described here overlaps with many areas of active research, the amount of related previous work is enormous. This section is an attempt to identify some of the most closely related work, with special emphasis on work that led to or inspired the current project. While all previous work on obstacle avoidance in autonomous agents is relevant to some degree, the work described here is concerned with behaviors that are: *evolved* (as opposed to hand-coded), *remote sensor based* (rather than using touch sensors, or global knowledge of the world), and based on *variable* sensor morphology (instead of static placement of a predefined number of sensors).

A classical work in behavioral modeling is by Braitenberg,[5] who touched on many of the ideas here, but as thought experiments whose implementation was somewhat fanciful. Many researchers (for example, Brooks[6] and Nehmzow[24]) have worked with simple obstacle avoidance for real robots based on touch sensors.

Obstacle avoidance techniques have been developed for synthetic actors in computer animation: see Amkraut,[2] Girard,[16] Reynolds,[28,29] and Ridsdale.[31,32] All of these were based on global knowledge about the environment, which, while suitable for their intended application, is an unnatural model for an autonomous agent.

In contrast, Renault[27] developed a hand-coded, vision-based obstacle avoidance technique for animation applications. In this system the agent examined its simulated environment by a variant of computer graphic *rendering,* producing a z-buffer (range image) in addition to a color perspective image. This allowed the agent to direct its motion based on information about the distance to various obstacles in its path. Renault's system made use of "object tags" which allowed it to unambiguously determine what object was visible at each pixel and how far away it was. This too can be seen as a form of global knowledge, or as imposing a constraint on the environment, such as requiring every object to be painted a unique color.

An example of non-evolved, fixed-sensor architecture, obstacle avoidance for real-time robotics based on video image processing is presented by Horswill.[21]

Evolved vision-based behaviors are central to PolyWorld (see Yaeger[39]), a conceptually vast, open-ended Artificial Life simulator. In PolyWorld, as in nature, creatures thrive to the extent that they are able to find food and to mate. Unlike a genetic algorithm system, there are no externally imposed fitness tests in PolyWorld, and no explicit goal other than survival. Simple behaviors like obstacle avoidance can appear in PolyWorld, but only as a side effect of the creature's quest for survival. The simulated world is rendered to provide color pixel data for a one-dimensional retina which is the primary input to the creatures behavioral controller.

An interesting example of another kind of evolved behavior (see Ventrella,[37]) is a legged critter whose interactive, physically based walking behavior was derived through evolution. The walker's fitness was measured by its ability to move toward a globally designated "food" location. In a similar vein, Maes[25] describes robotic leaning of coordinated leg motions, and de Garis[15] used a mutation-based technique to produce controllers for leg motion. In both cases, fitness was based on achieving forward motion.

The behavioral controllers described in this chapter are very much in the spirit of the subsumption architecture originally described by Rodney Brooks.[6] These reactive agents base their behavior directly on the world as perceived through their sensors. They have little or no higher cognition and do not bother with complex mental models of their environment, preferring to use "the world is its own map."[8]

The current work is very closely related to recent research by Randy Beer and colleagues. His earlier work (Beer[3]) involved painstakingly duplicating by hand the behavioral repertoire of an insect. His more recent work[4] derives weights for continuous-time recurrent neural net behavioral controllers using simulated evolution based on fitness criteria that measure achievement of the desired behavior. A detailed analysis of the complex dynamics of these controllers can be found in a paper by Gallagher.[14]

The work reported here was also strongly influenced by Dave Cliff's manifesto on *computational neuroethology*[9] as well as his hoverfly simulation.[10] While the models described here are not based on neurons, the principles of using a closed, grounded simulation to test behavioral models are fundamental to this project. Contemporaneous with the work reported here, the Evolutionary Robotics group[17] at the University of Sussex were engaged in some very closely related experiments, subsequently reported at SAB92. Harvey et al.[17] describe evolving dynamical, recurrent neural nets to control a simulated robot using touch sensors engaged in tasks such as "wandering" and "exploring" a space cluttered with obstacles. In work by Cliff et al.,[11] dynamical, recurrent neural nets are evolved to control a simulated vision-guided robot whose task was to move to, and remain at, the center of a circular room. These vision-guided robots had exactly two eyes, but various parameters such as placement and "zoom" were under the control of evolution. A significant difference between these projects and the work reported here is that the Sussex group used noise at each node in their neural nets, both to simulate real-world conditions and to require evolution to discover robust controllers that do not (can not) depend on coincidental or chance correlations in the simulated world.

At about the same time this author did some experiments, using a model essentially identical to the one described here, to investigate the evolution of coordinated group locomotion behavior.[30] In those experiments, in order to survive, groups of critters needed to avoid collisions with static obstacles, to coordinate with (not run into) each other, and to flee from a pursuing predator. The results obtained can at best be considered preliminary.

GENETIC ALGORITHMS, GENETIC PROGRAMMING, AND THE STEADY STATE

At the heart of the work described here is the notion of simulated evolution. The basic evolution model used here is the venerable Genetic Algorithm ("GA"), originally developed by Holland.[19] The Genetic Algorithm has been widely studied and applied to a myriad of practical problems in many different fields.[20] Over the years, many variations on the basic Genetic Algorithms have been proposed. A hybrid of two such variations are used to implement the simulated evolution described in this paper.

John Koza forged a link between the Genetic Algorithm and computer programming technology with his Genetic Programming paradigm.[22,23] Genetic Programming ("GP") is a technique for automatically creating computer programs (often, but not necessarily, in the LISP language) that satisfy a specified fitness criteria. There is a very strong analogy between the operation of the Genetic Algorithm and Genetic Programming; the main difference is in the representation of genetic information: bit strings in GA, fragments of LISP code in GP. The use of fitness-proportionate (or rank-based, or tournament) selection, reproduction, crossover, and mutation are all directly analogous.

One significant difference is that while classic Genetic Algorithms work on fixed-length bit strings, Genetic Programming deals with objects of inherently varying size. The complexity of programs created by GP tends to correspond to the complexity of the problem being solved. If simple programs do not satisfy the fitness function, the Genetic Programming paradigm creates larger programs that do. As a result, Genetic Programming does not require that the user know, or even estimate, the complexity of the problem at hand. This was an important consideration for the goal of coevolving sensor arrays and obstacle avoidance strategies. We did not want to specify, for example, how many sensors should be used. Genetic Programming did not require such a specification; instead that implicit parameter could be left to evolve its own preferred value.

Karl Sims independently developed the idea of using LISP code as genetic material. He used the concept in combination with a system much like The Blind Watchmaker ("BW"),[13] where fitness is based on a human's aesthetic judgment. An analogy which might help clarify the relationships is that GA is to GP as BW is to Sims' work. A discussion of crossover and mutation operations can be found in Sims.[33] These mutation operations were incorporated into the GP system described in this paper, but, following Koza's lead, the experiments described here did not use mutation.

Gilbert Syswerda described a variation on traditional GA he called *Steady-State Genetic Algorithms* ("SSGA") (see Syswerda,[35] Appendix A) An analysis of their performance can be found in papers by Syswerda[36] and Davis.[12] A similar technique had previously used in classifier systems (see Holland,[19] pp. 147–148). Darrell Whitley independently discovered this variation.[38] While the term "steady state" has apparently become accepted, the comparison of "traditional GA" versus "steady state GA" suggests terms like "batch" versus "continuous" to this author.

The basic idea of SSGAs is to do away with the synchronized generations of traditional GAs. Instead there is a continuously updated population ("gene pool") of evolving individuals. Each step of the SSGA consists of fitness proportionate (or rank-based or tournament) selection of two parents from the population, creation of new individual(s) through crossover and (occasional) mutation, removal of individual(s) from the population to make room for new individual(s), and finally insertion of the new individual(s) into the population. An additional requirement is that all individuals in the population are required to be unique. (The step of creating new individuals loops until unique offspring are found.) In the work reported here, the concept of SSGA was applied to GP to produce a system for "Steady-State Genetic Programming."

SSGAs are clearly more *greedy,* because they focus more intently on high-fitness members of the population. For "easy" problems at least, this implies that a SSGA will find a solution more quickly than a traditional GA. However, what "greediness" actually implies is faster convergence of the population. This is good news if the population happens to converge on a global maximum, but it is clearly bad news if the population converges prematurely on a local maximum. The likelihood of being stranded on a local maxima is probably what characterizes a "hard" problem.

OBSTACLE AVOIDANCE AS GENETIC PROGRAMMING

In order to solve a problem with Genetic Programming we must restate the problem in a canonical form. We must specify a list of *functions* and a list of *terminals.* The Genetic Programming paradigm will then evolve programs as hierarchical expressions with the functions applied to subexpressions which are either one of the terminals or (recursively) another such expression. These hierarchies are known variously as "s-expressions," "lists," "LISP fragments," "parse trees," and so on.

The terminals used in the evolved obstacle avoidance problem are just an assortment of numerical constants: 0, 0.01, 0.1, 0.5, and 2. The list of functions is:

```
+
-
*
%
abs
iflte
turn
look-for-obstacle
```

The functions `+`, `-`, and `*` are the standard Common LISP arithmetic functions for addition, subtraction, and multiplication (see Steel[34]). Each of them take an arbitrary number of arguments. The function `abs` is the standard Common LISP absolute value function which takes one argument. The functions % and `iflte` are suggested by Koza.[23] Koza calls "%" *protected divide*, a function of two arguments A and B, which returns 1 if B = 0 and A/B otherwise. The conditional `iflte` combines the standard Common LISP functions `if` and `<=` into "if less than or equal to." In the implementation described here, `iflte` is a LISP *macro* which makes this source-level transformation:

```
(iflte a b c d) → (if (<= a b) c d)
```

The functions `turn` and `look-for-obstacle` are specific to the obstacle avoidance problem. Each of them take a single argument, an angle relative to the current heading. Angles are specified in units of *revolutions,* a normalized angle measure: one revolution equals 360 degrees or 2π in radians. These functions will be explained more fully below, but basically: `turn` steers the critter by altering its heading by the specified angle (which is returned as the function's value) and `look-for-obstacle` "looks" in the given direction and returns a measure of how strongly (if at all) an obstacle is seen.

We must also provide the *fitness function* that the Genetic Programming paradigm will use to judge the fitness of the programs it creates. The fitness function takes one argument, an evolved program, and returns a numerical fitness value. In the implementation described here, fitness values are normalized to lie in the range between zero and one inclusively. A fitness of zero means "totally unfit," a category that can include programs that get errors during execution. A fitness of one signifies a perfect solution to the problem at hand.

Finally there are a few other parameters required to specify a Genetic Programming run. In these experiments the maximum size of programs in the initial random generation is set to 50. The size of the "steady-state gene pool" (which is roughly comparable to the population in a traditional "batch" generation GA) is 1000 individual programs. The mutation rate is zero.

THE CRITTER

The critter model used in these experiments is a computer simulation based on widely used principles of computational geometry and computer graphics. Its simplicity and abstraction make it an equally good (or by the same token, equally bad) model of a living creature, or a mechanical robot, or simply an abstract synthetic agent.

The critter moves on a two-dimensional surface. It is essentially equivalent to the LOGO *turtle* (see Abelson[1]). Its state consists of a position and an orientation. In the accompanying illustrations, the critter's body is depicted as a triangle to indicate its heading. For purposes of collision detection, however, its body is modeled as a disk of unit diameter. Besides the critter itself, the world includes various *obstacles.* In these experiments, the obstacles are represented as line segments. A *collision* is when the critter gets too close to an obstacle. Specifically, whenever the distance from its center point to the closest point on any obstacle is less than one half its body length, a collision has occurred and obstacle avoidance has failed. Collisions are considered to be fatal: a critter's life span is determined by how long it can avoid collisions with all obstacles in the world. Figure 1 shows the critter and the obstacle course named "Bilateral" which was used in all of the experiments described here.

There are two kinds of motor actions in the critter's repertoire: *move forward* and *turn*. In the problem being studied here, we will assume that forward motion is constant and innate. The critter will always move forward (that is, along its own heading) by one half of its body length each simulation step. This implies that we are actually dealing with the "*constant-speed*, vision-based obstacle avoidance" problem.

Beyond these fixed innate properties, the critter's *volition* comes from its evolved control program. Typically the controller will use the perception primitive `look-for-obstacle` to get information about the environment, do some conditional and arithmetic processing, and then a `turn` based on the result. Most of the effective evolved control programs contain many calls to both `look-for-obstacle` and `turn`.

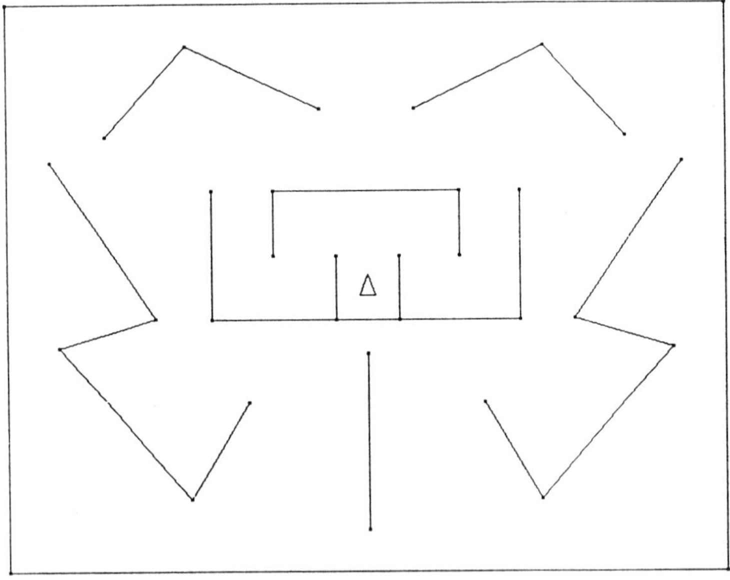

FIGURE 1 Critter in the obstacle course named "Bilateral."

Two kinds of limitation were placed on the critter's ability to `turn`. The first is that if the controller specifies an excessive turning angle it will be truncated to a "reasonable" range. In the experiments described here, "reasonable" is defined to be ± 0.05 revolutions (18 degrees or 0.31 radians). Secondly the critter will die ("from whiplash") if it turns too fast. This can happen despite the first limitation because the controller is capable of performing more than one `turn` operation per time step. The fatal `turn` threshold used in these experiments is ± 0.075 revolutions (27 degrees or 0.47 radians). The result of these limitations is to give the critter a non-zero turning radius. The motivation is to prevent them from spinning in place. Strictly speaking, spinning in place *is* a valid solution to the obstacle avoidance problem. Because spinning is a degenerate and not particularly interesting solution, it has been ruled out by these limitations on the turning angle.

PERCEPTION

The `look-for-obstacle` function simulates the critter's perception and so is the controller's only source of information about its world. All adaptive, time-varying behavior must be derived somehow from the variations of these perceptions as the critter moves through the world. When `look-for-obstacle` is called, a *ray-tracing* operation is performed. That is, a geometric ray ("half line") is constructed from the critter's position in the direction specified by the sum of the critter's heading and the argument to `look-for-obstacle`. The intersection (if any) of that ray with each obstacle is calculated. The obstacle whose ray intersection is closest to the critter's center is deemed the *visible obstacle.*

In order to provide the controller with an indirect clue about distances, we have postulated that the critter's world is a very foggy place. Visual stimuli are attenuated by distance. Certainly this aspect of the simulation can be criticized as being *ad hoc.* But the alternatives are daunting: without attenuation, a critter surrounded by obstacles is in a featureless environment; in whatever direction it looks it sees an obstacle. Postulating more complicated sensory mechanisms (such as stereo vision or texture recognition) seemed too complex and would have introduced a new crop of *ad hoc* details to be explained away.

The "foggy world" model is somewhat plausible: real fish in murky water face a very similar perceptual realm. The phenomenon known as "aerial perspective" refers to attenuation in air caused by dust and water vapor. Simulated aerial perspective is widely used (in classical painting, photo-realistic computer graphics, and cinematography of miniatures for special effects) to provide a sense of distance and scale. If desired, say for robotic experimentation, technology exists (in the special effects industry) for filling rooms with "smoke" of precisely controlled density. The last rationalization for the foggy world model is that it is only slightly different in effect from perception based on active sonar as used by bats and dolphins. A similar effect would be obtained for a real robot using narrow-angle photoreceptors (what photographers call "spot meters") measuring diffuse reflections from uniformly colored walls if the only light source was located on the robot itself. The $1/r^2$ falloff of illumination with distance would produce a result very similar to the "foggy world" model presented here.

The value returned from `look-for-obstacle` is a number between zero and one. A value of one would indicate that the obstacle is coincident with the critter, but this does not occur in practice since this would imply a collision had occurred. As the distance between the critter and the obstacle increases, the visual signal drops off quadratically in strength. At a certain threshold value (15 body lengths in these experiments) the signal will have reached zero. Hence a value of zero returned from `look-for-obstacle` indicates that the closest obstacle in the given direction (if any) is more than 15 units away.

In the original conception of this project, it was imagined that evolution would design a sort of retina consisting of a series of calls such as `(look-for-obstacle 0.1)` with various constant values used for the argument. Hence evolution would select the "field of view" and density distribution of the retina. As always, evolution turned out to have its own ideas. There was nothing in the problem statement that said the argument `look-for-obstacle` had to be a constant, and so the controllers soon evolved bizarre mechanisms based on calculating a dynamic value and using that to specify the direction in which to look. For example, one species of obstacle avoider seemed to be based on the puzzling expression `(look-for-obstacle (look-for-obstacle 0.01))`. In hindsight it becomes clear that the correct model is not that of a retina but rather a simple form of *animate vision* (see Cliff[10]) where the controller "aims" its visual sensor at the area of interest.

FITNESS

In order to test the fitness of a newly evolved control program, we place the critter in the world, start it running, and measure how long it can avoid collisions with the obstacles. Specifically, the critter is allowed to take up to 300 steps. The number of steps taken before collision, divided by 300, produces the normalized raw score.

The raw score is modified by multiplication by some "style points" (less objective criteria). To encourage critters to use generally smooth paths, they are penalized for "excessive turning." In order to encourage symmetrical behavior and discourage looping paths, the critters are penalized for "unbalanced turning." Statistics on turning are accumulated during the run, a sum of all `turn` angle magnitudes, a sum of signed `turn` angles, and a count of left versus right steps. A critter that made the maximum legal `turn` each time step would receive the "most harsh" excessive turn penalty of 0.5, whereas a critter that never turned would get value of 1.0 (i.e., no penalty). The penalty for "unbalanced turning" is analogous.

A single fitness test will produce a controller for solving the one exact situation represented by the fitness test. However, this solution may turn out to be very fragile and opportunistic. It may well have incorporated expectations of incidental properties of the specific fitness test. In the case of obstacle avoidance, this means that a controller would evolve that could nimbly negotiate a given obstacle course flawlessly, but could be utterly stumped by a slightly different obstacle course or by starting its run in a slightly different initial placement.

To strive for *robust* obstacle avoidance we need an alternative to a single fitness test. One solution would be to randomize some aspect of the world, such as obstacle placement or the critter's starting position. This is an appealing approach since fitness trials in nature are effectively randomized. When randomization was attempted in these experiments, it became clear that the noise injected into the fitness values made it very hard for both the experimenter and for simulated evolution to determine if progress was actually being made. Another approach is to have multiple obstacle courses for each critter to run, or the critter can make multiple runs through the same obstacle course from slightly different starting positions. These both have the effect of discouraging fragile solutions, but do not introduce uncorrelated noise into the fitness measure. Some other interesting approaches are discussed in the Future Work section below.

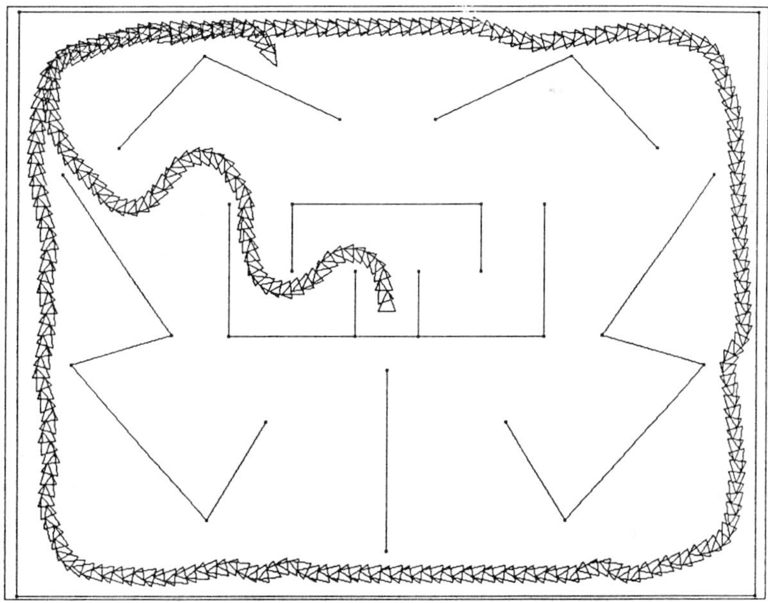

FIGURE 2 Trail of 300 steps of the best individual.

RESULTS

The end result of two runs and some analysis is presented here. One run (named "bilat-pool-1") is as described above but is based on a single (non-randomized) fitness test per individual. Figure 2 shows the "trail" of 300 steps of the best individual after about 22,000 fitness tests.

Note that while this critter is capable of moving through "wide open spaces," it displays a pronounced preference to get close to a wall and move along it. When observed in real animals, this behavior is called *thigmotaxis,* and the term seems equally appropriate here. The critter seems to depart from closely following the outside wall only when it passes the vicinity of an obstacle on its other side. Evolution has discovered that there is a relatively uncluttered path around the edge of the world. (A feature which was not intentionally designed in by the author.) This critter seems content to spend its lifetime circumnavigating the world. After some hand-simplifying of the control program, to removed redundant or ignored calculations, the circumnavigator is:

```
(turn (- 0.1
         0.02
         (% 0.02
            (look-for-obstacle 0.01))
         1
         (look-for-obstacle 0.1)
         (* (look-for-obstacle 0)
            (look-for-obstacle 0.01))
         0.1))
```

An interesting detail of this program is that it does not contain any conditionals. The GP function set used here contains the conditional primitive `iflte` but, in this case, evolution chose to solve the problem without it. If a human programmer was asked to write a controller for this problem, it is very likely that the resulting program would contain conditionals.

Because this individual avoided all collisions, its raw score was 1.0; its final score was 0.64 because it got an unbalanced turn penalty of 0.93 (presumably because it was turning right as it circled the world); and it got an excessive turn penalty of 0.69, which is quite severe. The reason for this can be seen in the "gait" this creature developed. Looking at its trail one can see that it jitters from side to side on almost every step. This behavior has been observed in many of the evolved critters. One possible explanation is that since the critter has a limited set of sensors, it has many "blind spots." A gait that causes its sensors to sweep back and forth should provide it with much better sensor coverage of its path. A plot of best-of-run fitness and population-average fitness after every 1000 individuals is shown in Figure 3.

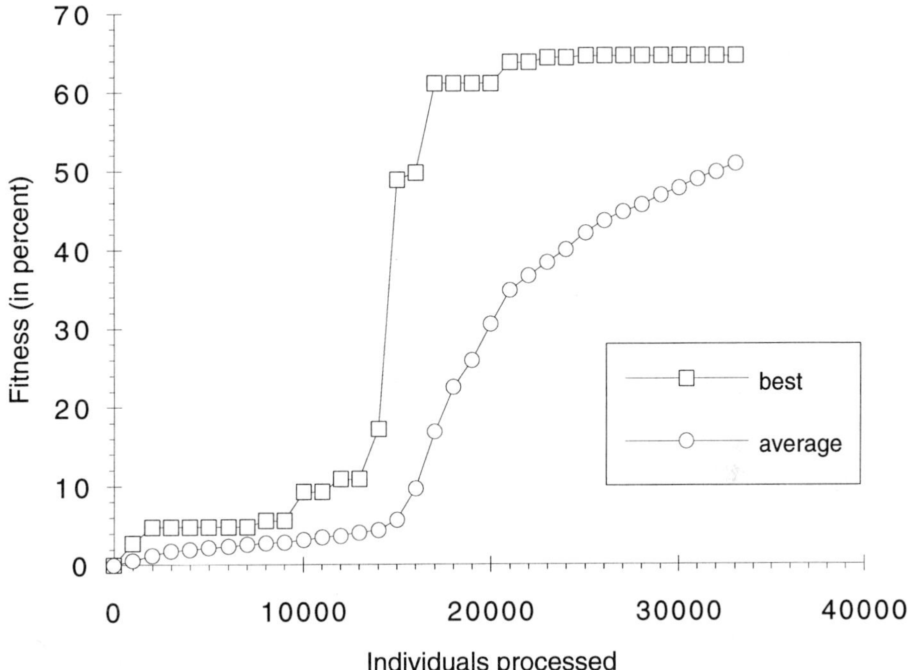

FIGURE 3 Plot of best-of-run fitness and population-average fitness, for a single run.

The second example (named "bilat-repeat-1") was identical except that its fitness function consisted of three separate runs through the Bilateral obstacle course with three slightly different starting positions. All three were within one-third of a body length of the original starting point. Before the run there was some concern that these starting position were too similar and would not present the controller with enough variation to promote robust behavior. In fact, experience showed that most controllers got markedly different fitness values for each of the three starting conditions. The final fitness for each individual was defined to be the average of its three runs.

The best individual found appeared after about 31,000 trials and had not been improved upon by 39,000 trials. The fitness of that individual was 0.22, the average of its three runs which scored 0.46, 0.19, and 0.02. Its average penalties were 0.69 for excessive turning and 0.71 for unbalanced turning. The three overlaid "trails" for this individual are shown in Figure 4. A plot of best-of-run fitness and population-average fitness after every 1000 individuals is shown in Figure 5. The hand-simplified code for the best-of-run individual is:

```
(iflte (look for obstacle (look for obstacle 0.01))
       (* 2
          (look for obstacle 2)
          (look for obstacle -0.001))
       (turn (* (look for obstacle -0.01)
                (look for obstacle 0.1)))
       (+ (turn (+ (look for obstacle 0.1)
                   (look for obstacle 0.1)))
          (turn 0.01)))
```

One odd feature of this run is that at one point the critter gets into a quasi-periodic looping behavior. Like spinning in place, looping is a valid solution to the fitness function as defined above, but it seems somewhat degenerate and does not fit our intuitive idea of "good" critter behavior. If we saw a real animal behaving like this, we might assume it was distressed or ill. From a dynamic systems point of view, the semi-stable nature of this behavior is interesting: it loops for several cycles, but then suddenly escapes. This suggests a sensitive dependence on certain aspects of the environment. The first five (?) times around that loop the controller decided to turn left, but the last time around, displaced only slightly from its state on the previous orbit, it decided to turn right.

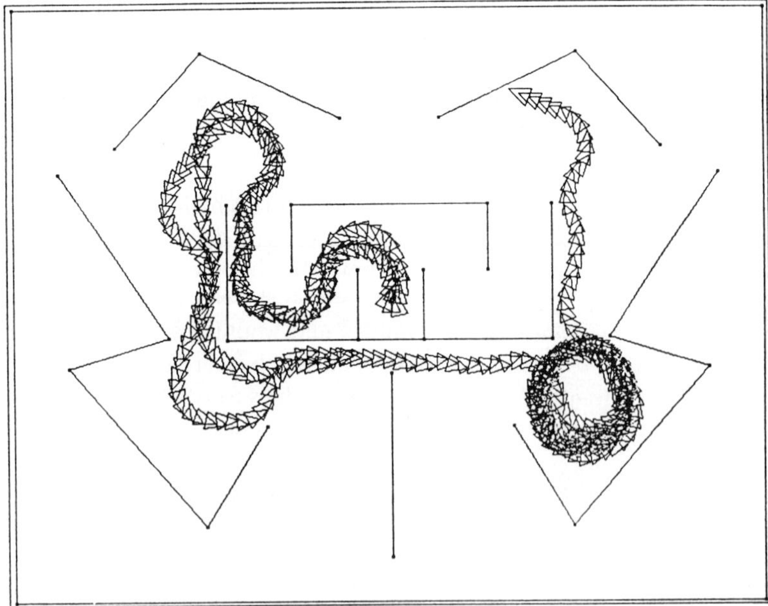

FIGURE 4 Three overlaid trails for the best individual.

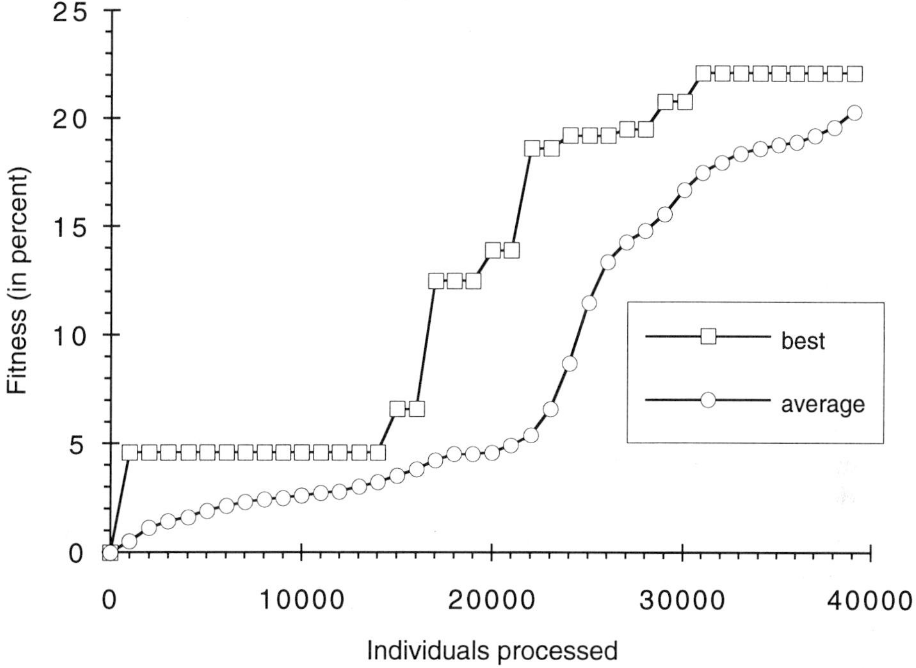

FIGURE 5 Plot of best-of-run fitness and population-average fitness, for runs per individual.

CONCLUSIONS

The preliminary results reported here represent only partial solutions to the problem of robust vision-based collision avoidance. None of the behaviors evolved so far are robust, general-purpose solutions to the full problem. Thoughts about how to proceed beyond these limitations are given in the next section.

It seems significant that while a nice solution was obtained for the first example described above ("bilat-pool-1"), the second example ("bilat-repeat-1") seemed to converge before a good solution was found. Similarly, the first example attained a fitness value almost three times higher than the second example, even though the latter was allowed to run for almost twice as many individuals. This suggests that using multiple fitness cases makes the problem significantly more difficult to solve. It seems reasonable to conclude that multiple fitness tests require increased generality in the controller, and that generality tends to preclude the "easy" solutions which are opportunistic and brittle. The results presented here do not fully justify these conclusions, and more work is required to better understand these issues.

On the other hand, there is no question that vision-based collision avoidance strategies have begun to emerge, given only the requisite primitives, an appropriate fitness measure, and the action of Darwinian evolution through natural selection and survival of the fittest.

FUTURE WORK

Preliminary results from a more recent set of experiments similar to those presented here tend to confirm the notion that adding fitness cases increases the apparent difficulty of the problem, as can be measured by the amount of "evolution work" (number of individuals processed) required to find a solution to the problem. Furthermore, attempting to add noise to the system, as suggested by Harvey,[17] seems to *significantly* increase the difficulty of the problem.

Future work in this area needs to investigate the effect of noise in controller programs. A systematic comparison should be made between the effect of using the average or the minimum of the multiple values obtained in fitness testing. This has been done in the neural net domain by Harvey,[17] and we would like to compare those results to what happens in the GP domain.

As was noted above, spinning or looping behavior is in some sense degenerate and undesirable. To incorporate this bias, an alternative measure of fitness would be to keep track of how much area the critter has covered before collision. This could be accomplished, for example, by overlaying a grid on the critter's world and setting a flag in each grid cell the critter visits. The number of flagged cells would be a quantized approximation of area covered. Using this measure instead of, or in addition to, the number of steps taken before collision would promote collision-free paths that do not cross over themselves repeatedly. This "*exploratory* obstacle avoidance" criteria may yield a more interesting kind of behavior.

While the obstacles used in these experiments are static, nothing in the model depends on that fact. The critters have no maps of the world nor memory of any kind. As a result, their obstacle avoidance strategies are completely *reactive* (see Brooks[7]), so we would expect them to be able to deal equally well with dynamic obstacles. An interesting variant of the current experiment would be to evolve avoidance in the presence of obstacles that move, or that appear and disappear "at random."

It is anticipated that better, more robust, obstacle avoidance strategies could be obtained by using a sequence of ever more challenging obstacle courses. But that approach puts a significant burden on the human designer of the obstacles. First, there is the work involved in creating some large number of increasingly complex obstacle avoidance test cases. Second, as the critters become proficient at obstacle avoidance, it would become increasingly difficult to come up with obstacle courses that would challenge them. These considerations suggest that perhaps the obstacles should be seen as *predators* or *parasites* that coevolve in competition with the critters. This kind of coevolution of problem-solvers and problem-poser has been examined by Hillis.[18] The "fitness" of an evolved obstacle would be judged by how well it could frustrate a critter's attempts at collision avoidance. In the confrontation between a given critter and obstacle, a collision early in the simulation would reflect poorly on the critter and well on the obstacle. Conversely, a collision late in the simulation (or the absence of a collision) reflects well on the critter and poorly on the obstacle.

ACKNOWLEDGMENTS

The author wishes to thank his current employer, Electronic Arts for sponsoring the preparation of the final draft of this paper, and for sponsoring ongoing research on related topics.

The following remarks refer to the original work performed in the summer of 1992:

> The research reported here was pure amateur science. It was not an officially sanctioned project of any corporation, university, or other funding agency. The individuals and companies listed below helped the author out of the goodness of their hearts and their interest in the subject matter. For generous grants of computer facilities, I am deeply indebted to Tom McMahon and Bart Gawboy of Information International Incorporated, Koichi Kobayashi of Nichimen Graphics Incorporated, and Richard Hollander of Video Image Associates. Special thanks go to Andrea Lackey whose workstation I repeatedly commandeered. Heartfelt thanks to Andy Kopra who cheerfully put up with my intrusions to his busy production schedule, and who helped brainstorm some of these ideas back at ALife II. Thanks to: Larry Malone, Dave Dyer, Jay Sloat, Dave Aronson, Joseph Goldstone, Glen Neufeld, DJ, and Antoine Durr. Thanks to Charles Taylor of UCLA for supplying the word "thigmotaxis." Finally, this work would not have happened without John Koza's trail-blazing work on the Genetic Programming paradigm. Many thanks to John and to James Rice for their helpful and encouraging comments during this project.

REFERENCES

1. Abelson, Harold, and Andrea diSessa. *Turtle Geometry: The Computer as a Medium for Exploring Mathematics.* Cambridge, MA: MIT Press, 1981.
2. Amkraut, Susan, and Michael Girard. "Eurhythmy." In *SIGGRAPH Video Review,* item 8, issue 52. ACM-SIGGRAPH, 1989.
3. Beer, Randall D. *Intelligence as Adaptive Behavior: An Experiment in Computation Neuroethology.* Boston, MA: Academic Press, 1990.
4. Beer, Randall D. and John C. Gallagher. "Evolving Dynamical Neural Networks for Adaptive Behavior." *Adapt. Beh.* **1** (1992): 91–122.
5. Braitenberg, V. *Vehicles: Experiments in Synthetic Psychology.* Cambridge, MA: MIT Press, 1984.
6. Brooks, R. "A Robust Layered Control System for a Mobile Robot." *IEEE J. Robot. & Automation* bf 2(1) (1986).
7. Brooks, R. "Elephants Don't Play Chess." In *Designing Autonomous Agents,* edited by Pattie Maes, 3–15.
8. Brooks, R. "Intelligence Without Representation." *Art. Int.* **47**(1991): 139–160.
9. Cliff, D. "Computational Neuroethology: A Provisional Manifesto." In *From Animals to Animats: Proceedings of the First International Conference on Simulation of Adaptive Behavior* (SAB90), edited by Meyer and Wilson. Cambridge, MA: MIT Press, 1991.
10. Cliff, D. "The Computational Hoverfly: A Study in Computational Neuroethology." In *From Animals to Animats: Proceedings of the First International Conference on Simulation of Adaptive Behavior* (SAB90), edited by Meyer and Wilson. Cambridge, MA: MIT Press, 1991.
11. Cliff, D., P. Husbands, and I. Harvey. "Evolving Visually Guided Robots." In *From Animals to Animats 2: Proceedings of the Second International Conference on Simulation of Adaptive Behavior* (SAB92), edited by Meyer, Roitblat, and Wilson, 374–383. Cambridge, MA: MIT Press, 1993.
12. Davis, L., ed. *Handbook of Genetic Algorithms.* New York: Van Nostrand Reinhold, 1991.
13. Dawkins, R. *The Blind Watchmaker.* Harlow: Longman, 1986.
14. Gallagher, John C., and Randall D. Beer. "A Qualitative Dynamical Analysis of Evolved Locomotion Controllers." In *From Animals to Animats 2: Proceedings of the Second International Conference on Simulation of Adaptive Behavior* (SAB92), edited by Meyer, Roitblat, and Wilson, 71–80. Cambridge, MA: MIT Press, 1993.
15. de Garis, Hugo. "Genetic Programming: Building Artificial Nervous Systems Using Genetically Programmed Neural Network Modules." In *Proceeding of the Seventh International Conference on Machine Learning,* edited by Porter and Mooney, 132–139. Los Altos, CA: Morgan Kaufmann, 1990.

16. Girard, Michael, and Susan Amkraut. "Eurhythmy: Concept and Process." *Visualization & Comp. Animat.* **1** (1990): 15–17.
17. Harvey, Inman, Philip Husbands, and Dave Cliff. "Issues in Evolutionary Robotics." In *From Animals to Animats 2: Proceedings of the Second International Conference on Simulation of Adaptive Behavior* (SAB92), edited by Meyer, Roitblat, and Wilson, 364–373. Cambridge, MA: MIT Press, 1993.
18. Hillis, W. Daniel "Co-Evolving Parasites Improve Simulated Evolution as an Optimization Procedure." In *Emergent Computation* (Special Issue of Physica D 42), edited by S. Forrest, 228–234. Cambridge. MA: MIT Press/North-Holland, 1990.
19. Holland, John *Adaptation in Natural and Artificial Systems.* Ann Arbor, MI: University of Michigan Press, 1975.
20. Holland, John. "Genetic Algorithms." *Sci. Am.* (1992): 66–72.
21. Horswill, I. "Proximity Detection Using a Spatial Filter Tuned in Three-Space." In *AAAI Fall Symposium on Sensory Aspects of Robotic Intelligence.* November, 1991.
22. Koza, John R. "Hierarchical Genetic Algorithms Operating on Populations of Computer Programs." In *Proceedings of the 11th International Joint Conference on Artificial Intelligence.* San Mateo, CA: Morgan Kaufmann, 1989.
23. Koza, John R. *Genetic Programming: On the Programming of Computers by Means of Natural Selection.* Cambridge, MA: MIT Press, 1992.
24. Nehmzow, Ulrich, John Hallam, and Tim Smithers. "Really Useful Robots." In *Intelligent Autonomous Systems 2,* 284–293. Amsterdam, 1989.
25. Maes, Pattie, and Rodney Brooks. "Learning to Coordinate Behaviors." In *Proceedings of the Eighth Nation Conference on AI* (AAAI-90), 796–802. AAAI, 1990.
26. Mataric, Maja J. "Minimizing Complexity in Controlling a Mobile Robot Population." In *Proceedings of the IEEE International Conference on Robotics and Automation,* 830–835. IEEE, 1992.
27. Renault, Olivier, Nadia Magnenat Thalmann, and Daniel Thalmann. "A Vision-Based Approach to Behavioral Animation." *Visualization & Comp. Animat.* **1** (1990): 18–21.
28. Reynolds, Craig W. "Flocks, Herds, and Schools: A Distributed Behavioral Model." (SIGGRAPH '87 Conference Proceedings) *Comp. Graph.* **21(4)** (1987): 25–34.
29. Reynolds , Craig W. "Not Bumping Into Things." In *Notes for the SIGGRAPH 1988 Course Developments in Physically-Based Modeling,* edited by Alan Barr, G1-G13. New York: ACM-SIGGRAPH, 1998.
30. Reynolds, Craig W. "An Evolved, Vision-Based Behavioral Model of Coordinated Group Motion." In *From Animals to Animats 2: Proceedings of the Second International Conference on Simulation of Adaptive Behavior* (SAB92), edited by Meyer, Roitblat, and Wilson, 384-392. Cambridge, MA: MIT Press, 1993.

31. Ridsdale, Gary J. "The Director's Apprentice: Animating Figures in a Constrained Environment." Ph.D. Thesis, Simon Fraser University. Also in Technical Report TR-87-2.
32. Ridsdale, Gary J., and Tom W. Calvert. "Animating Microworlds from Scripts and Relational Constraints." In *Proceedings of Computer Animation 1990.* Geneva, 1990.
33. Sims, Karl. "Artificial Evolution for Computer Graphics." (SIGGRAPH '91 Conference Proceedings) *Comp. Graph.* **25(4)** (1991): 319–328.
34. Steele, Guy L., Jr. *Common LISP: The Language,* 2nd ed. Bedford, MA: Digital Press, 1990.
35. Syswerda, G. "Uniform Crossover in Genetic Algorithms." In *Proceedings of the Third International Conference on Genetic Algorithms,* 2–9. San Mateo, CA: Morgan Kaufmann, 1989.
36. Syswerda, G. "A Study of Reproduction in Generational and Steady-State Genetic Algorithms." In *Foundations of Genetic Algorithms*, 94–101. San Mateo, CA: Morgan Kaufmann , 1991
37. Ventrella, J. "Walker" Video shown at Artificial Life III conference. Produced at Syracuse University, Syracuse, NY, 1990.
38. Whitley, D. "The GENITOR Algorithm and Selection Pressure: Why Rank-Based Allocation of Reproductive Trials is Best." In *Proceedings of the Third International Conference on Genetic Algorithms,* 116–121. San Mateo, CA: Morgan Kaufmann, 1989.
39. Yaeger, L. "Computational Genetics, Physiology, Metabolism, Neural Systems, Learning, Vision, and Behavior or PolyWorld: Life in a New Context" In *Artificial Life III,* edited by C. G. Langton. This volume.

David Zeltzer† and Michael McKenna‡
†Sensory Communication Group, Research Laboratory of Electronics, Massachusetts Institute of Technology, 50 Vassar St., Room 36-763, Cambridge MA, 02139;
e-mail: dz@irts.mit.edu
‡Medical Media Systems, 32 South Main Street, Hanover, NH 03755;
e-mail: mikey@media-lab.mit.edu

Simulation of Autonomous Legged Locomotion

INTRODUCTION

The idea of a technologically created artificial person has been with us for centuries—the synthesis of "natural" and "realistic" motions of human and animal figures has long been a goal in computer animation. In recent years, increases in computer performance have allowed the creation of representations of the real world that can be interacted with in real time. These *virtual environments* (VEs) allow us to unite the pursuits of animation, robotics, and human motor studie.

In this chapter, we will discuss work spanning nearly a decade by researchers at MIT concerning the simulation and animation of autonomous agents, with a focus on the simulation of legged locomotion of animals and people. A common theme seen throughout our work is autonomy—we have approached the problem of graphical simulation from a *task level.* High-level goals are given to our animated characters—such as "walk to the door," or "shake hands with the user," and the characters automatically perform the movements required to achieve the task. Over the years, our work has changed in focus from kinematic simulations, to dynamic, or physically based, simulations. Earlier work by Zeltzer[24] demonstrated a walking human skeletal model. Motion control was achieved by a hierarchy of finite state machines which kinematically controlled joint motions. Sims' work[21] allowed the interactive creation of animal forms, which could then be automatically animated to negotiate moderately uneven terrain using a variety of gaits. A kinematic model of hexapod locomotion was demonstrated by McKenna,[9] using a biologically inspired model of coordination based on coupled oscillators and reflexes. This autonomous "Roach" could be interacted with and commanded in the *bolio* virtual environment system.[26] The model was extended to incorporate a dynamic, or physically based model of motion, resulting in realistic and adaptive behaviors.[11] Our most recent research concerns a physically based model of human walking for computer animation. Preliminary results have focused on static balance and "passive" motions, which result from inherent system dynamics. We will conclude with a discussion of future work, including the addition of simulated skin and muscle to our rigid-body models, in order to create a more realistic and complete "artificial person."

SKELETON ANIMATION SYSTEM

In the early 1980's Zeltzer developed the *skeleton animation system* which was used to animate the motions of articulated figures.[24,25] His primary research concerned task-level interaction with a simulated human capable of walking over moderately uneven terrain, using a skeletal model. The simulation employed a hierarchy of finite state machines and kinematic motor programs to control joint motions. (See Color Plate 7.)

The simulations were kinematic in nature, using geometric rules to create motion. In other words, the motions were generated without regard to the forces that would be required to create them. Because the simulated motions were based on clinical data describing normal human gait,[6,20] the motions appear quite realistic for slow walking over planar terrain. However, rapid movements, or motion over uneven terrain, cannot be accurately simulated without at least accounting for the rigid-body dynamics of jointed figure motion. Thus, the strictly kinematic nature of the skeleton animation system limits its adaptability and restricts what can be learned from the simulation.

JOINTED FIGURE LOCOMOTION

Sims employed inverse kinematics and dynamic elements to simulate various adaptive gaits, over uneven terrain.[21] Sims' system allowed for the rapid, interactive creation of animal forms with varying numbers of legs and limb configurations. The animals could then be automatically controlled to walk, trot, run, hop, etc. (See Color Plate 8.)

Sims designed and implemented a *figure editor*, a visual tool for designing jointed figures.[21] The figure editor provides a MacDraw-style interface which allows the drawing of two-dimensional schematics of three-dimensional, jointed figures. (See Color Plate 9.) These two-dimensional representations are automatically transformed into three-dimensional, articulated figures—which can then be made to walk over uneven terrain using inverse kinematics and a gait sequencer.

Once a jointed figure has been created, the user simply selects the desired gait for the figure to use, and the system automatically generates the animated motions required to negotiate a given terrain. Different gaits are generated by specific functions which sequence the stepping of the legs, relative to each other. The leg motions are controlled using inverse kinematics, such that the target positions for the "feet" are specified, relative to the body, and the system computes the joint angles needed to reach that target.

KINEMATIC HEXAPOD

The *roach*, a real-time, kinematic simulation of a hexapod was developed by McKenna et al.[9] The user interacts with the hexapod in the virtual environment system *bolio*[26] (See Color Plate 10). Using a gestural interface, the user guides the walking behavior of the hexapod, issuing commands, specifying walking directions or target positions, etc. The hexapod controls its own low-level behavior, and can act autonomously in response to environmental "stimuli." The gait controller coordinates the stepping of the hexapod by triggering legs to step in patterns appropriate to the desired walking speed. An *oscillator*, or pacemaker, is associated with each leg. The oscillators rhythmically trigger the legs to step. Coupling between the oscillators creates the overall stepping pattern. This oscillator implementation is based on neurological mechanisms found in the cockroach and other insects, and is hypothesized to operate in mammals as well.[16,23] The stepping patterns generated by the coupled oscillators are virtually identical to the real patterns displayed by insects.

Different oscillator frequencies create different stepping pattern, and different walking speeds. The slow gaits are "wave" gaits, in which a wave of steps travels up each side of the body, from back to front (see Figure 1). The fastest gait generated by the coupled oscillator, is the tripod gait, in which a stable tripod of legs supports the body, while the other three legs step (see Figure 2). The coupled oscillator mechanism generates smooth gait changes, as the oscillator frequency is smoothly varied.

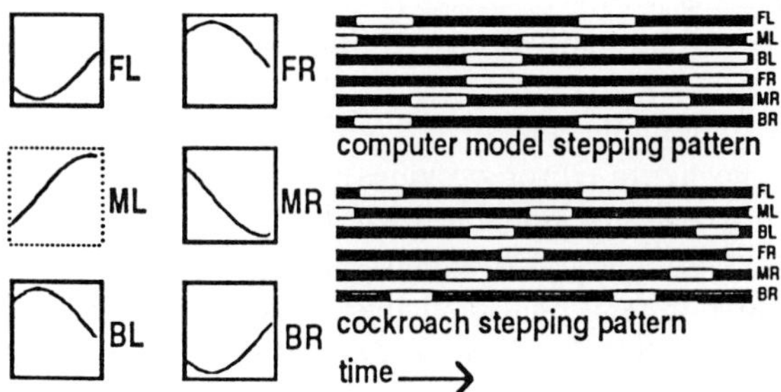

FIGURE 1 A wave gait stepping pattern, generated by the coupled oscillator mechanism, compared to a slightly different wave gait stepping pattern, exhibited by the cockroach. The white regions in the patterns denote stepping activity. The boxes and sine wave segments at the left depict the computational model of the oscillators over a short period of time. An oscillator will trigger stepping activity in its leg when the oscillation reaches its peak. The phase differences between oscillators can be seen in the diagram.

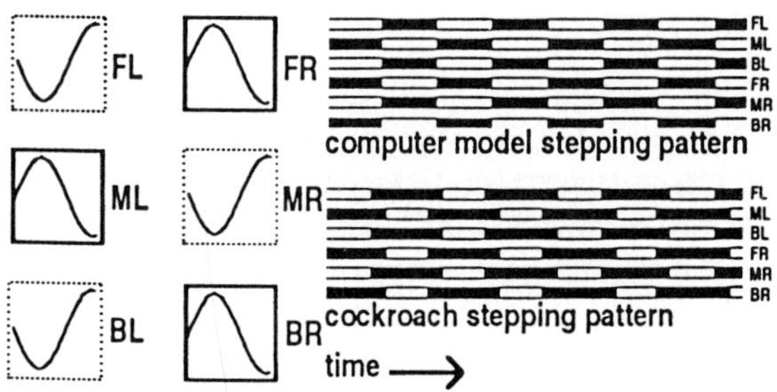

FIGURE 2 The tripod gait. The primary difference between the computer model and the cockroach stepping pattern is that the computer model produces longer step and shorter stance phases. Due to the idealization of the kinematic model, a leg can change from step to stance and take up the load instantaneously.

We have also experimented with *reflexive feedback* from the environment. A *step reflex* triggers stepping when the leg angle, relative to the body, exceeds a specified value. A simulated *"load-bearing"* reflex prevents a leg from stepping when an unstable leg configuration would result. These reflexes serve to reinforce the basic stepping pattern generated by the coupled oscillators, while making the system more robust.[9,10,11]

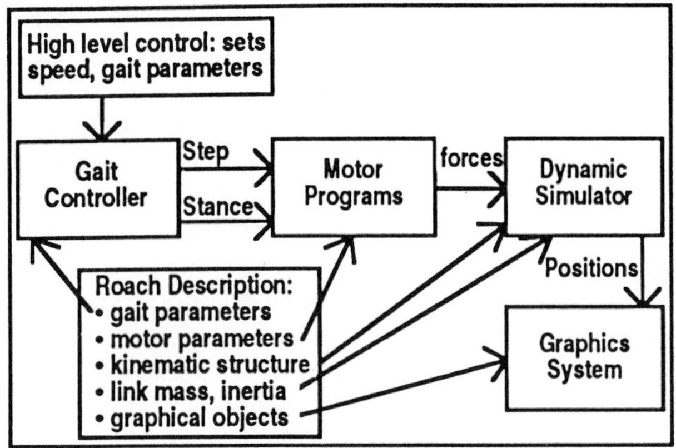

FIGURE 3 Block diagram of the dynamic hexapod control and simulation system, in the program *corpus*.

The gait controller triggers "step" and "stance" motor programs in the legs. The "step" program moves the leg up and forward, then down and forward, relative to the body. The "stance" program keeps the "foot" in place on the ground, as the body moves forward. Inverse kinematics is used to compute the leg joint angles from the specified foot location.

The *roach* software was used to create an animated character in the short animation *Cootie Gets Scared*[8] (see Color Plate 11). The control of the Cootie was "scripted" using high-level commands to control the walking speed and direction, as well as other character properties, such as head turning, etc. The leg and body motions were kinematically controlled; however, the head and antennae of the Cootie were dynamically simulated, to realistically respond to the force of gravity and the accelerations of the body.

DYNAMIC HEXAPOD

The hexapod locomotion research has been extended to a incorporate a dynamic model for motion simulation and control.[10,11] The kinematic structure of the dynamic roach is based on the general biomechanical properties of insects, in particular, the cockroach. The motions of the roach are simulated forward in time: all accelerations are the result of applied forces and torques. Locomotion is coordinated by the gait controller, as described above for the kinematic roach. Motor control for the legs is provided by dynamic motor programs and spring actuators, which deliver forces to the leg joints. Forward body motion is the result of traction at the ground, as the legs "push" backwards. The simulated roach displays stepping patterns very similar to those of real insects, as well as realistic walking behaviors.

The program *corpus* was developed by McKenna for the simulation and control of dynamic locomotion. The flow of control in *corpus* proceeds basically as follows (see Figure 3). At the highest level of control the animator, or another program, sets the desired speed and other parameters. The gait controller coordinates the stepping pattern based on the specified speed, and sends appropriate step and stance commands for each leg to the motor programs. The motor programs compute leg joint forces which are sent to the dynamic simulator. The simulator incorporates these forces with externally applied forces, such as gravity and contact forces, and computes the motion of the figure. The graphics system then renders the figure and its environment, based on the computed positions. Each module has several parameters that can be set in a simple scripting language. These parameters include such factors as stepping speed, spring stiffnesses, the kinematic structure of the figure, and link size, shape, and density.

The dynamic simulator employs the Articulated Body Method for dynamic simulation, based on the work of Roy Featherstone,[3,4] and is unfortunately too complex to describe in detail here. However, we will briefly discuss its operations, and its advantages and disadvantages.

The following is the equation of motion for a single rigid body:

$$\hat{f} = \hat{I}\hat{a} = \hat{p}^v \tag{1}$$

The carat ("ˆ") above the quantities denote that they are written in *spatial notation*, also developed by Featherstone, which unites the translational and rotational components of motion into single six-dimensional vector and matrix quantities. This allows for more compact equations, and faster solutions. The equation reads that "force equals inertia times acceleration plus a bias force." The bias force accounts for velocity-dependent forces. The equation of motion for a link, i, within an articulated figure is surprisingly similar to the basic equation of motion:

$$\hat{f}_i = \hat{I}_i^A \hat{a}_i + \hat{p}_i \tag{2}$$

The equation reads: "the force equals *articulated body inertia* times acceleration plus a bias force." The bias force, in this case, incorporates the velocity-dependent forces as well as forces transmitted through the joint. This equation establishes a linear relationship between the force applied to a body and its resulting acceleration, even though it is attached to, and therefore constrained by, other bodies.

The basic Articulated Body Method operates on arbitrary branching structures, without kinematic loops. Starting with the outermost, or "leaf," links in the figure, the articulated body inertia is computed inward to the base or "root" link. The acceleration of the root link is then solved, based on the applied forces, using its articulated body inertia. Alternately, the acceleration of the root can be constrained to any desired motion. The joint accelerations are then solved outward to the leaf links. The recursive solution makes the Articulated Body Method a very efficient method for dynamic simulation. Featherstone claims that it is the fastest method when the number of joints exceeds nine. The major drawback of this method is that loops cannot be simulated efficiently.

In order to derive the articulated figure we used in our locomotion experiments, we referred to diagrams of the insect Blatta and Periplaneta Americana, and descriptions of insect physiology written by entomologists.[5] We then derived the hexapod shown in Color Plate 12. This articulated figure is modeled as being 2.9 cm long, and has a mass of 2.1 gm. There are 38 unconstrained degrees of freedom in the figure.

The stepping pattern is generated using the coupled oscillator mechanism, described previously for the kinematic hexapod. The gait controller triggers motor activity, but the actual motion is controlled in *corpus* using dynamic motor programs. These functions operate at each joint at which the motion must be controlled, and can be grouped together to create more complex motor programs, such as "step" or "stance" for an entire leg.

The motor programs generate forces by modeling the force response of exponential springs, in combination with linear velocity dampers. As their name implies, exponential springs have an exponential relationship between the force they generate and the displacement of the joint from the spring rest angle, as in:

$$f = \text{sign}\ (x_s - x)\alpha e^{\beta(|x_s - x|)} - b\dot{x} \tag{3}$$

where f is the generated force, x is the joint angle, $\dot{x}$ is the joint velocity, x_s is the spring rest angle, α is the linear spring constant, b is the exponential spring constant, and β is the damping constant. In a sense, the exponential spring forces create a steep potential well, such that the controlled joint will likely stay near the rest angle of the spring.

In order to generate motion, the motor programs move the spring rest angles from their current angle to a target angle. This moves the potential well and, in effect, "drags" the joint along with the spring. In some sense, this method "keyframes" the controlling space of the springs.

The motor programs were progressively "tuned" by the animator, over the course of several simulations, using a trial and error method. For example, initially the posture was too low to avoid dragging the abdomen, so the joints were extended further, raising the posture. Also, the legs did not step high enough initially, so the motor programs were modified to retract the legs further during stepping.

The "springy" method of motor control, with trial and error calibration, may not be suitable for actions that require exact movements, such as grasping and manipulating objects. However, the fact that the springs create a compliant system can be of great benefit. For example, we have experimented with simulations over uneven terrain, which employed the level terrain motor programs, and the hexapod conformed to the uneven surface due to the springy compliance at the joints.

The dynamic roach and other simulated figures played key roles in the award winning computer animation, *Grinning Evil Death,* by McKenna and Sabiston[12] (see Color Plate 13).

DYNAMIC BIPED

Recently, McKenna and Zeltzer have begun development of a physically based model of bipedal walking. The mechanical properties of the simulated biped are based on human biomechanics, and dynamic simulation techniques are used to generate the walking movements, as with the dynamic *roach.* The control of walking is facilitated by both passive and active components. *Passive* control is provided by the inherent mechanical properties of the biped and the environment.[7] *Active* control is provided by force-generating actuators, which are based on a variable spring model. The actuator control parameters are regulated by motor programs, which effect active control by varying the spring parameters over time in order to accomplish particular movements. Walking and standing are controlled by a set of motor programs, which are calibrated using analyses of the motions executed by humans as they walk, coupled with inverse dynamics and inverse control to automatically derive the control parameters.

One of the most obvious, yet significant, differences between the hexapod research and this biped research is that balance becomes an important issue. In initial tests, the trial and error approach for "tuning" control parameters was used, as previously discussed concerning the dynamic hexapod. Over successive trials we attempted to adjust the joint spring parameters to achieve a stable, balanced posture for the biped. Although we were not especially serious about using such an approach, we were interested in its feasibility. After a dozen trials, the biped continued to fall over, demonstrating that dynamic motor programming is not an intuitive exercise, and is not well suited to interactive control when forces or spring controls are directly manipulated.[22]

However, automatic approaches can be used to determine the spring control parameters—achieving a stable posture. Using inverse dynamics, we calculate the joint forces which are required to achieve a given postural or motion goal.

By computing not only inverse dynamics, but also "inverse control," the control system (spring) parameters are determined from the computed force. (See Figure 4.) The springs provide a feedback loop, creating a compliance field about the calibrated posture. When an external force is applied, the posture deviates from its rest state by some amount. When the force is removed, the figure returns to the calibrated posture, unless the figure has toppled over in the meantime (caused by the center of gravity passing outside of the support region formed by the feet). The control system can also take into account the externally applied forces, and will attempt to counteract their effect.

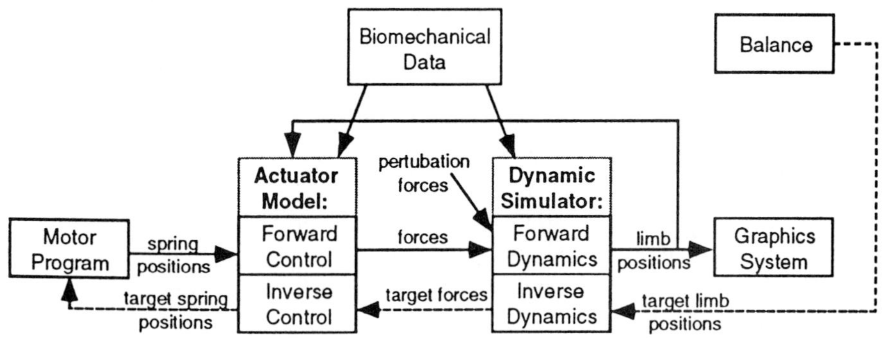

FIGURE 4 A control system for static balance. The dashed feedback lines only need to be computed once per posture.

Preliminary work from this research includes a simulation of a passive stepping motion. This simulation reproduces some of the results from Mochon and McMahon's "ballistic walking" experiments,[13,14] except with a fully three-dimensional model. A hybrid dynamics approach, mixing kinematic and dynamic control, was used for the motion simulation in order to simplify the problem. The stance knee was kinematically locked, and the stance hip and ankle were kinematically controlled to undergo a constant angular velocity rotation. Initial conditions were established for the joint angles and velocities of the step leg and for the initial overall body velocity. The system was then allowed to simulate forward, and a successful stepping motion resulted. The step leg acts as a passively swinging double pendulum, and the stance leg acts as a passive inverted pendulum. The coupling of these two types of motions results in a natural-appearing human step. The passive step is shown in Color Plate 14.

FUTURE WORK

Our work at MIT continues to focus on task-level interaction with autonomous virtual actors.[27] Future work includes research to enhance the realism of the simulated human model, with the addition of a physically based description of skin and muscle tissue. Chen has demonstrated a dynamic simulation of human skeletal muscle, using a finite-element model[1,2] (see Color Plate 15). The material properties of the simulated muscle were based on the biomechanical properties of muscle tissue. Based on these properties, nonlinear forces are applied at the finite element mesh nodes, and the motion and shape of the muscle is then simulated. In order to verify the accuracy of the simulated muscle, several key biological experiments, such as "tension length" and "quick release" experiments, were recreated in simulation. The results show a close match between the real and simulated experiments.

Pieper has developed a physically based model of human skin tissue, with an interface suitable for plastic and reconstructive surgery planning and simulation[18,19] (see Color Plate 16). The system simulates the flexible nature of the skin and allows for the simulation of surgical operations. Surgical procedures are interactively specified by the user on the graphical representation of the skin. The system then computes the post-operative shape of the skin tissue, using the physical properties of the skin to predict how the surface will be reshaped by the procedure.

Our long-term goal is to unify these models—rigid body simulation to compute gross motions, muscle simulation to compute active forces and muscle shape, and skin tissue simulation to compute the surface shape of the artificial person. All three systems would be interdependent, and the final motions and shapes would be dependent on all components.

We can envision many applications for such a complex system, from training in a virtual environment with realistic virtual actors, to personalized entertainment. One specific application might be a doctor or surgeon, examining a patient and augmenting reality with real-time computer models and scanned medical data. These graphics could appear to occupy the same three-dimensional space as the patient, using a head-mounted display or other tracked display system, allowing intuitive examination of the patient, with the ability to plan a surgery and perform "what if" types of experiments.

ACKNOWLEDGMENTS

This work has been sponsored, in part, by NHK (Japan Broadcasting Corp.), DARPA/Rome Laboratories Contract F30602-89-C-0022, and the National Science Foundation, with equipment grants from Hewlett Packard, Apple Computer, and Silicon Graphics.

The human skeletal model was created from medical data by Donald Stredney, under a grant from the National Science Foundation, Project No. MCS-7923670, at the Advanced Computing Center for the Arts and Design, Ohio State University.

Grinning Evil Death was co-directed by Mike McKenna and Bob Sabiston.

Members of the Computer Graphics and Animation Group, currently, and in the past, who have contributed to this work are: Cliff Brett, David Chen, Brian Croll, Steven Drucker, Paul Dworkin, Michael Johnson, Michael McKenna, Jim Puccio, Steven Pieper, Alex Seiden, Karl Sims, Steve Strassmann, David Sturman, and Clea Waite.

REFERENCES

1. Chen, David Tzu-Wei. "Pump It Up: Computer Animation of a Biomechanically Based Model of Muscle using the Finite Element Method." *Ph.D. Thesis,* Massachusetts Institute of Technology, 1991.
2. Chen, David T. "Pump It Up: Computer Animation of a Biomechanically Based Model of Muscle using the Finite Element Method." *Proceedings of SIGRRAPH, 1989*, 89–98. Boston, MA: Addison-Wesley,1989.
3. Featherstone, R. "The Calculation of Robot Dynamics Using Articulated-Body Inertias." *Robotics Research* **2(1)** (1983): 13–29.
4. Featherstone, R. *Robot Dynamics Algorithms.* THE NEDERLANDS: Kluwer Academic Publishers, 1987.
5. Hughes, G. M., and P. J. Mill. "Locomotion: Terrestrial." In: *The Physiology of Insecta*, edited by M. Rockstein, Vol. III, 335–379. New York, NY: Academic Press, 1974.
6. Inman, V. T., H.J. Ralston, and F. Todd. *Human Walking* Baltimore: Williams & Wilkins, 1981.
7. McGeer, Tad. "Passive Dynamic Walking." *The International Journal of Robotics Research* **9(2)** (1990): 62–82.
8. McKenna, Michael A. *Cootie Gets Scared.* Computer animation. Cambridge, MA: Computer Graphics and Animation Group, MIT Media Laboratory, 1988.
9. McKenna, M. A., S. Pieper, and D. Zeltzer. "Control of Virtual Actor: The Roach." In: *Proceedings of the 1990 Symposium on Interactive 3D Graphics*, 165–174. Snowbird, Utah: ACM Press, 1990.
10. McKenna, M. A. "A Dynamic Model of Locomotion for Computer Animation." *Master's Thesis* Massachusetts Institute of Technology, 1990.
11. McKenna, Michael, and David Zeltzer. "Dynamic Simulation of Autonomous Legged Locomotion." *Proceedings of SIGGRAPH 1990.* Dallas, TX, 1990. Also in: *Computer Graphics* **24(4)** (1990): 29–38.
12. McKenna, Michael A., and Bob Sabiston. *Grinning Evil Death.* Computer animation. Cambridge, MA: MIT Media Laboratory, 1990.
13. Mochon, Simon, and Thomas A. McMahon. "Ballistic Walking." *Journal of Biomechanics* **13** (1980): 49–57.
14. Mochon, Simon, and Thomas A. McMahon. "Ballistic Walking: An Improved Model." *Mathematical Biosciences* **52** (1980): 241–260.
15. Muybridge, E. *The Human Figure in Motion* New York, NY: Dover, 1955.
16. Pearson, K. "The Control of Walking." *Scientific American* **235(6)** (1976): 72–86.
17. Phillips, Cary B., and Norman I. Badler. "Interactive Behaviors for Bipedal Articulated Figures." In: *Proceedings of SIGGRAPH 1991.* Las Vegas, NV, 1991. Also in: *Computer Graphics* **25(4)** (1991): 359–362.

18. Pieper, Steven D. "CAPS: Computer-Aided Plastic Surgery." *Ph. D. Thesis.* Massachusetts Institute of Technology, 1991.
19. Pieper, Steven, Joseph Rosen, and David Zeltzer. "Interactive Graphics for Plastic Surgery: A Task-Level Analysis and Implementation." In: *Proceedings of 1992 Symposium on Interactive 3D Graphics.* Cambridge, MA, 1992.
20. Saunders, J. B., Inman, V. T., and Eberhart, H. D. "The Major Determinants In Normal and Pathological Gait." *Journal of Bone and Joint Surgery* **35A** (1953): 543–558.
21. Sims, K. "Locomotion of Jointed Figures over Complex Terrain." *M.S.V.S Thesis.* Massachusetts Institute of Technology, June 1987.
22. Wilhelms, J. "Using Dynamic Analysis for Realistic Animation of Articulated Bodies." *IEEE Computer Graphics and Applications* **7(6)** (1987): 12–27.
23. Wilson, D. M. "Insect Walking." *Annual Review of Entomology* **11** (1966): 162–169.
24. Zeltzer, D. "Motor Control Techniques for Figure Animation." *IEEE Computer Graphics and Applications* **2(9)** (1982): 53–59.
25. Zeltzer, D. "Representation and Control of Three Dimensional Computer Animated Figures." *Ph.D. Thesis.* Dept. of Computer and Information Science, Ohio State University, 1984.
26. Zeltzer, D., Pieper, S. and Sturman, D. "An Integrated Graphical Simulation Platform." *Proceedings of Graphics Interface 1989*, 266–274. London, Ontario, 1989.
27. Zeltzer, David. "Task Level Graphical Simulation: Abstraction, Representation and Control." In: *Making Them Move: Mechanics, Control and Animation of Articulated Figures* edited by N. Badler, B. Barsky and D. Zeltzer. San Mateo, CA: Morgan Kaufmann Pub., Inc., 1990.

Johndale C. Solem
Theoretical Division, Los Alamos National Laboratory, Los Alamos, NM 87544

The Motility of Microrobots

At the physical limits of technology, crude robots on the size scale of 10–100 μm may be possible. For many applications, especially in the biomedical and defense-related arenas, it will be necessary for these miniscule robots to move about on their own. But by what mechanisms will they achieve this motility? I address this question with exploratory calculations for microrobots traveling by air, land, and sea.

The Reynolds number for airborne robots is close to unity—the viscous forces dominate the inertial forces. I show that there is no sense to using a lifting airfoil; a microrobotic helicopter could fly by simply gripping the viscous air around it. A design using *wheel-wings* can hover at angular frequencies resembling mosquito pitch. I describe the scaling of microrobotic helicopters and show that exceedingly small craft would have to be externally powered. Since swimming robots encounter a higher Reynolds number, I explore a variety of propulsion mechanisms. The best propulsion appears to be a fan propellor using blades of rather unusual design or wheel propulsion similar to the helicopter. Surprisingly, the corkscrew-flagellum propulsion of the motile form of *Escherichia coli* is a good deal less efficient than fan or wheel propulsion. Nature is known for her parsimonious use of energy: perhaps she uses the flagellum because it is easy to fabricate from protein.

Hopping seems to be the most effective mode of transport for earth-bound robots. It is stealthy, predator-evading, and energy-efficient and provides mobility over many types of terrain. I calculate optimum hopping strategies as a function of weight and atmospheric viscosity.

Finally, I show various ways adhesion and electric fields can be used for walking on walls.

INTRODUCTION

There has been much informal speculation on the use of microrobotics. Some of that speculation has been devoted to the shape and size of these lilliputian automata: What would they look like? For many practical applications, the robot will be required to move about, and the actuators that provide its mobility will probably dominate its external appearance.

But what would those actuators look like? A lot depends on the robot size. On the human size scale, the actuator function has been well examined. But not much has been written about the exceedingly small scale. So how small can they get? This question can be answered in very general terms. Microfabrication depends on materials called *resists*. These are the photoetched materials used to form the actual shapes and mechanisms. The two best x-ray resists are polymethyl methacrylate and copolymer. When x-rays penetrate the resist, the concomitant electron showers break cross-links in the polymer and render the material locally less resistant to etching. The etchant is usually a strong base. The optimum wavelength[7,8,17,18] for fabricating from the polymer is about 50 Å. At shorter wavelengths the electron shower is too energetic and blurs the exposure owing to the increased transport length of the higher velocity electrons. At longer wavelengths, diffraction within the polymer itself causes blurring. Thus, the finest detail possible with x-ray resist technology is about 50 Å. Thus, if fabricated at the physical limit, a cube 10 μm on a side could contain 8×10^9 pixels, which would be quite enough for a crude robot. Because at this scale we enter the regime of microbes, it seems appropriate to use the biological term *motility* in lieu of mobility or locomotion.

The motility actuators would probably not be legs for a number of reasons: (1) walking machines are a rather formidable challenge even on the very large size scale, as the DoD knows from nearly 40 years' experience; (2) walking is always a rather slow way of getting about; and (3) legs are relatively inefficient. Wheels might be useful for fine adjustments to the robot's location, but they would not be very good for covering large distances. On the robot's size scale, all terrain is rough: A bread crumb may be an insurmountable obstacle. The robot would probably cover larger distances by flying or hopping. Aquatic robots, even more like their microbe counterparts, will undoubtedly be of some value in biomedical applications.

ARISTOTLE'S PHYSICS

A Reynolds number is used to describe the relative importance of *inertial* forces and *viscous* forces. It is always given for a specific geometric shape; and it scales with linear dimension x so that

$$\Re \sim \frac{\rho v x}{\eta}, \tag{1}$$

where ρ is the air density, v is the velocity through the fluid, and η is the fluid viscosity. Two systems with the same Reynolds number are said to be *dynamically similar*; thus, the flow around a half-scale model has the same dynamics as the full-scale object in a fluid of half viscosity. For exceedingly small objects, inertial forces are swamped by viscous forces. The physical world is well described by the teachings of Aristotle.[1] Each object falls according to its own shape and composition. The natural state is rest. When you stop pushing, it stops moving.

FLYING AROUND

Flying is an interesting challenge on the microrobot size scale. Some of the salient features of such locomotion can be extracted from very fundamental principles. Suppose the robot were constructed as a scaled-down version of a conventional helicopter. Specifically, it might have a pair of counter-rotating wings (rotorblades). The lifting force on a wing is given by the well-known[16] formula

$$F_L = \rho v L \Gamma, \tag{2}$$

where L is the wing length, and Γ is the *circulation*, defined by

$$\Gamma \equiv \oint \mathbf{v} \cdot d\mathbf{l}, \tag{3}$$

where the integral is taken on path $\mathbf{l}$ around the cross-sectional surface of the wing. Thus, for geometrically similar wings

$$F_L \propto \rho v^2 s, \tag{4}$$

where s is the surface area interacting with the fluid. So for a helicopter wing of radius R,

$$F_L \propto \rho R^4 \omega^2, \tag{5}$$

where ω is the angular velocity. From dimensional analysis it is easy to show that the viscous drag on an object characterized by a single linear dimension is proportional to $\eta v x$. Perhaps the best known example is Stokes' law, which gives the drag force on a sphere[13] as $6\pi\eta v r$, where r is the sphere's radius. So the drag on a helicopter wing at low Reynolds number is roughly described by

$$F_D \propto \eta R^2 \omega. \tag{6}$$

Some interesting results are derived directly. We note that the mass of the helicopter scales as R^3. So the angular velocity required to hover is

$$\omega_h \propto \frac{1}{\sqrt{R}}. \tag{7}$$

When hovering, all the power goes into overcoming drag,

$$\begin{aligned} P &\propto R^3 \omega_h^2 \\ &\propto R^2. \end{aligned} \tag{8}$$

A reasonable figure of merit is the specific power required to hover,

$$\frac{P}{m} \propto \frac{1}{R}. \tag{9}$$

Assuming most of the robot mass is fuel, the smaller the robot, the shorter its life. For protracted missions, the robots might have to be externally powered, perhaps by microwaves or lasers.

The *lift-over-drag* ratio[1] (sometimes called "L/D") is proportional to

$$\frac{\rho R^2 \omega}{\eta}, \tag{10}$$

which has the dimensions of Reynolds number. At a small enough Reynolds number, lift will become less than drag, and there is no sense to having a lifting airfoil at all. *The helicopter might as well climb up the air, gripping it by its own viscosity.*

[1] Lift-over-drag ratio used in this context is simply the lifting force divided by the drag force, without reference to the airframe of the vehicle.

THE BIRDS AND THE BEES

At what Reynolds number does viscous lifting become favorable? A more general form of Eq. (2) that encompasses all kinds of flow is

$$F_L = \frac{1}{2} C_L \rho v^2 s, \tag{11}$$

where C_L is called the *lift coefficient.* To be analogous to Eq. (11), the drag is usually written

$$F_D = \frac{1}{2} C_D \rho v^2 s, \tag{12}$$

where C_D is called the *drag coefficient.* Viscous lifting becomes favorable when $C_D > C_L$.

The most familiar wing cross section is a "lopsided teardrop." It is used on almost all subsonic aircraft. All wings with this standard shape possess a *Critical Reynolds Number*[9] usually between 10^4 and 10^5, below which the laminar flow separates from the top side of the wing, resulting in a sudden increase in drag and decrease in lift. A simple cambered plate, however, does not suffer from the critical transition. At low Reynolds number, the cambered plate will have more lift and less drag than the standard shape. Similarly, at high Reynolds number it will have less lift and more drag. This is apparently why large birds, such as the condor and albatross, have rounded leading edges and standard-shaped wings, while hummingbirds and bumblebees have sharp leading edges and relatively planar wings. Any robotic aircraft less than 10 cm would likely benefit from cambered plate wings. Thin plate wings will be mandatory for microrobots, even at high L/D.

At low Reynolds number, C_D will increase as Reynolds number decreases. For many shapes, C_D is roughly inversely proportional to $\Re$, owing to the argument of dimensional analysis given above. For a plate in tangential flow at low Reynolds number, it can be shown[10] that

$$C_D \simeq \frac{40\pi}{\Re[16 - 5\ln(\Re)]}, \tag{13}$$

which is good for $10^{-4} < \Re < 10^2$. Over the same region, $C_L \simeq 0.8$. Thus, viscous lifting becomes preferable for Reynolds numbers less than

$$\Re_{vl} \simeq 1, \tag{14}$$

which is about what we would have expected. Angular velocity of the wing (rotorblade) will be the most important limiting factor. Although there are ultimate limitations owing to stress and sound speed (see Appendix I), the practical limitation will likely have to do with lubrication. Say we have a maximum angular velocity

ω_m. Then the wing length R at which viscous lifting could become advantageous is given by

$$R \simeq \sqrt{\frac{\eta}{\rho\omega_m}}. \tag{15}$$

For air at 20°C, $\eta = 1.8 \times 10^{-4}\text{gm}\cdot\text{cm}^{-1}\cdot\text{sec}^{-1}$ and $\rho = 1.2 \times 10^{-3}\text{gm}\cdot\text{cm}^{-3}$. If $\omega_m = 10^4\text{sec}^{-1}$ (about mosquito pitch), then viscous lifting becomes advantageous for wing lengths less than $\sim 3.87 \times 10^{-3}$cm.

VISCOUS HELICOPTER

In Figure 1 is a conceptual drawing of a viscous-lift helicopter, using motor-driven wheel-wings suggestive of induction micromotors,[2] but more likely driven by electrostatic[5,6,11,19] or resonant-structure[14] motors, although their realization will likely invoke entirely new technologies. The viscous drag on each of the four wheels is $\sim \frac{8}{3}\eta R^2\omega$. The mass is $\sim 2\pi R^3\rho_m$, where ρ_m is the fabrication material. The angular velocity for this gadget to hover is

$$\omega_h \simeq \frac{3\pi\rho g R}{16\eta}. \tag{16}$$

If we make the helicopter out of silicon, $\rho_m = 2.33$ gm $\cdot$ cm^{-3}, and build it at a size $R = 10\ \mu m$, then it will hover at $\omega_h \simeq 7.5 \times 10^3\text{sec}^{-1}$.

If the wheels that serve as "wings" can also be operated at high torque and low angular velocity, they can be used for moving around on the ground. Because of their relatively large size, the wheels can be used on fairly rough terrain.

Depending on size and the details of the wheel-housing cavity, the viscous helicopter could be more or less efficient than the aerodynamic helicopter. A narrow wheel-housing cavity will result in a large power loss owing to laminar shear. This might be mitigated by a large cavity for all four wheels, which could support a circulating toroidal flow. It could be further mitigated by filling the cavity with a less viscous gas.

SWIMMING AROUND

It is reasonable to believe that some motile microrobots will need to swim. The design of a microrobotic submarine has a lot in common with the design of a microrobotic helicopter. Somewhat surprisingly, Reynolds numbers will be about an order of magnitude *larger* in water than in air. For the tiniest robots, we will still be in a world dominated by viscous forces.

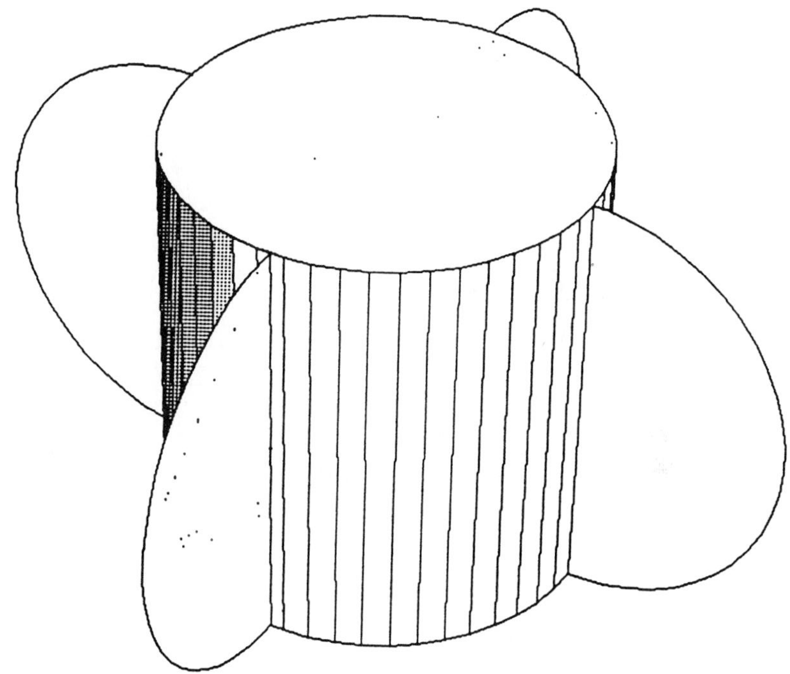

FIGURE 1 Conceptual drawing of a viscous-lift helicopter. The viscous drag on each of the four wheels is $\sim \frac{8}{3}\eta R^2\omega$. The mass is $\sim 2\pi R^3\rho_m$.

PADDLE PROPULSION

The most straight-forward propulsion device is a paddle. At low Reynolds number, the drag force on a thin circular disk[12] of radius R is given by

$$D_\perp = \delta_\perp \eta R v \tag{17}$$

for flow perpendicular or normal to the disk and

$$D_\parallel = \delta_\parallel \eta R v \tag{18}$$

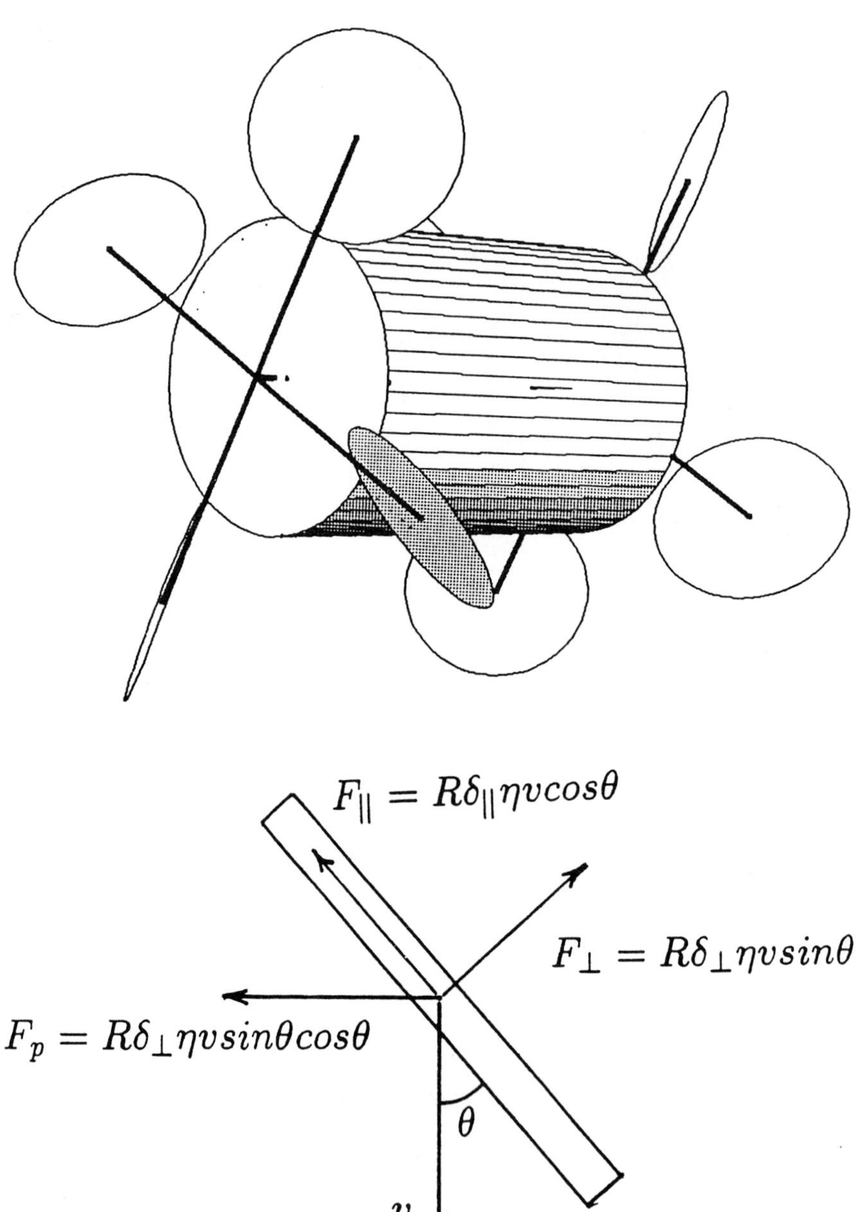

FIGURE 2 Fan Propulsion Submarine: (a) a propulsion system design using counter-rotating fan propellors; (b) drag force on a single blade is $F_d \simeq r_b\omega\eta R(\delta_{\perp} + \delta_{\|})$, while propulsive force is $F_p \simeq r_b\omega\eta R\delta_{\perp} \sin\theta \cos\theta$.

for flow parallel or tangential to the disk, where $\delta_\perp \simeq 16$ and $\delta_\parallel = 10.68$. The act of paddling consists of moving the disk back and forth, holding it parallel to the direction of travel on the *up stroke* and perpendicular to the direction of travel on the *power stroke*. Say the length of the stroke is l, and the velocity of the paddle is v. Of course, at least two paddles would have to be operated in unison to keep the vessel properly directed. The average propulsion force, averaged over a full cycle, will be

$$\langle F_p \rangle = \frac{1}{2}(\delta_\perp - \delta_\parallel)\eta R v, \tag{19}$$

and the average drag

$$\langle F_d \rangle = \frac{1}{2}(\delta_\perp + \delta_\parallel)\eta R v. \tag{20}$$

The propulsion efficiency is therefore

$$\varepsilon = \frac{\langle F_p \rangle}{\langle F_d \rangle} = \left(\frac{\delta_\perp - \delta_\parallel}{\delta_\perp + \delta_\parallel}\right)\eta R v \simeq 0.2. \tag{21}$$

The average power used for propulsion will be

$$P = \frac{1}{2}(\delta_\perp + \delta_\parallel)\eta R v^2; \tag{22}$$

the length of the stroke does not matter.

FAN PROPELLOR PROPULSION

In Figure 2(a) we show a propulsion system design using counter-rotating fan propellors. The fan propellors consist of four disk-shaped blades. They may not be the optimum design, but they are calculable. From Figure 2(b) it can be seen that the drag force on a single blade is given by

$$F_d \simeq r_b \omega \eta R(\delta_\perp + \delta_\parallel), \tag{23}$$

while propulsive force is given by

$$F_p \simeq r_b \omega \eta R \delta_\perp \sin\theta\cos\theta. \tag{24}$$

The propulsion efficiency is therefore

$$\varepsilon = \frac{\delta_\perp \sin\theta\cos\theta}{\delta_\perp + \delta_\parallel}. \tag{25}$$

The angle for maximum efficiency is easily found by setting $(d\varepsilon/d\theta) = 0$.

$$\theta_m = \frac{\pi}{4}. \tag{26}$$

So the maximum efficiency is

$$\varepsilon_m = \frac{\delta_\perp}{2(\delta_\perp + \delta_\parallel)} \simeq 0.3. \tag{27}$$

Surprisingly, the fan propellor is more efficient than the paddle.

WHEEL PROPULSION

The same wheel design of the *viscous helicopter* could be used for submarine propulsion. The viscous torque on the spinning wheel is given by

$$\tau = \delta_\theta \eta R^3 \omega, \tag{28}$$

where $\delta_\theta \simeq 8/3$ as for the helicopter case above. If the wheel housing is filled with air, then

$$\tau = \delta_\theta (\eta_{water} + \eta_{air}) R^3 \omega. \tag{29}$$

The viscous force averaged along the radius of the wheel for that portion in contact with the water is

$$f(\theta) = \delta_\theta \eta_{water} R^2 \omega \cos\theta. \tag{30}$$

The forward-directed viscous force used for propulsion is therefore

$$F_p = \frac{1}{\pi} \int_{-\pi/2}^{\pi/2} f(\theta)\, d\theta = \frac{\delta_\theta \eta_{water} R^2 \omega}{\pi}, \tag{31}$$

and the drag is

$$F_d = \delta_\theta (\eta_{water} + \eta_{air}) R^2 \omega. \tag{32}$$

The propulsion efficiency is

$$\varepsilon = \frac{\eta_{water}}{\pi(\eta_{water} + \eta_{air})} \simeq .032, \tag{33}$$

a little bit better than the fan propellor.

FLAGELLUM PROPULSION

Many microbes propel themselves with undulating flagella or by undulations of their bodies. The size of the microbe seems to relate to the complexity of the motion.

Spirillum volutans (20–40μm) uses a sinusoidal motion of both its body and flagella, *Chromatium okenii* (10–20μm) uses a sinusoidal motion of its flagellum, and a certain motile form of the famous *Escherichia coli* (1–2μm) actually uses an electric motor to turn a corkscrew as a propellor.[3]

To assess the effectiveness of this form of propulsion, we need to approximate the flagellum as a thin rod. The viscous force[12] on a unit length of circular-cylinder rod of radius R is

$$D_\perp = \delta_\perp \eta R v \tag{34}$$

for flow perpendicular or normal to the rod and

$$D_\parallel = \delta_\parallel \eta R v \tag{35}$$

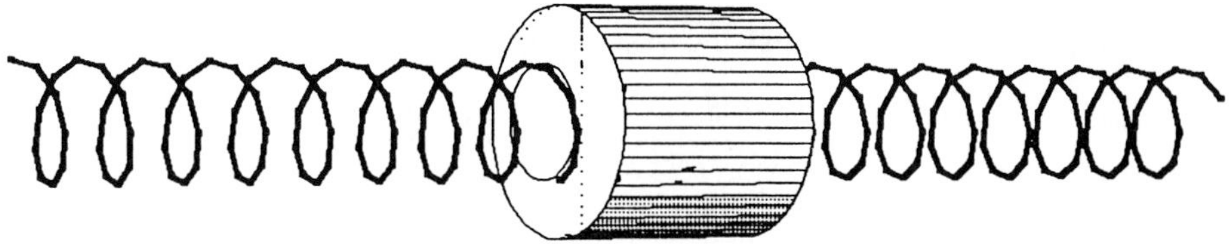

FIGURE 3 Conceptual Corkscrew Submarine. Design imitates a motile form of *Escherichia coli* that uses an electric motor to turn a corkscrew as a propellor.

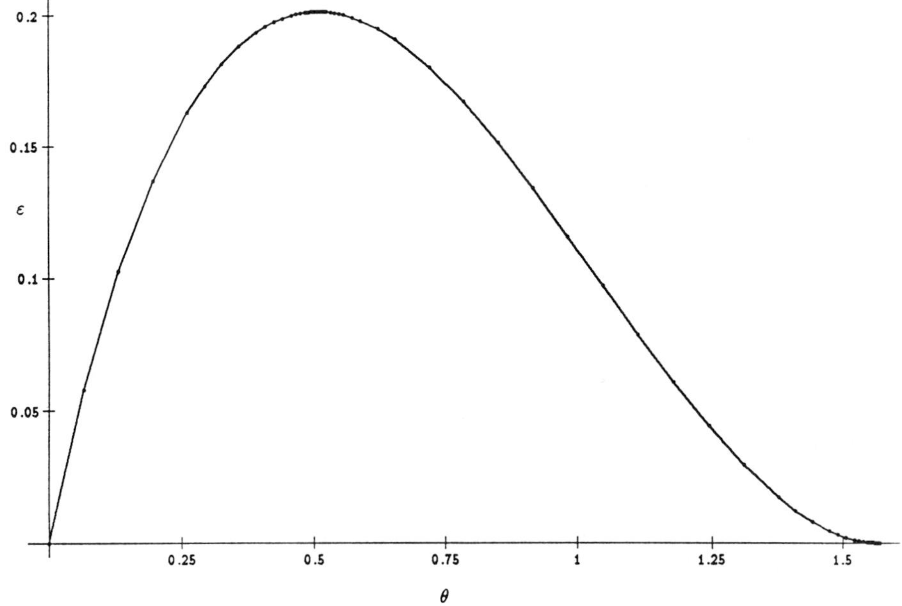

FIGURE 4 Corkscrew Efficiency. A plot of Eq. (37). With $\delta_\perp \simeq 2\delta_\parallel$ the pitch angle for optimum efficiency is $\theta_m \simeq 0.6$, which gives a maximum efficiency $\varepsilon_m \simeq 0.2$.

for flow parallel or tangential to the rod, where $\delta_\perp \simeq 22$ and $\delta_\parallel \simeq 12$. The rotating corkscrew seems to be considerably more easily adapted to a microrobotic submarine than the sinusoidally undulating flagellum. In Figure 3 we illustrate a conceptual vessel of this kind. A nontrivial exercise in geometry reveals that

$$F_d = 8\pi^3 R^2 \eta \frac{\delta_\perp \tan\theta + \delta_\parallel}{\omega},$$
$$F_p = 8\pi^3 R^2 \eta \frac{\delta_\perp - \delta_\parallel}{\omega} \sin\theta \, \cos\theta, \tag{36}$$

where R is the radius of the helical corkscrew and θ is its pitch angle. The efficiency is

$$\varepsilon = \frac{\delta_\perp - \delta_\parallel}{\delta_\perp \tan\theta + \delta_\parallel} \sin\theta \, \cos\theta. \tag{37}$$

Equation (37) is hard to optimize analytically, but is plotted in Figure 4. With $\delta_\perp \simeq 2\delta_\parallel$ the pitch angle for optimum efficiency is $\theta_m \simeq 0.6$, which gives a maximum efficiency $\varepsilon_m \simeq 0.2$. Clearly *E. coli* was not designed for maximum efficiency.[2] Its corkscrew is about as efficient as a paddle. However, the corkscrew is easy for nature to fabricate: it is essentially one protein whose tertiary structure is the helix.[4] Perhaps a microfabricated robotic submarine would use corkscrew propulsion for the same reason. A corkscrew is easy to make.

HOPPING AROUND

Hopping may be an effective means of transportation for many of the same reasons it is used by insects. Hopping is stealthy. It may be particularly useful for surveillance operations. The robot is not conspicuous until it hops, and it can choose a time when it is not being observed. Hopping offers a quick escape to avoid capture. Terminal velocity for microrobots is small enough that they can hop many body lengths without damage. Hopping provides mobility over many types of terrain. Hopping is relatively energy efficient.

A hopping strategy depends on the robot's size and the velocity of launch. If $\left\{ {x \atop y} \right\}$ are the $\left\{ {\text{horizontal} \atop \text{vertical}} \right\}$ and the initial conditions are $\left\{ {x_0 \atop y_0} \right\} = 0$, $\left\{ {\dot{x}_0 \atop \dot{y}_0} \right\} = v \left\{ {\cos\theta \atop \sin\theta} \right\}$, then the path described by a roughly spherical shaped hopping robot is

$$\begin{aligned} x &= \frac{mv \, \cos\theta}{6\pi\eta r} \left[1 - \exp\left(-\frac{6\pi\eta r}{m} t \right) \right], \\ y &= \frac{m}{6\pi\eta r} \left\{ \left[v \, \sin\theta + \frac{gm}{6\pi\eta r} \right] \left[1 - \exp\left(-\frac{6\pi\eta r}{m} t \right) \right] - gt \right\}, \end{aligned} \tag{38}$$

where v is the initial velocity and θ is the initial angle from the ground at which the robot launches its hop. One result is immediately apparent. The greatest distance that can be covered by the hop under any circumstances is

$$\frac{mv \, \cos\theta}{6\pi\eta r}. \tag{39}$$

The exact distance covered by the hop is $x(t_h)$, where t_h is obtained from the nontrivial solution $y(t_h) - 0$. The exact form is transcendental, but an approximate solution can be obtained for small η and large η.

[2] The relatively low efficiency of the cork screw was observed by Purcell,[15] whose work I discovered rather late in this investigation.

LOW VISCOSITY

We obtain the low-viscosity solution by expanding y in a Taylor series around $\eta = 0$

$$\begin{aligned} x &\simeq vt\ \cos\theta, \\ y &\simeq vt\ \sin\theta - \frac{gt^2}{2}. \end{aligned} \tag{40}$$

The optimum angle and maximum range are

$$\theta_m = \frac{\pi}{4}; \qquad x_m = \frac{v^2}{g}. \tag{41}$$

These are the well-known gunnery equations. The length of the hop is proportional to the energy expended. There is no optimum length of the hop, the energy expended in getting from one place to another is independent of the number of hops.

HIGH VISCOSITY

For high viscosity we take $6\pi\eta r \gg m$, in which case

$$y \simeq \frac{m}{6\pi\eta r}\left[v\ \sin\theta + \frac{\mathrm{gm}}{6\pi\eta r} - gt\right]. \tag{42}$$

Solving Eq. (42) for t_h and substituting in Eq. (38), it can be shown that

$$\begin{aligned} t_h &\simeq \frac{v\ \sin\theta}{g} + \frac{m}{6\pi\eta r}, \\ x(t_h) &\simeq \frac{mv\ \cos\theta}{6\pi\eta r}\left[1 - \exp\left(-\frac{6\pi\eta rv\ \sin\theta}{mg} - 1\right)\right], \end{aligned} \tag{43}$$

The maximum range is found by solving $dx(t_h)/d\theta = 0$. This is also a transcendental, but the solution can be approximated by taking the first term of the Taylor series of $dx(t_h)/d\theta$ around $\theta = 0$, which results in

$$\begin{aligned} \theta_m &\simeq \frac{6\pi\eta rvmg}{(e-1)(mg)^2 + (6\pi\eta rv)^2}, \\ x_m &\simeq \frac{(1-e^{-1})mv}{6\pi\eta r} + \frac{6\pi\eta rv^3 m}{(e-1)e(mg)^2 + (6\pi e\eta rv)^2}. \end{aligned} \tag{44}$$

A further simplification derives from recognizing that the second term in the denominator of Eq. (44) dominates at high viscosity. In this limit we find

$$\begin{aligned} \theta_m &\simeq \frac{mg}{6\pi\eta rv}, \\ x_m &\simeq \frac{1+e^2-e}{e^2}\frac{mv}{6\pi\eta r}. \end{aligned} \tag{45}$$

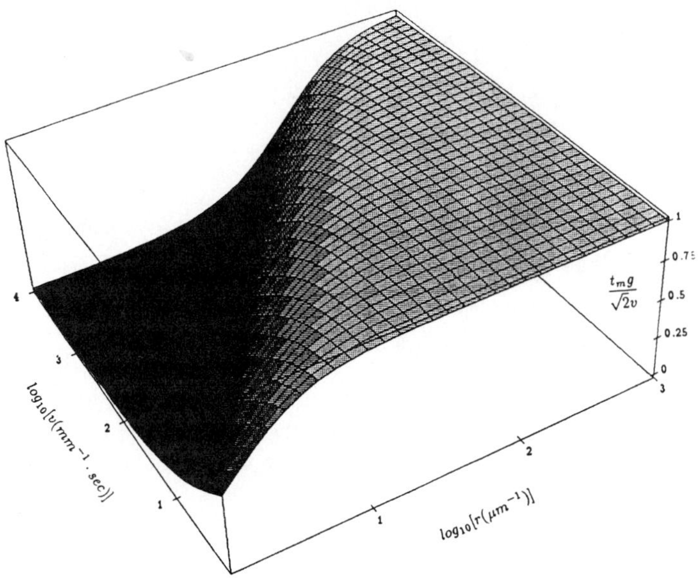

FIGURE 5 Optimum Hop Distance. Numerical solutions for the maximum range obtained from Eq. (38). The range is given in units of v^2/g, the low-viscosity range. The region around $v \sim 10^4$ mm $\cdot$ sec^{-1} and $r \sim 1$ μm is well described by Eq. (45). Newton's physics dominates in the regime $\log_{10}(v \text{ sec} \cdot \text{mm}^{-1}) < 2\log_{10}(\text{r } \mu\text{m}^{-1}) - 0.9$.

The length of the hop is proportional to the momentum at launch. The most energy-efficient robot will take very short hops. A more detailed treatment of hopping at low Reynolds number is given in Appendix II.

TRANSITION TO ARISTOTLE'S WORLD

As r gets smaller or v gets larger, the maximum range and the optimum angle of launch get smaller. The transition from Eq. (41) to a set of equations more or less like Eq. (45) occurs at a border determined by the parameters r and v. Figure 5 shows numerical solutions for the maximum range obtained from Eq. (38). The range is given in units of v^2/g, the low-viscosity range. The region around $v \sim 10^4$ mm $\cdot$ sec^{-1} and $r \sim 1$ μm is reasonably well described by Eq. (45). Newton's physics dominates in the regime

$$\log_{10}(v \text{ sec} \cdot \text{mm}^{-1}) < 2\log_{10}(\text{r } \mu\text{m}^{-1}) - 0.9. \tag{46}$$

It would be interesting to see how closely the dynamics of jumping insects approximate the optimal hopping strategies given here.

WALKING ON WALLS

The force with which a robot can cling to a surface, whether by adhesion or with a self-generated electric field, is proportional to the area of the robot in contact with the surface. Because mass is proportional to the cube of the robot's characteristic dimension and contact area is proportional to the square of the robot's characteristic dimension, smaller robots will more effectively walk on walls and ceilings.

Focussing our attention on clinging by self-generated electric fields, we note that the force between a plane conductor and the surface is given by

$$F = \frac{CV^2}{2d} = \frac{\epsilon A V^2}{2d^2}, \tag{47}$$

where d is the distance between the conducting plane and the wall surface, a is the area of the conducting plane, V is the voltage between the robot and the ceiling, and ϵ is the permittivity of the insulating dielectric between the conducting plane and the wall surface. If we assume a cubic shape for the robot, then the voltage between the robot and the ceiling required for the robot to cling is given by

$$V = d\sqrt{\frac{2\rho_m l g}{\epsilon}}, \tag{48}$$

where l is the cube edge. For $l = 100\ \mu m$, $\rho_m = 2.33$ gm $\cdot$ cm^{-3}, $\epsilon \simeq \epsilon_0 = 8.85 \times 10^{-12}$ (mks, the permittivity of air), and $d = 10\ \mu m$, we find $V \simeq 7.2$ volts.

In general, the stronger the adhesive, the fewer times it can be used. Adhesives such as that used on Post-it^{TM} notes are in the range 10^2 to 10^3 dyne $\cdot$ cm^{-2}. The characteristic dimension of a robot that could cling to the ceiling with an adhesive of strength S is given by

$$l = \frac{S}{\rho_m g}. \tag{49}$$

If $S = 10^2$ dyne $\cdot$ cm^{-2}, then the largest robot that could effectively hang from the ceiling has dimension $l \simeq 0.44$ mm.

An alternative method of clinging is provided by the surface tension of fluids that might be secreted from the robot's interior to wet the surface of the wall or ceiling. In this case, the clinging force will be proportional to the perimeter of the wetted surface, which is proportional to the characteristic dimension. For our cubical robot to cling, we have

$$l = \sqrt{\frac{4\sigma}{\rho_m g}}, \tag{50}$$

where σ is the surface tension. For water at room temperature, $\sigma \simeq 71$ dyne $\cdot$ cm^{-1}, so the largest robot that could cling has a dimension of $l \simeq 3.5$ mm. While this seems better than the weak adhesive, not all surfaces can be effectively wetted.

SUMMARY & CONCLUSION

The variety of locomotive means available to microrobots spans a far greater range of physical processes than for macroscopic automata. This variety is largely owing to viscous forces being large compared to inertial forces, and electrostatic and adhesive forces being large compared to the force of gravity. We have seen that microrobots could fly and swim using viscous effects alone, and that viscous forces are important in planning optimal hopping strategies. Microrobots have the unique ability to crawl up walls using either adhesion or electric fields. In several cases, microrobots can exercise more effective and efficient motility than their biological counterparts.

The research presented here is avant-garde, but may become useful when micromechanical technology reaches the projected level of competence.

APPENDIX I

STRESS AND SOUND-SPEED LIMITATIONS ON THE MICROROBOTIC HELICOPTER

Two ultimate limitations are introduced by limiting wing speeds: the wing may not exceed the sound speed, and it may not exceed the stress limit imposed by the wing's yield strength. The sound-speed limitation requires

$$r\omega < c, \tag{AI.1}$$

where c is the sound speed. Then to keep the lift greater than the drag, we need merely to have

$$r > \frac{\eta}{\rho c}. \tag{AI.2}$$

For air at 20° C, $\eta = 1.8 \times 10^{-4}$gm $\cdot$ cm$^{-1} \cdot$ sec^{-1}, $\rho = 1.2 \times 10^{-3}$gm $\cdot$ cm^{-3}, and $c = 3.3 \times 10^4$cm $\cdot$ sec^{-1}. This is not very restrictive; robots smaller than 5×10^{-6}cm would have to use viscous lifting. But that radius corresponds to $\omega \simeq 6.6 \times 10^9$sec^{-1}, which is probably impossible.

The yield-strength limitation derives from centrifugal force. Specifically, we must have

$$\rho_w \omega^2 r^2 < Y_w, \tag{AI.3}$$

where ρ_w and Y_w are the density and yield strength of the wing. Viscous lifting is mandatory if

$$r < \frac{\eta}{\rho}\sqrt{\frac{\rho_w}{Y_w}}. \tag{AI.4}$$

For steel, $Y_w/\rho_w \simeq 3.5 \times 10^8 \text{cm}^2 \cdot \text{sec}^{-2}$. Robot helicopters smaller than 8.06×10^{-6}cm would have to use viscous lifting. A robot helicopter would have to be exceedingly small to meet this criterion.

APPENDIX II

THE MECHANICS OF HOPPING ROBOTS

A hopping strategy depends on the robot's size and velocity of launch. Assuming the robot has a roughly spherical shape, the equations of motion in the horizontal (x) and vertical (y) directions are

$$\begin{aligned} \ddot{x} &= -\frac{6\pi\eta r}{m}\dot{x}, \\ \ddot{y} &= -g - \frac{6\pi\eta r}{m}\dot{y}. \end{aligned} \tag{AII.1}$$

With initial conditions $x(0) = y(0) = 0$, $\dot{x}(0) = v \ \cos\theta$, and $\dot{y}(0) = v \ \sin\theta$, the solutions to Eq. (1) are

$$\begin{aligned} x &= \frac{mv \ \cos\theta}{6\pi\eta r}\left[1 - \exp\left(-\frac{6\pi\eta r}{m}t\right)\right], \\ y &= \frac{m}{6\pi\eta r}\left\{\left[v \ \sin\theta + \frac{\text{gm}}{6\pi\eta r}\right]\left[1 - \exp\left(-\frac{6\pi\eta r}{m}t\right)\right] - gt\right\}, \end{aligned} \tag{AII.2}$$

where v is the initial velocity and θ is the initial angle from the ground at which the robot launches its hop. The exact distance covered by the hop is $x(t_h)$, where t_h is obtained from the nontrivial solution $y(t_h) = 0$. The exact form is transcendental, but an approximate solution can be obtained for small η and large η.

HIGH REYNOLDS NUMBER

We obtain the low-viscosity solution by expanding y in a Taylor series around $\eta = 0$,

$$\begin{aligned} x &\simeq vt\ \cos\theta, \\ y &\simeq vt\ \sin\theta - \frac{gt^2}{2}. \end{aligned} \tag{AII.3}$$

These are the familiar gunnery equations. Solving for the second zero-crossing of x,

$$\begin{aligned} t_h &= \frac{2v\ \sin\theta}{g}, \\ x(t_h) &= \frac{2v^2\ \cos\theta\ \sin\theta}{g}, \end{aligned} \tag{AII.4}$$

which gives a maximum range at

$$\begin{aligned} \theta_m &= \frac{\pi}{4}, \\ x_m &= \frac{v^2}{g}. \end{aligned} \tag{AII.5}$$

These equations are well known to freshman physics students and artillery officers.

LOW REYNOLDS NUMBER

For high viscosity we take $6\pi\eta r \gg m$, in which case

$$\begin{aligned} x &= \frac{mv\ \cos\theta}{6\pi\eta r}\left[1 - \exp\left(-\frac{6\pi\eta r}{m}t\right)\right], \\ y &\simeq \frac{m}{6\pi\eta r}\left[v\ \sin\theta + \frac{\mathrm{gm}}{6\pi\eta r} - gt\right], \end{aligned} \tag{AII.6}$$

where we have altered only the y equation. Solving for the second zero-crossing of y we have

$$\begin{aligned} t_h &\simeq \frac{v\ \sin\theta}{g} + \frac{m}{6\pi\eta r}, \\ x(t_h) &\simeq \frac{mv\ \cos\theta}{6\pi\eta r}\left[1 - \exp\left(-\frac{6\pi\eta rv\ \sin\theta}{mg} - 1\right)\right]. \end{aligned} \tag{AII.7}$$

The maximum range is found by solving $dx(t_h)/d\theta = 0$. This is also a transcendental, but the solution can be approximated by taking the first term of the Taylor series of $dx(t_h)/d\theta$ around $\theta = 0$, which results in

$$\begin{aligned} \theta_m &\simeq \frac{6\pi\eta rvmg}{(e-1)(mg)^2 + (6\pi\eta rv)^2}, \\ x_m &\simeq \frac{(1-e^{-1})mv}{6\pi\eta r} + \frac{6\pi\eta rv^3 m}{(e-1)e(mg)^2 + (6\pi e\eta rv)^2}. \end{aligned} \tag{AII.8}$$

If the weight is much more than the drag, $mg \gg \eta r v$, then θ_m increases with drag. If the weight is much less than the drag, $mg \ll \eta r v$, then θ_m decreases with drag. For robots in the size range of interest (not outrageously small), the first term in the denominator of Eq. (8) dominates the second term, so we can further approximate

$$\begin{aligned} \theta_m &\simeq \frac{6\pi\eta r v}{(e-1)mg}, \\ x_m &\simeq \frac{(1-e^{-1})mv}{6\pi\eta r} + \frac{6\pi\eta r v^3}{(e-1)emg^2}. \end{aligned} \qquad (AII.9)$$

The velocity components are

$$\begin{aligned} \dot{x} &= v \ \exp\left(-\frac{6\pi\eta r}{m}t\right) \ \cos\theta, \\ \dot{y} &= \left(v \ \sin\theta + \frac{mg}{6\pi\eta r}\right) \ \exp\left(-\frac{6\pi\eta r}{m}t\right) - \frac{mg}{6\pi\eta r}. \end{aligned} \qquad (AII.10)$$

So using Eq.(6), the impact velocity components are

$$\begin{aligned} \dot{x}(t_h) &\simeq v \ \exp\left(-\frac{6\pi\eta r v \ \sin\theta}{mg} - 1\right) \ \cos\theta, \\ \dot{y}(t_h) &\simeq \left(v \ \sin\theta + \frac{mg}{6\pi\eta r}\right) \ \exp\left(-\frac{6\pi\eta r v \ \sin\theta}{mg} - 1\right) - \frac{mg}{6\pi\eta r}. \end{aligned} \qquad (AII.11)$$

If we extend the approximation to $mg \ll 6\pi\eta r$, then

$$\begin{aligned} \dot{x}(t_h) &\simeq v \ \exp\left(-\frac{6\pi\eta r v \ \sin\theta}{mg} - 1\right) \ \cos\theta, \\ \dot{y}(t_h) &\simeq v \ \exp\left(-\frac{6\pi\eta r v \ \sin\theta}{mg} - 1\right) \ \sin\theta. \end{aligned} \qquad (AII.12)$$

The speed at impact is therefore given by

$$v(t_h) \simeq v \ \exp\left(-\frac{6\pi\eta r v \ \sin\theta}{mg} - 1\right). \qquad (AII.13)$$

Using Eq. (8) this becomes

$$v_m \simeq v \ \exp\left[-\left(\frac{1}{e-1}\right)\left(\frac{6\pi\eta r v}{mg}\right)^2 - 1\right]. \qquad (AII.14)$$

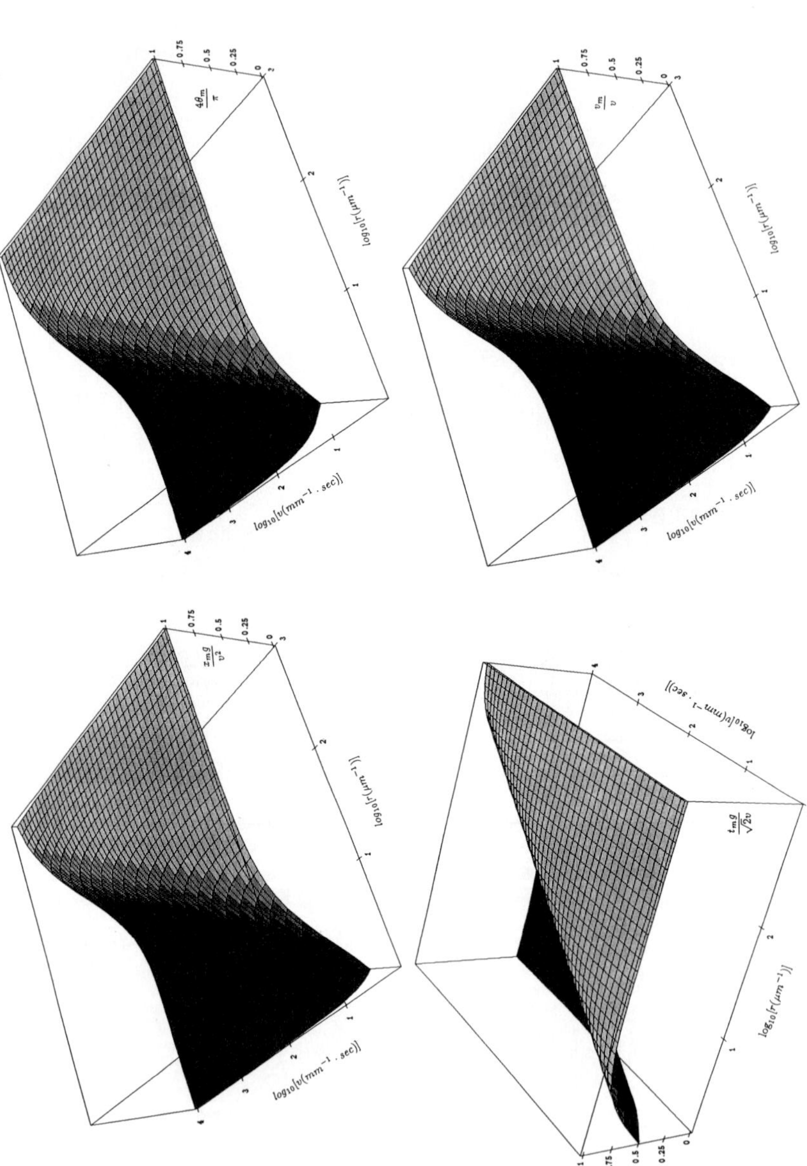

FIGURE 6 Optimum Hopping Strategies. Numerical solutions to Eqs. (2) and (10) for the maximum range and the corresponding angle of launch, time of flight, and impact velocity. Units are relative to the high Reynolds number case given in Eq. (5): (a) optimum distance; (b) optimum angle; (c) optimum time of flight; and (d) optimum impact velocity.

For most purposes when $r < 0.1$mm, we can approximate

$$\begin{aligned} \theta_m &\simeq 0, \\ x_m &\simeq \frac{(1-e^{-1})mv}{6\pi\eta r}, \\ v_m &\simeq v\, e^{-1}. \end{aligned} \tag{AII.15}$$

Performing a Taylor series expansion of the exponential in the y equation, we obtain

$$t_h \simeq \frac{2mv\ \sin\theta}{mg + 6\pi\eta r\ \sin\theta}. \tag{AII.16}$$

This gives a length of hop of

$$x(t_h) \simeq \frac{mv\ \cos\theta}{6\pi\eta r}\left[1 - \exp\left(-\frac{12\pi\eta r m v\ \sin\theta}{mg + 6\pi\eta r\ \sin\theta}\right)\right] \tag{AII.17}$$

and speed of impact of

$$v(t_h) \simeq v\ \exp\left(-\frac{12\pi\eta r m v\ \sin\theta}{mg + 6\pi\eta r\ \sin\theta}\right). \tag{AII.18}$$

Figure 6 shows numerical solutions to Eqs. (2) and (10) for the maximum range and the corresponding angle of launch, time of flight, and impact velocity. Units are relative to the high Reynolds number case given in Eq. (5). Specifically, $t_m = \sqrt{2}v/g$, $x_m = v^2/g$, $\theta_m = \pi/4$, and $v_m = v$.

REFERENCES

1. Aristotle. *Physics, Books III & V*, translated by W. Ross. *The Works of Aristotle.* Reprinted by University of Chicago & Encyclopedia Britanica, 1971. Library of Congress Number: 55-10318.
2. Bart, S., and J. Lang. "An Analysis of Electroquasistatic Induction Micromotors." *Sens. Actuators* **20** (1989): 97–106.
3. Berg, H. "Bacterial Behavior." *Nature* **254** (1975): 389–392.
4. Berg, H. "How Bacteria Swim." *Sci. Am.* **233** (1975): 36–44.
5. Fan, L., Y. Tai, and R. Muller. "IC-Processed Electrostatic Micromotors." *Sens. Actuators* **20** (1989): 41–47.
6. Fan, L., Y. Tai, and R. Muller. "IC-Processed Micro-Motors Design, Technology, and Testing." In *IEEE Micro Electro Mechanical Systems. An Investigation of Micro Structures, Sensors, Actuators, Machines and Robots*, 1–6. IEEE Cat. No. 89THO249-3. New York: IEEE, 1989.

7. Feder, R., E. Spiller, and J. Topalian. *Polymer Eng. & Sci.* **17** (1977): 385.
8. Feder, R., E. Spiller, J. Topalian, A. Broers, W. Gudat, B. Panessa, Z. Zadunaisky, and J. Sedat. *Science* **197** (1977): 259.
9. Hoerner. *Aerodynamic Drag.* Dayton, OH: Otterbein Press, 1951.
10. Hoerner. *Fluid-Dynamic Drag.* Brick Town, NJ: Hoerner Fluid Dynamics, 1965.
11. Jacobsen, S., R. Price, J. Wood, T. Rytting, and M. Rafaelof. "The Wobble Motor, an Electrostatic Planetary-Armature Microactuator." In *IEEE Micro Electro Mechanical Systems. An Investigation of Micro Structures, Sensors, Actuators, Machines and Robots*, 17–24. IEEE Cat. No. 89THO249-3. New York: IEEE, 1989.
12. Lamb, H. *Hydrodynamics.* New York: Dover, 1945.
13. Margenau, H., W. Watson, and C. Montgomery. *Physics Principles and Applications.* New York: McGraw-Hill, 1953
14. Pisano, A. "Resonant-Structure Micromotors." In *IEEE Micro Electro Mechanical Systems. An Investigation of Micro Structures, Sensors, Actuators, Machines and Robots*, 44–48. IEEE Cat. No. 89THO249-3. New York: IEEE, 1989.
15. Purcell, E. "Life at Low Reynolds Number." *Am. J. Phys.* **45** (1977): 3–11.
16. Sears, F. *Mechanics, Wave Motion, and Heat.* Reading, MA: Addison-Wesley, Reading, 1958.
17. Solem, J. "Imaging Biological Specimens with High-Intensity Soft X-Rays."
18. Solem, J., and G. Chapline, "X-Ray Biomicroholography." *Opt. Eng.* **23** (1984): 193–203 (1984).
19. Tai, Y., and R. Muller. "IC-Processed Electrostatic Synchronous Micromotors." *Sens. Actuators* **20** (1989): 49–55.

Harold J. Morowitz
George Mason University, 207 East Building, Fairfax, VA 22030

Artificial Biochemistry, Life Before Enzymes

It is my purpose to explore certain aspects of artificial life from the perspective of physical biochemistry. One purpose of this approach is to ask which features of what we recognize by the term "life" are tied to the one chemical embodiment we know, life on the planet Earth, and which aspects can be abstracted and conceptually removed from this context. My own sense is that prokaryotic life is tied to its chemical reification in very profound ways and these should be understood before proceeding.

To scientists interested in biogenesis on Earth, life is an emergent property of systems of carbon, hydrogen, nitrogen, oxygen, sulfur, and phosphorus, subject to an influx from a high frequency (high temperature source) and an efflux to a low frequency (low temperature sink). From the point of view of physical biochemistry, the statement by Perrett[2] stands out:

> Life is a potentially self-perpetuating open system of linked organic reactions, catalyzed stepwise and almost isothermally by complex and specific organic catalysts which are themselves produced by the system.

This statement focuses on the complex catalysts and presents one of the enigmas of biogenesis. How can the system function without complex catalysts, and how can the catalysts exist without the system? It is the pursuit of questions of this nature that lead to considering artificial biochemistry, those reactions of biochemical significance that can be carried out in the absence of enzymes, which I take to mean specific macromolecular catalysts specified by a cellular genome.

I would like to move to the grand empirical generalization of biochemistry. There is a universal chart of intermediary metabolism applicable to all contemporary organisms. This chart for autotrophs includes glycolysis, the pentose phosphate shunt, the itric acid cycle, oxidative phosphorylation, amino acid biosynthesis, nucleatide biosynthesis, and synthesis of macromolecules. Each species has its own metabolic chart which includes a subset of the universal chart plus a set of taxon-specific reactions which are usually designated secondary metabolism. The chart consists of a connected set of reactions which are highly interconnected. The chart in any species is determined by the genome, although only certain parts of the chart may be used in any tissue in a multicellular organism.

This graph is so crucial to studying biology at all hierarchical levels that studying any branch of biology without a consideration of the metabolic chart is like pursuing chemistry without knowledge of the periodic table.

In all likelihood, present-day metabolism emerged about 3.9 billion years ago with the universal ancestor. It is the oldest fossil and is perhaps older than any extant rocks. Because of the principle of continuity, the chart of intermediary metabolism is the best way to look back into prebiotic chemistry.[1]

All of metabolism can be represented by three general types of chemical reactions: isomerizations, binary reactions, and condensation-splitting reactions. They are represented as Figure 1 and appear to represent an irreducible set of reaction types for all of metabolism. Using the line and dot representation of these in Figure 1, we can convert the chart of intermediary metabolism into a type of graph. Note this differs from ordinary graph theory where there are only points and lines. The present type of graph demands three kinds of connections. Associated with each type of connection are conservation rules such as the conservation of atoms between reactants and products.

The dots represent chemical species, and a number associated with each dot is the concentration. If we are dealing with a compartmentalized system, a dot represents a chemical species in a specific cellular compartment. Some of the isomerization steps then represent transport between compartments. The operations also have a direction from a higher Gibbs free energy to a lower free energy.

I would now like to introduce the term "artificial biochemistry," which is "artificial" in a different sense than in most of the other contributions in this volume. An artificial metabolic operation is one from the graph of intermediary metabolism that can take place in aqueous solution or at aqueous nonpolar interfaces in the absence of enzymes.

The importance of artificial biochemistry, apart from possible industrial applications, is in discussions of the origin of life. If some subset of reactions of intermediary metabolism can take place in the absence of enzymes, then the precursors

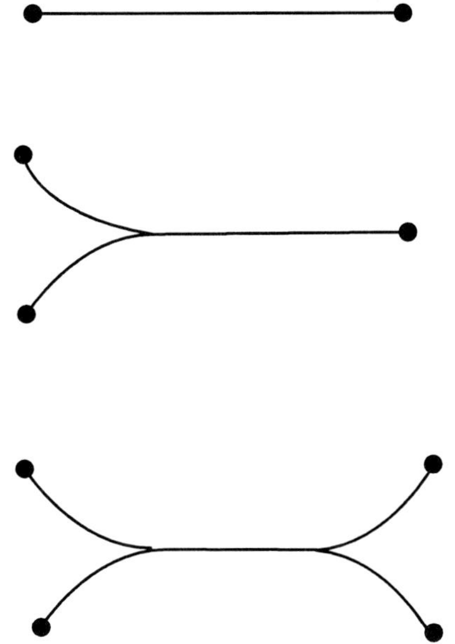

FIGURE 1 Three general types of chemical reactions: isomerizations, binary reactions, and condensation-splitting reactions.

of macromolecular complexity can derive from the simplicity of aqueous organic chemistry. The conjecture is that a primitive metabolic system led to catalysts that improved the functioning of the system. A second set of reactions arose to conserve the catalysts by genetic encoding.

We can distinguish between the initial energy metabolism and the subsequent information-specific metabolism. There is a general feature of the chart revealing that energy metabolism involves low-molecular-weight molecules of carbon, hydrogen, oxygen, and phosphorous, while the entry of nitrogen into the system leads to catalytic molecules, such as combinations of amino acids, and information molecules, such as nucleotides.

At this point the chart of metabolism reveals a universal feature. In autotrophs (plants and autotrophic bacteria), there is a universal pathway of entry of nitrogen into the biosphere. Since all heterotrophs have to rely on nitrogenous compounds made from autotrophs, this becomes a general port for entry of nitrogen.

Thus, the amination of energy metabolism compounds leads to information metabolism. The ammonia group enters at two reactions only, hence the biological universality of these reactions. First, of course, nitrogen in the environment must be converted to ammonia, NH_3, if it is in a more oxidized form. The most common oxidized forms are dinitrogen, N_2, in the atmosphere and nitrite NO_3^- and nitrite NO_2^-, in the soil. The entry of ammonia into the biosphere is by the following two reactions:

$$NH_3 + \alpha\text{ketoglutarate} + NADPH_2 \rightleftharpoons \text{L-glutamate} + NADP + H_2O. \qquad (1)$$

$NADP_2$ is reduced Nicotinamide adenine dinucleotide phosphate and NADP is the oxidized form of that molecule. The enzyme is L-glutamate-dehydrogenase.

$$NH_3 + \text{glutamate} + \text{ATP} \rightleftharpoons \text{glutamine} + \text{ADP} + P_i. \quad (2)$$

ATP is adenosine triphosphate, ADP is adenosinne diphosphate, P_i is the orthophosphate, and the enzyme for the reaction is glutamine synthetase. The first reaction is the critical step in converting inorganic nitrogen in ammonia to nitrogen in the amino acid glutamic acid. The task of artificial biochemistry as I have defined it is to carry out reactions without an enzyme and without the nitrogen-containing cofactor $NADPH_2$. As a further set of constraints, we require the reaction to take place in aqueous solution at plausible temperatures for the earth's surface.

Again, note that this reaction links energy metabolism with catalysis and information metabolism. If complexity is to arise from simplicity, there must be a nonenzymatic pathway from keto acids to amino acids. Equation (1) can be generalized.

$$NH_3 + \text{keto acid} + \text{reducing agent} \rightleftharpoons \text{amino acid} + H_2O + \text{oxidant}. \quad (3)$$

In a series of experiments carried out with Sherwood Chang and Eta Peterson, we attempted the artificial biochemistry shown in Eq. (3). The details will be reported elsewhere. For the reducing agent, we introduced formic acid, which is perhaps the simplest of organic reductants.

$$\overset{\displaystyle O}{\overset{\|}{HC}}\text{–OH} \rightleftharpoons CO_2 + 2[H].$$

Thus αketoglutaric acid was mixed with aqueous ammonium formate at room temperature. The mixture, after one hour at room temperature, produced substantial amounts of DL glutamic acid as determined by cellulose thin layer chromatography, column chromatograph and GC–MS. Thus reaction can occur in the absence of enzymes. The synthesis has also been carried out with ammonia and sodium borohydride. Other keto acids have been tried. Thus glycine has been synthesized from glyoxylic acid and alanine from pyruvic acid. This points to the full generality of reaction 3.

The next step in this approach, currently under study, is to find other key reactions and reduce them to artificial biochemistry. One of the most important of these involved carbon chain extension. In normal biochemistry, this proceeds by

$$\text{acetyl coenzyme A} + CO_2 + \text{ATP} \rightleftharpoons \text{malonyl coenzyme A} + \text{ADP} + P_i. \quad (5)$$

This reaction is currently done with the enzymes acetyl CoA carboxylase, biotin carbonyl carrier protein, and carboxyl transferase. We must seek nonenzymatic analogues for these reactions. Carbon chain extension is so important in biogenesis that we must seek to carry out some form of the reaction

$$CO_2 + \text{acetate} \rightleftharpoons \text{malonate}.$$

Another feature of artificial biochemistry has long been known but not assigned its full role in biogenesis. Amphiphiles are a class of molecules consisting of a polar and a nonpolar portion. In an aqueous environment the nonpolar parts cluster, giving rise to a family of colloids. The structures that form minimize the nonpolar water interactions and represent local Gibbs free-energy minima. They are equilibrium structures in the time domain of intermolecular interactions. A class of these structures consists of planar biomolecular leaflets or membranes. This allows all the nonpolar parts to be in contact and all the polar parts to be in contact with water. These membranes seal their loose ends to form a class of vesicles known as liposomes. The existence of vesicles immediately makes possible the following properties of cells:

1. Cellularity
2. Three-phase system
3. Interior-exterior concentration difference
4. Transmembrane electrical potential difference
5. Transmembrane pH difference
6. Transmembrane redox difference
7. Light-driven electrical potentials
8. Chemosmotic photophosphorylation
9. Binding of ligands on the membranes, after which they act as catalysts

Note that this is all artificial biochemistry. The amphiphiles are relatively low-molecular-weight molecules, and no macromolecules are involved in their assembly.

I now want to argue that in a watery world, phase behavior is dominated by differences in polarity. This is generally measured by dielectric constant going from values of 80 for a very highly polar substance like water to 3 for very nonpolar substances such as alkanes. All of aqueous-phase colloidal chemistry and colloidal structures depend on these differences. Polarity which depends on symmetry and charge mobility is deeply embedded in the structure of matter.

In cellular systems there are two fundamental methods of structure:

1. Macromolecules which are precise, depend on enzymes, and are extremely sensitive to changes in monomers.
2. Membranes which are variable in composition, form as a thermodynamic local minimum, and do not depend on enzymes for formation.

Membranes clearly fit our definition of artificial biochemistry and are logical candidates for prebiotic structures if we assume that biogenesis moves from the simple to the complex. Again, there is the implication that artificial biochemistry can be related to prebiotic chemistry.

Next we move to the question of how much of the biology we know is specific to the chemical details of the chemistry of carbon, hydrogen, nitrogen, oxygen, phosphorous, and sulfur and how much can be generalized in more abstract terms. To do this consider a small balanced microecosystem (microbiosphere) closed to the flow of matter but open to the flow of energy. If sunlight flows in and heat flows out, the system is optically pumped and maintains itself above the thermodynamic ground state of a system of the same atomic composition and at the same temperature. For most biological systems this amounts to about five kilocalories per gram of dry weight of biomass. This energy is largely stored in covalent bond energy. Tables 1

TABLE 1 Energies of Bonds of C, H, N, and O

Bond	Enthalpy of Formation	Enthalpy/electron	Bond Type[1]
H–O	-111	-55.5	T
H–H	-104	-52.0	T
H–C	-99	-49.5	T
H–N	-93	-46.5	T
C=O	-174	-42.5	T
C–O	-84	-42.2	E
C–C	-83	-41.5	E
N≡N	-226	-37.7	T
C=C	-147	-36.8	E
N=O	-143	-35.8	T
N=C	-140	-35.0	E
N–C	-69	-34.5	E
N≡C	-207	-34.5	T
C≡C	-192.1	-32.0	E
N–O	-53	-26.5	E
N=N	-100	-25	E
O=O	-96	-24.0	T
N–N	-38.4	-19.2	E
O–O	-34.9	-17.2	E
H•	0	0	T
O•	0	0	T
N•	0	0	T
C•	0	0	T

[1] T stands for chain terminating, E stands for chain extending.

TABLE 2 Energy grouping of bonds of C, H, N, and O

	(LECT) Low energy chain terminating	(IECE) Intermediate Energy chain extending	(HECT) High energy chain terminating	Diatomic oxygen
	H–O	C–O	H.	0=0
	H–H	C–C	O.	
	H–C	N=C	N.	
	H–N	C=C	C.	
	C=O	N–C		
	N≡N	C≡C		
	N=O	N–O		
	N≡C	N≡N		
		N–N		
		O–O		
Average energy/ electron	-44.25	-31.01	0.0	-24.0

and 2 indicate the covalent bond energies and classification for bonds of carbon, hydrogen, nitrogen, and oxygen.

Next, place the ecosystem in contact with an isothermal reservoir and shut off the flow of energy. Allow it to decay for a long time to approach chemical equilibrium. It will then be dominated by CO_2, H_2O, and N_2 and, if sufficiently reducing, there will be CH_4 and NH_3. The equilibrium or lowest free energy will be dominated by small gas phase molecules.

Comparing the equilibrium state with the living state, the latter is dominated by chain-extending bonds or chemical complexity. The major chain-terminating bond in the living state is the relatively high energy 0=0. At very high energy the chain-terminating free radials take over, and the system again moves toward chemical simplicity. The living state is characterized by large, stable, covalently bonded structures and oxygen. This is a consequence of the chemical energetics of molecules made of CHNO. The use of oxygen as the released gas in optically pumping the system to higher energy levels also seems implied in this elementary energy analysis. Thus, oxygen as the product of photosynthesis seems tied to the energetics.

There are thus a number of features characteristic of life, such as those just discussed which seem closely related to the chemical and thermochemical properties of the atomic components. It would seem well to incorporate these chemical constraints into the theoretical structures of artificial life.

ACKNOWLEDGMENTS

I would like to thank the National Aeronautics and Space Administration for support of work reported in this paper.

REFERENCES

1. Morowitz, H. J. *The Beginnings of Cellular Life: Metabolism Recapitulates Biogenesis.* New Haven, CT: Yale University Press, 1992.
2. Perrett, J. *New Biology* **12** (1952): 6.

Kurt Fleischer and Alan H. Barr
Computation and Neural Systems, Mail Stop 350–74, California Institute of Technology, Pasadena, CA 91125

A Simulation Testbed for the Study of Multicellular Development: The Multiple Mechanisms of Morphogenesis

This paper presents a simulation framework and computational testbed for studying multicellular pattern formation. The approach combines several developmental mechanisms (chemical, mechanical, genetic, and electrical) known to be important for biological pattern formation. The mechanisms are present in an environment containing discrete cells that are capable of independent movement (cell migration). Experience with the testbed indicates that the *interactions* between the developmental mechanisms are important in determining multicellular and developmental patterns.

Each simulated cell has an artificial genome whose expression is dependent only upon its internal state and its local environment. The changes of each cell's state and of the environment are determined by piecewise continuous differential equations. The current two-dimensional simulation exhibits a variety of multicellular behaviors, including cell migration, cell differentiation, gradient following, clustering, lateral inhibition, and neurite outgrowth (see Color Plates 1–6).

We plan to perform simulated evolution on developmental models as part of a long-range goal to create artificial neural networks that solve problems in perception and control.[8] The testbed is a step on the path towards this goal.

1. INTRODUCTION

The testbed system described in this paper is part of a larger project to generate a new class of artificial neural networks. We are studying the problem of generating artificial neural networks that share some properties of biological neural networks, in particular:

- **problem-specific geometric structure:** some biological neural circuits solve a problem largely by assuming a particular geometric configuration. For example, the owl auditory localization circuit[7] uses intercalated axons as delay lines to compute the difference between time of arrival of auditory signals.
- **asymmetric topological connectivity:** most artificial neural networks have a prescribed regular connectivity. Biological systems exhibit a variety of connection topologies.
- **heterogeneous neural types:** most neural circuits involve multiple neural types that have different morphology and function.

These properties are evident in real neural networks and are believed to be closely related to the functions they compute. To capture these properties in an artificial neural network, we have chosen to perform simulated evolution on a developmental model.[1]

Why evolve a *developmental* model? Developmental models have two properties that may make simulated evolution more fruitful:

- robustness—the process of development can compensate for deleterious changes to the genome, and
- developmental gain—a small change in the genome can make a large change in the organism (e.g., add another layer, add another segment).

The simulation testbed presented in this paper was created to find a simple developmental model that can create neural networks with arbitrary topological connectivity and a large degree of geometric complexity. The testbed must be flexible so that we can explore different strategies for development and evolution.

We use the term *modeling framework* to refer to our modeling abstraction, which we distinguish from the *testbed* (our implementation of the framework).[2] For example, our framework contains the concept of an environment with diffusing chemicals. The testbed currently implements this as a discretized grid in two dimensions, and the cells' sensors are implemented as smooth interpolations on this

[1] A *developmental model* is a model that captures some aspect of the changes that occur as an organism grows to achieve its mature shape and function.

[2] This distinction between abstraction and implementation is a useful tool in applied mathematics and other fields.[3]

grid. We may later decide to change the implementation or to extend it to three dimensions, but our abstraction (the modeling framework) would remain the same.

From developmental biology, we know that the gross structure of a multicellular organism arises from the expression in each cell of preprogrammed genetic information in a particular environment. Each cell inherits its state from its parent, in the form of its genome as well as other cellular materials and structures (e.g., organelles, mRNA, etc.). Although each cell only has access to local information, the development process robustly creates successful organisms, even in the presence of fairly severe perturbations. What sort of local behaviors can account for this?

Previous mathematical models of morphogenesis have shown that chemical effects can account for some behaviors,[16,30] mechanical effects for others,[19] and cell-lineage control of the geometry of cell division for yet other shapes.[5,15] We combine these factors in one modeling system, to explore how the *interaction* between these factors can determine developmental patterns.

Our modeling framework consists of discrete cells that are capable of independent movement and that are controlled by an artificial genome model. They move about in a simulated environment comprised of chemical, mechanical, and electrical elements. The individual cells are modeled as physical entities within this environment, subject to mechanical collisions, adhesive forces, and drag. Each cell's activities are determined by its internal state and the expression of its genome within its local environment. The genome is a set of differential equations that depend on the cell's current state and its local environment. The changes in the environment are governed by differential equations which implement the mechanical effects and the diffusion of extracellular chemicals. The current testbed is two-dimensional; extension to three dimensions is straightforward, although computationally more expensive.

Our first results show that the testbed is able to model several simple multicellular patterns. We show examples of following gradients, clustering, cell differentiation, pattern formation, and network generation (see Color Plates 1–6).

1.1 OVERVIEW OF THE PAPER

Several of the elements of our model are based on previous work in developmental biology. This work and other related work is briefly described in Section 2.

Section 3 describes the modeling framework, and its relation to previous work. The testbed implementation is presented in Section 4 (more details appear in Appendix A). Sections 5 shows some experiments we have run with the testbed and Section 6 discusses what we have learned from our experience with the system.

2. RELATED WORK

2.1 PREVIOUS DEVELOPMENTAL MODELS

In this section we describe some previous models of development, each of which focuses on a particular developmental mechanism (chemical, mechanical, genetic, or electrical). Recently, some of these researchers have enhanced their models to include elements of other mechanisms, as we advocate.

CHEMICAL FACTORS. In his 1952 paper "The Chemical Basis of Morphogenesis," Turing proposed a mathematical theory of cell-to-cell interaction via chemical substances (*morphogens*). He showed that these *reaction-diffusion* systems could exhibit stable patterns, and proposed this as a possible mechanism for pattern formation in development. This has formed the basis of much work on developmental modeling using reaction-diffusion equations, such as that by Meinhardt,[16] the chemotaxis models of *dyctyostelium* slime molds,[13] and the creation of patterns such as those of the zebra coat[2] and some butterfly wings.[18]

MECHANICAL FACTORS. In "The Mechanical Basis of Morphogenesis," Odell et al.[19] discuss how a mechanical model can account for gastrulation, neural tube formation, and eversion behaviors such as that observed in Volvox.[19] In their system, the cell membrane is modeled as several springs with variable spring constant and rest length. These parameters are modified based on interactions between adjacent cells, and this gives rise to the various behaviors. Later work in this area uses more detailed mechanical models and incorporates some chemical signalling to model cell intercalation[20] and other phenomena.

GENETIC FACTORS: L-SYSTEMS AND GRAMMARS. Grammar-based techniques such as L-systems[14] are convenient for describing cell lineage and genetic control of cell division. These systems use rewrite rules to sequentially modify strings which represent organisms, and are capable of creating realistic-looking models of biological structures. L-systems have been particularly successful for modeling plants[15,22] and have also been applied to modeling cell layers,[5,6,22] topology of neural networks,[3] and other systems.

Some grammar-based models incorporate environmental influences and cell-cell interactions using context-sensitive languages.[14,15] Other grammar-based models have been enhanced to include computation of physical forces for the modeling of cell layers.[6,22] Between cycles of cell division, adjacent cells apply forces to each other, changing shape until equilibrium is achieved. The cell walls are modeled as linear springs (similar to Odell's model[19]), and the cells expand and contract due to an approximation of osmotic pressures.

[3] A more detailed comparison of our differential equation approach to evolving neural networks versus the grammar based approach of Kitano[12] appears in Fleischer.[8]

This sort of hybrid system, which incorporates both grammatical elements and differential equation elements, seems to be an effective way to make computationally feasible models which incorporate physical behaviors.[17,23]

ELECTRICAL ACTIVITY. Electrical activity affects the development of neural structure in many ways. For example, it is thought that correlated firing between neighboring axons can affect their destinations[9] and that synapse formation can be strengthened in some cells when the firing of the input and output cells are correlated.[10]

Fraser and Perkel proposed a developmental model of the neural map between the retina and the tectum in the visual system of lower vertebrates. This model incorporates several different mechanisms, involving modulation of cell and neurite adhesion, competition for space, and activity-dependent processes (which depend on the firing of neurons), and can reproduce a variety of observed experimental data.

2.2 OTHER RELATED WORK

Several researchers have independently determined that it is now computationally feasible to compute medium-scale developmental simulations for a variety of applications. We mention here two systems that bear some similarity to our model.

THE CONNECTIONIST MODEL. Mjolsness et al.[17] have noted the similarities between gene regulation and standard neural net dynamics. They use differential equations to describe dynamics of gene interactions at a short time scale, combined with a grammar-based model to describe cell state changes at a longer time scale (e.g., mitosis,[4] interphase, post-mitotic).

The connectionist model uses a type of fixed differential equation to describe the detailed dynamics. The differential equations are of the standard connectionist form[11]: $\tau_a(dv_i^a/dt) = g_a(\sum_b T^{ab}v_i^b + h^a) - \lambda_a v_i^a$. The v_i^a are the components of the state vector for each cell i, g_a is a sigmoidal threshold function, T^{ab} are the components of a connection matrix, and h^a are offsets.

This model is currently being successfully applied to the early developmental genetics of *Drosophila*,[25] by optimizing the parameters to match biological data (the optimized parameters are T^{ab}, h^a, and parameters associated with synthesis, decay, and diffusion rates).

[4] cell division.

THE CELL PROGRAMMING LANGUAGE. The "Cell Programming Language"[1] makes more simplifying assumptions (discrete time and space, and direct interactions between cells), which enable it to compute simulations with a few thousand cells. In the discretized spatial model, cells exist only on grid points. A single cell can cover several adjacent grid points, creating a nontrivial shape. Each cell's genome consists of a set of states for the phases of the cell (pre-mitotic, post-mitotic, etc.), and associated with each state is a sequential list of instructions. Time is also discrete and, during each time step, every cell executes its instruction list. The state of each cell is directly available to neighboring cells, so they can modify or react to the state of their neighbors. It has been applied to modeling cellular sorting by differential adhesion, aggregation in slime molds, and other phenomena.

3. THE MODELING FRAMEWORK

We propose a multiple mechanism model for cellular development, based on the chemical, mechanical, genetic, and electrical models discussed in the Previous Work section (Section 2.1). Several others have been moving in this direction as well. The modeling of cell layers[22] incorporates a grammar-based model for cell lineage and a mechanical model for cell-cell interactions. The cell intercalation models of Oster et al.[20] combine detailed mechanical models with simple chemical signaling models. Fraser and Perkel's retino-tectal model,[9] Mjolsness' connectionist model,[17] and the Cell Programming Language[1] are all models that use a combination of mechanisms.

Our modeling framework combines multiple mechanisms within a single simulated environment (see Table 1). The environment contains diffusing and reacting chemicals, mechanical barriers, adhesive substances, etc. Cells move about within this environment, interacting with each other and with the environment.

Cell migration is an important aspect of the neural network development, and we explicitly model discrete cells that are free to move continuously within the environment. This distinguishes our work from most of the previous work in reaction-diffusion systems and grammar-based system that do not have this capability. Other models that do allow cell migration often do not include the other mechanisms (chemical, genetic) that are critical to developmental pattern formation.

The inclusion of multiple mechanisms enables many forms of interaction between cells. Interactions between cells can occur directly from one cell to another or indirectly mediated by the environment between the cells. Examples of direct interaction are collision (applying a force) and contact recognition (changing the amount of a cell's artificial membrane protein that is in a bound state). Indirect cell interactions occur when a cell changes the state of the environment, which in turn is sensed by another cell. For example, one cell can emit a chemical into the environment which will then diffuse spatially. Another cell some distance away can sense and respond to that emitted chemical, thus reacting to the actions of the first cell in an indirect manner. Genetic factors such as cell lineage are also important in forming developmental patterns. Some simple grammar-based systems that only model cell lineage are capable of making biologically relevant patterns.[22] The capabilities of these models motivate us to include genetic mechanisms, although our representation is somewhat different from that of L-systems which generally have pre-defined cell types and which determine the types of the children directly from the parent via a production rule. Our cells inherit their state from their parent, and then cells with

TABLE 1 The Modeling Framework and Its Implementation.

Modeling Framework (abstraction)	Testbed (current implementation)
Discrete cells (allows cell migration)	
cell geometry	two-dimensional circles
cell substructures	none
growth cones	modeled as small cells
neurites	path of growth cone and communication link between cell and growth cone
Genetic/cell lineage	
genetic control of cell operations	parallel ODEs w/conditions
inherit state from parent cell	yes
control over orientation of cell divisions	yes
asymmetric cell division	not implemented yet
Extracellular environment	
chemical	two-dimensional reaction-diffusion grid
mechanical	mechanical barriers, viscous drag
Cell-cell interactions	
mechanical	collisions and adhesion between cells
chemical (membrane proteins)	adhesion and contact recognition
electrical (gap junction, synapse)	not implemented yet
Cell-environment interactions	
chemical	emit, absorb, sense values in grid
mechanical	cell-env collisions and adhesion

different behavior or composition can be interpreted as different types. Our model is thus at a slightly lower level of abstraction than the L-systems models, with cell type derived rather than specified.

The *connectionist model* mentioned previously[17] is a hybrid system that uses grammar rules to model state changes within individual cells, and differential equations to describe the continuous behavior within a state. This system is similar to ours in many ways, the major differences being:

- They use a fixed type of differential equation, whereas we have a more arbitrary form (user-specified function; see Sections 4 and A.2.1).
- They have grammar rules for state changes, whereas we use conditional terms on our ODEs (see Sections 4 and A.2.1).
- With regard to goals, they are matching biological data, whereas we are making artificial neural networks (although both systems are probably general enough to be applied in either domain).

4. TESTBED IMPLEMENTATION

In this section, we describe in broad terms the implementation of each feature of the model sketched in Table 1. The current implementation is a subset of the entire modeling framework. The electrical model is as yet incomplete, so we leave its description to a future paper. Appendix A contains a more detailed description of the data structures, mathematical methods, and numerical solutions.

The testbed meets our design goals of:

- faithfully implementing the modeling framework,
- showing the ability to generate neural networks with asymmetric structure and heterogeneous cell types, and
- computing simulations with hundreds of cells.

The testbed is designed to be flexible and exttensible, enabling us to search for a genomic language suitable for simulated evolution (as discussed in the introduction).

We model discrete cells as containing two classes of proteins: those in the cytoplasm of the cell and those in the membrane. These proteins form the *state* of our cells. Each protein is represented by a floating point state variable ($state[i]$) describing the amount of that protein currently in the cell (or membrane). Conversion between amounts and concentration is done by using cell volume.

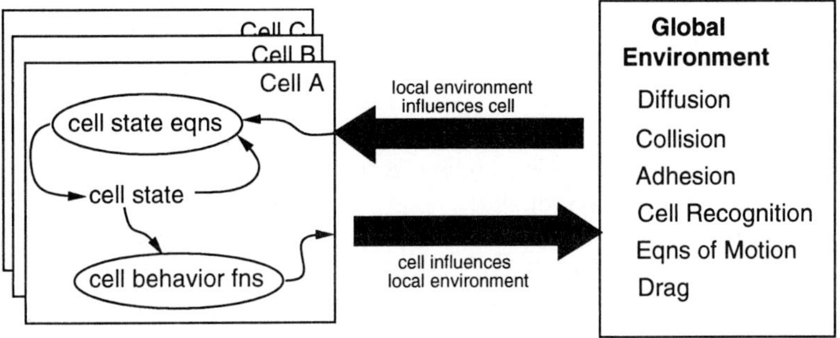

FIGURE 1 Each cell's state is modified by the *cell state equations* (the genome), which have access to the cell's local environment. The local environment information is computed from the processes in the global environment. The cell's state determines the cell's behavior via the *cell behavior functions.* The behavior can then affect the global environment (e.g., by releasing a diffusable chemical).

The cells change state continuously via an artificial genome (the *cell state equations*; see Figure 1 and Section A.2.1), which encodes the change of amount of protein over time ($dstate[i]/dt$). This derivative depends on the current state of the cell and on information available in the cell's local environment (locally measured values of diffusable chemicals, amount of contact with neighbors, etc.). To allow the artificial genes to regulate each other (turn each other on and off), we include a conditional term to each differential equation. For a cell c, a single artificial gene for a state variable i has the form:

$$\text{IF } \mathbf{Condition}_c(state_c, env_c), \text{ THEN } \frac{dstate_c[i]}{dt} = \mathbf{Consequent}_c(state_c, env_c).$$

Several artificial genes may code for the same artificial protein, and their contribution is summed compute the total change in state. We implement the conditions as continuous functions which vary from zero to one in a sigmoidal fashion, which are multiplied by the consequent to give the actual contribution. Using a continuous function avoids signalling a discontinuity every time a condition fires.

$$\frac{dstate_c[i]}{dt} = \sum_S (\mathbf{ContinuousCondition}_c^S(state_c, env_c))(\mathbf{Consequent}_c^S(state_c, env_c)).$$

The state of a cell determines its behavior via the *cell behavior functions* (Section A.2.2). In real biochemical systems, a protein or group of proteins has a specific function in the cell, which is a consequence of the molecular structure of the proteins involved. We use a mathematical expression (the *cell behavior function*) to describe how a protein or group of proteins acts to perform some function: e.g., produce a locomotor force, release a diffusable chemical, adhere to another molecule, etc.

Cells also exhibit discontinuous behaviors (events), such as cell division, emitting a growth cone, and dying. We determine the timing of each event with a cell behavior function; when the function crosses zero, the specified event occurs (the cell divides, emits a growth cone, or dies).

The continuous and discontinuous behaviors of the cells form a system of piecewise continuous ordinary differential equations.[3] These are solved similarly to ordinary differential equations (ODEs), except that, when an event occurs (as signalled by an event function crossing zero), the solver is briefly halted and structures are created or destroyed (e.g., during cell division or cell death). Thus the solver must also do root finding on the event functions while it is integrating the ODEs forward in time.[5]

The testbed supports a broad range of experiments, in which a simulated organism is allowed to undergo development according to its artificial genome and the given environmental conditions. An experiment is described in a file containing the cell state equations (artificial genome), cell behavior functions, parameter settings, boundary conditions, and initial state (an example file appears in Appendix B). Thus the user can change both the state equations (genome) and the behavior equations. In practice, we tend to change only the genome and use the same behavior functions between experiments.

5. RESULTS

Snapshots from the animations generated by the testbed are shown in Color Plates 1–6 (see insert). Two of them are also included in black and white with the text, for convenience (Figures 5(d) and 6(d)). These simulations show the ability of our testbed to exhibit some basic multicellular behaviors. The cells are represented as circles, and the diffusing chemicals are rendered as graded colors. Neurites are shown as white lines (in Color Plate 6).

The simulations were computed on Hewlett Packard 9000 series 800 and 700, IBM RS6000, and DEC Alpha computers. The running times range from a several seconds to a few hours.

[5]The piecewise ODEs manage state changes, performing a similar function to the grammar rules in the *connectionist model*.[17]

PLATE 1: NEURITE PATH FINDING. In this experiment, a growth cone from a cell on the left climbs the gradient of the red chemical, pushing through the barriers in its path until it reaches the far cell (which is emitting the chemical). Note the robustness exhibited here: the connection is made despite the presence of barriers. If the barriers were removed or differently located, the connection is still likely to be made.

PLATE 2: CELL DIFFERENTIATION. The initial cell divides under the control of a cytoplasmically inherited factor (a protein that the cell inherits from its parent). When this is diluted beyond a threshold, the cells stop dividing and begin to emit diffusable chemicals. One of the chemicals is a fast-diffusing inhibitory factor (green) and the other is slow-diffusing and excitatory (red). The combination of lateral inhibition and local positive feedback gives rise to patterns of differentiated cells, as in Meinhardt's work.[16] Those turned "on" are emitting both red and green. The other cells' activity is suppressed by the inhibitory green chemical. Initial conditions and environmental differences account for which cells are selected, creating a heterogeneous population.

PLATES 3(A–B): CYCLIC BEHAVIOR. Size regulation (creating and maintaining the proper number and types of cells) is an important function in multicellular organisms. In every experiment we conduct, we must deal with size regulation, avoiding explosive nonterminating growth (which simply fills our simulated Petri dish).

We include this example to emphasize that it is not trivial to predict what a given artificial genome will do, nor is it easy to concoct by hand a genome to create a particular pattern.

In Color Plates 3(a) and 3(b), we see an attempt to regulate size using diffusable chemicals. Instead of regulating the size, this experiment unexpectedly resulted in cyclic behavior. There were initially two cell types: one emits and seeks red, the other emits and seeks green. Growth and splitting for each type limited by the presence its respective chemical. The amount of chemical emission is regulated by the presence of the chemical. This specification forms clusters, however, they are not stable. Once the clusters form, the cells in the center of the cluster begin to reduce their emission of the chemical, which causes a local depression. The gradient then leads the cells out from the center, forming a ring. The ring pattern is not stable either, and the cells tend to cluster together again, cyclically.

PLATES 4(A–D): CHAINS OF CELLS This experiment shows how the ratio of forces due to two different mechanisms can be effective in creating a pattern. The two opposing forces are an attractive force due to cell adhesion and a repulsive force due to a diffusable chemical. The cells are held together by the adhesive force but attempt to move away from each other due to the repulsive chemical. This interaction leads to a pattern of chains of cells, or very small clumps.

The cells divide when the concentration of the chemical is low, so the ones at the end of a chain tend to divide more frequently. This also serves to regulate the size of the chains. (The interpreted code for this experiment appears in Appendix B.)

PLATES 5(A–D): SKELETON. This sequence begins with a single cell which divides for several generations. These first cells emit a diffusing chemical (shown as yellow in the color plate) and move about until they sense a certain level of the chemical. This forms a pattern of roughly equally spaced cells. When this pattern is stable, some of the cells change state and begin to exhibit the chaining behavior (as in Color Plate 4). The second wave cells avoid large concentrations of the (yellow) diffusing chemical but are attracted to it if the concentration is too small. Thus they intercalate between the original cells but do not move too far from the cluster.

PLATES 6(A–D): NETWORK. Two cells of different types give rise to a small network in this simulation run. One cell divides a few times to create a small cluster of green emitting cells. At a later time (determined by the dilution of an inherited factor), these cells emit growth cones[6] which seek the red chemical. Meanwhile, the other cells have been dividing and emitting the red chemical while waiting to be contacted by the growing neurites.

When a growth cone initially contacts a red-emitting cell, it adheres via a certain surface factor (an artificial membrane protein). Upon recognizing the contact, both the cell and the growth cone undergo a state change. They both begin to express a different adhesive surface factor and stop expressing the first. In addition, the cell stops emitting red and the growth cone stops moving. Thus the cell will no longer attract growth cones (since it has stopped emitting red), and later-arriving growth cones will no longer adhere to it (since it has stopped expressing the first surface factor).

Growth cones that have not made contact continue to search for red-emitting cells. Eventually, the growth cones that fail to find a target will die, as will their neurites. The process of excess neurites dying off can be seen in the difference between the number of white neurites in Color Plates 6(c) and 6(d). This "pruning" of excess neurites is a phenomenon observed in biological networks.[4,24]

[6] A growth cone is a structure at the tip of a growing neurite, which is "the driving force behind neurite extension."[24]

6. DISCUSSION

6.1 WHAT HAVE WE LEARNED FROM THESE EXPERIMENTS?

- Size regulation (creating and maintaining the size and shape of a multicellular organism) is critical and nontrivial.
- Using the combination of multiple mechanisms to specify a pattern can be more robust than using a single mechanism.
- Specifying cell lineage (via cytoplasmically inherited factors) is a useful way of describing a developmental pattern (this is the mechanism used by L-systems; see Section 2.1).
- It is difficult to design an artificial genome that develops into a particular pattern or, conversely, to predict the pattern that a particular artificial genome will create.
- Developmental models are adaptive (robust). They tend to generate similar patterns under perturbations of environment or initial conditions.

SIZE REGULATION. Size regulation is a difficult problem for multicellular organisms. We separate this into two related problems: limiting growth, and clustering (keeping the cells together). Each of these may be solved using chemical, mechanical, or genetic operations. For limiting growth, we tried two methods:

- using a threshold on the concentration of a diffusable chemical to turn cell division on and off, and
- using cytoplasmically inherited factors. Cells keep subdividing until the supply of an irreplaceable chemical is exhausted. Since the chemical is not regenerated, each cell division reduces the total amount per cell and, as the cells grow in size, the concentration diminishes.

For clustering, we also tried two methods:

- cells move up the concentration gradient of a diffusable chemical, and
- using cell adhesion to keep cells together.

The most robust size regulation behavior was produced using a combination of approaches (as in Color Plates 4, 5, and 6).

RELATION TO L-SYSTEMS Control via cytoplasmically inherited factors is an effective way to regulate a developmental process. The combination of asymmetric cell division and inherited factors can emulate L-systems[22] see Section 2.1). Each cell division is like a production rule ($parent \rightarrow child_a\ child_b$), with the children differing as specified by the asymmetric cell division. As discussed in Section 2, this approach can be seen as a model at a lower level of abstraction than that of L-systems. Asymmetric cell division was not used in the simulations shown in this paper; however, we did use inherited factors to control cell division and state changes.

HARD TO MAKE THE GENOMES BY HAND. As we mentioned in the Results section (Section 5), it is difficult to construct by hand an artificial genome that will give rise to a particular pattern. This is not surprising, since we are writing a specification in a very indirect language, and the resulting simulated organism will be affected by many factors during its development. However, it is not our intention that the system be easily programmed by humans. A more pertinent issue is whether the artificial genome will be appropriate for simulated evolution.

IS THIS GENOME REPRESENTATION APPROPRIATE FOR SIMULATED EVOLUTION? The experiments reveal that developmental models adapt well to changes in environment and initial conditions. It seems likely that they will also have robustness with respect to genomic variation (as mentioned in the introduction). Hence our initial experiments lead us to believe that our artificial genome is well suited to simulated evolution. We have started performing simulated evolution on these developmental models and will report those results when they are complete (the preliminary results seem promising).

6.2 USEFUL TECHNIQUES FOR DEVELOPMENTAL SIMULATION

During the construction of the testbed, we focused on getting qualitatively biological behavior while aiming for computational efficiency. The following techniques were effective in achieving this balance:

- Artificial Genome (see Sections 4 and A.2.2):
    - use a condition to allow regulation of genes and groups of genes
    - compute condition as a continuous function (avoid a discontinuity at the firing of every condition)
    - sum the contribution of multiple artificial genes
- use one sensor with a time-varying location instead of many fixed sensors (see Section A.3.1)
- use simplest stable solver (adaptive Euler solver; see Section A.6)
- use viscous dynamics ($F = kv$) for cell motion (see Section A.2.3)
- use penalty method for sloppy collisions (see Section A.3.2)

Noise can play an important role in dynamical systems, knocking a system off of an unstable point or popping it out of a local minimum. It also can be used as part of a stochastic estimation process. For instance, some bacteria move up a nutrient gradient by the strategy of moving randomly in various directions, but with a smaller likelihood of changing direction if there seems to more nutrient. We incorporate noise by adding it at the cell's sensors (user can specify the amount of noise).

7. CONCLUSION

We have presented a developmental model which captures sufficient biological detail to produce patterns with asymmetric structure and heterogeneous cell types. In fact, asymmetry is the rule rather than the exception. Yet, in Color Plate 5, we see a fairly regular global pattern which can arise despite local disorder. Heterogeneity is also evident in the multiple cell types.

The modeling framework is based on several mechanisms known to be important in biological development (chemical,[30] mechanical,[19] electrical,[9] and genetic/cell lineage[14] elements). The interaction between mechanisms can lead to heterogeneous patterns; also, patterns can be redundantly specified via different mechanisms.

This is work in progress towards a long-range goal of performing simulated evolution on developmental models. The experiments performed to date suggest that developmental models can be robust to changes in the environment or initial conditions. Despite perturbations, they tend to form similar patterns. This property is likely to be important for simulated evolution, where adaptation to small changes in the genome may help the organism survive mutations that lead to new structures.

ACKNOWLEDGEMENTS

The authors would like to thank all members of the Caltech Graphics Group for collaboration, support, and encouragement. Some of the code used in the testbed was written by Ronen Barzel, Tim Kay, David Laidlaw, Mike Meckler, John Snyder, and Adam Woodbury. Thanks also to Dan Fain, David Kirk, David Laidlaw, Erik Winfree, and an anonymous reviewer for proofreading and helpful suggestions.

This work was supported in part by grants from Apple, DEC, Hewlett Packard, and IBM. Additional support was provided by NSF (ASC-89-20219), as part of the NSF/DARPA STC for Computer Graphics and Scientific Visualization. All opinions, findings, conclusions, or recommendations expressed in this document are those of the authors and do not necessarily reflect the views of the sponsoring agencies.

APPENDIX A: DETAILED IMPLEMENTATION

We begin with definitions of variables and objects, then proceed with an explanation of the equations that control the behaviors of the cells and environment. The last two subsections describe the implementation of growth cones and neurites, and the numerical solution methods.

- Definitions
- Cell Equations (State Equations, Behavior Functions, Equations of Motion)
- Environment Equations (Diffusion, Collision)
- Model of Membrane Proteins and Cell Adhesion
- Neurites and Growth Cones
- Numerical Computation

A NOTE ON THE PROTEIN MODEL. The model was initially implemented as described in Section 4. Since the cell state variables encoded protein amounts, they were constrained to be greater than or equal to zero. However, we found that our cell state equations and cell behavior functions often used a difference of the state values to create a signed value ($state[i] - state[j]$). This suggested a computational speedup by doing a change of variables and by allowing our state variables to represent the difference of a *pair* of proteins. This removes the constraint that the state variables be non-negative, and reduces the number of variables by half.

A.1 DEFINITIONS

cell: A cell is modeled as a geometric shape (currently a circle, with optional neurites) with a given response to applied forces, as well as an array of *cell state variables.*

continuous cell behaviors: Cells exhibit several continuous behaviors, determined by the cell behavior functions (see Section A.2.2):

- attempt to move in some direction (may be limited by collisions, adhesion, or drag)
- attempt to grow in size
- emit or absorb chemicals from the environment
- change the amount of particular proteins in the membrane (e.g., cell adhesion proteins, which mediate how much this cell will adhere to another cell)

discontinuous cell behaviors (events): The cell provides functions that determine the timing of the following events. An event is a discontinuity in the solution, which stops the solver and may create or destroy data structures. The timing of events is determined by cell behavior functions which are described in Section A.2.2 below.

- split (cell division)
- die
- emit neurite with growth cone

cell state variables ($\mathbf{state}_c[\,]$): An array of variables that loosely represent the amounts of proteins within the cell (or differences, as noted above). The values of these variables affect the cell's movements, the timing of events, and the cell's interaction with the environment.

environment: All of the simulated cells interact within a single global environment. The environment contains diffusing, reacting chemicals, as well as physical barriers. Within the simulation, cells access information about their environment locally through an array of *local environment variables.*

local environment variables ($\mathbf{env}_c[\,]$): An array of variables that represent the local environmen of a cell. The values available to the cell as a function of time, and they depend on the extracellular environment. Since each cell is in a different location, in general the local environments of two cells will differ. These variations can then lead to different behavior for the cells, even though their genomes may be identical.

LOCAL ENVIRONMENT VARIABLES. The local environment of a cell can be accessed via the array of local environment variables. These include:

- amount and gradient of diffusable chemicals at a local sensor (with noise) amount of cell membrane proteins in a bound state (for detecting contact with other cells—see section below on Model of Membrane Proteins and Cell Adhesion; Section A.4)
- cell size

Cell size is included as an environmental variable since the equations of motion and growth of cells are actually computed in a global process, and then propagated back to the cells. A cell does not have access to its absolute location, but it does know its size.

In the simulations shown, the local environment variables were defined to contain the values and gradients of the chemicals around each cell. For each diffusable or nondiffusable chemical in the environment ($i \in 0, \ldots, m$), there are three variables: one for the value ($chem_i$) and two for the components of the two-dimensional gradient ($\partial chem_i/\partial x, \partial chem_i/\partial y$).

For each membrane protein ($j \in 0, \ldots, n$), there is a value (mem_j) that approximates the amount of the protein that is bound to a matching protein on an adjacent cell (in our model, the cells must be in contact for their membrane proteins to bind—see Section A.4 below for details).

At present the local environment array looks like:

$$env_c[\,] = (size,\ chem_0,\ \frac{\partial chem_0}{\partial x}, \frac{\partial chem_0}{\partial y},$$
$$chem_1,\ \frac{\partial chem_1}{\partial x}, \frac{\partial chem_1}{\partial y}, \cdots, chem_{n-1},\ \frac{\partial chem_{m-1}}{\partial x}, \frac{\partial chem_{m-1}}{\partial y},$$
$$mem_0,\ mem_1,\ \cdots,\ mem_{n-1}).$$

A.2. CELL EQUATIONS

The cell state variables indirectly control a cell's behavior within its environment. This is accomplished via three categories of functions: the *cell state equations* (genome), *cell behavior functions* (protein structure to function model), and the *cell equations of motion* (see Figure 1).

cell state equations (the genome): The state equations modify the cell's internal state variables based on the local environment and the cell's current state.

cell behavior functions (protein structure to function model): The cell's current state determines what it is trying to do: the forces it is applying, events such as cell division, etc. The behavior functions compute all of the cell's forces and events from the state variables.

cell equations of motion: given the cell's behavior functions which describe what the cell is *trying* to do, these equations will compute the end results. For example, the cell may be applying forces to move right, but the collision forces may counteract that movement, producing a net movement to the left or right.

Cells can only access local information such as the local concentration of chemicals (see Section A.1). They cannot directly access their absolute position and orientation in the world. Nor can they directly change their position, but they can do so only by applying forces that may be counteracted by other forces (e.g., collision with a wall).

A.2.1 CELL STATE EQUATIONS (THE GENOME). The cell state equations are the model of the cell's genome (see Section 4). These equations encode how the cell changes state based on its local environment and its current state. The state equations implement a crude model of protein synthesis. In biological cells, a gene encodes a protein. In our model, a conditional differential equation determines the change of a variable related to protein amount. (Examples of cell state equations appear in Appendix B.)

This artificial genome was designed to be amenable to simulated evolution. We focus on the following properties:

- allow regulation of genes by other gene products (to switch on and off single equations)
- allow groups of genes to be regulated together (to switch on and off groups of equations)
- if there are multiple genes for the same protein, just make more of it

The conditional element models the regulation of a gene or group of genes, enabling us to switch on or off groups of equations based on the state of the cell or its local environment.

It is possible to have multiple contributions to the same state variable (multiple genes for the same protein). The differential equation for state variable $state[i]$ is formed from the contribution of all of the consequents pertaining to that state variable. The condition, implemented as a continuous function (rather than a zero-to-one step), that is multiplied by the consequent. This is both more "biological" (rates of protein production turn on and off with some probability) and more efficient (no need to do root finding for the exact time when the condition changes).

$$\frac{dstate_c[i]}{dt} = \sum_S (\mathbf{ContinuousCondition}_c^S(state_c, env_c))(\mathbf{Consequent}_c^S(state_c, env_c)).$$

A.2.2 CELL BEHAVIOR FUNCTIONS. The behavior functions compute the cells' attempted behaviors based on their current state. Both continuous and discontinuous behaviors are handled; continuous behaviors are simply continuous functions of the state variables (e.g., **MotiveForce**, **GrowthForce**, **Spew**) and discontinuous behaviors are events triggered by the zero-crossing of a behavior function (e.g., **TimeToSplit**, **TimeToDie**).

This example illustrates the behavior functions used to compute the simulations shown in Section 5. These may be changed to arbitrary C mathematical expressions that depend on the state and local environment.

$$\begin{aligned}
\mathbf{MotiveForce}_c(state_c, env_c) &\equiv (state_c[0], state_c[1]);\\
\mathbf{GrowthForce}_c(state_c, env_c) &\equiv state_c[2];\\
\mathbf{TimeToSplit}_c(state_c, env_c) &\equiv \min(\mathrm{thresh}(\mathrm{r}_0, env[\mathrm{radius}]),\\
&\qquad \mathrm{thresh}(\mathrm{split}_0, state_c[\mathrm{split}])) - 0.5).
\end{aligned}$$

The definition of **TimeToSplit**() is a continuous version of the condition ((radius > r_0) & ($state_c$[split] > s_0)); i.e., a cell splits when it is large enough (bigger than r_0) and has accumulated enough of the $state_c$[split] protein (more than s_0). The other event functions **TimeToDie**() and **TimeToEmitNeurite**() are defined similarly.

For each chemical $a \in 0, 1, \ldots, nchems$, this function defines the amount of a diffusable chemical a being emitted into or absorbed from the environment by cell c. In the simulations shown, the rate at which chemical a is emitted by the cell is determined a single state variable $state_c[i_a]$, where i_a is an index into the state array.

$$\mathbf{Spew}_{a,c}(state_c, env_c) \equiv state_c[i_a].$$

As mentioned before, this is an example of the equations used in the simulations shown; other functions of the state and environment variables can easily be defined. For instance, a useful alternative to defining the components of the motive force directly via state variables is to use state variables for the magnitude and direction of motion, then transform them to obtain the xy components of the force: $\mathbf{MotiveForce}_c(state_c, env_c) \equiv (state_c[0] \cdot \cos(state_c[1]), state_c[0] \cdot \sin(state_c[1]))$.

A.2.3 EQUATIONS OF MOTION FOR THE CELLS. The motion and growth of the cells is determined by the forces they generate, and the forces applied from collisions and other extracellular effects. In the low Reynolds number domain of small objects in viscous fluids, we determine the velocity v from balancing the drag force $F = kv$ with the applied forces ($k = k_{\text{drag}} \times \text{area}(c)$). The **CollisionForce** and **AdhesionForce**, computed in the global environment, are described in detail below (under Environment Equations).

$$\begin{aligned}\sum_{\text{forces}} &= \mathbf{CollisionForce}_c + \mathbf{AdhesionForce}_c + \mathbf{MotiveForce}_c - kv \\ &= 0.\end{aligned}$$

Unlike the drag force, the collision forces do not depend on velocity (see Section A.3.2). Adding a dependence on velocity is straightforward and leads to a set of simultaneous equations of motion that can be solved to find cell velocities.

A.3. ENVIRONMENT EQUATIONS

A.3.1 DIFFUSION. The diffusion of each chemical is governed by a partial differential equation for f_a, the concentration of chemical $a \varepsilon A$. The $\mathbf{R}_a(\mathcal{A})$ function computes reactions occurring naturally between chemical a and all other chemicals $\mathcal{A}$ as they mix in the extracellular matrix. For each chemical a, at a particular location:

$$\frac{\partial f_a(x,y)}{dt} = -\nabla^2 f_a(x,y) - \text{dissipation}_a + \mathbf{R}_a(\mathcal{A}) + \mathbf{SourcesAndSinks}_a(x,y,t).$$

Each cell can emit or absorb chemicals locally and, thus, contributes to $\mathbf{SourcesAndSinks}(x,y,t)$. We have experimented with two models for the location of the sources/sinks on the cells:

- several locations along the cell perimeter and
- a single location at the center of the cell.

We primarily use the single location since it gives qualitatively similar results and is computationally more efficient. For this case, the function for chemical a can be specified as:

$$\mathbf{SourcesAndSinks}_a(x,y,t) = \sum_{c \in \text{cells}} \delta(x - c_x, y - c_y)\ \mathbf{Spew}_{c,a}(state_c, env_c)$$

where the location of cell c is (c_x, c_y), and $\delta(x,y)$ is the Dirac delta function (a spike at $x = 0, y = 0$).

Color Plates

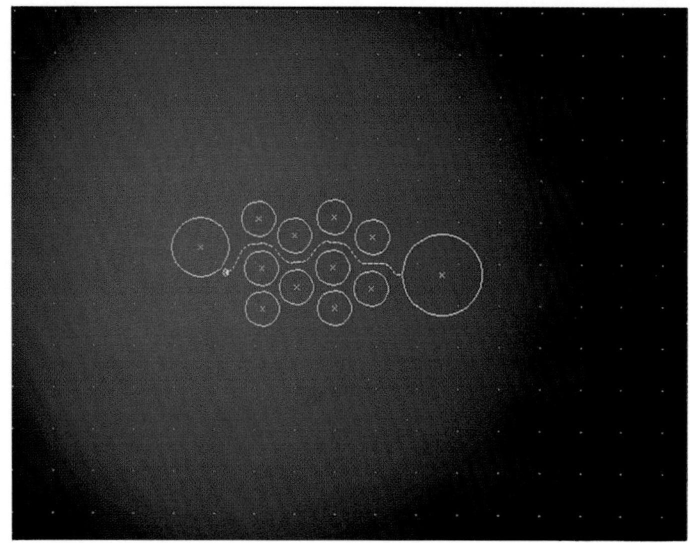

Plate 1: Neurite grows past barriers (towards red).
[© Fleischer]

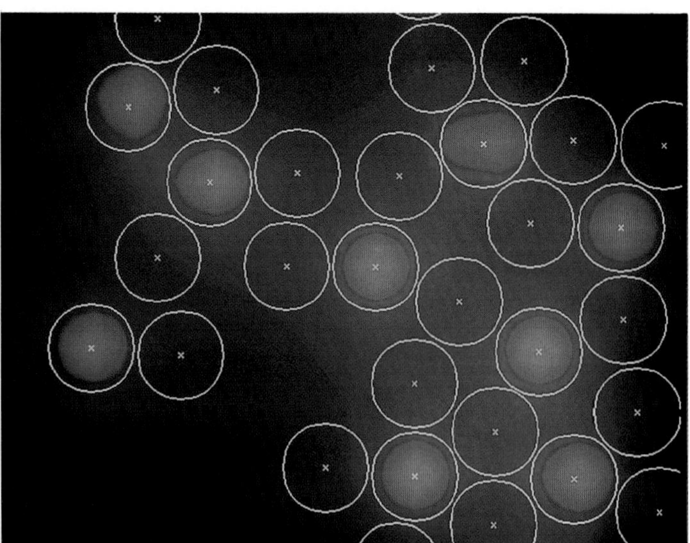

Plate 2: Cells of initially similar types differentiate into two types. [© Fleischer]

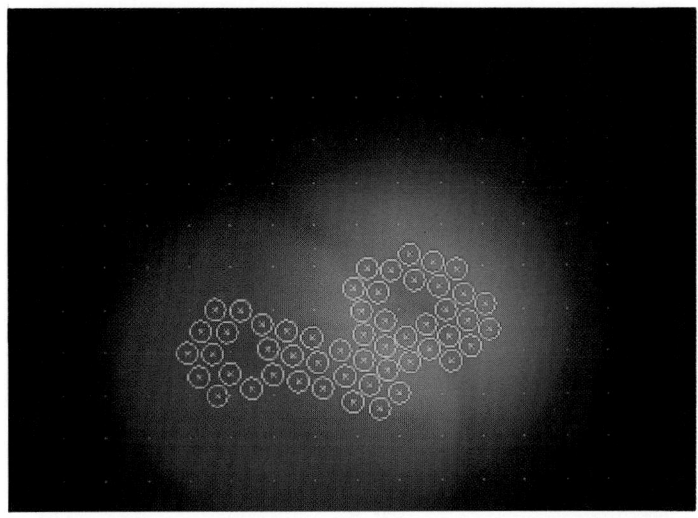

(a)

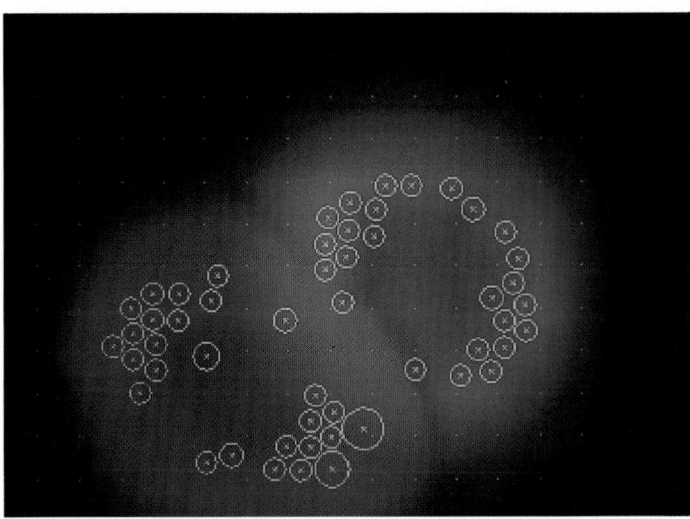

(b)

Plate 3: Cells clump together and then expand outwards, cyclically. [© Fleischer]

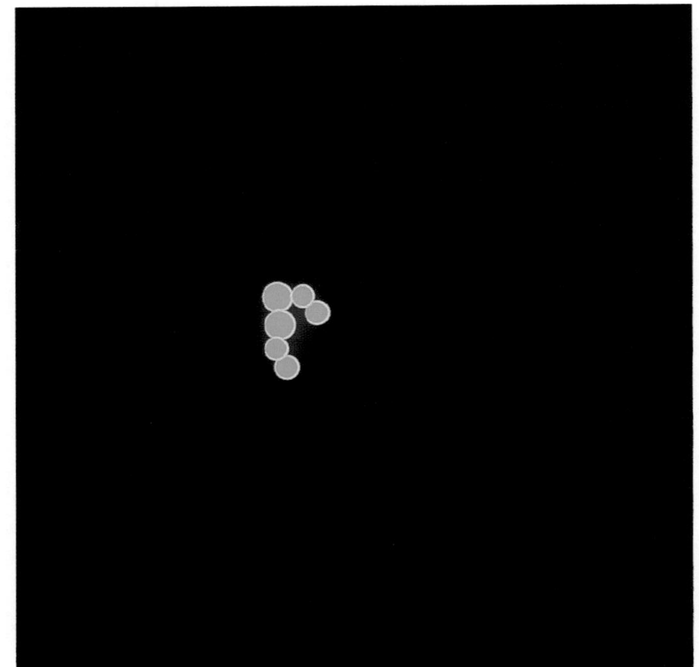

(a)

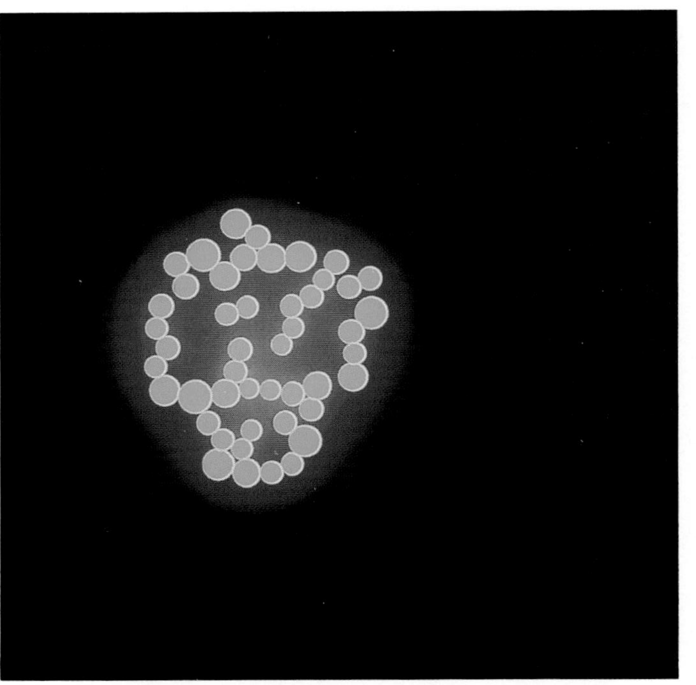

(b)

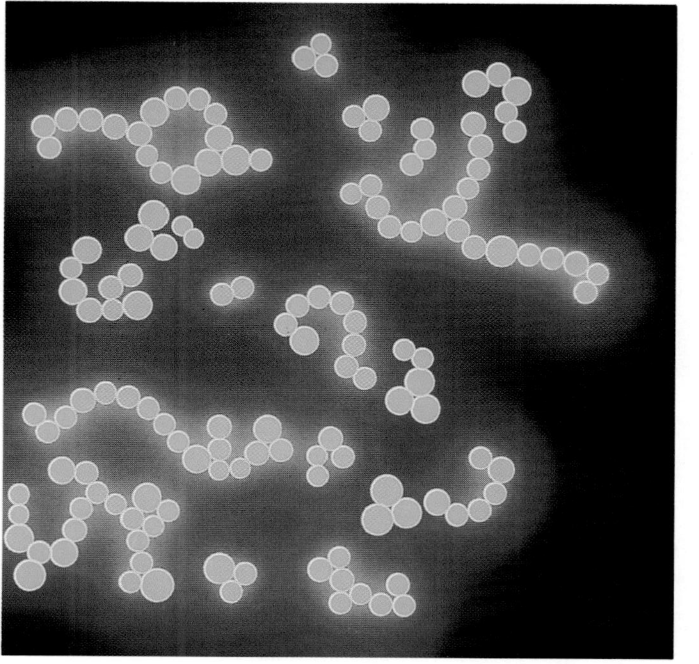

(c)

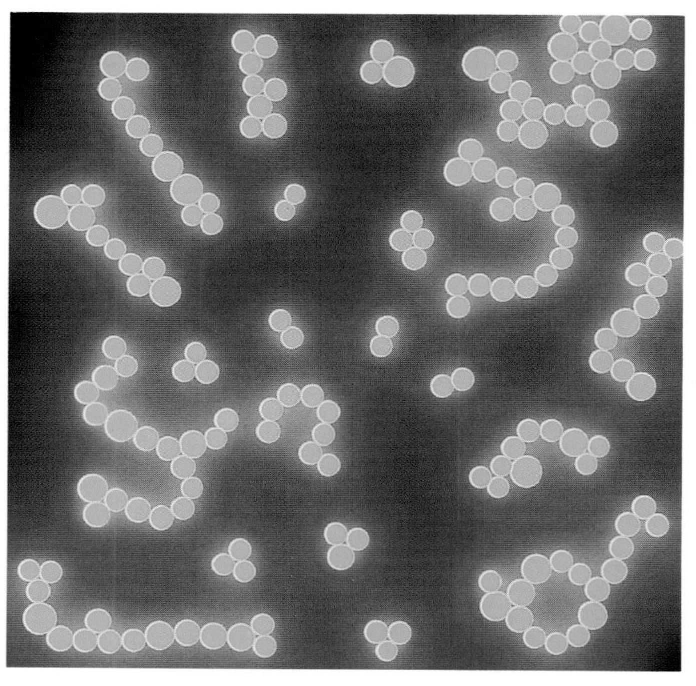

(d)

Plate 4: A combination of cell adhesion and a repulsive chemical makes chains of cells. [© Fleischer]

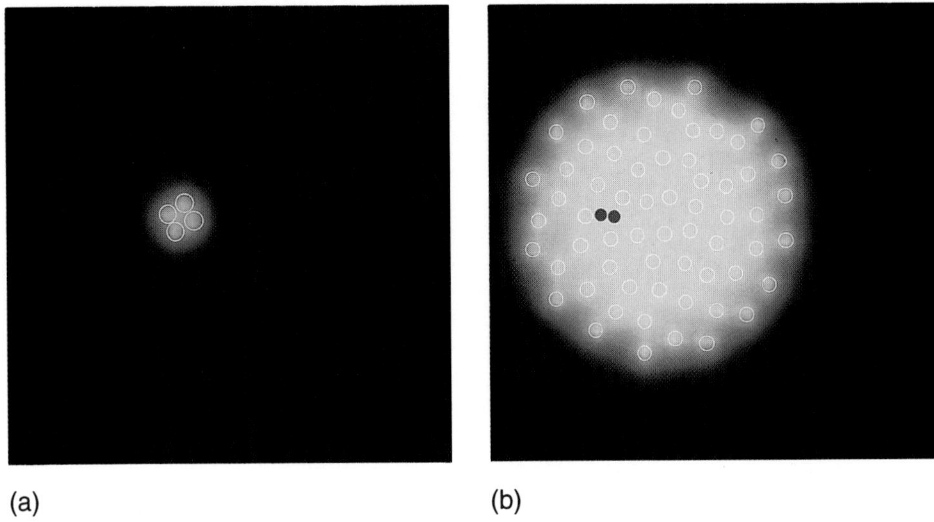

(a) (b)

Plate 5: Initial division and separation of yellow-emitting cells is followed by differentiation and proliferation of adhesive purple cells, forming chains between the yellow cells. [© Fleischer]

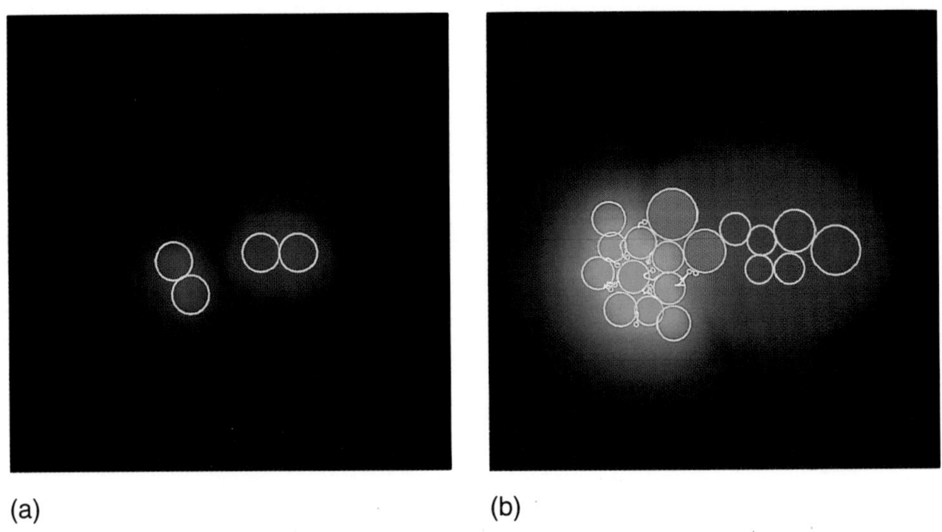

(a) (b)

Plate 6: Neurite Growth. After the network is complete, excess neurites die (note the difference in white neurites between (c) and (d)). [© Fleischer]

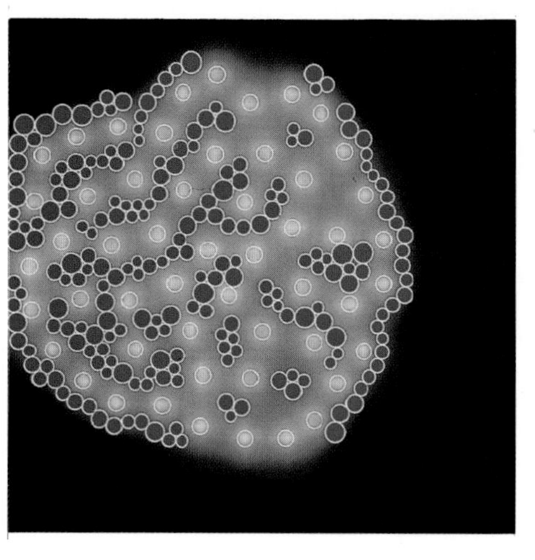

(c)

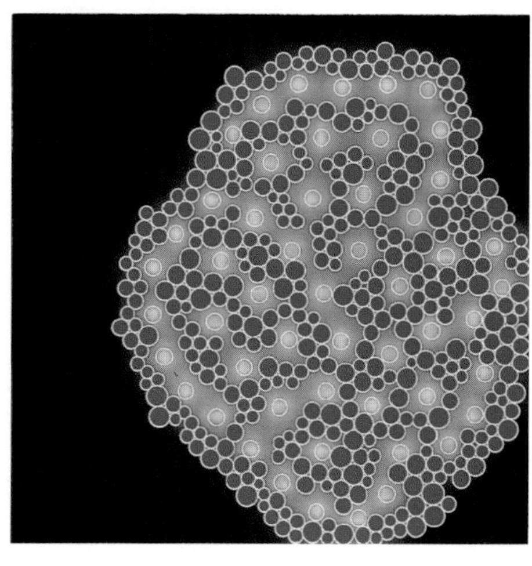

(d)

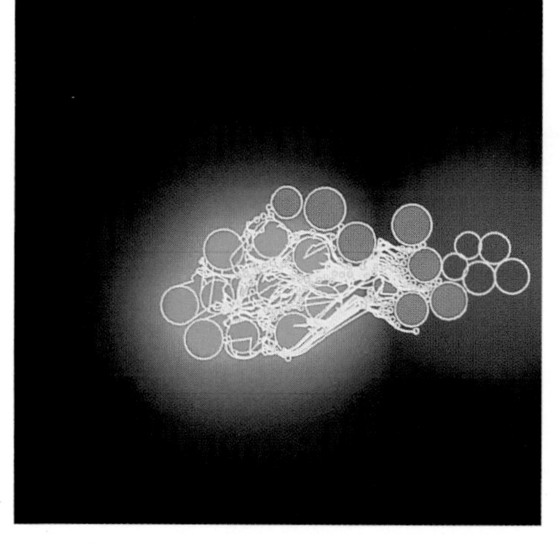

(c)

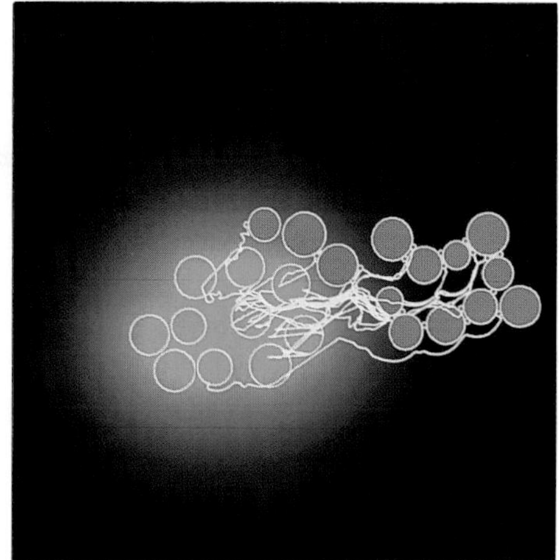

(d)

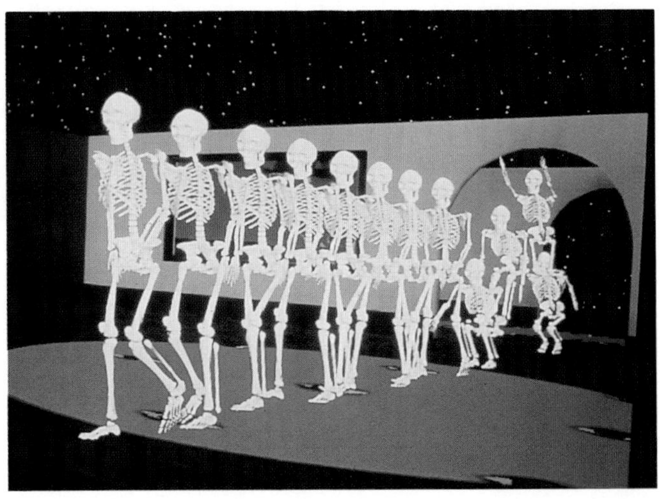

Plate 7: A walking human skeleton. The motions are controlled by a hierarchy of finite state machines. (From *IEEE Computer Graphics and Applications* **2(9)** (1982); reprinted by permission of the author.) [Zeltzer & McKenna]

Plate 8: Two walking quadrupeds negotiate uneven terrain. (From M.S.V.S. Thesis by K. Sims, Massachusetts Institute of Technology, 1987; reprinted by permission of the author.) [Zeltzer & McKenna]

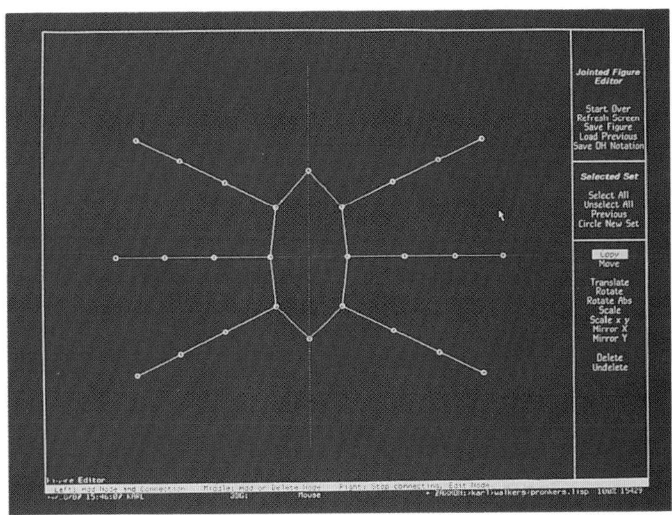

Plate 9: *Figure editor* for designing articulated structures. Circles represent hinge joints, and line segments stand for rigid links. Limbs are attached to the body via ball joints. (From M.S.V.S. Thesis by K. Sims, Massachusetts Institute of Technology, 1987; reprinted by permission of the author.) [Zeltzer & McKenna]

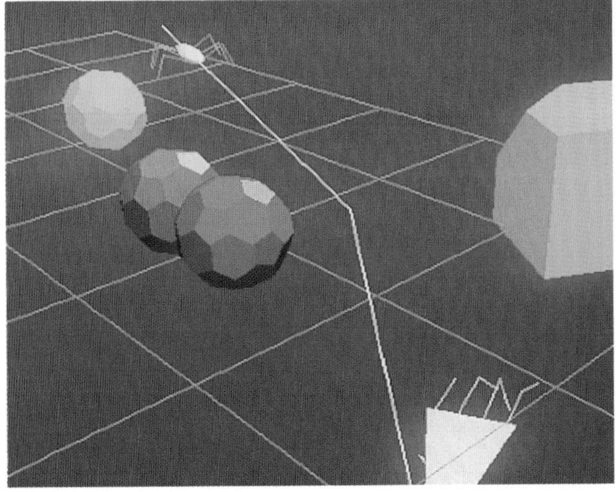

Plate 10: The kinematic roach follows a collision-free path in the *bolio* virtual environment system. (From *Computer Graphics* **24(2)** (1990); reprinted by permission of the author.) [Zeltzer & McKenna]

Plate 11: The hexapod, "Cootie," from the computer animation *Cootie Gets Scared.* The Cootie's high-level actions were "scripted" by the animator. The motions of the legs, body, and antennae were automatically generated by the software. The Cootie's movements were kinematically controlled, in general, with the exception of the head and antennae, which were dynamically simulated. (From *Cootie Gets Scared,* Media Laboratory, MIT (1988); reprinted by permission of the author.) [Zeltzer & McKenna]

Plate 12: The parametrized dynamic roach. The figure begins to step with three legs using the tripod gait in this illustration. It has 5 DOFs per leg, and a head and abdomen joint. The entire system has 3 translating and 3 rotating DOFs, for a total of 38. [© Zeltzer & McKenna]

Plate 13: Scene from the animation *Grinning Evil Death.* As the Roach walks forward, it generates collision forces against the wires. (From *Grinning Evil Death,* Media Laboratory, MIT (1990); reprinted by permission of the author.) [Zeltzer & McKenna]

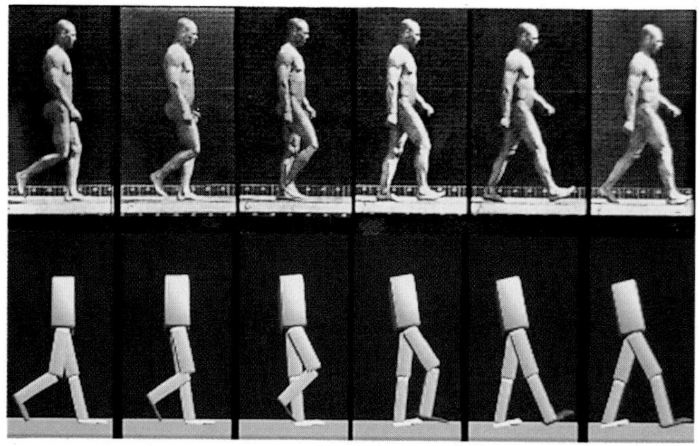

Plate 14: Passive stepping motion, compared to Muybridge sequence (from Dover, 1955; reprinted by permission of the author). This preliminary simulation from the dynamic biped project, reproduces some of the results from Mochon and McMahon's "ballistic walking" experiments (*J. Biomech.* **13** (1980) and *Math. Biosci.* **52** (1980)). After initial conditions have been established, the system is forward-simulated, without any changes in the control state. The step leg swings with the motion of a double pendulum, as the stance leg pivots on the ground as an inverted pendulum. [Zeltzer & McKenna]

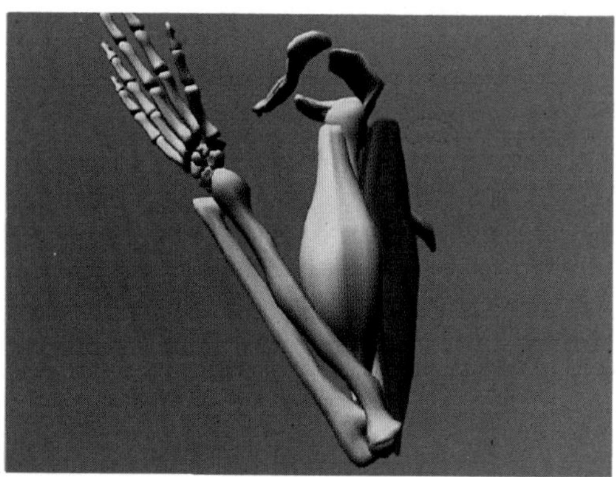

Plate 15: A simulation of the human biceps muscle. The shape of the muscle was determined by the material properties of the simulated muscle, with active and passive nonlinear forces generated within the muscle, as it actively contracts. The skeleton of the arm was attached to the muscle using inverse kinematics. (From *Computer Graphics* **26(4)** (1992); printed by permission of the author.) [Zeltzer & McKenna]

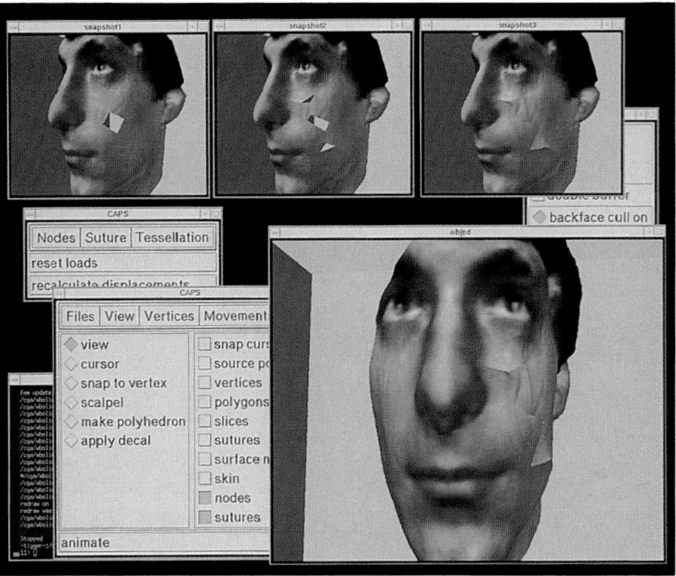

Plate 16: A simulation of a plastic surgery procedure on the face. Top left image: a small tumor is to be removed from the skin. The procedure is planned—the lines of the incision are defined on the skin, and the area to be excised is designated. Next frame: the tumor is removed, along with a square of surface skin. Next frame: the excised region is closed by moving flaps of skin; the upper right segment moves down, and the lower left segment moves up. Bottom image: the incisions are closed. Note how the skin has deformed as the skin is pulled taut in some areas, and loosened in others. (From MIT Press, 1986; reprinted by permission of the author.) [Zeltzer & McKenna]

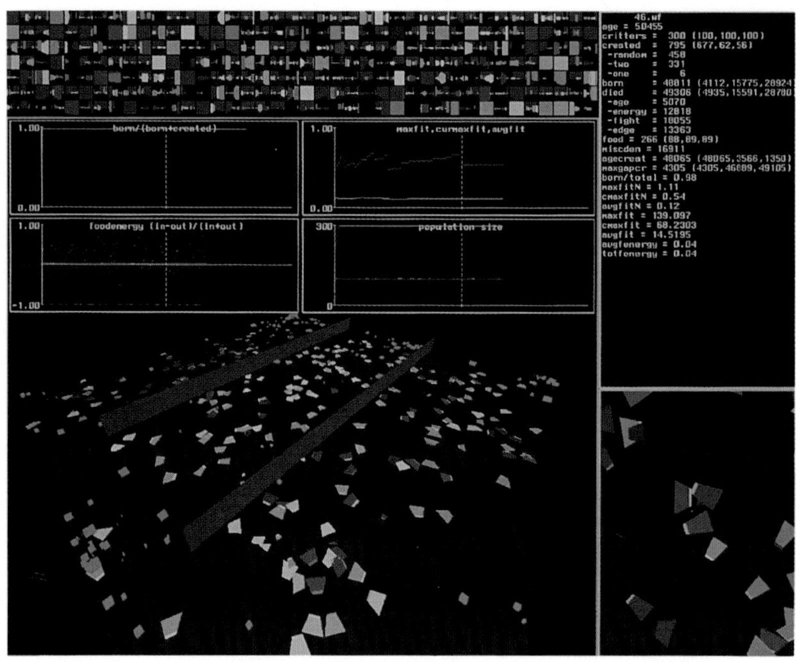

Plate 17: Screen shot of PolyWorld Ecological Simulator. [© Yaeger]

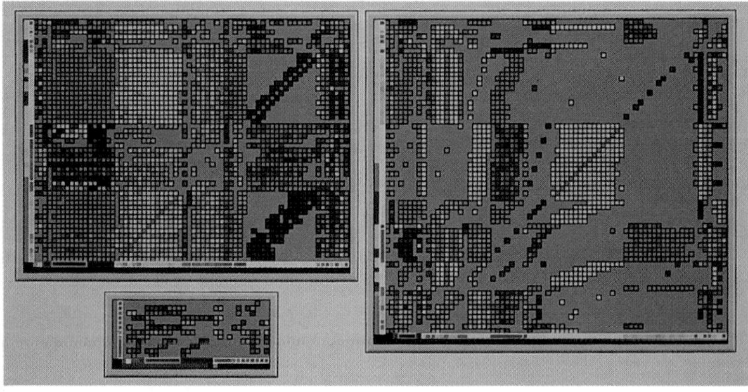

Plate 18: Three different evolved neural architectures. [© Yaeger]

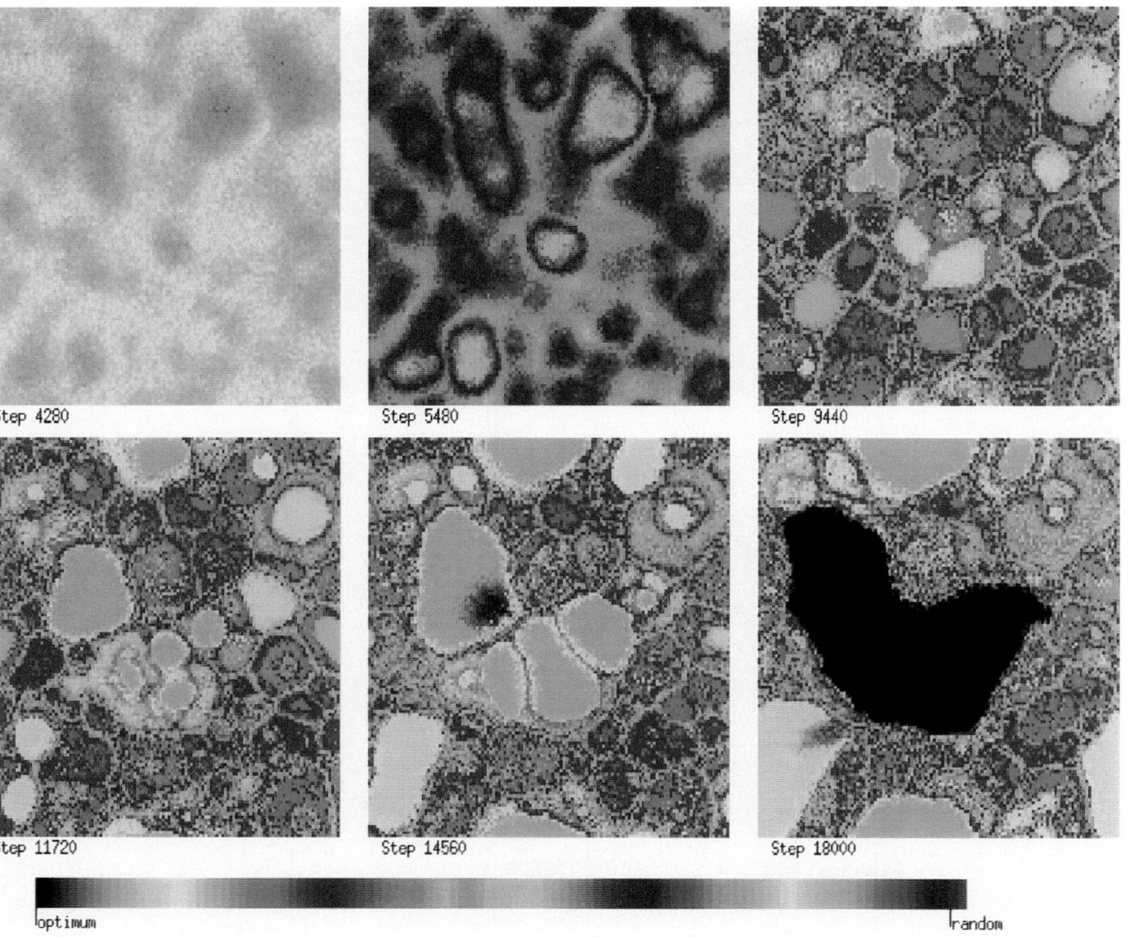

Plate 19: Stages in distributed Lamarckian evolution (see text for more information). [© Ackley]

The function $f_a(x, y, t)$ is discretized on an $n \times n$ grid in two dimensions to give n^2 ODES. The notation f^{ij} indicates the value of the discretized variable at node (i, j) in the two-dimensional grid.

$$\begin{aligned}\frac{df_a^{ij}}{dt} = &-(f_a^{i+1,j} + f_a^{i-1,j} + f_a^{i,j+1} + f_a^{i,j-1}) - 4f_a^{ij} \\ &- \text{dissipation}_a + \mathbf{R}_a(f_a, f_b, f_c, \ldots) + \mathbf{SourcesAndSinks}_a^{ij}(t).\end{aligned}$$

The discrete version of the **SourcesAndSinks**() function is computed by partitioning the components of a cell's emission/absorption between the adjacent grid points using bilinear extrapolation.

In addition to modifying the information in the diffusion grid (via the **SourcesAndSinks**() function), a cell can sense the values and gradient of a chemical locally via sensors. We have implemented several sensor strategies to date:

- multiple sensors on the cell's periphery (sensing value only, gradient is computed by differences),
- a single sensor at the cell's center (sensing value and gradient directly), and
- a single sensor at a random location on the cell (sensing value and gradient).

We have found the randomly moving sensor to be the most effective of the three. The randomly moving sensor computes an approximation to the value at the center of the cell. It is more efficient than using multiple sensors, and it avoids some problems that occur when using a stationary sensor and source that are both located at the center of the cell.

An alternative strategy for implementing the sensors is to represent the amount of sensor proteins in the cell membrane. The strength of the sensor signal then depends on this amount, similar to the contact recognition computation discussed below in Section A.4.

The diffusion equation for moving point sources/sinks can be solved in closed form using an integral expression. It is possible that this approach could speed up the computation, although the dependence of the integral on the history of cell movement may make this undesirable. Also, using a closed form solution would put more restrictions on the extracellular reaction function **R**().

A.3.2 COLLISION. Collisions are computed using a penalty term,[21] which is introduced into the equations of motion for each cell as a force. For every colliding pair of objects, the collision manager computes equal and opposite forces. Note that the objects for the collision computation are not restricted to being cells. For instance, collisions are also computed with the boundaries of the environment, and other objects can be introduced as well (bone, fibers, or other substances in the environment).

In the case of colliding circles, the forces are very simple, but more complex collisions are possible within our framework. For every pair of objects b, c, we denote the function computing their maximum overlap as $\mathbf{d}(b,c)$. $\mathbf{p}$ is a unit vector in the direction of the maximum overlap. Then we have:

$$\mathbf{CollisionForce}_c = \sum_{b \in \text{objects}} k\ (\mathbf{d}(b,c) - \mathit{offset})^n\ \mathbf{p}.$$

This method of collision computation is somewhat inaccurate and requires choosing parameters k, n, and *offset* arbitrarily. However, accurate collisions are not critical to our application, and the cells we are modeling are not rigid objects so some amount of overlap is acceptable. In practice, object interpenetration has only occasionally been troublesome, and the penalty method has been computationally efficient when combined with our solution methods (see Section A.6).

A.4. MODEL OF MEMBRANE PROTEINS AND CELL ADHESION

Real cells have many proteins in their membranes which perform a variety of functions. We propose a simple model of membrane proteins (which we refer to as surface factors). Our model captures a few of the major functions:

- cell recognition (cells recognize that they are in contact) and
- cell adhesion (cells physically bind together).

In both cases, our model contains surface factors that are both homophilic (like binds to like) and heterophilic (a complementary pair binds together).

A single state variable state[j] directly controls the amount a given surface factor in the cell's membrane. The surface factor that is bound to its complement on an adjacent cell is reported in the mem_j variables in the local environment array. This suffices for cell contact recognition; cells can determine that they are in contact with another cell that expresses a certain surface factor. For cell adhesion, there is a force computed that depends on the size of the region of contact, and the amounts of surface factors available on the contacting cells.

Note that this model does not allow for asymmetric expression or recognition along the cell membrane. Cells cannot express a surface factor on just one side, not can they tell which side has been contacted. This limitation may be important, and we are considering various models to enable specifying a spatial distribution of surface factors without adding too much computational burden.

All of the functions of the membrane protein model depend on the amount of a surface factor which is bound. This is computed from the contact area between two cells and the amount of the complementary factors in their membranes. Let $(\mathbf{Conc}_c(a))$ return the concentration of a on cell c. For a particular surface factor a with complement a', we have:

$$\mathbf{AmountBound}_c(a) = \sum_{b \in \text{cells}} \mathbf{ContactArea}(c,b) \times \mathbf{Conc}_c(a) \times \mathbf{Conc}_b(a') \times p_{\text{bind}},$$

where p_{bind} is a multiplicative constant that roughly corresponds to the probability that two proteins a and a' will bind. **ContactArea**(c, b) estimates the contact area between two cells as the chord length of their overlap. The recognition factors report **AmountBound** in the appropriate local environment variable (the mem_i mentioned in Section A.1).

The adhesion force on a cell is the sum of the forces from all the adhesive surface factors on all of the cells in contact:

$$\mathbf{AdhesionForce}_c = \sum_{j \in \text{adhesion factors}} \mathbf{AmtBound}_c(j)^* \mathit{stickiness}_j.$$

Biological cells also use membrane proteins as sensors to detect diffusable chemicals in the environment, and as channels for emitting diffusable chemicals. We could model these effects similarly, but instead we have chosen to directly measure/change the chemicals in the local environment (discussed above in the **Diffusion** section, A.3.1). This is done for efficiency.

A.5. NEURITES AND GROWTH CONES

We model growth cones as small cells that are connected to the parent cell by a neurite. They have the same capabilities as cells, except that they die if the parent cell dies, and they cannot emit growth cones of their own. Branching neurites can be implemented as splitting of growth cones, analogous to cell division. Growth cones and cells communicate via a set of state variables which are held in the neurite. The growth cones and cells can modify and sense the levels of these state variables in the neurite.

Although the growth cone computes collisions with the cells, collisions between neurites and other objects are not computed. The geometry of the neurites is simply a record of the path of the growth cone (Color Plate 6). In biological systems, adhesion and mechanical interactions between neurites are known to be factors in neural development. However, implementing the neurite-cell collisions in two dimensions seems too restrictive on the cells' movements. We have opted for computational expedience and do not compute collisions between neurites and other objects.

A.6. NUMERICAL COMPUTATION

We have combined the equations arising from chemical, mechanical, and electrical sources into one large system of ordinary differential equations. Since discontinuous events occur in the ODE system (representing cell division, collisions between cells, etc.), the numerical implementation is based on a piecewise-continuous ordinary differential equation (PODE) solver.[3] The PODE solver allows the addition/deletion of variables at discontinuities, which occur when cells split or die during the course of a simulation run.

There is a tradeoff regarding the choice of numerical method for solving the simulation equations and the solution's computation time, stability, and accuracy. The solver needs to work well in the context of the three dominating effects encoded in the differential equations: diffusion of chemicals on a grid, forces that cause the cells to migrate, and forces induced by cell-to-cell contacts.

We have chosen one of the simplest numerical solvers that produces stable and qualitatively accurate solutions. After experimenting with several ODE solution techniques (variable-order, variable-step Adams method; Runge-Kutta;...); we have settled on a type of adaptive Euler solver that greedily increases its step size but is limited by a function that looks for signs of instability and other undesirable behaviors. We were able to eliminate the more advanced but computationally more expensive ODE modules, without compromising qualitative accuracy.

The solver rejects differential equation steps that do not meet the following criteria:

- Per ODE step, cells cannot move more than a specified distance (typically 5% of a cell diameter). This prevents tunneling of cells through one another and prevents gross instabilities when there are many cells in close contact with one another, with very large forces acting on them.
- The largest diffusion rate (α) and the spacing between points in the diffusion grid (h) impose a maximum step size to ensure stability (in two dimensions, we require $\alpha \Delta t / h^2 < 1/4$).

If an ODE step is rejected, the solver tries again with a smaller step size.

With these restrictions, the simple Euler solver is sufficiently accurate and efficient, particularly when dissipation effects (such as from chemical diffusion and from viscous drag on the cells) dominate the equations. The main advantage of this approach is that gross instabilities are eliminated despite the use of a very simple ODE solution method. The solutions produced by the simpler method approaches the numerical solution from the more advanced solvers as the step size is reduced.

Another reason for using the adaptive Euler solver is that it copes better with discontinuities such as random noise added to the system. More sophisticated solvers do not handle this well.

APPENDIX B: EXAMPLE EXPERIMENT FILE

This is an example of the interpreted file used to generate the simulation shown in Color Plate 4.

```
nchems = 1;  /* How many diffusable chemicals in environment */
user_state = 1; /* Number of state variables user wants */

/* diffusion and dissipation coefficients for all chemicals */
diffusion[0] = 4.0;
dissipation[0] = 1.0;

setcolor(0, 0.8, 0.2, 0.0);  /* color of diffusing chem 0 */

/* The values fx, fy, fsize, etc. are integers
*  which are used as array indices into the state array
*
* Array indices for the state array (z[])
*
*  System defined (defined for every experiment):
*   fx, fy, fsize  -- forces on x, y, radius
*   split, die, emitgc -- event indices
*   homophilic -- homophilic cell surface chemical,
*      sticks to same chemical on adjacent cells
*   spew -- rate at which to emit chemical 0
*
*  User defined (just for this experiment):
*   rundown -- this indexes a state variable
*     which decreases over time, diluted by cell division
*     as well as consumed inside the cell. Controls how
*     many generations of cell division occur.
*
* Array indices for local environment array (env[])
*  rvl, rdx, rdy  -- chemical 0 value and gradient
*  radius -- size of cell
*  stuck -- how much of the homophilic cell adhesion molecule
*   on our surface is bound.
*/
int rundown = nstate+0;

/* addstmt(condition, index, consequent);
* adds a 'gene' to the artificial genome which does:
* if (condition(state, env))
* then dstate[index]/dt += consequent(state, env)
*/
beginprogram();
addstmt("(TRUE)", fx, "-state[fx]");
addstmt("(TRUE)", fy, "-state[fy]");
addstmt("(TRUE)", fsize, "-state[fsize]");   /* Grow  */
addstmt("(TRUE)", split, "-state[split]");   /* Split */

/* just dilute down to 0 as we grow and divide */
addstmt("(1)", rundown, "-1");
```

```
/* initially, split if we aren't stuck to too many neighbors */
/* (until rundown runs down) */
addstmt("((state[rundown] > 30) && (env[stuck] < 0.6)) ||
        (env[radius] > 1.0)", split, 4.3);
addstmt("(state[rundown] > 30) && (env[stuck] < 0.6)", fsize, 0.1);

/* later, split if there is not much of chemical 0 */
addstmt("(env[rvl] < 0.2) || (env[radius] > 1.0)", split, 4.3);
addstmt("(env[rvl] < 0.2)", fsize, 0.1);

/* Spew until we see enough of chem 0 in the local environment */
addstmt("(1)", "spew", "5*(0.5 - env[rvl])");

/* if we are not stuck to enough neighbors, move towards chem 0,
*  else if too many, move away */
addstmt("(env[stuck] < 0.2)", fx, "7*env[rdx]");
addstmt("(env[stuck] < 0.2)", fy, "7*env[rdy]");
addstmt("(env[stuck] > 0.6)", fx, "-7*env[rdx]");
addstmt("(env[stuck] > 0.6)", fy, "-7*env[rdy]");

/* set the concentration of adhesion molecule in the membrane */
addstmt("1", "homophilic", "0.4-state[homophilic]");

addcell("FirstCell", 5.0, -0.48); /* initial cell name and pos */

/* set initial value for state[rundown] */
/* changing values here makes different numbers of cells */
initcell("FirstCell", rundown, 1200);
```

REFERENCES

1. Agarwla, Pankaj. "The Cell Programming Language." Presented at the Symposium on Pattern Formation, Harvey Mudd College, Claremont, CA, February 12, 1993.
2. Bard, Jonathan B. L. "A Model for Generating Aspects of Zebra and Other Mammalian Coat Patterns." *J. Theor. Biol.* **93** (1981): 501–531.
3. Barzel, Ronen. "A Structured Approach to Physically Based Modeling." Ph.D. Thesis, California Institute of Technology, 1992. Also Roman Barzel, as *Physically-Based Modeling: A Structured Approach.* Cambridge MA: Academic Press, 1992.
4. Brown, M. C., W. G. Hopkins, and R. J. Keynes. *Essentials of Neural Development.* Cambridge, MA: Cambridge University Press, 1991.
5. de Boer, Martin. "Analysis and Computer Generation of Division Patterns in Cell Layers Using Developmental Algorithms." Ph.D. Thesis, University of Utrecht, November 20, 1989.
6. de Boer, Martin, David Fracchia, and Przemyslaw Prusinkiewicz. "Analysis and Simulation of the Development of Cellular Layers." *Artificial Life II*, edited by C. G. Langton, C. Taylor, J. D. Farmer, and S. Rasmussen. Santa Fe Institute Studies in the Sciences of Complexity, Proc. Vol. X, 465–483. Redwood City, CA: Addison-Wesley, 1991.
7. Carr, Catherine E., and Masakazu Konishi. "Axonal Delay Lines for Time Measurement in the Owl's Brainstem." *Proc. Natl. Acad. Sci. USA* **85** (1988): 8311–8315.
8. Fleischer, Kurt W. "Generating Artificial Neural Networks Using Simulated Evolution of a Developmental Model." Ph.D. Thesis, California Institute of Technology, expected 1993.
9. Fraser, Scott, and Donald Perkel. "Competitive and Positional Cues in the Patterning of Nerve Connections." *J. Neurobiol.* **21(1)** (1990): 51–72.
10. Hebb, Donald O. *The Organization of Behavior.* New York: Wiley 1949.
11. Hopfield, J. J. "Neurons with Grade Response have Collective Computational Properties like Those of Two-State Neurons." *Proc. Natl. Acad. Sci. USA* **81** (1984): 3088–3092.
12. Kitano, Hiroaki. "Designing Neural Networks Using Genetic Algorithms with Graph Generation System." *Complex Systems* **4** (1990): 461–476.
13. Lin, C. C., and L. A. Segel. *Mathematics Applied to Deterministic Problems in the Natural Sciences.* Philadelphia, PA: SIAM, 1988. This is just one reference of Dictyostelium (slime mold) modeling which has been described by many authors.
14. Lindenmayer, Aristid. "Mathematical Models for Cellular Interaction in Development, Parts I and II." *J. Theor. Biol.* **18** (1968): 280–315.
15. Lindenmayer, Aristid, and Przemyslaw Prusinkiewicz. "Developmental Models of Multicellular Organisms: A Computer Graphics Perspective." *Artificial*

Life, edited by C. G. Langton. Santa Fe Institute Studies in the Sciences of Complexity, Proc. Vol. VI, 221–249. Reading, MA: Addison-Wesley, 1989.

16. Meinhardt, Hans. *Models of Biological Pattern Formation.* London: Academic Press, 1982.
17. Mjolsness, Eric, David Sharp, and John Reinitz. "A Connectionist Model of Development." *J. Theor. Biol.* **152** (1991): 429–453.
18. Murray, J. D. "On Pattern Formation Mechanisms for Lepidoptera Wing Patterns and Mammalian Coat Markings." *Phil. Trans. Roy. Soc. (B)* **295** (1981): 473–496.
19. Odell, Garrett M., G. Oster, P. Alberch, and B. Burnside. "The Mechanical Basis of Morphogenesis." *Devel. Biol.* **85** (1981): 446–462.
20. Oster, G., W. Weliky, and S. Minsuk. "Morphogenesis by Cell Intercalation." *1989 Lectures in Complex Systems*, edited by E. Jen. Santa Fe Institute Studies in the Sciences of Complexity, Lect. Vol. II. Reading, MA: Addison-Wesley, 1990.
21. Platt, John. "Constraint Methods for Neural Networks and Computer Graphics." Ph.D. Thesis, California Institute of Technology, 1989.
22. Prusinkiewicz, Przemyslaw, and Aristid Lindenmayer. *The Algorithmic Beauty of Plants.* New York: Springer-Verlag, 1990.
23. Prusinkiewicz, Przemyslaw, Mark Hammel, and Eric Mjolsness. "Animation of Plant Development Using Differential L-Systems." *Computer Graphics* **27** (July 1993): to appear.
24. Purves, Dale, and Jeff W. Lichtman. *Principles of Neural Development.* Sunderland, MA: Sinauer, 1985.
25. Reinitz, John, Eric Mjolsness, and David H. Sharp. "Model for Cooperative Control of Positional Information in *Drosophila* by *Bicoid* and Maternal *Hunchback*." Research Report YALEU/DCS/RR-922, Yale University, 1992 or Unclassified Report 92-2942, Los Alamos National Laboratory, Los Alamos, NM, September 1992.
26. Sims, Karl. "Interactive Evolution of Dynamical Systems." In *Proceedings of the First European Conference on Artificial Life*, held in Paris, 1991. Cambridge, MA: MIT Press, 1991.
27. Sims, Karl. "Artificial Evolution for Computer Graphics." *Computer Graphics* **25(4)** (1991): 319–328.
28. Snyder, John. "Generative Modeling: An Approach to High-Level Shape Design for Computer Graphics and CAD." Ph.D. Thesis, California Institute of Technology, 1991.
29. Stork, David G., Bernie Jackson, and Scott Walker. "Non-Optimality via Pre-Adaptation in Simple Neural Systems." *Artificial Life II*, edited by C. G. Langton, C. Taylor, J. D. Farmer, and S. Rasmussen. Santa Fe Institute Studies in the Sciences of Complexity, Proc. Vol. X. Redwood City, CA: Addison-Wesley, 1991.
30. Turing, Alan. "The Chemical Basis of Morphogenesis." *Phil. Trans. B.* **237** (1952): 37–72.

Mark M. Millonas
Complex Systems Group, Theoretical Division, and Center for Nonlinear Studies, MS B258, Los Alamos National Laboratory, Los Alamos, NM 87545, USA and Santa Fe Institute, 1660 Old Pecos Trail, Suite A, Santa Fe, NM 87501, USA

Swarms, Phase Transitions, and Collective Intelligence

A spacially extended model of the collective behavior of a large number of locally acting organisms is proposed in which organisms move probabilistically between local cells in space, but with weights dependent on local morphogenetic substances, or morphogens. The morphogens are in turn effected by the passage of an organism. The evolution of the morphogens and the corresponding flow of the organisms constitutes the collective behavior of the group. Such models have various types of phase transitions and self-organizing properties controlled both by the level of the noise and other parameters.

The model is then applied to the specific case of ants moving on a lattice. The local behavior of the ants is inspired by the actual behavior observed in the laboratory, and analytic results for the collective behavior are compared to the corresponding laboratory results.

It is hoped that the present model might serve as a paradigmatic example of a complex cooperative system in nature. In particular, swarm models can be used to explore the relation of nonequilibrium phase transitions to at least three important issues encountered in artificial life. Firstly, that of emergence as complex adaptive behavior. Secondly, as an exploration of continuous phase transitions in biological systems. Lastly, to derive behavioral criteria for the evolution of collective behavior in social organisms.

1. INTRODUCTION

1.1 THE APPEAL OF SWARMS

The swarming behavior of social insects provides fertile ground for the exploration of many of the most important issues encountered in artificial life. Not only do swarms provide the inspiration for many recent studies of the evolution of cooperative behavior,[6,19,20] but the action of the swarm on a scale of days, hours, or even minutes manifests a nearly constant flow of emergent phenomena of many different types.[1,2,8,11,14,17,26,27] Models of such complex behavior range from abstract cellular automata of models[21] to more physically realistic computational simulations.[7,12] The notion that complex biological behavior, from the molecular to the ecological, can be the result of parallel local interactions of many simpler elements is one of the fundamental themes of artificial life.[21] The swarm, which is a collection of simple locally interacting organisms with global adaptive behavior, is a quite appealing subject for the investigation of this theme.

When one includes an evolutionary dimension, the appeal becomes even more robust, since we have, in many ways, a much better notion of the ultimate purpose or utility of insect behavior than we have of many other types of emergent phenomena in nature. The notion of utility provides a link between the emergent *behavior* of swarms and the *evolution* of cooperative social behavior.

An additional impetus to this type of study which is lacking in many areas of artificial life research is the contact with experiments so necessary for the healthy growth of science. By this I mean not just computer simulations, but actual work with real organisms.[1,8,26] In many ways the physical motivations behind the types of models discussed here are inspired by such experiments. The knowledge that many kinds of controlled investigations can be performed on systems which closely resemble the models described here not only informs the interpretation of the results, but suggests new types of laboratory studies.

In the end perhaps the most pervasive appeal of swarms centers on a kind of emotional attractiveness of the subject. Undoubtedly all of the above considerations play a role in this, but probably the main reason is hidden within in the human psyche. More than a paradigm, swarms are almost, at times, an archetype.

1.2 BASIC PRINCIPLES OF SWARM INTELLIGENCE

It is useful to list some broad behavioral categories that might be classified as collective intelligence, or swarm intelligence. These may be thought of as evolutionary principles of selection, and are not intended to be definitive.

The first is the **proximity principle**. The group should be able to do elementary space and time computations. Since space and time translate into energy expenditure,[29] the group should have some ability to compute the utility of a given response to the environment in these terms. Computation is understood as *a direct behavioral response to environmental stimuli which in some sense maximizes the utility of some type of activity to the group as a whole.* Although the kinds of activity may vary greatly, depending on both the type and complexity of the organisms, some typical activities include the search and retrieval of food, the building of nests, defense of the group, collective movement, and, in the case of higher organisms, the interaction necessary for many social functions.[17]

Second is the **quality principle**. The group should be able to respond not only to time and space considerations, but to quality factors, for instance, to the quality of foodstuffs the or the safety of a location.

Third is the **principle of diverse response**. The group should not allocate all of its resource along excessively narrow lines. It should seek to distribute its resources along many modes as insurance against the sudden change in any one of them due to environmental fluctuations. It is clear that a completely ordered response to the environment, even if possible, may not even be desirable.

Fourth is the **principle of stability**. The group should not shift its behavior from one mode to another at every fluctuation of the environment, since such changes take energy, and may not produce a worthwhile return for the investment. The other side of the coin is the **principle of adaptability**. When the rewards for changing a behavioral mode are likely to be worth the investment in energy, the group should be able to switch. The best response is likely to be a balance between complete order and total chaos and, therefore, the level of randomness in the group is an important factor. Enough noise will allow a diverse response, while too much will destroy any cooperative behavior.

The behavior of many complex adaptive systems would probably fall into some version of these principles. It is amusing to note the resemblance of these rules to many good economic decision-making principles, or folk maxims like *time is money, only buy the best, don't put all your eggs in one basket, better safe than sorry, a bird in the hand is worth two in the bush, invest for the future,* etc.[13]

1.3 INSPIRATION: THE BEHAVIOR OF REAL ANTS

The techniques used in some recent experiments with ants[7,8,14,26] in many ways inspired the approach used here. Typically, ants in the laboratory are exposed to a set of bridges connecting two or more areas where the ants explore, feed, and go about their business. As the ants wander, they discover and cross the bridges. As they move on the effectively one-dimensional paths, they come to junctions where they choose a new branch and continue on their way. In Figure 1 we show a picture of what such a bridge might look like. Since the ants both lay and follow scent as they walk, the flow of ants on the bridges changes as time passes. For instance, in the case shown in Figure 1, most of the flow will eventually concentrate on the one of the branches. In this case the swarm is said to have chosen a branch of the bridge. The types of emergent behaviors of the ants can then be studied in a controlled manner by observing their response to various situations. In addition, many interesting mathematical models and computer simulations have been studied which capture some of the behavior observed in the laboratory.[1,7,27]

The experimental results indicate the type of environmental, or **emergent, computation** that real ants perform. The term environmental computation here again refers to the fact that the ants collectively perform information gathering and processing on the local environment. Both the information gathering and processing happen simultaneously and without centralized controls. For example, both the location of a food source and its utilization are *computed* by the self-organization of a column of ants between the nest and the food source. These experimental observations form a basic set of tasks that any model needs to explore. The following results were obtained using a few different species of ants.[8,14,26]

1. When ants are exposed to two paths of unequal length, the ants will choose the shortest path.
2. If a shorter path is offered after the ants have chosen, they are unable to switch to the new path.

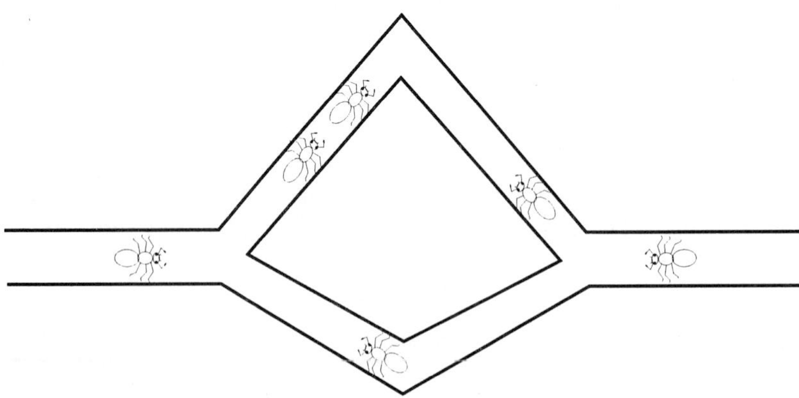

FIGURE 1 The binary bridge experiment.

3. The ants will break symmetry and chose one path, even when both paths are equal.
4. If ants are offered two unequal food sources, they will usually choose the richest source. If a richer food source is offered after the ants have chosen, some species can switch to this new source, and others are unable to.
5. If two equal food sources are offered, the ants exploit the source unequally, breaking symmetry.

In addition, the rules for the effect of the pheromone density on the motion of the ants can be determined experimentally from a setup such as the one pictured in Figure 1. It has been found from such experiments that ants choose a branch of the bridge in proportion to the function $(\alpha + \sigma)^{\beta}$, where σ is the density of pheromone on the branch measured in unit of the average density layed down during th e passage of a single ant, and α is a fixed parameter with the same units. β is some dimensionless fixed parameter. For one type of ant, the argentine ant, values of $\alpha = 20$ and $\beta = 2$ were obtained.

1.4 GOALS

The collective behavioral characteristics of a group of organisms must, of course, be encoded in the behavior of the individual organisms. We will be interested in *how* the collective behavior is encoded by the individual behavior. In particular, we explore the idea that complex adaptive behavior is the result of the *interactions between* organisms as distinct from behavior that is a direct result of the actions of *individual* organisms.

In line with the ideas proposed above, the most important modeling constraint is **the principle of locality**. In the models we will study, the behavior of the individual organisms will be determined solely by local influences. This means that the individual organisms will not have any memory, nonlocal navigational skills, or any type of behavior that involves storage of internal information. Any information flow must then be a product of the collective behavior. Wilson[32] introduced the concept of mass communication to designate information transmission that cannot be communicated by the individual. Here we extend the designation to include transmission of information that the individuals cannot even possess. Of course, that the actual situation in nature will almost always be more complicated than this, but we may hope to achieve some level of understanding by studying such restricted models and by comparing them to real situations. In this way it may be possible to isolate many of the collective emergent processes that are taking place.

1.5 PREVIEW

In Section 2 the Swarm network model is presented, and its analogy with other connectionist models is discussed. We show that the distribution of organisms on the network always relaxes, on the short time scale, to a unique stationary distribution which is independent of the initial configuration. This allows us to write down the general deterministic learning rules for the network. In Section 3 we introduce a type of dynamics, inspired by experiment, that allows us to make an analogy with a thermodynamic system. The quite general conditions for self-organization (symmetry breaking) are derived, and the various attractors and phase transitions of the Swarm network system are explored. In Section 4 a specific model of ant swarms is presented. The behavioral function of the ants is taken from the experimental results and a theoretical model. I will particularly focus on the role of phase transitions, which are of relevance to the study of artificial life in at least three important ways. The first and most obvious role is that phase transitions control the global behavior of the swarm. In Section 5 a few simple cases are compared to the observed behavior of ants in the laboratory. Secondly, since it has recently been suggested[22] that second-order phase transition might have an important influence on the complex adaptive and computational properties of biological systems, these transition are worthy of study in their own right. Here we provide an example inspired by real-life biology which can be studied. Finally, not all possible phase transitions have significance on a behavioral level. Phase transitions on the variation of the behavioral parameters of the organisms, which are fixed on the behavioral time scale, can provide *evolutionary criteria* for the development of cooperative behavior in social organisms. We thus hope to illuminate this issue as well.

2. SWARM NETWORKS

In this section we present the basic structure and dynamics of swarm networks. The dynamics of the organisms is discussed, and the deterministic learning rule for a swarm is derived.

2.1 CONNECTIONIST MODELS

Connectionist models[9] have three basic earmarks. Firstly, their **structure** consists of a discrete set of *nodes*, and a specified set of *connections* between the nodes. For example, neural networks, the archetypal connectionist systems, are composed of neurons (nodes), and the neurons are usually linked by synapses (connections). Secondly, there are the **dynamics** of the *nodal variables*. In the case of neural networks the nodal variables are the firing rates of each neuron. The dynamics are controlled by the connection strengths, and the input-output rule of the individual neurons. The dynamics of the whole system is the result of the interaction of all the neurons. Lastly there is **learning.** In its most general sense, learning describes how the connection strengths, and hence the dynamics, evolves. In general, there is a separation of time scales between dynamics and learning, where the dynamical processes are much faster than the learning processes.

In addition to neural networks, there are many other type of connectionist type models, such as autocatalytic chemical reactions, classifier systems, and immune networks,[10,28] to mention just a few. Swarm networks are just another such example.[24]

2.2 NETWORK STRUCTURE

The structure of the network is a representation of the physical space on which the organisms move, and possibly of certain environmental factors and constraints. In this paper we will not seek details on arbitrarily small scale, but will divide the physical space into discrete elements, or **nodes**, which will be labeled by an index $i = 1, \ldots, m$. In certain cases this discretization may be a natural reflection of the actual physical setup of a laboratory experiments such as those discussed above, or another naturally occurring discrete structure such as an existing network of path segments. In other cases the discretization may represent a more abstract division of the physical space. For instance, positions on a plane may be divided up into a chessboard of discrete cells. Additionally, each discrete position in space might have additional discrete divisions representing the possible orientations of an organism at that point. Each discrete position and orientation will be specified by a single node. A network is then a kind of discretized phase space on which the organisms move.

In addition to nodes, we must specify **connections**. Connections express the possibility of an organism moving from one node to another, and are specified by an ordered pair of nodes, one of which is the initial node and one of which is the final node. Obviously, most transitions will not be possible. The set of possible final nodes for an initial node are said to be the nodes *local* to the initial node.

Lastly, we must specify boundary conditions on the network. These will take the form of **adsorbing sites** and **input sites**. Adsorbing sites are nodes from which an organism may leave the network entirely. Input sites are nodes that receive an input of organism from the world outside the network. For the purposes of this paper we will consider only closed networks.

2.3 DYNAMIC AND PARAMETRIC NETWORK VARIABLES

A number of organisms will be allowed to move on the network. Most of what follows can easily be extended to the case where there are a number of different *types* of organisms, but for simplicity we will only deal with the case where there is one type. Each node has a given volume or measure, μ_i, with units depending on the dimension of the nodes. We will denote the number of organisms at node i at time t by $n^i(t)$, so the density of organism at i is $\rho^i(t) = n^i(t)/\mu_i$.

In addition, there is a quantity of a morphogenetic substance, or morphogen, at each node. I have adopted this terminology from Turing's famous paper, "The Chemical Basis of Morphogenesis,"[30] where the term is used to indicate the form producer. While in the present case there is not a one-to-one relationship with Turing's morphogen concept, the basic idea is the same. The term is adopted mainly to avoid the misunderstanding that the processes modeled here depend on any specific realizations or mechanisms. Much of what follows is widely applicable.

The density of morphogen on node i will be denoted σ^i. We will, for simplicity, consider the case where only one *type* of morphogen is present. This may be thought of as a chemical substance that the organisms both emit, and respond to, but it is possible in certain circumstances that it might have another physical meaning. For instance, such networks could model the formation of animal trails, or routes in the undergrowth. The density of plant growth at a given point would be analogous to a morphogenetic substance. The route that animals take while moving from place to place is influenced by the density of undergrowth. In addition, the undergrowth is affected by the passage of animals, which tend to crush young plants and otherwise impede the growth in that area. One might expect the result, over time, to be the formation of trails. This type of consideration is also know to play a role in the formation of more or less permanent network of *physical* trails by insect colonies.[17]

ρ^i and σ^i are essentially the dynamical variables of the model. However, we can, if we wish, make a distinction between **dynamical** variables and **parametric** variables under certain conditions. For practical purposes we will assume that the σ^i are more or less constant for time scales typical of the variables ρ^i. We will then treat the σ^i as parameters that determine the dynamics of ρ^i. The ρ^i variables are said to be *slaved* to the variables σ^i. They are parametric *variables* because they are endowed with a dynamics of their own. This dynamics is usually known in connectionist terminology as *learning*. This separation of time scales, and the resulting distinction of dynamic versus parametric variables, is a common theme running through connectionist theories. For some types of behavior we may not be able to strictly separate time scales, but we will view these situations as complications on the base of a model where the time scales are well separated. In fact, it is the passage of the swarm through a supercritical point of its global dynamics which makes possible the amplification of small nonequilibrium signals, such as time delay feedback.[8] A good example of this will be our analysis of the choice of a closer food source in Section 5.

In addition to dynamic and parametric variables, there will also be **fixed parameters** which can be considered constant on both dynamical and learning time scales. The network will possess certain parameters such as the measure of each node, μ_i, or the orientation of each node at the junction if we are modeling a network of paths. In addition the organism may have internal parameters that determine their response to, and emission of, the morphogenic substances. The choice of these parameters by natural selection may be considered an additional type of dynamics of the system. Lastly, there may be one or more parameters that describe the evolution of the parametric variables, such as the decay rate of the morphogen.

2.4 DYNAMICS: THE FLOW OF ORGANISMS

In addition to reacting to the morphogens, the organisms may in general interact with each other. For the purpose of this paper we will consider situations where the organism do not interact directly, but only through the medium of the morphogenetic substance. In this case the the flow on a network is described by

$$\mu_i \frac{d\rho^i}{dt} = \sum_j \left(v_0 \omega_{ij} \rho^j - v_0 \omega_{ji} \rho^i \right) - v_0 a^i \rho^i + f^i, \tag{1}$$

where ω_{ij} is the probability that organisms leaving node j will choose node i; and a^i is an adsorption probability, that is, the probability that organisms leaving i will leave the network entirely. v_0 is a rate constant that can be thought of as the speed of the organisms. The left-hand side of the equation is just the rate of change of the number of organisms on node i. The first term on the right-hand side represents the flow of organisms into the node from all of the connecting nodes. The second term is the flow out of the node to all the local nodes. The third term represents the loss of organisms to the outside world at an adsorbing site, and the last term is the number of organisms entering node i per unit time from some outside source. Clearly $\omega_{ij}, a^i, f^i(t) \geq 0$. Additionally, probability considerations lead to the following restriction

$$a^i + \sum_j \omega_{ji} = 1, \tag{2}$$

which can be read as, the probability that an organism on node i goes somewhere is one. ω_{ji} will generally depend on the parametric variables, and also on the fixed parameters of the network, and a^i and f^i may be taken to be fixed, or given some special dependence on the parametric variables depending on the situation we are trying to model.

The distribution of organisms on the network described above in nearly all physically reasonable situations relaxes to a stationary state that is independent of the initial distribution. The state is known as a stationary state because the distribution is constant in time. Since in general the flow of organisms at any given point on the network does not vanish, the state is not an equilibrium state. The more specific case of equilibrium states, and the conditions necessary to produce them, will be discussed in detail in the next section.

2.5 LEARNING: THE PARAMETRIC DYNAMICS

The learning rule is a prescription for the way the morphogen distribution changes with time. It is not arbitrary, but depends on the details of the mechanism by which the morphogens evolve. For our purpose we will make use of the following simple and sensible rule. As time passes, the morphogen density decays in proportion to the amount present with decay rate κ. In addition, each organism at i lays down a density η_i of the morphogen as it passes a point. The evolution of the morphogen distribution on the network is then described by the equation

$$\frac{d\sigma^i}{dt} = -\kappa\, \sigma^i + \eta_i\, \rho^i, \tag{3}$$

where σ^i is the density of the morphogen on node i, and ρ^i is the density of the organism on node i. Equations (1) and (3) then completely describe the dynamics once the dependence of ω_{ij} on σ^i is specified.

As discussed before, we will assume the of separation of time scales. In this case the flow of organisms relaxes to a stationary state on time scales short compared to the typical times scales of Eq. (1). The **deterministic approximation** then consists in replacing ρ^i by their stationary values $\rho_s^i(\sigma)$ and in *ignoring the fluctuations* about these values. This leads us to the generalized learning rule for a swarm network

$$\frac{d\sigma^i}{dt} = -\kappa\, \sigma^i + \eta_i\, \rho_s^i(bf\sigma). \tag{4}$$

It is of interest that even though we have strictly local couplings between nodes, the learning rule is a globally coupled system of equations, since $\rho_s^i(\sigma)$ depends in a complicated way on the parametric variables σ^i. This is made possible by the separation of time scales. On the dynamical time scale the system *feels out* its entire space of configurations, and relaxes to a state that is determined by global considerations, even though the dynamics is strictly local. This globally determined state then determines the dependence of some of the variables in the local learning rule. *Globally coupled parametric dynamics in strictly local models via separation of time scales* is an important aspect of connectionist models, and is one of the central reasons why connectionism is such an attractive model of adaptive complex systems.

3. THE THERMODYNAMIC ANALOGY

3.1 DETAILED BALANCE AND THE THERMODYNAMIC ANALOGY

For the rest of this paper we will consider the case where the transition matrix takes the form $\omega_{ij} \propto f\left(\sigma^i\right) \Omega_{ij}$, where f is some weighting function describing the effect of σ^i on the motion of the organisms; and $\Omega_{ij} = 1$ if i connects to j, and zero otherwise. When properly normalized, the transition matrix is given by

$$\omega_{ij} = \frac{f(\sigma^i)\Omega_{ij}}{\sum_k f(\sigma^k)\Omega_{kj}}. \tag{5}$$

Transition matrices of this type obey the detailed balance relations $\omega_{ij} f(\sigma^j) = \omega_{ji} f(\sigma^i)$. The property of detailed balance allows us to determine all of the statistical properties of the quasi-stationary particle field. In addition a one-to-one analogy with a thermodynamic system with energy $U(\sigma)$ and temperature $T = \beta^{-1}$ can be made if we set $f(\sigma) = \exp\left(-\beta U(\sigma)\right)$, where any parameter T that affects f can be regarded as a temperature parameter if $f(\sigma; T)$ scales like $f(\sigma; \alpha\, T) = f^{-\alpha}(\sigma; T)$. Statistical quantities of interest can then be calculated from the one-particle partition function

$$Z = \frac{1}{V} \sum_i \mu_i \exp\left(-\beta U(\sigma^i)\right) \tag{6}$$

according to the usual prescriptions,[18] where V is the total volume of the system. The N-particle partition function is $Z_N = Z^N$.

Let us partially evaluate the partition function over the volume μ_ϵ of the network with a given energy ϵ,

$$Z = \frac{1}{V} \sum_{U \neq \epsilon} \mu_i \exp\left(-\beta U(\sigma^i)\right) + \frac{\mu_\epsilon}{V} \exp(-\beta\epsilon). \tag{7}$$

The mean equilibrium particle density in the energy state ϵ is then given by[18]

$$\rho_e^\epsilon = \frac{N}{VZ} \exp(-\beta\epsilon). \tag{8}$$

3.2 THE ORDER PARAMETER EQUATION

The equations for the pheromonal field densities can be written as

$$\frac{ds^i}{d\tau} = -s^i - \rho_e^i(\mathbf{s}), \tag{9}$$

where we have introduced the new variables $\tau = \kappa\, t$, and $s^i = \kappa\sigma^i/\eta$. We have adiabatically eliminated the organisms from the picture by replacing ρ^i with ρ_e^i. We introduce the bimodal variables

$$s^+ = \frac{1}{\mu^+} \sum_{i; s^i > \gamma} s^i, \qquad s^- = \frac{1}{\mu^+} \sum_{i; s^i < \gamma} s^i, \tag{10}$$

where $\gamma = N/V$ is the mean density of organism on the network, and

$$\mu^+ = \sum_{i;s^i>\gamma} \mu^i, \quad \mu^- = \sum_{i;s^i<\gamma} \mu^i. \tag{11}$$

We will make use of the mean-field approximation $s^i = s^\pm$, which leads to the mean-field bimodal equations

$$\frac{ds^+}{d\tau} = -\ s^\pm + \rho_e^\pm\left(s^+, s^-\right). \tag{12}$$

The quasi-stationary distribution $\rho_e^\pm(s^+, s^-)$ is calculated from the master equation for the particle density. In the case of a closed systems with detailed balance,

$$\rho_e^\pm(s^+, s^-) = -\frac{N}{\beta\mu^\pm}\frac{\partial \ln Z}{\partial \epsilon^\pm}. \tag{13}$$

It will be helpful to introduce the dimensionless parameter $\upsilon = \mu^-/\mu^+$, the ratio of the volume in the $-$ state to the volume in the $+$ state. For compactness we also define the function

$$R(s^+, s^-) = f(s^+)/f(s^-). \tag{14}$$

The equations can then be put in the form

$$\frac{ds^+}{d\tau} = -\ s^+ + \frac{\gamma(1+\upsilon)R^\beta}{\upsilon + R^\beta}, \tag{15}$$

$$\frac{ds^-}{d\tau} = -\ s^- + \frac{\gamma(1+\upsilon)}{\upsilon + R^\beta}. \tag{16}$$

Instead of working in the variables $s^\pm$, we will chose the new order parameter variables

$$m = s^+ - s^-, \quad s = \frac{s^+ + \upsilon\, s^-}{1+\upsilon} - \gamma. \tag{17}$$

The second of these is proportional to the difference between the total density of pheromone present and the value at which the total density of pheromone equilibrates. The equation for the evolution of s is given by $ds/d\tau = -s$, which has solutions $s \propto e^{-\tau}$. No self-organization will occur until the pheromone density has built up sufficiently. Thus the pheromone density will initially evolve uniformly over the configuration space, there will be little feedback, and m will be small on time scales where $s \to 0$. We can thus make the adiabatic approximation $s = 0$. The deterministic equation for the order parameter m is then given by

$$\frac{dm}{d\tau} = -m + F(m), \tag{18}$$

where

$$F = \gamma(1+\upsilon)\frac{R^\beta - 1}{R^\beta + \upsilon}, \tag{19}$$

and where F is determined as a function of m by

$$R(m) = R\left(\gamma + \frac{\upsilon\, m}{1+\upsilon}, \gamma - \frac{m}{1+\upsilon}\right). \tag{20}$$

In addition the longtime evolution of the order parameter will only depend on the longtime limit $s \to 0$ and is not affected by the validity of the adiabatic approximation.

The deterministic dynamics is then described by the equation

$$\frac{dm}{d\tau} = -\ \Psi'(m), \tag{21}$$

where the deterministic potential $\Psi(m)$ is given by

$$\Psi(m) = \frac{1}{2}m^2 - \int^m F(x)\ dx. \tag{22}$$

3.3 PHASE TRANSITIONS

We are interested in the various phase transitions in the collective behavior of the swarm. These are the points where the behavior changes abruptly upon variation of some parameter. As mentioned earlier, these points determine the behavior of the swarm when the varied parameter is an external one. In this case the basins of attraction of Ψ will determine the *behavior* of the swarm, so a phase transition represents a behavioral transition. Phase transitions are also significant in establishing evolutionary criteria when the varied parameter is a *fixed* behavioral parameter.

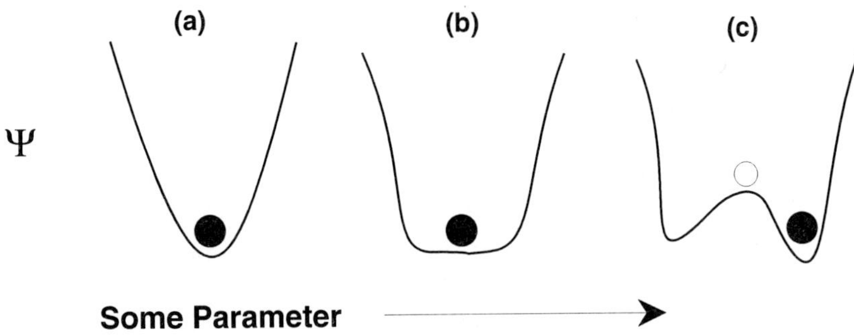

FIGURE 2 A phase transition.

The values of the order parameter m_k are determined from the condition

$$\Psi'(m_k) = 0, \tag{23}$$

and their stability by

$$\Psi''(m_k) > 0, \tag{24}$$

that is, by the the minima of Ψ. When a parameter is varied, the shape of Ψ varies. A phase transition occurs when there is a transition, such as the one shown in Figure 2, from one type of shape to another. In this case the various macroscopic states of the swarm may change in an abrupt way. The point of phase transition is known as a critical point, illustrated by the shape in Figure 2(b).

To include some basic terminology, we usually distinguish two types of phase transitions, **first order** and **second order**. First order designates phase transitions where the macroscopic states change in a discontinuous way upon passage through the critical point, and second order designates phase transitions where the states change in a continuous way. Second-order phase transitions usually have some kind of symmetry which is broken when passing through the critical point. This symmetry breaking is caused by the natural fluctuations in the system which we neglected in the deterministic approach. It is at the critical point where we would expect the fluctuations to become the most pronounced and our deterministic analysis to have the greatest problems.

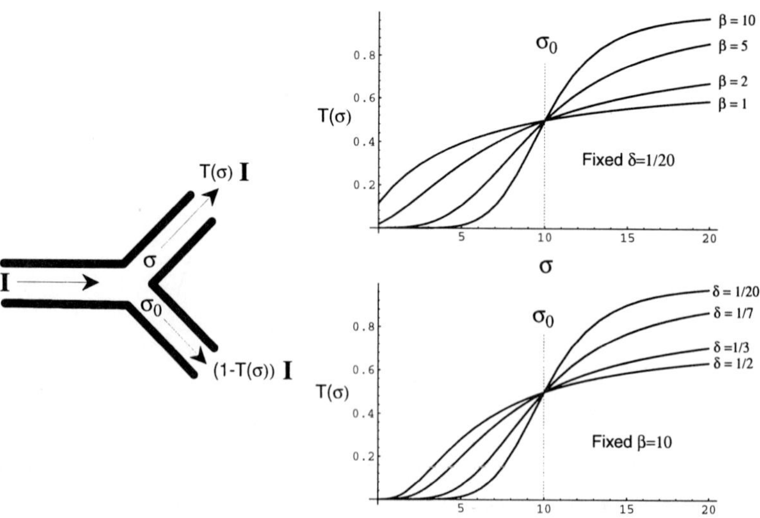

FIGURE 3 Transition functions for varying β and δ.

When a system passes through a second-order transition, it may be left sitting on top of the hill in the center of Figure 2(c). At this point the system is like a pencil balanced on its end. We cannot tell which direction it will fall, but a small perturbation can send it falling in a certain direction. It is the sensitivity of the system at this point to external influences which makes them of interest as a kind of *information amplifier* in biological systems. Typically, susceptibilities, the quantities that measure the response of a system to external influence, become infinite at the critical point.

In addition, many macroscopic quantities scale in a characteristic way near a critical point, according to **critical exponents**. These exponents have, in many cases, universal values which are independent of the details of the given universality class of models being studied.

From conditions (23) and (24) we can determine the stability condition for the homogeneous state $s^i = \gamma$:

$$\gamma\, \chi(\gamma) < T \qquad (25),$$

where $T = 1/\beta$ is the temperature parameter, and where χ is known as the chemotatic factor, or force $\chi(x) = -U'(x)$.

4. ANT SWARMS

4.1 MICROSCOPIC ANT BEHAVIOR

The microscopic dynamics of ants can be described by the pheromone energy function

$$U(\sigma) = -\ln\left(1 + \frac{\sigma}{1+\delta\sigma}\right), \qquad (26)$$

where σ is the pheromone density, and δ is a dimensionless behavioral parameter.[24] The temperature parameter $T = 1/\beta$ describes the internal randomness of the response of the ants to the pheromonal field. This function is based approximately on a model for Osmotropotaxi (scent gradient following)[3,4,5] and on experimental observations of actual ants.[8] For the case where the density of ants is low, and hence the pheromone density is low ($\sigma << 1/\delta$), we can make use of the approximate energy function $U_0(\sigma) = -\ln(1{+}\sigma)$. The constant $1/\delta$ will be known as the capacity. When σ approaches $1/\delta$, the ants respond less accurately to pheromone gradients.

An illustration of this effect is shown in Figure 3. A given current of organisms I flows into a junction from the left. On the lower branch the pheromone density is fixed at σ_0, and on the upper branch σ is allowed to vary. $\mathcal{T}(\sigma)$, the proportion of the current that flows into the upper branch, is given by the sigmoidal function

$$\mathcal{T}(\sigma) = \left[1 + \exp\left(\beta U(\sigma)/U(\sigma_0)\right)\right]^{-1}. \qquad (27)$$

The plots on the right of Figure 3 shows $\mathcal{T}(\sigma)$ for varying values of β and δ. The upper plot, where δ is fixed, shows the influence of increasing the temperature (lowering β). As the temperature increases, the threshold response becomes less and less pronounced. In the opposite limit $\beta \to \infty$, $\mathcal{T}(\sigma)$ would be a step function $\Theta(\sigma - \sigma_0)$. In this limit all of the ants would choose the branch with the greatest pheromone density. In the lower plot the noise level is fixed, and the capacity $1/\delta$ is varied. It is interesting to note that the effects of decreasing the capacity with fixed temperature are similar to the effects of increasing the temperature with fixed capacity. When the density of the ants increases, the pheromone density increases up to and beyond the capacity, and the qualitative effects on the behavior of the ants is the same *as if the temperature was increased.* This gives the swarm roughly the ability to modulate its temperature by modulating its numbers.

This can be made clearer by defining an effective temperature factor $\theta(\sigma)$ through the relation $f(\sigma) = \exp(-\beta U_0(\sigma)/\theta(\sigma))$. $\theta(\sigma)$ roughly measures the effective change in temperature as a function of the pheromonal field when compared to the case where $\delta = 0$, which correspond to the energy function U_0. The effective temperature is then given by $\theta(\sigma)T$ where

$$\theta(\sigma) = \frac{\ln\left(1 + \frac{\sigma}{1+\delta\sigma}\right)}{\ln(1+\sigma)}. \tag{28}$$

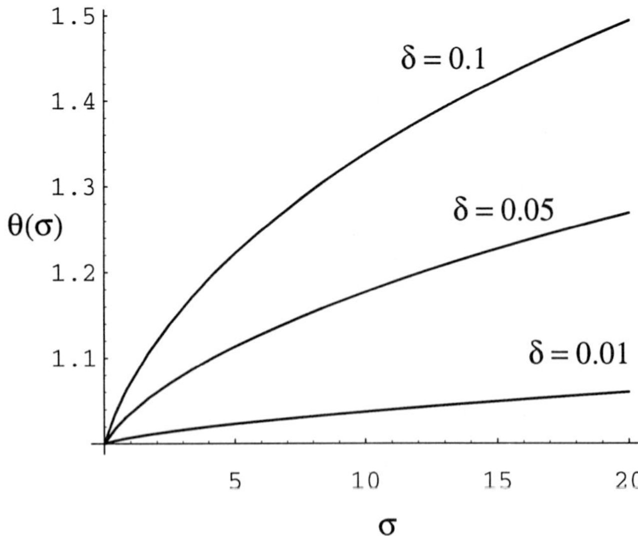

FIGURE 4 Effective temperature factor.

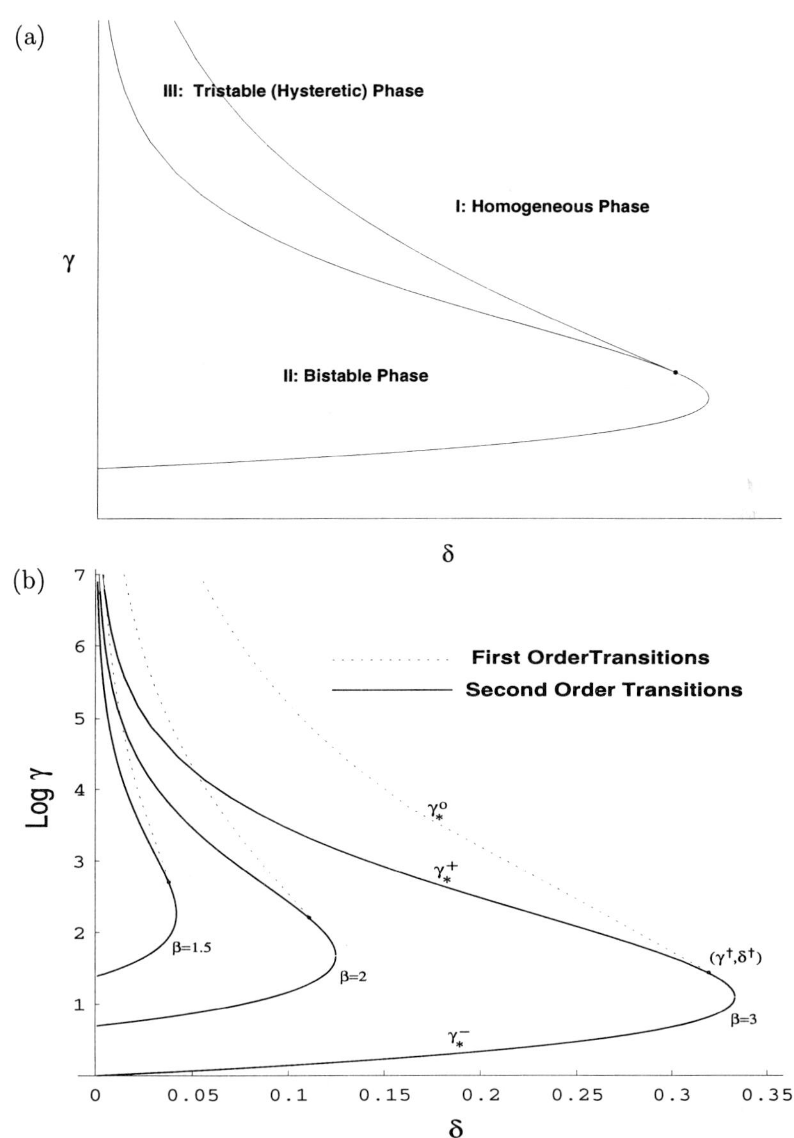

FIGURE 5 $\delta - \gamma$ phase diagrams.

In Figure 4 we illustrate the increase in the effective temperature with increasing σ for three different values of δ. Since increasing the temperature tends to decrease stability, we might expect any organized behavior to break down when the number of participants grows too large. It is this ability of the swarm to self-modify its temperature that allows it, in a sense, to traverse its various phase transition boundaries. Such boundaries are of crucial importance in self-organization and emergent phenomena, and it has been proposed that the ability to self-organize at or near these boundaries is at the heart of adaptive, emergent biological systems.[22]

4.2 THE PHASE DIAGRAM FOR THE SWARM

The condition for stability of the homogeneous phase is independent of υ as could be expected, since, in the homogeneous phase, υ does not really exist. The critical points of the homogeneous phase are given by

$$\gamma_*^{\pm} = \frac{1}{2\delta(1+\delta)} \left\{ \beta - 2\delta - 1 \pm \sqrt{\beta^2 - 2\beta - 4\beta\delta + 1} \right\}, \tag{29}$$

where γ_*^- is the value of γ where the symmetric phase becomes unstable as γ is increased, and γ_*^+ is the value of γ where the symmetric phase again becomes stable. These critical points themselves are the result of a bifurcation controlled by δ. This bifurcation only occurs for δ below the critical point

$$\delta < \delta_* = \frac{(\beta - 1)^2}{4\beta}. \tag{30}$$

When $\delta > \delta_*$, no symmetry breaking is possible, irrespective of γ.

In addition, there is a region of tristability where either the inhomogeneous phase or the homogeneous phase is possible. Which is chosen will depend on the initial conditions, and hysteresis (multiple values of the order parameter for the same values of the state variables) is possible. For a given δ this region extends from $\gamma_*^+ < \gamma < \gamma_*^0(\upsilon)$, where γ_*^0 marks the location of a first-order transition. For a certain case $\gamma_*^0(\upsilon)$ as a function of δ can be calculated analytically, but usually we will have to resort to Newton's method, or some other numerical scheme. All of this information can be illustrated by plotting the critical points $\gamma_*^{\pm}$ and $\gamma_*^0(1)$ as a functions of δ. The resulting phase diagram shown in Figure 5 illustrates the regions of symmetric phase, bistability, and tristability.

In general the various inhomogeneous states labeled by υ will become unstable at *different values of* γ. In this case we can have a quite complicated sequence of ordering transitions as γ is increased. This ordering can be used to explore some other experiments on ants but, due to the complexity of the subject, this discussion is best taken up elsewhere.

4.3 BRANCHES OF STABILITY

We can also investigate the behavior of the various branches of solutions of swarm networks. In Figure 6 we show five diagrams, known as bifurcation diagrams, which illustrate the types of behavior as the parameter γ is varied. A bifurcation point is the point were new stable equilibria come into being, that is, a phase transition using our terminology.

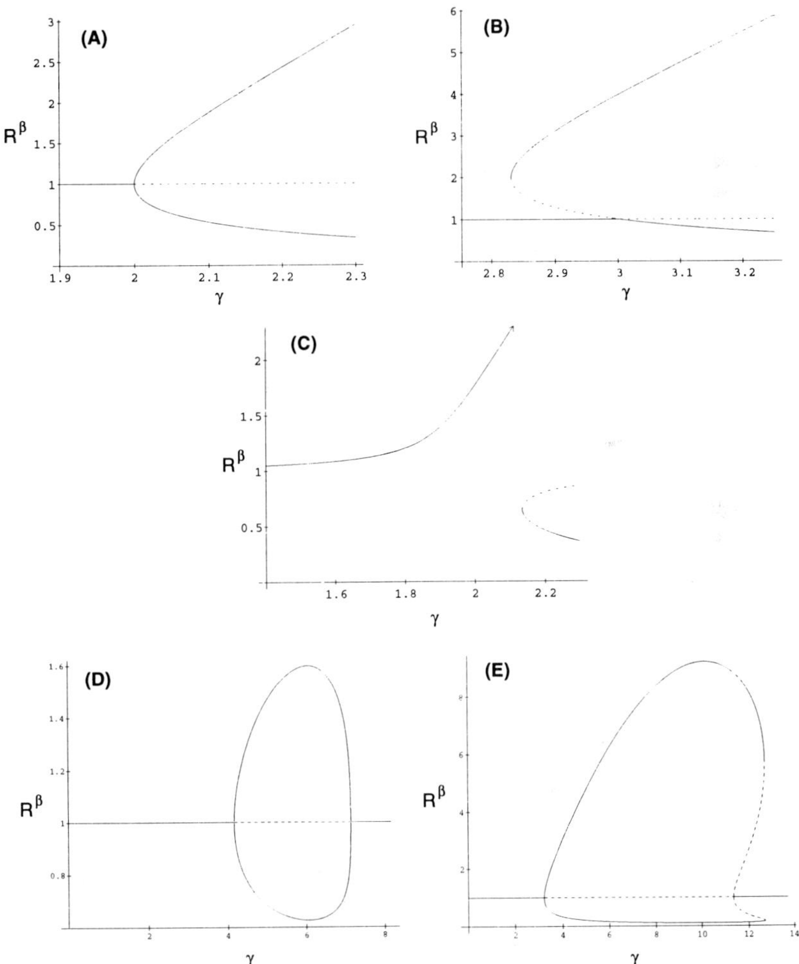

FIGURE 6 Bifurcation diagrams.

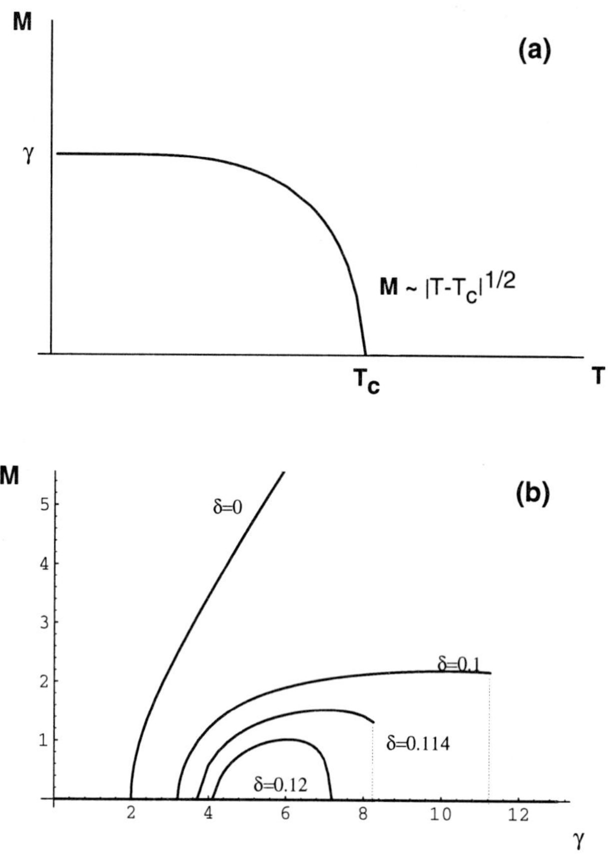

FIGURE 7 Plots of (a) the order parameter as a function of temperature, and (b) the equations of state for varying values of δ.

Figures 6(a)–(c) represent the case where $\delta = 0$. In Figure 6(a) we show the supercritical bifurcation for $\upsilon = 1$, the symmetric case. In Figure 6(b) we show a subcritical bifurcation for $\upsilon = 2$. Figure 6(c) represents a special case which will be discussed in the last section of the paper. It shows a transcritical bifurcation for the case where the ants choose between different qualities of food. It is this transcriticality that makes the choice of the better quality food source a more robust aspect of the behavior of swarms than the choice of a shorter segment, as shown by the experimental results mentioned at the end of Section 1.3.

In Figures 6(d)–(e) we show bifurcations for two different value of δ. These should be compared to the equations of state, shown in the next section, to get a feel for their significance. Both diagrams are for $\upsilon = 1$. Note that the homogeneous state in Figure 6(e) becomes stable before the broken symmetry states undergo a reverse bifurcation and disappears.

Letting $T = 1/\beta$ be the temperature, we can illustrate the phase transition at the critical temperature T_c and the resulting emergence of the order parameter in Figure 7(a). The critical temperature is given, in terms of the other parameters, by

$$T_c = \frac{\gamma}{1+\gamma+2\,\delta\,\gamma+\delta\,\gamma^2+\delta^2\,\gamma^2}. \tag{31}$$

Figure 7(a) is very reminiscent of the magnetization of a substance near its critical temperature.

We can also plot the order parameter as a function of γ, or the equations of state, shown in Figure 8(b), clearly illustrating both the second- and first-order transitions.

Very close to the critical points, the order parameter scales according to critical exponents which are independent of the particular parameters of the system. We obtain the classical critical behaviors

$$M \sim |T - T_c|^{1/2}, \tag{32}$$

$$\chi \sim |\gamma - \gamma_c|^{-1/2}, \tag{33}$$

where $\chi = \partial M/\partial\gamma$ is the susceptibility. In Figure 9 we illustrate this critical behavior.

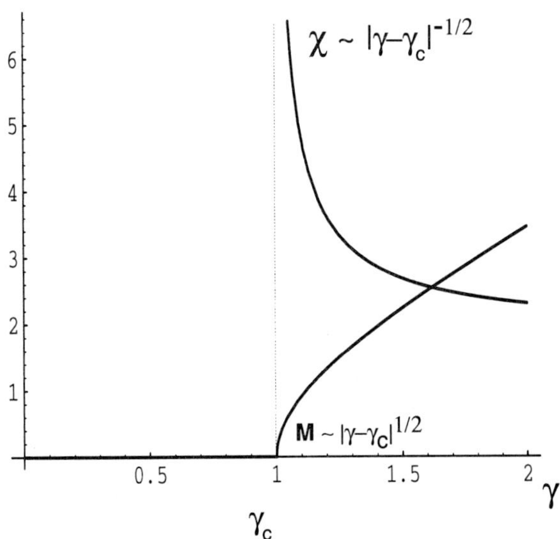

FIGURE 8 Susceptibility near second-order phase transition.

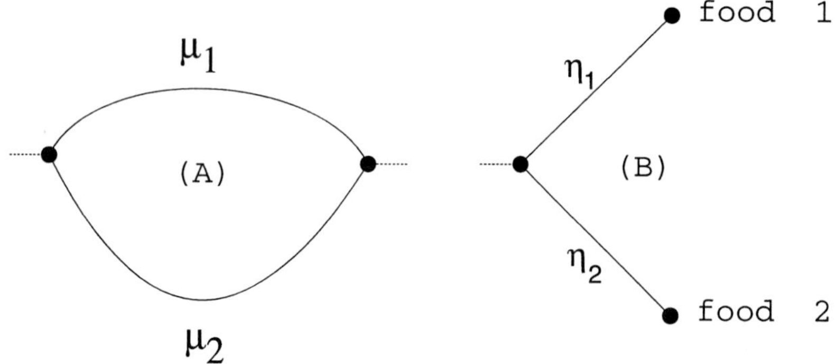

FIGURE 9 Binary node architecture.

5. A SIMPLE EXAMPLE OF SWARM BEHAVIOR

As a basic example of how the attractor structure of the model determines the behavior of a swarm, we will consider the simple cases corresponding to the binary bridge experiments with ants[8,14] discussed at the beginning of this paper, and compare these cases to the corresponding laboratory experiments.

5.1 BINARY NODE NETWORKS

Two basic types of two-segment architecture are illustrated in Figure 9.

In case (A) $\mu_1 \neq \mu_2$ and $\eta_1 = \eta_2$. In case (B) we let $\mu_1 = \mu_2$, and $\eta_1 \neq \eta_2$, where we assume, in agreement with experiment, that the ants returning from each food source lay scent at different rates depending on the quality of the food source.[15,17]

There are four interesting situations which correspond roughly to the cases (1–6) discussed at the beginning of this paper.

a. Asymmetric double bridge—(A) with $\mu_1 \neq \mu_2$ which models cases 1 and 2.

b. Symmetric double bridge—(A) with $\mu_1 = \mu_2$, which models case 3.

c. Asymmetric food sources—(B) with $\eta_1 \neq \eta_2$, which models cases 4 and 5.

d. Symmetrical food sources—(B) with $\eta_1 = \eta_2$ which models case 6.

These cases have all been previously explored experimentally, both with actual ants, and with some simple computer simulations.

For this analysis we set $\delta = 0$ and $\beta = 2$ in agreement with the experimentally observed values of these parameters. The fixed points in terms of the ratio of densities on the segments, $S_1/S_2 = R^2$, in various situations are plotted as a function of γ in Figures 6(a)–(c). The stable fixed points are shown by the solid lines, and the unstable ones by dashed lines. These plots, or bifurcation diagrams, show the values of S_1/S_2 corresponding to the fixed points in scent space. If there is only one stable branch for a given value of γ, then the ant densities will always evolve to a given configuration on the network. If, however, there are two stable branches, the system will be forced to choose one or the other. In the absence of noise the system that starts out on one side of the dashed line or the other will always evolve towards the stable state on that side of the line. The natural fluctuations of the density will introduce an element of randomness in the choice of stable state, particularly in those systems which start out on or near the dashed line.

5.1.1 THE BINARY BRIDGE. Symmetry breaking for the binary bridge occurs when $\gamma = 2$. Near a critical point where the $R = 1$ branch becomes unstable, one would expect to see a great increase in the fluctuations of the densities. In Figure 10(c) we show an example of these critical point oscillations.

In Figures 6(a)–(b) we show the bifurcation diagrams for the symmetric and asymmetric cases, respectively. In Figure 6(a) we illustrate situation **b** described above. The $R = 1$ becomes unstable when $\gamma > 2$ and, even though the system is completely symmetric, the ants will spontaneously break this symmetry and choose one of the bridges.

In Figure 6(b) we illustrate what will happen in situation **a**. There are three regions of interest. When $\gamma < 2\sqrt{v}$, only the branch $R = 1$ is possible. When $2\sqrt{v} < \gamma < 1 + v$, $R = 1$ and R_+, which represents the choice of the shorter branch, are stable. When $\gamma > 1 + v$, $R_\pm$ are stable and $R = 1$ is unstable. This says that while the choice of the shorter segment becomes possible before the longer segment, in the absence of noise, the ants should still choose the segment which gets the advantage first.

A full analysis of the effect of noise on the system would be necessary to determine whether the ants would choose the shortest segment with greater probability in this model. It can be shown that noise plays a constructive role in this case.[25] At the point $\gamma(v + 1) = 3$, the $R = 1$ becomes unstable. At this point, only R_+ is stable. If the number of ants builds up sufficiently slowly, the fluctuations in the system will cause a choice of the shorter segment before, and especially upon, passage through the point $\gamma(1 + v) = 3$.

The choice of the shorter segment can still be understood as a kind of time-delayed auto-feedback, as analyzed by Deneubourg et al.[8] The reasoning proceeds as follows. Ants that take the shorter route reach the food sooner. When they turn around and head for the nest, they follow the scent, and they find the scent slightly greater on the route they arrived on, and are thus more likely to choose this route and lay even more scent. Soon the equilibrium densities, which depend on the scents, will be reached, and the auto-feedback stops. However, in the short time this has taken, the shorter branch has built up a slightly greater scent. This system then effectively starts off above the $R = 1$ line in Figure 6(a). In this case, the ants will usually tend to attract to the shorter branch. From this argument we see how some simple nonequilibrium network properties can be taken into account within the framework of our equilibrium approximations.

This choice of the shorter segment depends on the short-term evolution of the system. However, our calculations show that in the long term, once an equilibrium has been established on a given route, the appearance of a shorter route will not cause the ants to shift to this new route, in agreement with experiment.

5.2 TWO FOOD SOURCES

In Figure 6(c) we illustrate the transcritical behavior for the asymmetrical two-food-source situation **c**. For a certain range of γ there is only one fixed point, and the ants will always choose the better food source. In this case they will be able to switch to a better food source should one become available. Depending on the scent-laying ratio η_1/η_2, there can be quite a large range of γ for which this type of switching is possible. If γ is large enough, the choice of the poorer food source also becomes a stable possibility. In this case the ants will be unable to switch to a better food source if offered. It is important to note that if the number of ants on the network slowly builds up from zero, which is what usually happens experimentally, the system will move along the upper branch in Figure 6 and will always choose the better food source, no matter what final value γ takes. The case of equal-quality food sources is equivalent to the symmetric double bridge case. Provided γ is great enough, the ants will choose one of the two sources at random.

5.3 MONTE CARLO SIMULATIONS

Monte Carlo simulations of these process were performed to test all of the above predictions, and in order to view the effects of fluctuations on the evolution of the ant densities. In the computer simulations a certain number of ants are allowed to move on the networks. During a time step each ant either moves forward one step on the segment it is on or, if it is at the end of a segment (the segments are all of integer length), chooses a new segment and moves onto it. The scent densities at the ends of each segment are then updated and used to compute the new transition probabilities. The initial configuration of ants and the scent densities on segments are set at the beginning of the program. This representation of the swarm is a probabilistic cellular automata.

In the simulations represented by Figure 10(a) and Figure 10(b), the ants were placed randomly on the network, and the initial pheromone density set to zero. In the situation shown in Figure 10(b) all of the ants started out on segment 2. In these simulations γ was varied by varying the total number of ants on the network. The numbers varied between approximately 400 and 550 ants for the three simulations shown.

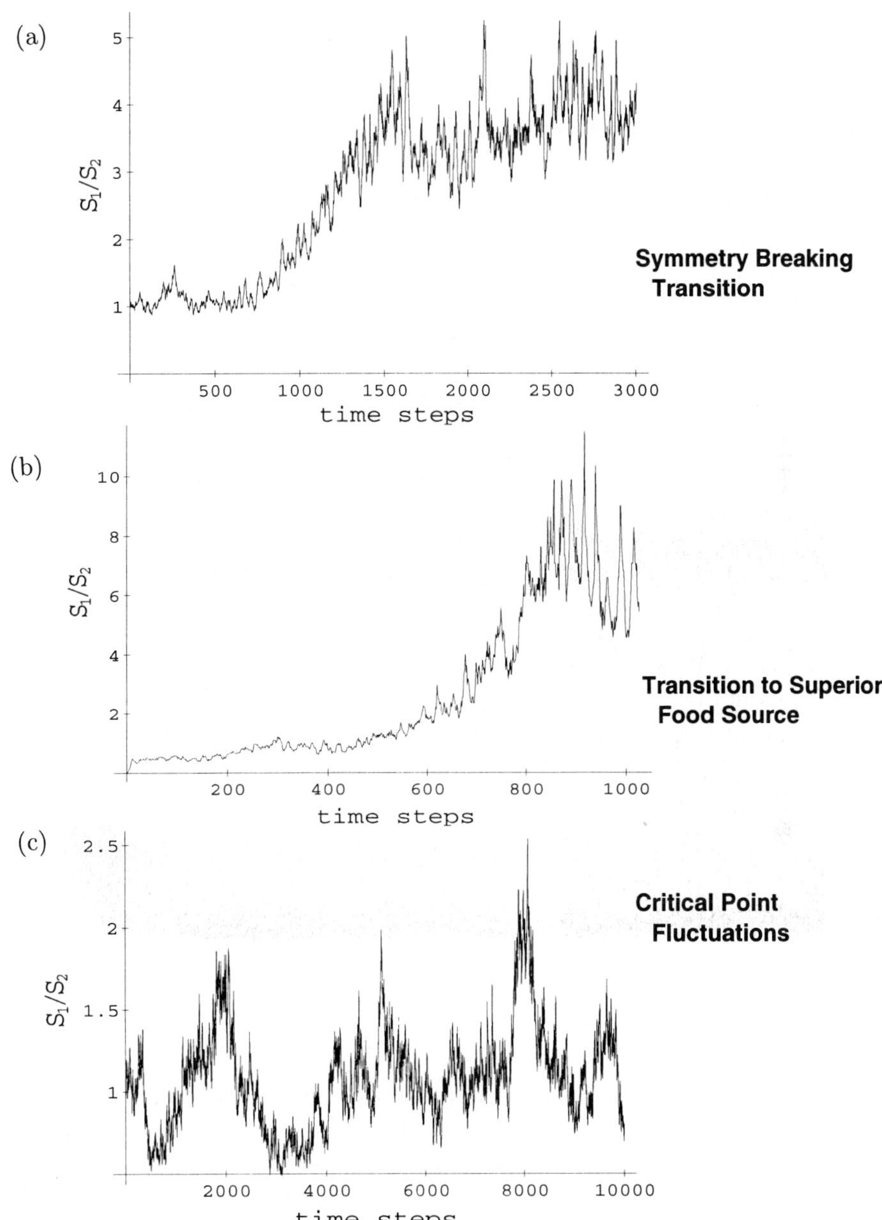

FIGURE 10 Monte Carlo simulations using a probabilistic cellular automata. (a) Symmetry breaking transition. (b) Transition to superior food source. (c) Critical point fluctuations.

In Figure 10(a) is a simulation of the binary double bridge, which shows the evolution of the density ratio S_1/S_2 on the network. In this case $\gamma = 2.5$. This is above the critical point $\gamma_c = 2$, so the ants must eventually break symmetry and attract to one of the segments. In this case the ants attract onto segment 1. S_1/S_2 hovers around the unstable fixed point at $S_1/S_2 = 1$ until a large enough fluctuation comes along and sends S_1/S_2 towards one of the attractive fixed points. Our previous calculations predict that S_1/S_2 should reach a mean value of 4.00 for this value of γ, and it can be seen that S_1/S_2 reaches a plateau at about this value.

In Figure 10(b) we illustrate the ability of ants to switch to a better food source. Initially, all the ants are on segment 2 which leads to the poorer food source. The scent-laying rate on segment 1 is set about five percent greater than on segment 2; γ is chosen to be just below the point where the bifurcation occurs (see Figure 7). In this case the ants initially find themselves in an unstable configuration, and it is only a matter of time before they shift to segment 1 where the better food source is located. In the simulation illustrated in Figure 10(b), S_1/S_2 becomes greater than 1 around $t = 500$, and reaches a new plateau around $t = 900$.

In Figure 10(c) we again illustrate the symmetric double bridge, this time at the critical point, $\gamma = 2$. The simulation was run for a long time in order to illustrate the long-range critical point fluctuations. In most of the simulations, S_1/S_2 reaches its fixed point mean value in 1000 to 2000 time steps. Note that when the system is at a critical point, it can never make a decision. S_1/S_2 shows fluctuations on an extremely long time scale. These long-range correlations are typical of systems at a critical point.

The approach used here is perhaps a little unusual for the study of artificial life. With the availability of computers the usual method of attack is straight simulation. Here, simulations have been performed more or less as illustrations of the theoretical understanding which has been gained. It is hoped that a theoretical attack will not only contribute to understanding in itself, but prove a road map for later more extensive simulations.

6. SUMMARY

In this paper I have tried to show how it is possible to understand the properties of swarms using some tools from statistical physics. Of course the main goal, that of complete understanding of swarms, was left unfinished. However, the more modest goal of setting up a structure that will serve to support such a study, I think, has been achieved. We have taken a tour through the mathematical devices that can be used for this purpose, and have introduced a number of new ways to look at the problem in terms of the vocabulary and physics of phase transitions. The analysis of one type of model gave an indication of the types of things that are possible, as well as ideas for new experiments involving ants. Finally, at the very end, we got a brief glimpse at the complexities that might arise when we eventually seek to go beyond the deterministic approximation, and that include a full understand of the complex effects of fluctuations.[25]

ACKNOWLEDGMENTS

The author benefited greatly from numerous conversations with L. E. Reichl, and from her support, both financial and spiritual. This work was supported by the Welch Foundation under grant no. F-1051. Much of this work was done at the Center for Statistical Mechanics and Complex Systems at the University of Texas at Austin.

REFERENCES

1. Beckers, R., J.-L. Deneubourg, S. Goss, and J. M. Pasteels. "Collective Decision Making Through Food Recruitment." *Insectes Soc.* **37(3)** (1990): 258–267.
2. Buser, M. W., C. Baroni Urbani, and E. Schillenger. "Quantitative Aspects of Recruitment to New Food by a Seed-Harvesting Ant (*Messor capitatus latreille*)." In *From Individual to Collective Behavior in Social Insects*, edited by J. M. Pasteels, and J.-L. Deneubourg. Basel: Birkhäuser Verlag, 1987. Also *Experientia Supplementum* **54** 139–154.
3. Calenbuhr, V., and J.-L. Deneubourg. "A Model for Trail Following in Ants: Individual and Collective Behavior." In *Biological Motion*, edited by W. Alt and G. Hoffmann, 453–469. Berlin: Springer-Verlag, 1990.
4. Calenbuhr, V., and J.-L. Deneubourg. "A Model for Osmotropotatic Orientation (I)." *J. Theor. Biol.* (1991): In press.
5. Calenbuhr, V., L. Chrétien, J.-L. Deneubourg, and C. Detrain. "A Model for Osmotropotatic Orientation (II)." *J. Theor. Biol.* (1991): In press.
6. Collins, D., and D. Jefferson. "Ant Farm: Towards Simulated Evolution." In *Artificial Life II*, edited by C. G. Langton, C. Taylor, J. D. Farmer, and S. Rasmussen. Santa Fe Institute Studies in the Sciences of Complexity, Proc. Vol. X. Reading, MA: Addison-Wesley, 1991.
7. Deneubourg, J.-L., S. Goss, N. Franks, and J. M. Pasteels. "The Blind Leading the Blind: Modeling Chemically Mediated Army Ant Raid Patterns." *J. Insect Behav.* **2(3)** (1989): 719–725.

8. Deneubourg, J.-L., S. Aron, S. Goss, and J. M. Pasteels. "Self-Organizing Exploratory Pattern of the Argentine Ant." *J. Insect Behav.* **3(2)** (1990): 159–168.
9. Farmer, J. D. "A Rosetta Stone for Connectionism." *Physica D* **42** (1990): 153–187.
10. Farmer, J. D., N. H. Packard, and A. S. Perelson. "The Immune System, Adaptation and Machine Learning." *Physica D* **22** (1986): 187–204.
11. Franks, N. R. "Army Ants: A Collective Intelligence." *Am. Sci.* **77(2)** (1989): 138–145.
12. Franks, N. R., N. Gomez, S. Goss, and J.-L. Deneubourg. "The Blind Leading the Blind in Army Ant Raid Patterns: Testing a Model of Self-organization (Hymenoptera: Formicidae)". *J. Insect Behav.* **4(5)** (1991): 583–607.
13. Genakopolos, G., and L. Gray. "When Seeing Farther is Not Seing Better." *The Bulletin of the Santa Fe Institute* **6(2)** (1991): 4–7.
14. Goss, S., R. Beckers, J.-L. Deneubourg, S. Aron, and J. M. Pasteels. "How Trail Laying and Trail Following Can Solve Foraging Problems for Ant Colonies. " In *Behavioral Mechanisms of Food Selection*, edited by Hughes. NATO ASI Series, vol. G20. Berlin-Heidelberg: Springer-Verlag, 1990. **G20**
15. Hangartner, W. "Structure and Variability of the Individual Odour Trail in *Solenopsis germinata* Fabr. (Hymenoptera, Formicidae)." *Z. Vergl. Physiol.* **62** (1969): 111–120.
16. Hebb, D. O. *The Organization of Behavior.* New York: Wiley, 1949.
17. Hölldobler, B., and E. O. Wilson. *The Ants.* Cambridge: Belknap, 1990.
18. Huang, K. *Statistical Mechanics.* New York: Wiley, 1963.
19. Jefferson D., D. Collins, C. Cooper, M. Dyer, M. Flowers, R. Korf, C. Taylor, and H. Wang. "Evolution as a Theme in Artificial Life: The Genesis/Tracker System." *Artificial Life II*, edited by C. G. Langton, C. Talor, J. D. Farmer, and S. Rasmussen. Santa Fe Institute Studies in the Sciences of Complexity, Proc. Vol. X. Reading, MA: Addison-Wesley, 1991.
20. Koza, J. R. "Genetic Evolution and Coevolution of Computer Programs." *Artificial Life II*, edited by C. G. Langton, C. Taylor, J. D. Farmer, and S. Rasmussen. Santa Fe Institute Studies in the Sciences of Complexity, Proc. Vol. X. Reading, MA: Addison-Wesley, 1991.
21. Langton, C. G. "Studying Artificial Life with Cellular Automata." *Physica D* **22** (1986): 120–149.
22. Langton, C. G. "Life at the Edge of Chaos." In *Artificial Life II*, edited by C. G. Langton, C. Taylor, J. D. Farmer, and S. Rasmussen. Santa Fe Institute Studies in the Sciences of Complexity, Proc. Vol. X. Reading, MA: Addison-Wesley, 1991.
23. Leuthold, R. H. "Orientation Mediated by Pheromones in Social Insects." In *Pheromones and Defensive Secretions in Social Insects*, edited by C. Noirot, P. E. Howse, and J. LeMasne. Dijon: University of Dijon Press, 1975.

24. Millonas, M. M. "A Connectionist-Type Model of Self-Organized Foraging and Emergent Behavior in Ant Swarms." *J. Theor. Biol.* **159** (1992): 529.
25. Millonas, M. M. "Nonequilibrium Statistical Field Theory of Swarms" (in preparation).
26. Pasteels, J. M., J.-L. Deneubourg, and S. Goss. "Self-Organzation Mechanisms in Ant Societies (I): Trail Recruitment to Newly Discovered Food Sources." In *From Individual to Collective Behavior in Social Insects*, edited by J. M. Pasteels, and J.-L. Deneubourg. Basel: Birkhäuser Verlag, 1987. Also *Experientia Supplementum* **54** 155–175.
27. Pasteels, J. M., J.-L. Deneubourg, and S. Goss "Transmission and Amplification of Information in a Changing Environment: The Case of Ants." In *Laws of Nature and Human Conduct*, edited by I. Prigogine and M. Sanglier. Brussels: Gordes, 1987.
28. Perelson, A. S. "Immune Network Theory." *Immunol. Rev.* **110** (1989): 5–36.
29. Stephens, D. W., and J. R. Krebs. *Foraging Theory.* Princeton: Princeton University Press, 1986.
30. Turing, A. M. "The Chemical Basis of Morphogenesis." *Phil. Trans. Roy. Soc.* **237** (1951): 37–72.
31. Van Kampen, N. G. *Stochastic Processes in Physics and Chemistry.* North-Holland, 1971.
32. Wilson, E. O. *The Insect Societies.* Cambridge: Harvard University Press, 1971.

Jeffrey O. Kephart
High Integrity Computing Laboratory, IBM Thomas J. Watson Research Center, P.O. Box 704, Yorktown Heights, NY 10598

How Topology Affects Population Dynamics

INTRODUCTION

How does an environment's topology affect the population dynamics of its inhabitants? Lack of attention to this simple, fundamental question has seriously hampered our theoretical understanding of population biology and epidemiology. For example, in this heyday of HIV, we are admonished daily by educators about the dangers of promiscuous activity; yet until recently[1] there were no quantitative theoretical studies of how the spread of a disease depends upon the detailed network of contacts among individuals. As critical as this oversight is for biology, it is even more so for artificial life. We can exert considerably more control over the design of *digital* organisms and their habitats than we can over that of natural or genetically engineered biological organisms, and the range of topologies that we can imagine and experiment with is perhaps greater than what nature and culture have realized. If we are to understand and control the behavior of digital organisms, we absolutely must gain a clearer quantitative understanding of how topology influences population dynamics.

First, I shall illustrate what I mean by topology by reflecting briefly on the variety of topologies that can be found in natural and artificial systems. Then, I shall select two simple ecological systems for quantitative study. In the first system, I study the dynamics of a single species; in the second, I study a particular brand of predator-prey interaction between two species. In both cases, I have been careful to scale the models in such a way that topological effects can be separated unequivocably from all other parameters and effects. I shall conclude with a speculative generalization of my findings.

THE TOPOLOGY OF BIOLOGICAL AND DIGITAL WORLDS

What do I mean by *topology*? Imagine that the world is composed of cells which can be inhabited by "creatures"[1] which can replicate into neighboring cells under specified conditions. (This can be thought of as a rather general type of cellular automaton.) The *topology* of a world is a description of which cells are neighbors of one another. What topologies do we typically find in biological and digital worlds?

One's first impulse might be to classify most biological habitats as being purely spatial—either two or three-dimensional. After all, an influenza virus infecting a human body has a three-dimensional arena in which to live and spread, as do the T cells and B cells which proliferate in response to it. A coconut washed ashore on a treeless desert island can launch a two-dimensional propagating wave of palm trees. However, at other levels of description, one can often discern important physical or virtual topological structures quite different in nature from those which are most immediately apparent. If one focuses on the proliferation of viruses, T cells, and B cells within one individual, the fractal nature of the body's vascular systems might turn out to be more relevant than the three-dimensional space in which they are imbedded. If one focuses instead on the spread of influenza or some other contagious disease among a population of individuals, the relevant substrate for the infectious agent is the network of social interactions, which could have an arbitrary topology. Interestingly, sociologists have been quite active in measuring the topological structure of social networks (e.g., Fararo and Sunshine[2]), but dynamical theories of cultural replicating entities such as ideas,[3] rumors,[4] and other memes rarely incorporate this structure.

Of course, the topology of *digital* habitats can be completely arbitrary, being entirely up to the designer. Included among numerous examples of spatial topologies are cellular automata,[5] Hillis' two-dimensional world of co-evolving parasites,[6] and coupled-map lattices.[7] Fontana's Turing gas[10] is a homogeneous mixture, which as we shall see can be thought of as a fully connected graph. Ray's *Tierra*[11] contains aspects of both one-dimensional and homogeneous worlds. The dynamical properties of Kauffman's Boolean networks have been studied on random graphs[8] and two-dimensional lattices.[9] These are just a few examples of models that probably

[1]or "particles," in the parlance of mathematicians who study "contact processes."

merit adaptation to other topologies, and I hope that one by-product of this paper will be to encourage researchers in artificial life to expand their topological horizons.

Of course, not all digital organisms live in idyllic worlds created just for them. Computer viruses and worms are busy establishing footholds in environments which came into existence long before such creatures were conceived. In the future, we can anticipate that more beneficial digital lifeforms will occupy a variety of ecological niches in the manifestly nonspatial web of electronic links and diskette exchanges that connects the world's computers.[12] Already, database consistency across a large network is being maintained by the action of replicating digital organisms.[13] Large-scale distributed computations are being made possible by processes which spawn themselves into as many processors as is physically or economically feasible.[18] Soon, crude analogs of the mammalian immune system will emerge as countermeasures against computer viruses and worms.

In order to make some sense out of this panoply of topologies, I have selected a few representatives on the basis of their simplicity, generality, variety, and relevance. The simplest topology is, of course, no topology—commonly referred to by population biologists as the "homogeneous mixing" model. It is all-too-commonly taken as gospel in theoretical epidemiology, where it reduces to the assumption that every individual in the world is equally likely to infect or be infected by any other individual. In this work it serves as a useful reference to which other topologies can be compared. In order to investigate the effect of limited connectivity, I have chosen random graphs of low degree. As a simple representative of spatial models, I have chosen a two-dimensional lattice in which each cell is connected only to its eight nearest neighbors. Finally, the hierarchically clustered random graph serves as an example of a local, nonspatial topology that may capture some of the essence of clustering in social and computer networks. I shall have occasion to mention other related topologies in passing.

SINGLE REPLICATING SPECIES

Consider the following very simple model of a single species A propagating in some given environment. Suppose that each cell in the world is either occupied by A or empty. If A occupies a particular cell, it will try to replicate itself into a randomly chosen neighboring cell. The replication attempt succeeds only if the neighbor is not already occupied by A. The average total rate at which an occupied cell attempts replication (summed over all neighboring cells) is the *birth rate* β. Occasionally, at some *death rate* δ, an occupied cell will become empty. Births and deaths associated with various cells are uncorrelated; they occur asynchronously at random times generated according to an exponential distribution. This is well known to epidemiologists as the SIS (susceptible $\rightarrow$ infected $\rightarrow$ susceptible) model.[14] To epidemiologists, A is some infectious agent, and the cells are hosts among which it spreads. Upon being cured, (i.e., after a death), a host immediately becomes susceptible again (i.e., there is no immunity). Of course, we need not think of A as a biological or computer virus; A could be a palm tree or a variety of other biological or digital replicators. Nevertheless, for the sake of consistency I shall refer to cells occupied with A as "infected" cells.

I shall now examine the behavior of this model (which I shall refer to as "Model 1") on several different topologies. Since my specification of the propagation rules does not refer explicitly to the topology, the topological effects can be isolated very cleanly, as will be seen.

HOMOGENEOUS MIXING

In the homogeneous mixing scenario, each cell is a neighbor of every other cell. Such a world can be pictured as a fully connected graph in which the nodes represent cells and the arcs represent connections (neighbor-relationships) between them. This topology and numerous variations on it have been favored by theorists for many decades—mainly for reasons of analyzability rather than realism. In this topology, the question of *which* nodes are infected has no bearing on the overall population dynamics; all that matters is *how many* are infected. Thus, as was stated earlier, topology is virtually eliminated as a consideration.

If the number of nodes is very large, stochastic effects can be ignored, and the problem can be treated deterministically. In this case, the fraction of infected nodes as a function of time, $a(t)$, can be obtained by solving a simple nonlinear differential equation[1]:

$$\frac{da}{dt} = \beta a(1-a) - \delta a,$$

the solution to which is

$$a(t) = \frac{a_0(1-\rho)}{a_0 + (1-\rho-a_0)e^{-(\beta-\delta)t}},$$

where $\rho \equiv \delta/\beta$ and $a_0 \equiv a(t=0)$.

This solution reveals that, when the death rate δ exceeds the birth rate β ($\rho > 1$), the fractional population of A decays exponentially from its initial value. Thus extinction is inevitable. However, when the birth rate exceeds the death rate, the fraction of infected nodes grows exponentially at first and then saturates at the value $1-\rho$. An example of such behavior is illustrated in Figure 1, which depicts $a(t)$ for $\beta = 5.0$ and $\delta = 1.0$, starting from an initial condition $a(0) = 0.0001$. (If the initial fractional population exceeds the equilibrium value, $a(t)$ decays to the equilibrium at an exponential rate.)

Thus there is a sharp "epidemic threshold" such that the population survives if the birth rate exceeds the death rate and is driven to extinction otherwise.[2] The

[2]Of course, if we are fond of A and wish to think of it as a cute furry animal rather than some evil virus, we might want to refer to this boundary as a "fitness" rather than an "epidemic" threshold.

existence of the epidemic threshold was first derived about 60 years ago, and has been perhaps the most powerful paradigm in theoretical epidemiology.[14]

In systems of finite size, probabilistic analysis reveals that, even above the threshold, the population has some chance of becoming extinct, but only if the initial number of infected nodes is small enough to be vulnerable to statistical fluctuations.[1] For example, starting from an initial condition in which one cell is occupied by A, the probability that the population will survive is $1 - \rho$. Having established these basic facts about homogeneous mixing in Model 1, we can now turn to other topologies for comparison.

RANDOM GRAPH

It is difficult to think of real ecosystems that have topologies resembling the fully connected graph described in the previous sub-section. In most realistic systems, a cell's neighbors constitute a very small fraction of the total number of cells. This leads very naturally to the consideration of various types of *random graph* topologies, in which the average number of neighbors is an adjustable parameter.

Here I consider a particular type of random graph which is generated as follows. Suppose that the desired average number of neighbors per node (also referred to as the *degree*) is d. Then, for each node i in the system, the number of neighbors, $d(i)$, is chosen according to a Poisson distribution with mean d. Finally, $d(i)$ other nodes are selectly randomly to be neighbors which are reachable from i.[3]

Exact analysis of the behavior of Model 1 on random graphs appears to be very difficult, if not impossible. In the homogeneous topology, no matter how many nodes are infected with A, each sees exactly the same environment. In other words, each infected node has the same number of uninfected neighbors, and each uninfected node can be infected by the same number of infected neighbors. This permits the full dynamics of a system with N nodes to be expressed in terms of the number of infected nodes—a quantity which can only take on $N + 1$ different values. By contrast, each node in a random graph sees a different environment. An exact analysis would require us to keep track of exactly which nodes are infected and which are not—2^N possible states in all. This is barely feasible (numerically) if N does not exceed 15 or 20. Unfortunately, even if we succeed in analyzing one particular randomly generated graph, our goal is to understand the behavior of Model 1 on the *class* of graphs generated by the algorithm described in the previous paragraph. This requires us to take a weighted average over $2^{N(N-1)}$ graphs, with a weighting function that depends on d and the number of connections in each graph. I have done this calculation for $N = 3$ and $N = 4$, and can imagine that some clever trickery might help push this limit to $N = 5$ or 6. However, the results

[3]Note that the connections between cells are asymmetric, i.e., i can "infect" j does not imply that j can "infect" i. A minor modification of this algorithm would render the connections symmetric, in which case the main qualitative features of the population dynamics turn out to be similar to those presented here.

are only useful in an academic sense, and fail to bring us anywhere close to the interesting large-N limit. Although I have experimented with some approximations, it is difficult to assess their validity. The only way to obtain believable results is to simulate the model.

Along with the deterministic, homogeneous solution discussed in the previous subsection, in Figure 1 I illustrate typical simulation runs on randomly generated graphs of average degree $d = 8.0$ and $d = 2.0$. In order to facilitate comparison with the deterministic solution, which is only valid in the large-N limit, the graphs contained 10,000 nodes each. The birth and death rates were taken to be the same as in the homogeneous case (i.e., the birth rate along each link is β/d). To match the homogeneous initial condition of $a(0) = 0.0001$, the simulations were started with just one of the 10,000 nodes being infected with A. Thus all of the parameters of the population model itself are identical; the three curves are distinguished solely by their topology. Taking a slightly odd point of view, one can consider the homogeneous curve to represent a 10,000-node random graph with $d = 9999$, which fits it neatly into the same framework as the two random-graph simulations.

The similarity in shape between the $d = 8$ and $d = 9999$ curves in Figure 1 is striking. The initial growth is quite similar in form, and the equilibrium fractional population is just slightly less for the $d = 8$ random graph. The only significant difference is a delay in when the population explosion occurs for the $d = 8$ simulation run. This apparent difference is illusory. By running several more simulations on the same graph (or different random graphs with $d = 8$), one can obtain a fairly wide spread in "ignition" times—some occurring before that of the homogeneous curve, and some occurring later. However, beyond fairly minor statistical fluctuations, the basic shape is unaltered. A more exact statistical analysis of the homogeneous topology reveals that the wide variance in ignition times is to be expected for $d = 9999$ as well. Thus the shift between the $d = 9999$ and $d = 8$ curves in Figure 1 is merely a consequence of trying to compare an expected value with a single sample of a distribution with a large variance. An average over 2500 simulations gives a survival probability of 0.77 and a fractional population equilibrium of approximately 0.77 for $d = 8$, as compared to 0.80 for both of these quantities for $d = 9999$. Further simulation shows that, for these values of the birth and death rates, the behavior of Model 1 on random graphs with $d \geq 5$ is substantially similar to that observed for the homogeneous topology in every respect. If the birth rate is decreased to bring the system closer to the epidemic threshold, this lower limit on d is increased.

This recent finding[1] is significant because it establishes the validity of the homogeneous mixing assumption over a much broader range than might have been expected. Thus diseases for which a typical individual has many infectious contacts (which may be true for influenza, for example) are well described by the homogeneous mixing approximation.

However, when the average degree d is sufficiently small, the equilibrium population is suppressed substantially, as illustrated by the $d = 2$ simulation run in Figure 1. The initial growth of the population is noticeably slower. (Unfortunately,

I have not yet established the functional form of this growth; it could be exponential with an exponent of approximately half the homogeneous value.) An average over 2500 simulations gives a survival probability of 0.58 and a fractional population equilibrium of approximately 0.58 for random graphs with $d = 2$. It seems to be more than mere coincidence that, for all three cases, the survival probability is equal to the fractional equilibrium population. This equivalence can be proven in the homogeneous case, but it is surprising that it seems to generalize to random graphs of arbitrary degree.

Thus, sparse connectivity not only slows the spread of A, but limits it in the sense of decreasing the survival probability and reducing the equilibrium population even when A does survive. Sparse connectivity is probably typical of sexually transmitted diseases, at least in many communities. The much more detailed results reported by Kephart and White[1] are perhaps the first to quantify the dangers of promiscuity from a global perspective.

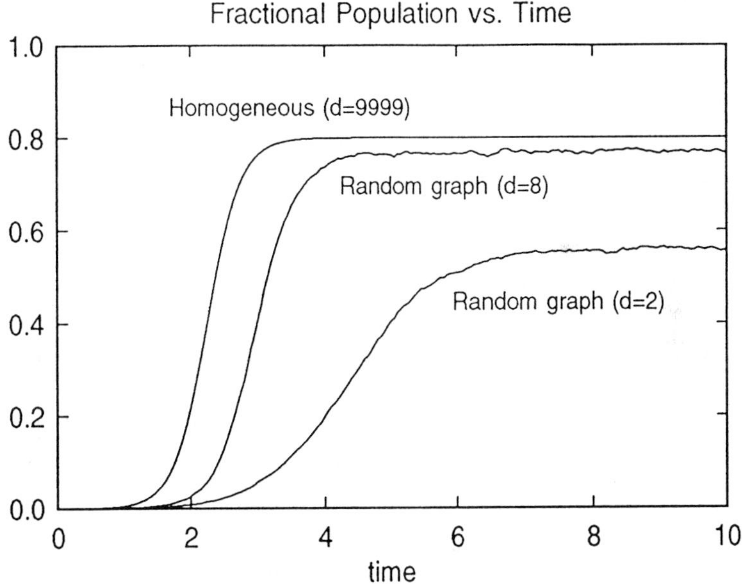

FIGURE 1 The effect of various topologies on the population dynamics of species A: (a) homogeneous mixing model, (b) 10000-node random graph with $d = 8.0$, and (c) 10000-node random graph with $d = 2.0$. The birth and death rates are $\beta = 5.0$ and $\delta = 1.0$ in all cases. The homogeneous mixing curve is given by a differential equation solution, in which the initial fractional population of A is taken to be $a_0 = 0.0001$. The random graph curves were obtained from typical simulation runs in which initially just one cell (out of 10000 total) was infected with A.

LOCAL TOPOLOGIES

Now suppose that the cells are arranged in a two-dimensional square lattice in which each cell is connected only to its eight nearest neighbors. How does this influence the population dynamics of the model?

In Figure 2 I compare a typical simulation run on a 100×100 square lattice to the previously discussed homogeneous behavior. (Note the change in time scales between Figures 1 and 2.) Again, the birth and death rates are 5.0 and 1.0, respectively, and initially just one of the 10,000 nodes is infected. In contrast to the rapid, exponential growth of populations in homogeneous systems, the population growth on the square lattice is much slower, and quadratic in form. Eventually, the population saturates at the same equilibrium as in the homogeneous case. Likewise, the survival probability is essentially the same as for the homogeneous topology, as I have established by observing many simulation runs.

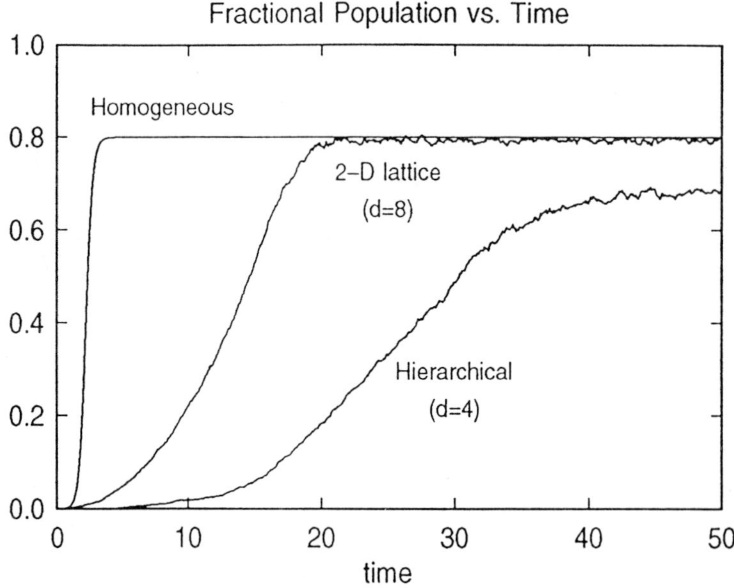

FIGURE 2 The effect of various local topologies on the population dynamics of species A: (a) homogeneous mixing model (for comparison), (b) 100×100 square lattice (wrapped around in both dimensions to form a torus), (c) 6561-node hierarchically clustered random graph with $d = 4$. The birth and death rates are $\beta = 5.0$ and $\delta = 1.0$ in all cases. In both simulation runs, just one cell was infected with A at $t = 0$. Note that the time scale is 5 times that of Figure 1.

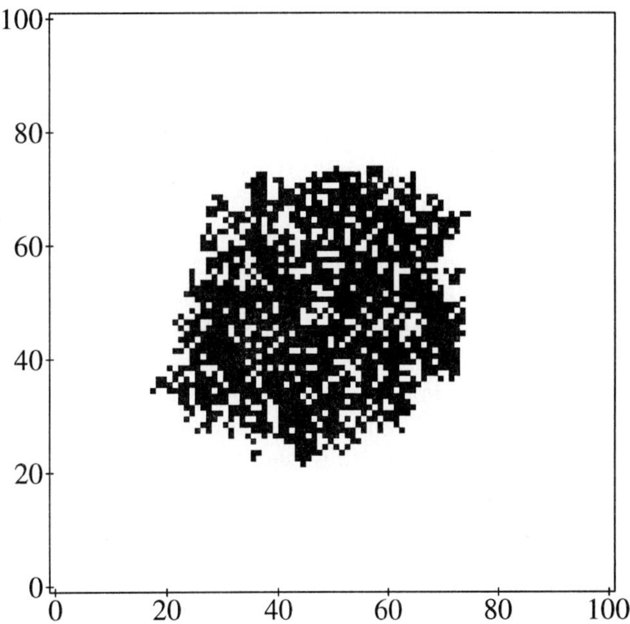

FIGURE 3 State of simulated epidemic in Figure 2(b) at time $t = 10$. Each node can infect the eight neighbors lying within a 3-by-3 square centered on itself. Black and white squares represent infected and uninfected individuals, respectively. The boundary of the expanding infected region is roughly circular (despite the fact that the infection neighborhood is square) and fairly sharp.

In Figure 3 I illustrate why the initial growth of the population is quadratic in this topology. Starting from a single infected cell, A tends to spread outwards in an expanding, roughly circular pattern, with a radius that grows linearly in time. Within this radius, the system is essentially at the homogeneous equilibrium. Thus the fractional population is proportional to the area of the infected region, which grows quadratically. See Kephart and White[1] for a further description of this effect, and Durrett[15] for a mathematical analysis.

It is relatively easy to do a deterministic analysis of the population dynamics of this model in D-dimensional spaces in which the cells are merged into a continuous medium rather than being distinct from one another.[1] The solution $a(\vec{x}, t)$ to the resulting partial differential equation exhibits behavior similar to that found in the discrete case. Thus, in both continuous and discrete d-dimensional spaces, the rate of population growth is polynomial ($\propto t^d$) rather than exponential.

Intuitively, the slower growth rate can be attributed to the fact that A wastes much of its replication effort in areas in which it is already dense, failing to take advantage of the vast unoccupied territory lying beyond the leading edge of growth.

In a sense, this is obvious. It might also seem obvious that this "locality" effect can only hold in spatial topologies. However, further reflection on the underlying reasons for the qualitatively slow population growth found in spatial topologies leads to a generalized concept of local topologies, of which spatial topologies are just a subset.

The basic reason why A keeps trying to infect already-infected cells in spatial topologies is that there is a significant amount of overlap between a node's neighbors and its neighbors' neighbors. Suppose that a single node i in the square lattice is infected with A. This one node has eight infectible neighbors ($d = 8$). Now suppose that i manages to infect all eight of these neighbors. Then there are nine infected nodes, but the degree of overlap between i's neighbors and its neighbors' neighbors is so large that they have among them just 16 infectible nodes (the ones located on the periphery of a 5×5 square centered on i). Thus nine times as many nodes have only twice as many nodes to infect. Now consider a random graph with $d = 8$. Again, node i would be able to infect eight neighbors. However, the next step in the chain of infection is quite different. Since in a random graph there is no correlation between neighbors and neighbor's neighbors, the chances that none of the nine infected nodes have any overlap among them are extremely high. Thus the number of infectible nodes at this point is $9 \times 8 = 72$. In general, the number of infected nodes and the number of infectible nodes remain proportional to one another until the population starts to saturate.

If we loosely define locality as the presence of significant overlap between a typical node's neighborhood and the neighborhoods of its neighbors, we can construct nonspatial local topologies. The hierarchically clustered random graph is a particularly interesting example because it captures some of the essence of both social and computer networks. Consider computer networks, for example. We can imagine that users within one research group exchange software frequently among themselves, are somewhat less likely to do so with other members of their department, and are even less likely to do so with users in other companies, universities, or countries. Thus it is not too far-fetched to suppose that the typical environment inhabited by a computer virus might consist of nested clusters, with some amount of cross-linkage among them. On the basis of the above discussion we would expect this to slow the spread of computer viruses profoundly.

Here is a recipe for generating a hierarchically clustered random graph. Imagine a virtual tree in which the root has b branches, each of which has b branches, and so on for L levels. There are b^L leaves at the bottom level of this virtual tree. These are our nodes. For each node i, generate the list of nodes which can be infected by it as follows: For each node $j \neq i$, let j be infectible by i with probability $p(\ell)$, where ℓ is the number of levels one must go up in the virtual tree before a common ancestor of i and j can be found. Typically, one would choose $p(\ell)$ to be a monotonically decreasing function, normalized such that the expected number of neighbors for each node is the desired degree d. A convenient choice is the geometrically decreasing $p(\ell) = \alpha r^\ell$, where the parameter r can be used to control the amount of localization.

A simulation run on a 6561-node ($b = 3$, $L = 8$) hierarchically clustered graph is compared with the square-lattice simulation and the homogeneous solution in

Figure 2. Although the functional form of the population growth has not yet been determined, it appears to be strongly subexponential, as one would expect. By varying the localization parameter r, one can increase or decrease the growth rate at will. In the limit $r \to 0$, the graph consists of isolated nodes or node clusters; as $r \to 1$, the topology approaches that of the standard random graph. In this particular example, the equilibrium population on the clustered graph was noticeably less than the homogeneous value because it was somewhat sparsely connected (d was only 4).

TWO INTERACTING SPECIES

As a convenient fiction for segueing into the two-species model (Model 2), suppose that we have just learned a grisly secret: A did not die of natural causes! In fact, the death rate δ is entirely attributable to a predator B lurking in the shadows, conducting brutal random killings of unsuspecting As.

Now suppose that, whenever a B kills an A, it occupies the cell that A had occupied. In addition, suppose that a B occupying a particular cell can replicate into neighboring cells that contain A. Denote the average total rate at which it attempts to do this as the predator birth rate β_B. Similarly, there is a predator death rate δ_B at which a cell occupied by B becomes empty. The characteristics of A remain the same as in the previous section; it can only replicate into empty cells. As before, all births and deaths associated with various cells are uncorrelated and asynchronous.

Model 2 shares some features with the mammalian immune system. Some time after infiltration of the body by an antigen A, the immune system manages to create T cells and B cells (represented by B) capable of locating and killing A, and successful destruction of an A stimulates further replication of B. Of course, this is a drastic oversimplification of the rich interplay of biological processes comprising the immune system, many of which are still little understood or unknown. One can imagine an analog of this process in the computer world, in which A represents computer viruses and B represents a hypothetical anti-virus virus. A third alternative is for A to represent rabbits introduced to Australia by English settlers in 1859; B represents the myxomatosis virus deliberately introduced into the Australian rabbit population in 1950 for the purpose of ridding the continent of what had become a troublesome pest.[17] Regardless of its possible applications to various natural or artificial ecosystems, Model 2 will prove to be a convenient device for comparing several topologies to one another.

HOMOGENEOUS MIXING

As was the case for the one-species model, placement of the two-species model just described on a fully connected graph effectively eliminates topology from the problem. Deterministic analysis (valid for sufficiently large systems) leads to the following coupled pair of nonlinear differential equations:

$$\frac{da}{dt} = \beta_A a(1 - a - b) - \delta_A a - \beta_B ab,$$
$$\frac{db}{dt} = \beta_B ab - \delta_B b + \delta_A a.$$

Model 2 can be solved numerically to yield $a(t)$ and $b(t)$, the fractional populations of A and B, respectively.

Analysis of the solution shows that the epidemic threshold for A is unaffected by B's birth and death rates; it remains at $\beta_A = \delta_A$. B has no intrinsic epidemic threshold; it can survive as long as there are As upon which to feed, regardless of the relative values of B's birth and death rates.

Figure 4(a) illustrates the population dynamics of A and B for a particular set of birth and death rates chosen such that A is above the epidemic threshold: $\delta_A = 1.0$, $\beta_A = 5.0$, $\delta_B = 0.1$, and $\beta_B = 0.5$. Thus $\rho_A = \rho_B = 5.0$, but the life cycle of B is ten times slower than that of A. When the initial populations are very small, they increase exponentially at first. At intermediate times, there may be some damped predator-prey oscillations—the existence, frequency, and decay rate of which depend upon the birth and death rates. Eventually, the populations settle to stable values determined by the birth and death rates. The presence of B can strongly suppress the equilibrium population of A. For the set of parameters illustrated in Figure 4(a), the equilibrium fractional population of A would increase by a factor of roughly 16 if B were removed, as can be seen by comparing the curve for A with that in Figure 1.

RANDOM GRAPH

In Figure 4(b) I show the populations of A and B on a 10000-node random graph in which a typical node has eight neighbors. The birth and death rates are the same as in the homogeneous case depicted in Figure 4(a). The dynamics are substantially similar to the deterministic solution for the homogeneous mixing model displayed in Figure 4(a). Note that even the amplitudes of the decaying predator-prey oscillations are quite similar. Furthermore, an average over 100 simulation runs shows that the probability that A and B will survive for a long time (beyond $t = 1000$) is approximately 80%, as in the homogeneous case.

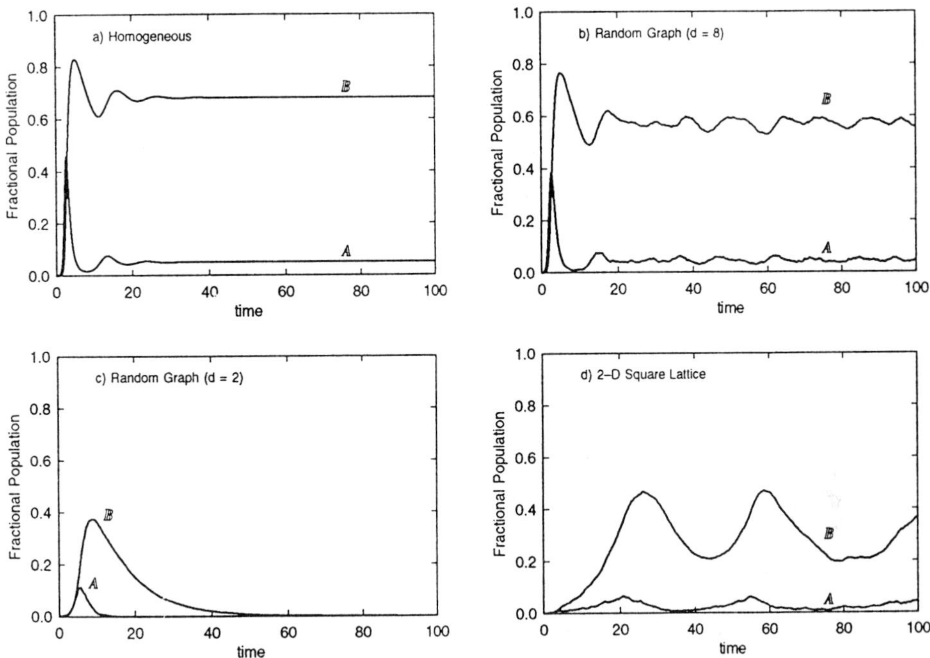

FIGURE 4 The effect of various topologies on the population dynamics of species A and B: (a) homogeneous mixing model, (b) 10000-node random graph with $d = 8.0$, (c) 10000-node random graph with $d = 2.0$, and (d) 100×100 square lattice wrapped around to form a torus. The birth and death rates are $\delta_A = 1.0$, $\beta_A = 5.0$, $\delta_B = 0.1$, and $\beta_B = 0.5$ in all cases. The homogeneous mixing curves are numerical solutions of a coupled pair of differential equations in which the initial fractional populations of A and B were 0.0001 and 0.0, respectively. The two-dimensional square lattice and the random graph curves were obtained from typical simulation runs in which initially just one cell (out of 10000 total) was occupied with A and none with B. (Recall that a B is born whenever an A dies.)

In Figure 4(c), the only parameter which has been changed from those of Figure 4(c) is the average number of neighbors, which has been decreased from 8 to 2.[4] After a short-lived growth spurt, the population of A becomes extinct near $t = 16.4$. Their supply of A having run out, the population of B decays exponentially due to the death rate δ_B, and becomes extinct near $t = 89.1$. In all 200 simulation runs that were conducted under these conditions, A and B never came close to surviving up to the time limit $t = 1000$. Thus it appears that the sparsity of the graph has

[4] Since the average total birth rates summed over all neighbors are held constant, this means that the birth rates along each individual link are quadrupled to compensate for the reduction in the number of neighbors.

plunged the system below the epidemic threshold. More extensive simulation studies show that, for values of d intermediate between 2 and 8, the survival probability and the stable equilibria for surviving populations increase monotonically with d, rising very dramatically in the range $3 < d < 5$.

SPATIAL TOPOLOGY

In Figure 4(d) I show the populations of A and B for a typical simulation run on a 100×100 square lattice (wrapped around in both dimensions to form a torus). Although the birth and death rates are identical to those used in the other three topologies presented in Figure 4, the population dynamics are remarkably different. As might be expected from our experience of the previous section, the initial growth in the populations of A and B is quadratic rather than exponential. However, a completely unexpected difference is also strongly evident—the populations exhibit large undamped oscillations centered about markedly lower equilibria. Simulations indicate that, after many oscillations, the fractional populations of A and B usually settle to values substantially less than half of the homogeneous-mixing equilibria. About 20% of the time, the oscillations persist beyond $t = 1000$.

By viewing a graphical display of the simulation as it is running, it is possible to get some insight into the nature of the oscillations. Initially, A expands in a roughly circular pattern (as in the simple model of the previous section). Soon, B follows the trail blazed by A, resulting in a fairly solid core of Bs surrounded by a ring of As. After a while, the Bs in the core near the site of the cell initially occupied with A start to die, having exhausted their supply of As. This situation is illustrated in Figure 5(a), which shows a snapshot of the system at time $t = 18.1$ for the simulation run of Figure 4(b). The first peaks in the populations of A and B in Figure 4(b) occur roughly when the expanding circle has fully wrapped around and filled the space. A is now hemmed in on all sides by B, which quickly consumes A, leaving just a few small pockets of surviving As. The safest haven for A is near the originally occupied cell, where most of the B have already died. A fringe of the original population explosion wraps around and ekes its way through a phalanx of Bs into this central area, giving birth to a new expanding circle which emerges from it, and a second population cycle begins. In Figure 5(b) I display the situation at $t = 47.4$, when the populations of A and B are headed towards their second peaks. Note the similarity to Figure 5(a)—the confinement of A to the edges of the expanding region and the empty inner core, which will eventually serve as a breeding ground for A and the third population cycle. Note also in Figure 5(b) the shards of B left over from the first population explosion.

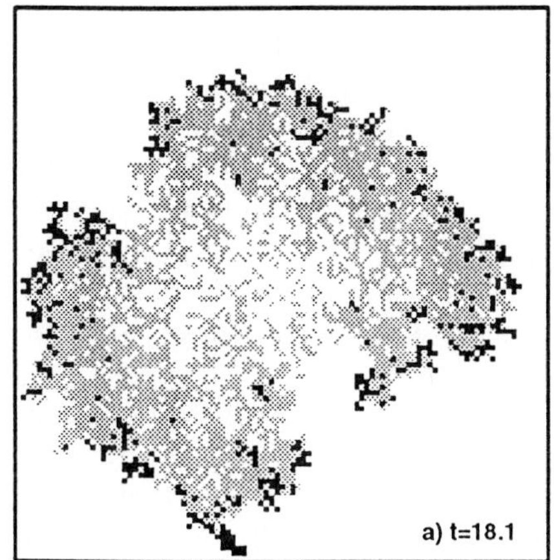

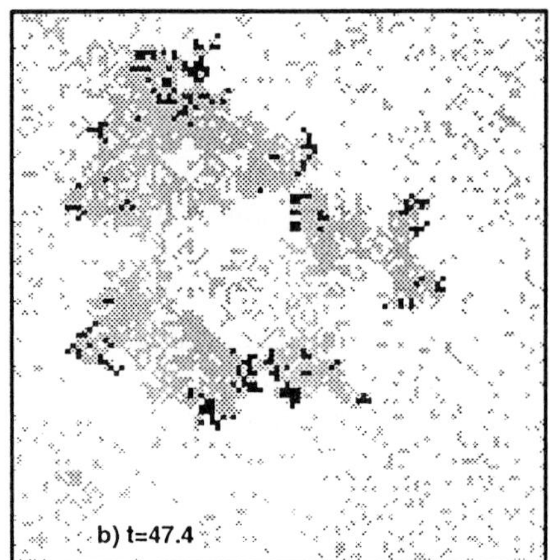

FIGURE 5 Snapshots of spatial distribution of A (black cells) and B (gray cells) at (a) $t = 18.1$ and (b) $t = 47.4$, taken from the simulation run portrayed in Figure 4(d).

CONCLUSIONS

In both population models, the homogeneous mixing (fully connected graph) topology fosters more rapid growth and supports higher levels of population than the spatial and random-graph topologies. This is so because, unlike *any* other topology, it adds no propagative restrictions to the intrinsic rules of the model.

Nearest-neighbor spatial topologies and other local topologies such as the hierarchically clustered random graph *do* introduce additional constraints on propagation. The maximum propagation rate in any direction can not exceed the largest birth rate of any creature in the system (the so-called "speed of light"[16]). By contrast, in a fully connected graph, each cell is just one step away from every other cell. Another way to view the constraints that local topologies place on propagation is that they force replicators to waste much of their effort on futile attempts to replicate in areas that are already densely occupied. We can expect a similar slowing of population growth in *any* topology which possesses a reasonable degree of locality. The slowing effects of local constraints can engender qualitatively different phenomena such as the population oscillations depicted in Figures 4(d) and 5. Essential to the cyclic behavior was the formation of an unoccupied core in the center of Figure 5(a) and 5(b), which would not have existed if cells far from the core had been permitted to replicate into it.

The constraints placed by sparsely connected random graphs upon propagation are of a different nature. Consider a node x with two neighbors y and z, both of which are unoccupied. As soon as x replicates into y, the rate at which it can successfully replicate in the future is immediately halved. In contrast, in a fully connected graph of N nodes, replication of the initially occupied cell into one of its $N-1$ neighbors only reduces the field of eligible targets by the negligible fraction $1/N-1$. Bottlenecks in sparsely connected graphs not only slow population growth; they diminish the equilibrium populations and increase the likelihood that the population will become extinct. Similar effects can be expected in *any* topologies in which each cell has only a few neighbors. In fact, I have observed this phenomenon in the hierarchical topology for the one-species model[1] and in a variant of the two-dimensional square lattice with reduced connectivity (i.e., 4 nearest neighbors instead of 8) for the two-species model.

Many interesting phenomena remain to be discovered through further exploration of other topologies and models. I hope that this paper will help to make potential creators of new artificial worlds more aware of the effects that their choice of topology might have on the behavior that they observe, and that it will encourage researchers in artificial life to adapt some of the very interesting worlds that they have developed to new topologies. In particular, the population models discussed here generalize quite naturally to a class of cellular automata on arbitrary topologies, in which a node's future state is determined (perhaps probabilistically) by its neighbors' states. Finally, I hope to convince biologists, sociologists, and computer scientists of the importance of understanding the topological structure of their respective worlds.

REFERENCES

1. Kephart, J. O., and S. R. White. "Directed-Graph Epidemiological Models of Computer Viruses." *Proceedings of the 1991 IEEE Computer Society Symposium on Research in Security and Privacy,* 343–359. Oakland, California, May 20-22, 1991.
2. Fararo, T. J., and M. H. Sunshine. *A Study of a Biased Friendship Net.* Syracuse, NY: Syracuse University Press, 1964.
3. Goffman, W., and V. A. Newill. "Generalization of Epidemic Theory, An Application to the Transmission of Ideas." *Nature* **204** (1964): 225–228.
4. Pittel, B. "On a Daley-Kendall Model of Random Rumours." *J. Appl. Prob.* **27** (1990): 14–27.
5. Wolfram, S. "Universality and Complexity in Cellular Automata." *Physica D* **10** (1984): 1–35.
6. Hillis, W. D. "Co-Evolving Parasites Improve Simulated Evolution as an Optimization Parameter." *Physica D* **42** (1990): 228–234.
7. Kaneko, K. "Spatiotemporal Chaos in One and Two-Dimensional Coupled Map Lattices." *Physica D* **37** (1989): 60–82.
8. Kauffman, S. A. "Requirements for Evolvability in Complex Systems: Orderly Dynamics and Frozen Components." *Physica D* **42** (1990): 135–152.

9. Lam, P. M. "A Percolation Approach to the Kauffman Model." *J. Stat Phys.* **50** (year): 1263–1269.
10. Fontana, W. "Functional Self-Organization in Complex Systems." In *1990 Lectures in Complex Systems*, edited by L. Nadel and D. Stein. Santa Fe Institute Studies in the Sciences of Complexity, Lect. Vol. III, Redwood City, CA: Addison-Wesley, 1991.
11. Ray, T. "An Approach to the Synthesis of Artificial Life." In *Artificial Life II*, edited by C. Langton. Santa Fe Institute Studies in the Sciences of Complexity, Proc. Vol. X. Redwood City, CA: Addison-Wesley, 1991.
12. Huberman, B. A. ed. *The Ecology of Computation,* New York: North-Holland, 1988.
13. Demers, A., D. Greene, C. Hauser, W. Irish, J. Larson, S. Shenker, H. Sturgis, D. Swinehart, and D. Terry. "Epidemic Algorithms for Replicated Database Maintenance." *Oper. Syst. Rev.* **22** (1988): 8–32.
14. Bailey, N. T. J. *The Mathematical Theory of Infectious Diseases and Its Applications*, 2nd ed. New York: Oxford University Press, 1975.
15. Durrett, R. *Lecture Notes on Particle Systems and Percolation.* Pacific Grove, CA: Wadsworths & Brooks/Cole, 1988.
16. Gardner, M. "The Fantastic Combinations of John Conway's New Solitaire Game 'Life'." *Sci. Amer.* **223(4)** (1970): 120–123.
17. McNeill, W. H. *Plagues and Peoples.* New York: Anchor Books (Doubleday), 1976.
18. Waldspurger, C., T. Hogg, B. A. Huberman, J. O. Kephart, and W. S. Stornetta. "Spawn: A Distributed Computational Economy." *IEEE Transactions on Software Engineering* **18** (1992): 103–117.

Andrew Wuensche
Santa Fe Institute, 1660 Old Pecos Trail, Suite A, Santa Fe, NM 87501, USA; and University of Sussex, School of Cognitive and Computing Science, Brighton BN1 9RH, UK; e-mail: wuensch@santafe.edu, andywu@cogs.susx.ac.uk
Contact address: 48 Esmond Road, London W4 1JQ, phone 081 995 8893, fax 081 742 2178, e-mail: 100020.2727@compuserve.com

The Ghost in the Machine: Basins of Attraction of Random Boolean Networks

This paper examines the basins of attraction of random Boolean networks, a very general class of discrete dynamical systems, in which cellular automata (CA) form a special subclass. A reverse algorithm is presented which directly computes the set of pre-images (if any) of a network's state. Computation is many orders of magnitude faster than exhaustive testing, making the detailed structure of random network basins of attraction readily accessible for the first time. They are portrayed as diagrams that connect up the network's global states according to their transitions. Typically, the topology is branching trees rooted on attractor cycles.

The homogeneous connectivity and rules of CA are necessary for the emergence of coherent space-time structures such as gliders, the basis of CA models of artificial life. On the other hand, random Boolean networks have a vastly greater parameter/basin field configuration space capable of emergent categorization.

I argue that the basin of attraction field constitutes the network's memory; but not simply because separate attractors categorize state space—in addition, within each basin, subcategories of state space are categorized along transient trees far from equilibrium, creating a complex hierarchy of content-addressable memory. This may answer a basic difficulty in explaining memory by attractors in biological networks where transient lengths are probably astronomical.

I describe a single-step learning algorithm for re-assigning pre-images in random Boolean networks. This allows the sculpting of their basin of attraction fields to approach any desired configuration. The process of learning and its side effects are made visible. In the context of many semi-autonomous weakly coupled networks, the basin field/network relationship may provide a fruitful metaphor for the mind/brain.

INTRODUCTION

Recent work in unravelling the global dynamics of discrete dynamical systems, such as cellular automata[31] and, more generally, of random Boolean networks,[32,33,34] allow their basins of attraction to be explicitly portrayed. These are diagrams that connect up the network's global states according to their transitions—typically, the topology is branching trees rooted on attractor cycles. The diagrams are efficiently constructed with a reverse algorithm that directly computes a state's set of pre-images (if any).

Following Hopfield,[8] I argue that *attractors* constitute the networks "content addressable" memory, but not simply because separate regions of state space flow to energy minima—in addition, state space is categorized hierarchically along transient trees far from equilibrium.

In continuous deterministic dynamical systems, all possible time series make up the vector field which is represented by the system's phase portrait, an idea introduced by Poincaré. This is the field of flow imposed on phase space by the systems' dynamical rule. A set of attractors, be they fixed point, limit cycles, or chaotic, attract various regions of phase space in the *basin of attraction field.* Analogous concepts apply to discrete deterministic dynamical systems, such as cellular automata (CA), and the more general case, random Boolean networks, which are noise free and update synchronously. An important difference, however, is that transients can merge onto one successor state far from equilibrium in these discrete systems, whereas in continuous system they cannot.

Neither does Hopfield's model[8] support deterministically merging transients because his updating method is randomly asynchronous and thus nondeterministic. It is open to debate whether synchronous or asynchronous updating in a local network is more or less biologically plausible. However, synchronous random networks have greater potential as content-addressable memory systems because not only attractors categorize state space. Subcategories of state space are also categorized by a reliable time series of unique states along each transient tree far from equilibrium, creating what is effectively a complex hierarchy of content-addressable memory.

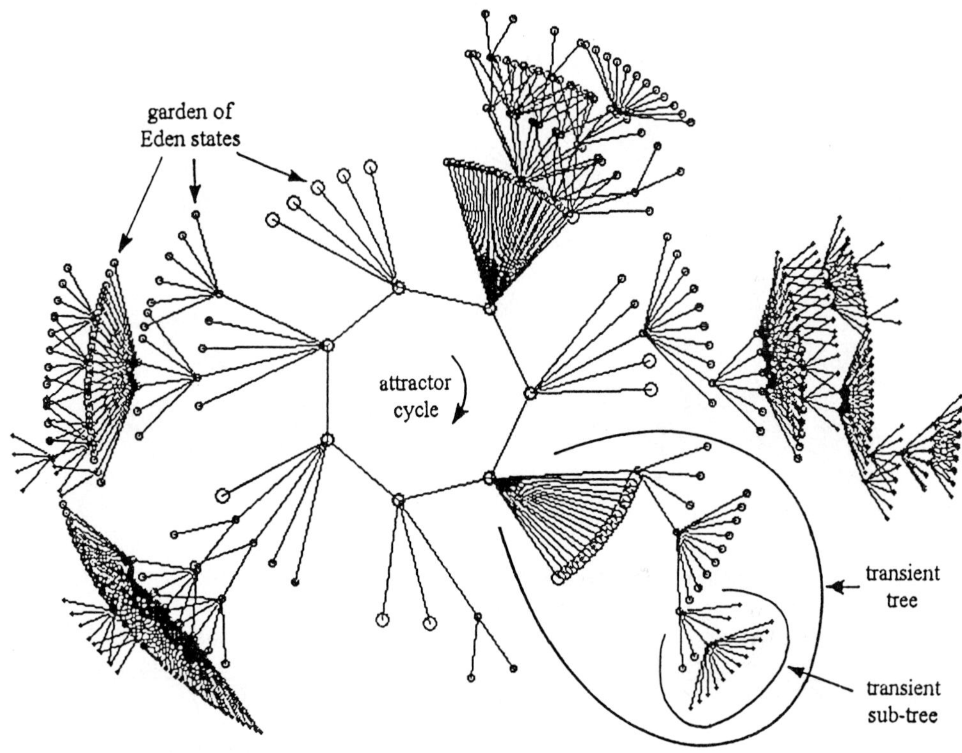

cell	*wiring*	*rule,*	*-table*
1	3,12,6	86,	01010110
2	7,11,4	4,	00000100
3	3,3,1	196,	11000100
4	11,3,9	52,	00110100
5	8,7,5	234,	11101010
6	1,8,1	100,	01100100
7	12,4,13	6,	00000110
8	8,6,8	100	01100100
9	9,2,6	6,	00000110
10	5,1,1	94,	01011110
11	2,7,1	74,	01001010
12	7,8,4	214,	11010110
13	1,4,7	188,	10111100

FIGURE 1 Above: A basin of attraction of a random Boolean network ($N = 13$, $K = 3$). The basin links 604 states, of which 523 are Garden of Eden states. The attractor has period 7. The direction of time is inwards from Garden of Eden states to the attractor, then clockwise. The basin is one of 15, and is indicated in the basin of attraction field in Figure 2. Left: The random Boolean network wiring/rule parameters. Wiring and rules were assigned at random, except that the neighborhood $000 \rightarrow 0$.

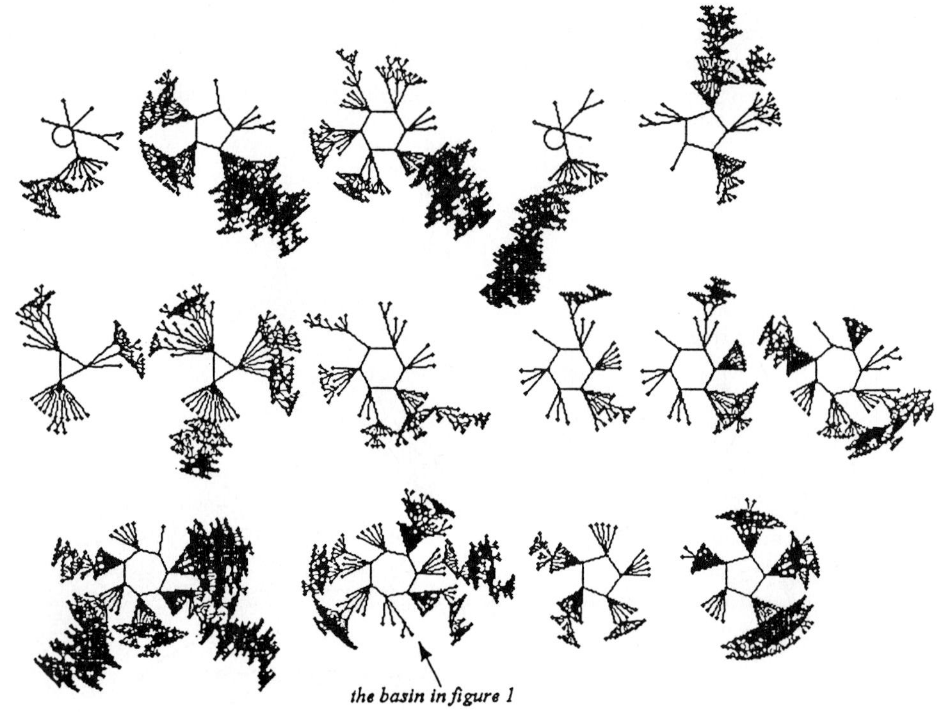

FIGURE 2 The basin of attraction field of a random Boolean network ($N = 13$, $K = 3$). The $2^{13} = 8192$ states in state space are organized into 15 basins, with attractor period ranging from 1 to 7. The number of states in each basin is: 68, 984, 784, 1300, 264, 76, 316, 120, 64, 120, 256, 2724, 604, 84, and 428. Figure 1 shows the arrowed basin in more detail, and the network's wiring/rule scheme.

The range of topologies of basins of attraction, and the potential for emergent complex categorization of network states, suggests that the basin of attraction field, a mathematical object in space time, is the network's cognitive substrate—*the ghost in the machine.*[32] A basic difficulty in explaining memory by attractors in biological networks has been the probably astronomical transient lengths needed to reach an attractor in large networks, whereas reaction times in biology are extremely fast. The answer may lie in the notion of memory far from equilibrium along merging transients.[34]

CA (of whatever dimension) may be regarded as a special random network subclass with homogeneous connectivity and rules. Evidence is presented that this local architecture is necessary for the emergence of coherent space-time structures such as gliders, the basis of CA models of artificial life. Random network architecture breaks these two basic premises: the *wiring/rule* scheme may be arbitrary and different at each cell, though divergence from CA architecture is a question of degree. An arbitrary wiring rule scheme implies a vastly greater parameter space, and thus basin field configuration space, than for CA. Perhaps any basin of attraction field configuration is possible. The process of adaptation and learning modifies the network's parameters, its wiring/rule scheme or size/connectivity, resulting in a modified basin of attraction field. The stability of the field under small perturbations to parameters is noteworthy.

This paper describes CA and random network architecture, and contrasts their dynamics in terms of space-time patterns and basins of attraction. The reverse algorithm for computing pre-images is explained. I suggest that random networks may provide a component for a biological model; in the context of many semi-autonomous weakly coupled networks, the basin field/network relationship may provide a fruitful metaphor for the mind/brain.

I describe learning algorithms that automatically re-assign pre-images in a single step. New attractors can be created and transient trees and subtrees transplanted, *sculpting* the basin of attraction field to approach any desired configuration. The effects and side effects of learning become immediately apparent by redrawing the modified basin of attraction field, or some fragment of it. Such *visible learning* may lead to useful applications as well as helping to clarify the process of memory and learning in a variety of artificial neural network architectures.

BASINS OF ATTRACTION

CA and random networks are both examples of discrete deterministic dynamical systems made up from many simple components acting in parallel. The dynamics is driven by the iteration of a constant global updating procedure (the *transition function*) resulting in a succession of global states, the system's *trajectory*. Given a noise-free, deterministic transition function within an *autonomous* system (cut off from outside influence), any global state imposed on the network will seed a determined trajectory (though it may be unpredictable). In fact the system may be regarded as *semi-autonomous*, in the sense that a global initial state must be imposed or perturbed from outside to set the system going along a new trajectory. The system also needs a channel to communicate its internal state to the outside.

A trajectory is one particular path within a *basin of attraction*, familiar from continuous dynamical systems. In a finite network of size N and value range V, there are V^N global states. Any path must inevitably encounter a repeat. When this occurs the system has entered and is locked into a state cycle (the *attractor*). Many trajectories typically exist leading to the same attractor. The set of all such trajectories, including the attractor itself, make up a basin of attraction. This is composed of *merging* trajectories linked according to their dynamical relationships, and will typically have a topology of branching transient trees rooted on the attractor cycle (though this may be a stable point—an attractor cycle with a period of 1).

Separate basins of attraction typically exist within state space. A transition function will, in a sense, crystallize state space into a set of basins, the *basin of attraction field*—a mathematical object in space time which constitutes the dynamical flow imposed on state space. If represented as a graph, the field is an explicit portrait of the network's entire repertoire of behavior. It includes all possible trajectories.

Basins of attraction are portrayed as computer diagrams in the same graphic format as presented in *The Global Dynamics of Cellular Automata.*[31] Various other names are sometimes used, for example, flow graphs, state transition graphs, and networks of attraction. Global states are represented by nodes, or by the state's binary or decimal expression at the node position. Nodes are linked by directed arcs. Each node will have zero or more incoming arcs from nodes at the previous time step (*pre-images*) but, because the system is deterministic, exactly one outgoing arc (one *out degree*). Nodes with no pre-images have no incoming arcs and represent so-called *Garden of Eden* states. The number of incoming arcs is referred to as the *degree of preimaging* (or *in degree*).

In Figure 1 I show a typical basin of attraction of a random Boolean network (it is part of the basin of attraction field shown in Figure 2). In Figure 4 I show the basin of attraction field of a CA where many symmetries are evident, a major difference between the topologies of the two systems.

In the graphic convention, the length of transition arcs decreases with distance away from the attractor, and the diameter of the attractor cycle asymptotically approaches an upper limit with increasing period. The forward direction of transitions is inward from Garden of Eden states to the attractor, which is the only closed loop in the basin, and then clockwise around the attractor cycle.

Typically, the vast majority of nodes in a basin of attraction lie on transient trees outside the attractor cycle, and the vast majority of these states are Garden of Eden states. A transient tree is the set of all paths from Garden of Eden states leading to a particular state on the attractor cycle. A transient subtree is the set of all paths from Garden of Eden states leading to a state within a transient tree, as indicated in Figure 1.

Computing transient trees or subtrees, and basins of attraction, poses the problem of finding the complete set of pre-images of any global state. The trivial solution, exhaustive testing of the entire state space, rapidly becomes intractable in terms of computer time as the network's size increases beyond modest values. A *reverse algorithm* for one-dimensional CA, that directly computes the pre-images of a global state, with an average computational performance many orders of magnitude faster than exhaustive testing, was recently introduced.[31] In Section 4 I set out a *general* direct reverse algorithm[32] for random Boolean networks (which includes CA of arbitrary dimension), and which may be generalized for random networks with a greater value range.

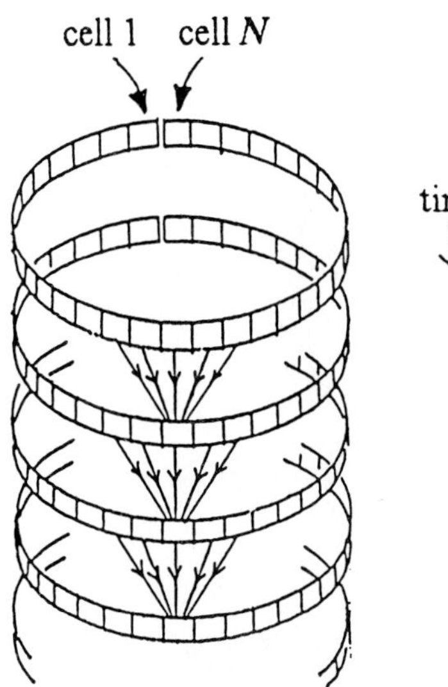

FIGURE 3 One-dimensional finite CA architecture, $K = 5$, each cell has the same wiring template and rule. Boundary conditions are periodic by definition. The network is synchronously updated in discrete time steps.

CELLULAR AUTOMATA

A CA is sometimes described as a discretized artificial universe with its own local physics.[14] Space is a lattice of cells with a particular geometry; each cell contains a variable from a limited range (often just 0 or 1). All cells update synchronously as time advances in discrete steps. The updating rule is the same for all cells, and depends only on local relations, usually a closed symmetrical neighborhood.

Conversely, one could say that the homogeneous neighborhood template defines a given space and, if the CA is finite, implies periodic boundary conditions (i.e., a circle of cells for one-dimensional architecture, a toroidal surface for two-dimensional). Finite one-dimensional CA architecture is illustrated in Figure 3 where cells are arranged in a circle. Time steps are shown in sequence from the top down.

The neighborhood template (the *wiring* scheme) extends to the previous time step only; there are no connections to the more distant past.

Consider a periodic one-dimensional lattice with N cells and an odd number of connections, K. $[K/2]$ is the integer neighborhood radius. The time evolution of the ith cell is given by

$$C_i^{(t+1)} = f\left(C_{i-[K/2]}^{(t)}, \ldots C_{i-1}^{(t)}, C_i^{(t)}, C_{i+1}^{(t)}, \ldots C_{i+[K/2]}^{(t)}\right)$$

to satisfy periodic boundary conditions, for $x < 1$, $C_x = C_{N+x}$; for $x > N$, $C_x = C_{x-N}$.

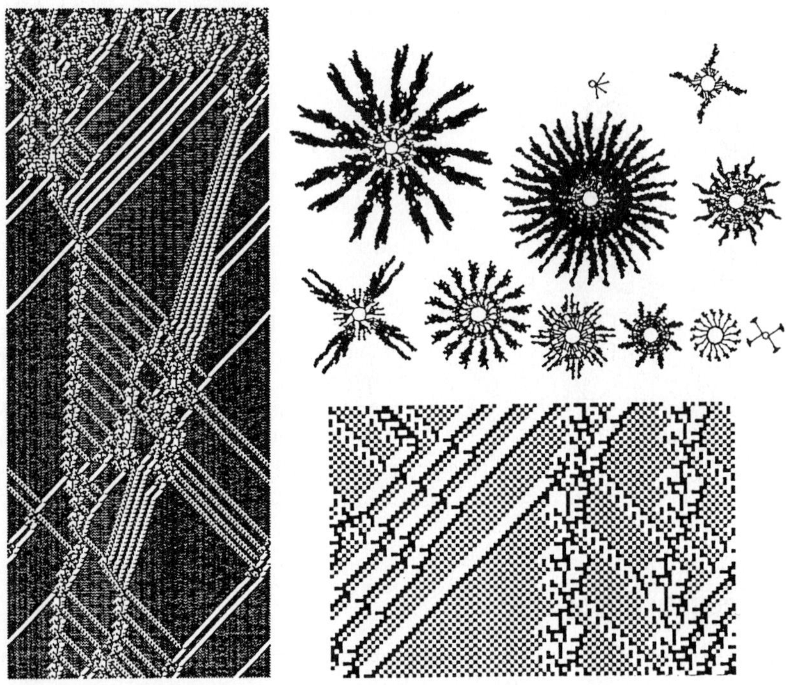

FIGURE 4 An example of one-dimensional, $K = 5$ complex CA dynamics; the rule number is 906663673. Left: the space-time pattern from a random initial state, $N = 200$, 480 time steps; gliders (and a glider gun) emerge against a checkerboard background after an initial sorting out phase. Top right: The basin of attraction field for the same rule, $N = 16$. The 2^{16} states in state space are organized into 17 basins of attraction; only the 11 nonequivalent basins are shown. The number of states in each basin, and the number of each type, is as follows (from left to right), 30928(1), 6(1), 220(4), 23808(1), 1136(2), 1972(2), 2064(1), 568(1), 448(2), 144(1), and 26(1). Note the basin symmetry due to equivalent transient trees. Bottom right: A detail of typical glider interactions.

If the value range $= V$, there are V^K permutations of values in a neighborhood of size K. The most general expression of the function f is a look-up table (the *rule table*) with V^k entries, giving V^{V^K} possible rules. There are also subcategories of rules that can be expressed as simple algorithms, Boolean derivatives,[22] totalistic rules, or threshold functions. The number of effectively different rules is reduced by symmetries in the rule table.[24,31] By convention[29] the rule table is arranged in descending order of the values of neighborhoods, and the resulting bit string converts to the decimal rule number. For example, the rule table for rule 30 ($V = 2, K = 3$) is

	111	110	101	100	011	010	001	000	...neighborhoods
rule table	0	0	0	1	1	1	1	0	...outputs (0 or 1)

Some CA support coherent space-time structures. Periodic self-sustaining configurations, known variously as particles, solitary waves, and gliders, may emerge and propagate across the lattice, as in Conway's two-dimensional *game of life*.[3] For simplicity I will call all such configurations *gliders* though their velocity may vary between zero and the system's "speed of light," equal to the neighborhood radius per time step.

The emergence of gliders characterizes complex rules, which are said to occur at a *phase transition* in rule space balanced between simple and chaotic behavior,[14] the so-called *edge of chaos*. From an initial random state, a limited number of glider types emerge after an initial sorting out phase, and continue to interact for an extended period of time. Collisions between gliders may produce new gliders that collide in their turn. So-called *glider guns* may eject other gliders at periodic intervals. These complex interactions can encode logical operations supporting universal computation.[3,14,30] von Neuman's original self-reproducing automaton[19] and subsequent examples of glider reproduction have lead to the notion of artificial life.[13,15]

Examples of such *edge-of-chaos* dynamics can be found in the simplest CA, for example, the much-studied $K = 3$, one-dimensional rule 110.[17] An example of a complex $K = 5$, one-dimensional CA rule (from amongst many others[35]) is shown in Figure 4. One-dimensional CA dynamics is conveniently represented as a space-time pattern diagram. The cylinder in Figure 3 is split between cells 1 and N and flattened out. A time series of global states is represented as in Figure 4, with space across (rows of black and white squares) and time running from the top down.

Previous work on the structure and topology of CA basins of attraction[31] have shown that there are a number of general constraints to CA dynamics, which are reflected in their basin of attraction fields. They do not necessarily apply to random networks. The constraints relate to various symmetries and hierarchies within state space and rule space, summarized below.

1. Rotation symmetry (the number of repeating segments in the bit pattern) is conserved. In a transient, rotation symmetry cannot decrease over time; in an attractor cycle, rotation symmetry must remain constant. In symmetric rules the same principles apply to bilateral symmetry.
2. Rotation equivalent states (that differ only by any rotation of the circular lattice) are embedded in equivalent behavior. This results in basins of attraction with identical topology, but rotated states. Symmetries within basins occur if a sequence of rotation equivalent states repeat in the attractor cycle; transient trees with identical topology, but rotated states, will be rooted on the repeats.
3. Rule space can be divided into symmetry categories by transformations within rule tables.

4. Equivalence classes and rule cluster relationships exist in rule space due to transformations between rule tables.

A parameter, Z, has been developed[31] to capture correlations between a CA's rule table and its behavior/basin field topology. The Z parameter is measured directly from the rule table and relates to Langton's λ parameter[13,14] and the equivalent concept of *internal homogeneity* introduced earlier by Walker.[23] Whereas λ simply counts the fraction of 1's in a binary rule table, Z takes into account the allocation of rule table values to subcategories of related neighborhoods. Behavior predictions on the basis of Z avoid the exceptions characteristic of λ. The Z parameter predicts various interrelated aspects of basin field topology such as the degree of pre-imaging (the convergence of state space) and the density of Garden of Eden states in state space. These, in turn, relate to attractor cycle and transient length, and the number and size of separate basins. The table below shows varying aspects of behavior of increasingly large CA arrays as the Z parameter changes from 0 to 1 (Wolfram's behavior classes[29,30] are shown in brackets).

	simple (class 1 & 2) →	*phase transition* complex (class 4)	→ chaotic (class 3)
Garden of Eden density	converges to 1	balanced	converges to 0
transient, attractor length	very short	moderately long	very long
Z parameter (0–1)	$\approx 0 - 0.6$	$\approx 0.6 - 0.8$	$\approx 0.8 - 1$

The variation of Garden of Eden density against the Z parameter for one-dimensional CA exhibits a strong inverse correlation.[35] Complex rules are rare, but are most likely to occur in a distinct but broad band of Z. Related correlations may exist for random Boolean networks.

RANDOM BOOLEAN NETWORKS

The idea of randomly connected multi-function networks as dynamical systems with a corresponding field containing all lines of behavior can be traced back to Ross Ashby, in his book *Design for a Brain.*[2] Random Boolean networks (sometimes called Kauffman's model) have been investigated for a considerable time by Stuart Kauffman in theoretical biology and complex systems,[9,10,11] specifically to model gene regulation underlying embryonic development. Others have studied variations of random Boolean networks in the context of memory in the immune network, for instance,[28] and in complex systems in general.[16,23,24,25,26] The studies have built statistical data on network dynamics from many separate *forward* simulations. The focus of interest has often been to gain an insight into the topology of basin of attraction fields in relation to a range of possible system parameters.

Random Boolean network architectures are in many ways similar to weightless (or logical) neural networks,[1] where standard memories (RAMs) hold each cell's look-up table. Classical neural network architectures use weighted connection and threshold functions. A random Boolean network may be regarded as a discrete generalization of a sparsely connected classical neural network. Connections with higher weights may simply be replaced by multiple couplings, and the threshold function applied. However, a threshold function is a tiny subclass of the V^{V^K} possible CA rules.

Random Boolean networks may be viewed as generalized (*disordered*) CA,[32] breaking two basic premises of CA architecture by allowing arbitrary wiring and/or rules at each cell. The effect on behavior of deviating from either or both of these premises by degrees will be discussed below (see Figures 8, 9, and 10). Not surprisingly, coherent space-time patterns and emergent complex structures such as *gliders*, characteristic of CA, are progressively degraded. A relatively small number of basins with low period attractors typically emerge.

Random Boolean networks have a vastly greater parameter space, and thus behavior space, than CA. The various symmetries and hierarchies that constrain CA dynamics previously described need no longer apply. Consequently it might be conjectured that any arbitrary basin of attraction field configuration is possible given the right set of parameters. There is no limit to the speed of propagation across the network. The notion of space and the "speed of light" lose significance as the homogeneous wiring/rule scheme of CA is progressively scrambled, though this can occur by degrees.

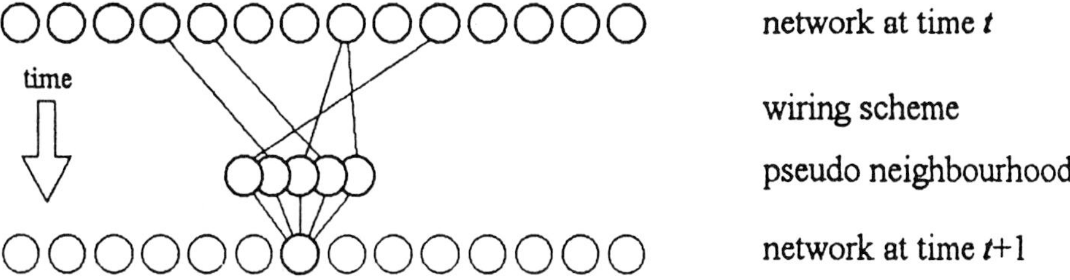

FIGURE 5 Random network architecture. Each cell in the network synchronously updates its value according to the values in a pseudo-neighborhood, set by single wire couplings to arbitrarily located cells at the previous time step. Each cell may have a different wiring/rule scheme. The system is iterated.

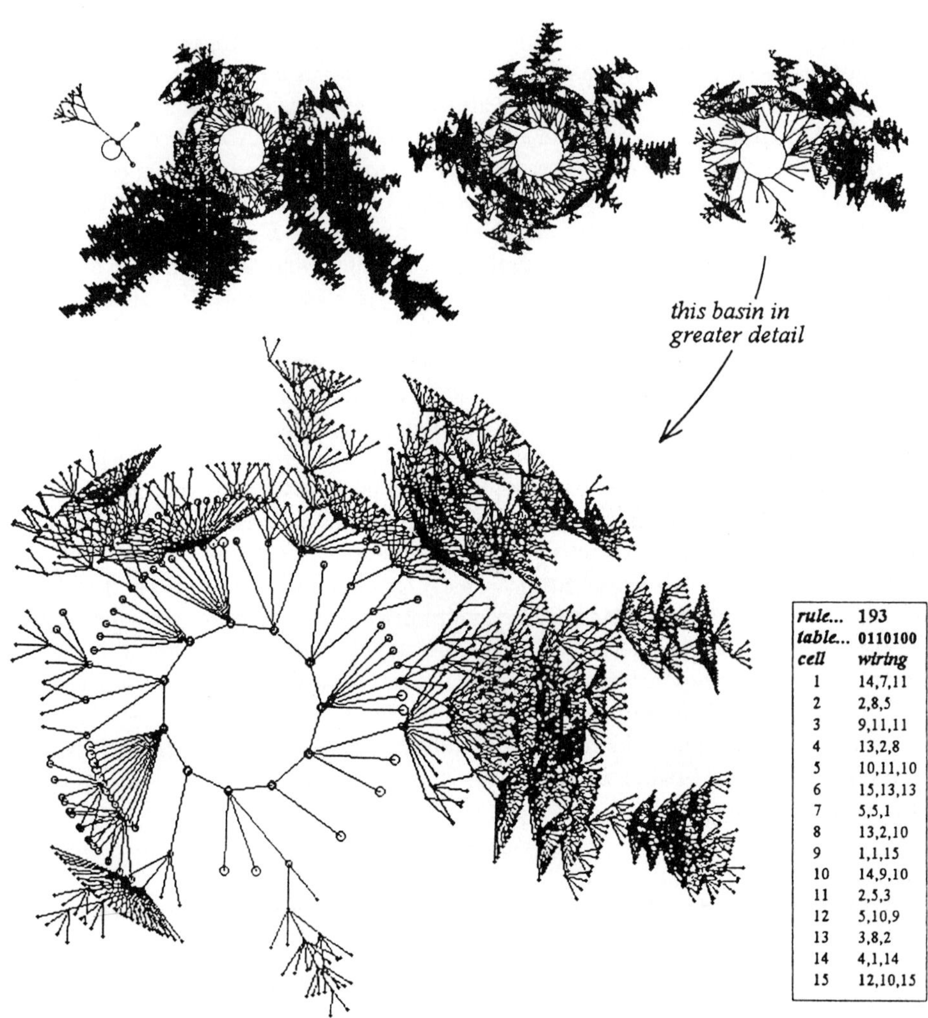

rule...	193
table...	0110100
cell	wiring
1	14,7,11
2	2,8,5
3	9,11,11
4	13,2,8
5	10,11,10
6	15,13,13
7	5,5,1
8	13,2,10
9	1,1,15
10	14,9,10
11	2,5,3
12	5,10,9
13	3,8,2
14	4,1,14
15	12,10,15

FIGURE 6 Top Left: The basin of attraction field of a randomly wired, single-rule network (non-local CA). $N = 15, K = 3$, rule 193. The $2^{15} = 32768$ states in state space are organized into four basins of attraction. The total number of states in each basin is as follows (with attractor period in brackets): 24 (1), 26926 (22), 3498 (17), and 2320 (11). The field has 27,057 Garden of Eden states; Garden of Eden density = 0.823. Bottom left: The last basin shown in greater detail (Garden of Eden density = 0.823). Bottom right: The pseudo-neighborhood wiring scheme.

A random Boolean network implies a value range of 2 (0 or 1), but in principle the arguments in this paper could equally apply to a network where cells have more than two values. As in CA, the global state of a network of N cells is the pattern resulting from values assigned to each cell, from a finite range of values V (usually $V = 2$). Each cell synchronously updates its value in discrete time steps. The value of a cell C_i at $t+1$ depends on its particular CA rule, f_i, applied to a notional or pseudo-neighborhood, size K. Values in the neighborhood are set according to single wire couplings to arbitrarily located cells in the network at time t. The system is iterated. The system's parameters are set by specifying the pseudo-neighborhood wiring and CA rule for each of the N cells. Each cell may have a different *wiring/rule* scheme (but not necessarily). Once set, the network's wiring/rule scheme is fixed over time.

The ith cell C_i has its neighborhood wiring connections chosen as $w_{i1}, w_{i2}, \ldots,$ w_{iK}. Connections are assigned to any of the N cells in the network, including C_i itself. Duplicate connections are allowed, giving N^K possible alternative wiring options. The ith cell's rule f_i is chosen from V^{V^K} alternatives in rule space. The time evolution of the ith cell is given by

$$C_i^{(t+1)} = f_i\left(C_{w_{i1}}^{(t)}, C_{w_{i2}}^{(t)}, \ldots, C_{w_{iK}}^{(t)}\right).$$

The number of alternative wiring/rule schemes that can be assigned to a given network turns out to be vast even for small networks, and is given by

$$S = (N^K)^N \times (V^{V^K})^N,$$

for example, a network where $V = 2$, $N = 16$, $K = 5$; $\quad S = 2^{832}$.

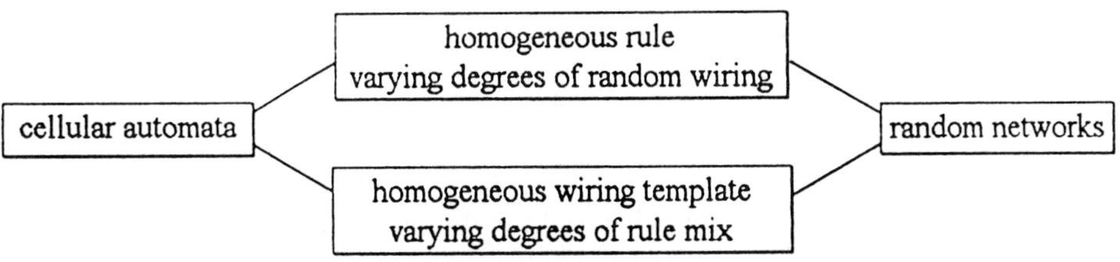

FIGURE 7 Intermediate architectures between CA and fully random networks.

INTERMEDIATE ARCHITECTURE

A spectrum of intermediate architectures is possible between CA and fully random networks as indicated in Figure 7. The two main categories are a homogeneous rule but random wiring, or homogeneous wiring but a random rule mix. Within each category there are may possible variations.

HOMOGENEOUS RULE—RANDOM WIRING

A variety of constraints can be imposed on random wiring; for instance, in Hopfield's model[8] only symmetric wiring is allowed. Connections to a cell may be constrained to be distinct. Duplicate connections would correspond to unequal weights on single input lines.[10]

An example of the space-time patterns of a network with a homogeneous rule but fully scrambled wiring is given in the lower half of Figure 8(a). Space-time patterns appear random until the onset of periodic behavior at the attractor. This is not surprising because totally random wiring destroys the continuity of space—geometrically the network becomes N-dimensional. However, the pattern density (the density of 1's) tends to settle and fluctuate at a characteristic mean-field level, though bi-stability is also possible as in Figure 9(a). The mean-field pattern density probably relates to the rule's internal homogeneity or λ parameter.

Fluctuations are sometimes large with extended periods. The trace of the pattern density (as in Figure 8(b)) is somewhat reminiscent of the EEG (electroencephalogram) measure of the mean excitatory state of a patch of neurons in the brain.

Networks with partially and totally random wiring (non-local CA) have been studied by Li.[16] He found that complex rules may result in a form of edge-of-chaos dynamics in these systems, by the emergence of cooperations among a cluster of components, which he calls *coherent clusters.* He also found evidence that the magnitude of density fluctuations becomes smaller for larger systems.

Walker[23,24,25,26] has studied basins of attraction by statistical methods for a $K = 3$ neighborhood where only the two outer wires were randomized (each cell is connected to itself). The system retains some notion of space. A spectrum of constraints on random wiring could be imposed in a similar way to retain a degree of spacial information in a network of given dimension and geometry. A varying proportion of wires from the neighborhood could be set at random. Random wiring could be confined to a local periodic zone of a certain radius; this seems to produce sub-attractors as in Figure 9(b). Connecting each cell to itself could be mandatory, or excluded. Kauffman's random Boolean network models[11] generally do not constrain random wiring, except that the number of connections per cell, K, is constant, and often set at $K = 2$.

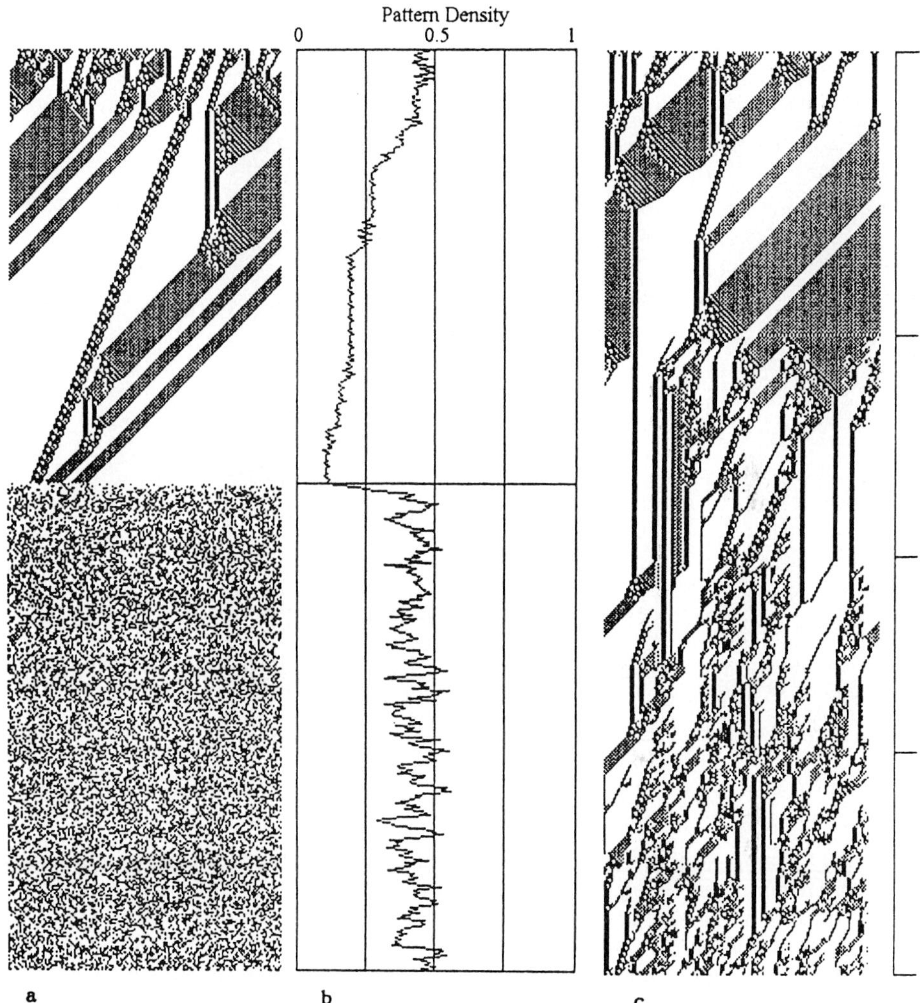

FIGURE 8 Single rule, local and random wiring (non-local CA), $N = 150, K = 5$, rule 3162662612. (a) The space-time pattern of a complex rule with local wiring from a random initial state. After about 240 time steps, the wiring scheme has been totally randomized. (b) The pattern density (density of 1's) in successive global states analogous to an EEG. (c) The space-time pattern of the same rule from another random initial state. At the times indicate, 2% of the wires (15 out of 750) are cumulatively randomized.

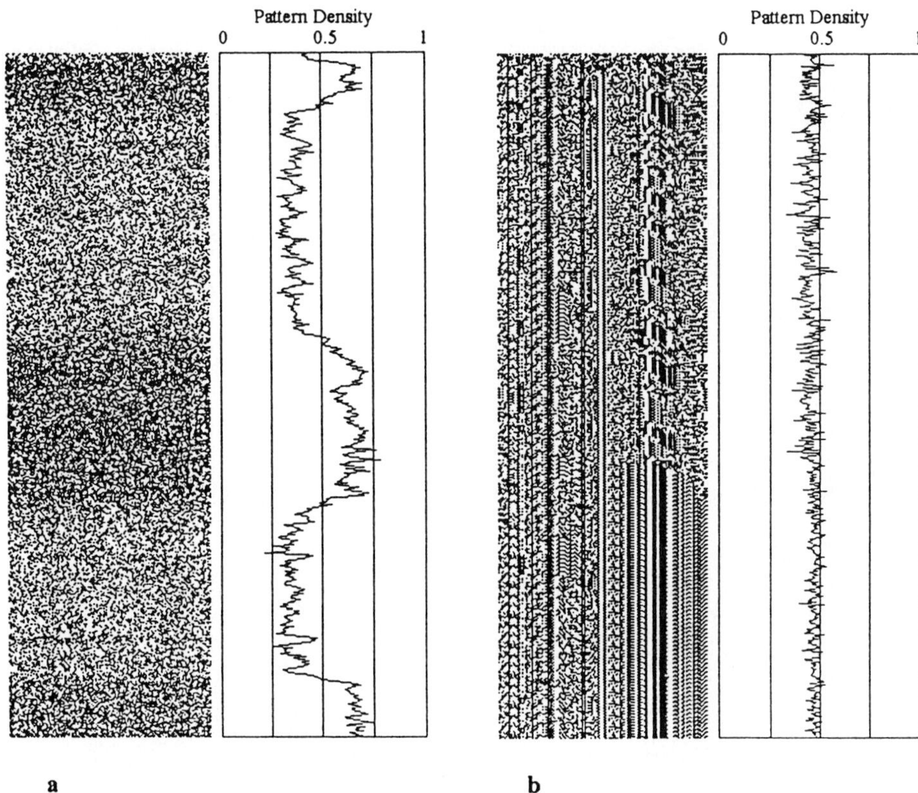

FIGURE 9 Examples of the space-time patterns and pattern density plots in non-local CA (single rule/random wiring), $N = 150$. (a) $K = 5$ rule 2129193089 with random wiring. This is a threshold rule set at 0.5 (majority rule) but with the end bits flipped, i.e., 11111 $\rightarrow$ 0 and 00000 $\rightarrow$ 1. Note the bi-stable pattern density. (b) $K = 3$ rule 193 with random wiring constrained to a ten cell wide local periodic zone. Sub-attractors emerge.

To illustrate how coherent space-time structures are degraded by randomizing *local* CA wiring, in Figure 8(a) I show the space-time pattern of a complex CA rule, $K = 5$, $N = 150$. Gliders emerge from a random seed. After approximately 240 time steps, the wiring is totally randomized (but not the rule). All coherent space-time structure appear to be destroyed. In Figure 8(b) I show the pattern density against time; irregular periodic fluctuations are apparent.

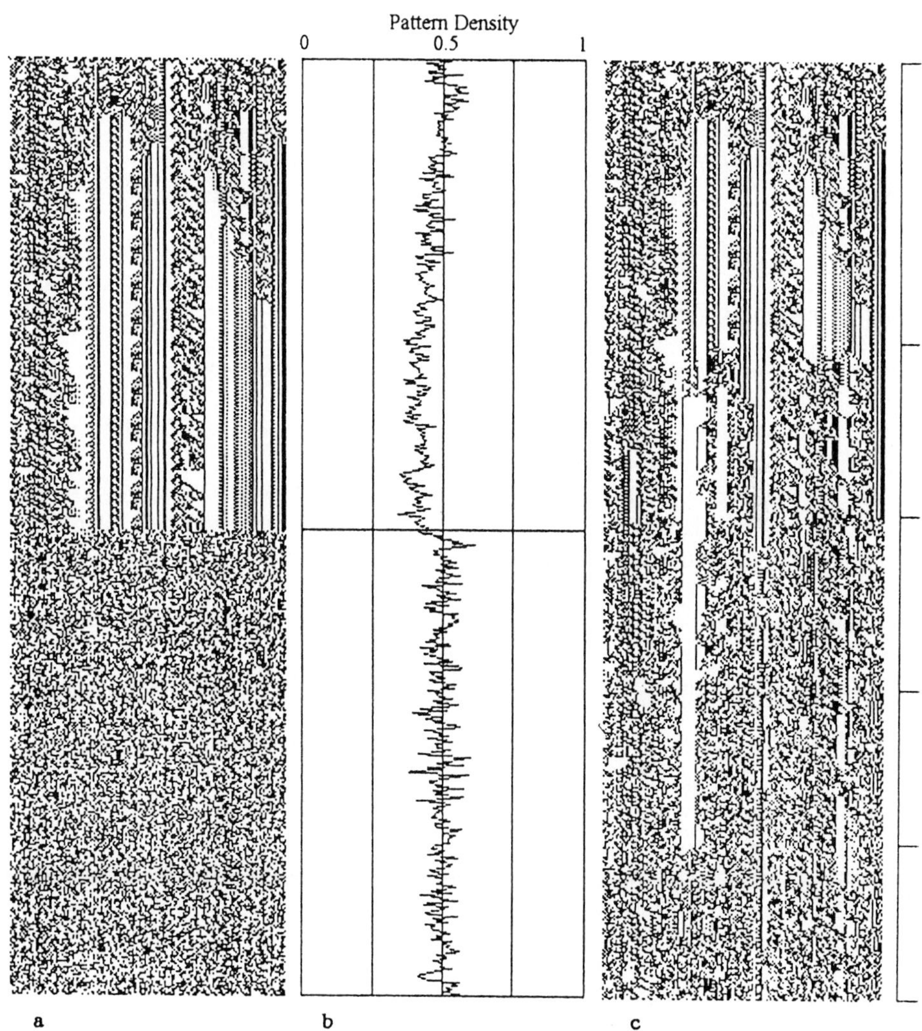

FIGURE 10 $K = 5$ mixed rule network with local and random wiring, $N = 150$. (a) The space-time pattern with local wiring from a random initial state; sub-attractors (vertical features) emerge. After about 240 time steps, the wiring scheme is totally randomized. The space-time pattern becomes chaotic but some vertical features persist. (b) The pattern density (density of 1's) in successive global states analogous to an EEG. (c) The space-time pattern of the same rule from the same random initial state. At the times indicated, 4% of the wires (30 out of 750) are cumulatively randomized.

Randomizing the wiring of the CA in stages will progressively transform a structured space-time pattern to a seemingly random pattern. In Figure 8(c) I show the space-time pattern of the same complex CA from another initial random state. At the intervals indicted, 2% of all available wires (randomly selected) are cumulatively randomized (15 wires out of 750). Coherent structure is progressively eroded. Eventually, the space-time pattern will look like the lower half of Figure 8(a).

HOMOGENEOUS WIRING—RANDOM RULE MIX

A network's rule mix may be assigned at random, or the choice may be restricted to any combination of subcategories of rules from rule space, for instance, rules with a particular setting of the Z parameter, only *additive* rules,[29] only threshold function, or only *canalizing* functions.[10]

In homogeneous wiring/mixed rule networks, periodic structures confined within vertical bands will rapidly emerge within the CA space (frozen islands and isolated islands of variable elements[10]). A random rule mix tends to compartmentalize space, depending on the degree of rule heterogeneity; the vertical features are local sub-attractors. In random networks (with both random wiring and mixed rules) local vertical features rapidly emerge for $K = 2$ rules due to frozen islands, but become less dominant as K increases. At $K = 5$, space-time patterns appear chaotic, but with some residual vertical features (see Figure 10(a)). The mean-field pattern density probably relates to the *mean* internal homogeneity or λ parameter of all the rules in the network, corrected to allow for the proportion of output couplings from each cell.

Figure 10(a) shows the space-time pattern of a homogeneous wiring/mixed rule network. Frozen islands and local sub-attractors emerge from a random initial state. After approximately 240 time steps, the wiring has been totally randomized (but not the rule scheme), resulting in the loss of all coherent structure, though some vertical features are evident. Figure 10(b) shows the pattern density. Figure 10(c) shows the space-time pattern of the same homogeneous wiring/mixed rule network from the same initial random sate. At the intervals indicated, 4% of all available wires (randomly selected) are cumulatively randomized (30 wires out of 750). Frozen island structure is progressively eroded. Eventually, the space-time pattern will look like the lower half of Figure 10(a).

COMPARATIVE GLOBAL DYNAMICS, CA $\leftrightarrow$ RANDOM NETWORKS

Although there is much work to be done given the enormous behavior space open to investigation, a few general observations can be made on the basin of attraction field topology of random Boolean networks, and intermediate architectures, on the basis of many computer runs to reconstruct basins of attraction to date.

The various symmetries and hierarchies that dominate CA basin field topology,[31] as in Figure 4, are absent in random Boolean networks; though some symmetries are still evident in small randomly wired networks with a single homogeneous rule or with a limited rule mix. This lack of constraint on the basin of attraction field topology, and their enormous parameter space, suggest that random networks (or a network of networks) are an ideal vehicle for emergent *categorization* and might be described as having the potential for *brain-like* behavior.

On the other hand, CA are discrete approximations of physical systems,[15] characterized by local connections. Just as the physical world has the potential for the

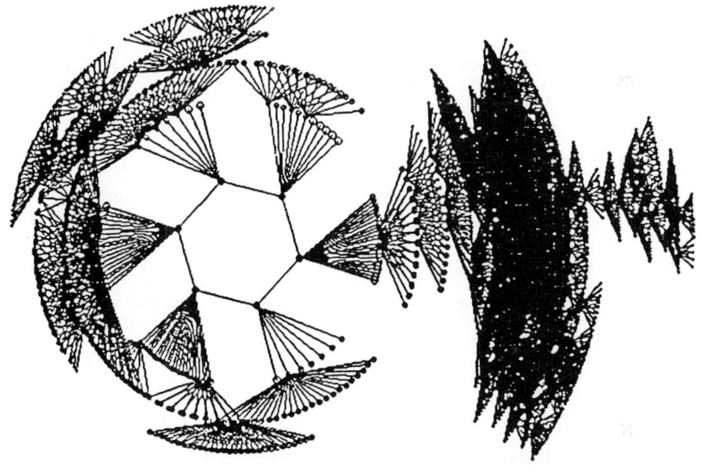

cell	wiring	rule,	-table
1	8,13,10	51,	00110011
2	4,4,3	55,	00110111
3	13,10,6	11,	00001011
4	8,12,8	166,	10100110
5	3,3,3	77,	01001101
6	1,9,9	182,	10110110
7	13,11,2	122,	01111010
8	6,2,11	188,	10111100
9	4,2,2	207,	11001111
10	9,12,6	112,	01110000
11	2,9,6	6,	00000110
12	11,13,9	63,	00111111
13	6,10,6	56,	00111000

FIGURE 11 The basin of attraction field of a random Boolean network with only one attractor, period 6. $N = 13, K = 3$. The $2^{13} = 8192$ states in state space are all linked into one basin of attraction. The network's wiring/rule scheme is set out in the table.

spontaneous emergence of life, under certain *edge-of-chaos* conditions, CA seem to support the spontaneous emergence of analogous complex dynamical phenomena such as gliders. In turn, a key property of life is the emergence of complex networks with *non-local* connections; biological systems are replete with non-local connections, from brains to economics.

The basin structure of random Boolean networks is extremely varied, but for parameters set at random, the number of basins and their attractor periods is generally very small and increases only slowly with system size, whereas in some categories of CA the growth may be exponential. Examples of just one attractor with a short period taking up the entire state space are not uncommon, as shown in Figure 11 for a $N = 13, K = 3$ network.

The surprisingly small number of attractors with small periods ties in with Kauffman's studies[11] for large random Boolean networks (for N up to 10,000). He reported that for $K = 2$ networks, attractor cycles and the number of alternative attractors increase as $\sqrt{N}$, and at a increasing rate with greater K.

A systematic investigations of the topology and structure of basins of attraction for random Boolean networks and various intermediate architectures, using the tools now available, has yet to be done.

COMPUTING PRE-IMAGES IN RANDOM BOOLEAN NETWORKS

The construction of a single basin of attraction poses the problem of finding the complete set of pre-images of a given network state. The trivial solution is to exhaustively test the entire state space. Every state that is linked together in the basin would require exhaustive testing. This obviously becomes intractable in terms of computer time as the network's size increases beyond modest limits. To overcome this problem, methods have been invented for computing pre-images directly, without exhaustive testing. The network's dynamics can, in effect, be run backwards in time; backward trajectories will, as a rule, diverge. A *reverse algorithm* that directly computes pre-images for one-dimensional CA was presented in *The Global Dynamics of Cellular Automata.*[31] I set out below a *general* direct reverse algorithm for random Boolean networks.[32]

Consider a random Boolean network with N cells, with a pseudo-neighborhood size K. The algorithm will be demonstrated for $N = 10, K = 3$. An equivalent algorithm applies for larger N and K. For convenience, the system is represented as a one-dimensional array, $A_1, A_2, \ldots, A_i, \ldots, A_N$. Each cell, A_i (value 0 or 1), has a preset wiring/rule scheme (possibly selected at random). The wiring scheme is given by A_i (w_1, w_2, w_3) where w_1 is a number between 1 and N signifying the position of the wire connections from the first branch of the pseudo-neighborhood, and so on. The rule scheme is given by $A_i(T_7, T_6, \ldots, T_0)$, the $K = 3$ rule table. In the example below, the wiring scheme for A_4 is $w_1 = 7, w_2 = 6$, and $w_3 = 3$.

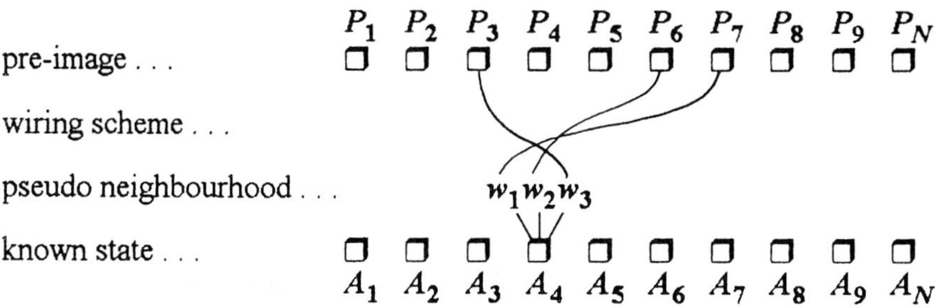

To derive the pre-images of an arbitrary global state, consider a candidate pre-image as an *empty* array, consisting of *empty* cells. The cells are empty because their values are unknown, and unallocated as either 0 or 1. Empty cells are denoted by the wild card symbol $\star$; known cells (with values established as 0 or 1) are denoted by the symbol $\Box$.

Consider a known network state, $A_1, A_2, \ldots, A_N$, and the empty pre-image state $P_1, P_2, \ldots, P_N$.

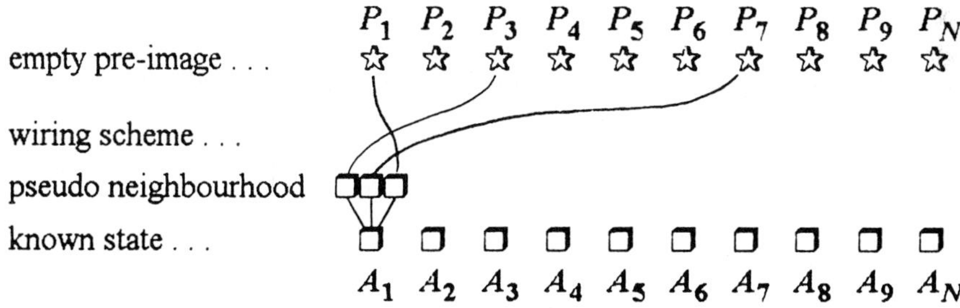

Starting with the first cell of the known state, A_1, the valid pseudo-neighborhood values consistent with the value of A_1 are assigned to separate copies of the empty pre-image according to the wiring scheme $A_1, (w_1, w_2, w_3)$. As there will be a mix of 0's and 1's in the rule table, only some of the eight possible pseudo-neighborhoods will be valid. If, say, three are valid, three *partial pre-images* (with some cells *known* and some *empty*) will be generated. For example, consider the $K = 3$ rule 50 at A_1, with a rule table as follows:

	111	110	101	100	011	010	001	000	...neighborhoods
rule table...	0	0	1	1	0	0	1	0	...outputs (0 or 1)
	T_7	T_6	T_5	T_4	T_3	T_2	T_1	T_0	

If $A_1 = 1$, then only three outputs match A_1, T_5, T_4 and T_1, corresponding to the neighborhoods 101, 100, and 001. These valid neighborhoods are allocated to three empty arrays according to the wiring scheme, say, $A_1(3,7,1)$. Each of the three arrays now have some of their cells allocated as 0's or 1's, and are termed *partial pre-images*, as illustrated below.

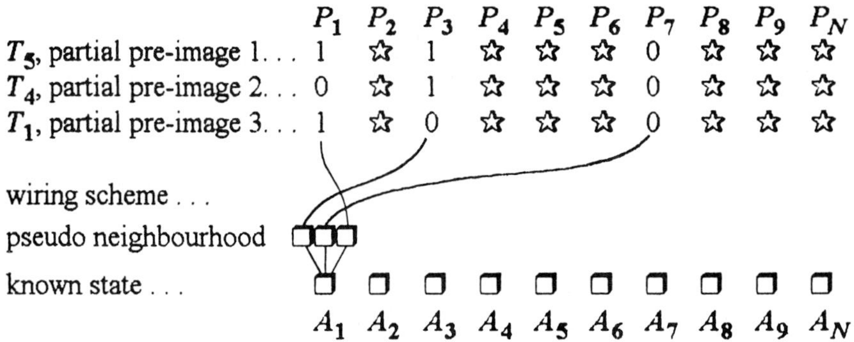

The procedure continues with the next cell of the known state, A_2 (though the order may be arbitrary). Say that the value of A_2 has 5 (out of 8) valid pseudo-neighborhoods in its rule table, $A_2(T_7, T_6, \ldots, T_0)$. The pseudo-neighborhoods are allocated to five copies of each of the partial pre-images that were generated at A_1, according to the wiring scheme $A_2(w_1, w_2, w_3)$.

If the allocation of a value to a given cell conflicts with the value already assigned to that cell, then the partial pre-image is rejected. Otherwise, the partial pre-image is added to the partial pre-image stack. The allocation will be valid if it is made to an empty cell, or to a known cell with an equal value. Valid allocation increases the size of the partial pre-image stack; conflicts reduce the size of the stack.

This procedure is repeated in turn for the remaining cells, $\ldots A_3, A_4, \ldots, A_N$. At each successive cell, more partial pre-images may be added to the stack, but also rejected. The size of the stack will typically vary according to a Gaussian distribution. If the stack size is reduced to zero at any stage, then the known state $A_1, A_2, \ldots, A_N$ has no pre-images; it is a *Garden of Eden* state. In general, the vast majority of states in state space turn out to be Garden of Eden states.

Note that the order in which cells in $A_1, A_2, \ldots, A_N$ are taken is entirely arbitrary. For the most efficient computation that minimizes the growth of the partial pre-image stack, the order should correspond to the greatest overlap of wiring schemes.

When the procedure is complete, the final stack may still have empty cells, signifying that these cells are not sampled by any wiring couplings. Final stack arrays with empty cells are duplicated so that all possible configurations at empty cell position are represented. The resulting pre-image stack is the complete set of pre-images of the given state, without duplication. An equivalent, but extended, procedure is used for $K = 5$ rules. In this case the wiring/rule scheme is specified for each cell in the array $A_1, A_2, \ldots, A_N$ as follows,

$$\text{the wiring scheme.} \;\ldots A_i(w_1, w_2, w_3, w_4, w_5)$$
$$\text{the rule scheme.} \;\ldots A_i(T_{31}, T_{30}, \ldots, T_0)\,.$$

The general reverse algorithm for computing pre-images works for fully random networks or any degree of intermediate architecture, which of course includes CA of any dimension—the wiring scheme is simply set accordingly. In principle, the algorithm will work for any size of pseudo-neighborhood, K, and any value range, V. Provided that $K < N$, the algorithm is many orders of magnitude faster than the exhaustive testing of state space, the only previous method available.

This information is used to construct the pre-image fan, from the given state to its set of pre-images (if any). The pre-image fan for each pre-image is then computed, and so on, until only Garden of Eden states remain. In this way transient trees (or subtrees) may be constructed. Basins of attraction (or the complete basin of attraction field) are constructed by first running the network forward to reveal the attractor cycle, then computing each transient tree in turn. The method, and the graphic conventions, are explained more fully in *The Global Dynamics of Cellular Automata.*[31]

MEMORY, FAR FROM EQUILIBRIUM

Memory far from equilibrium along merging transients may answer a basic difficulty in explaining memory by attractors in biological neural networks. A view of the brain as a complex dynamical system made up of many interlinked specialized neural networks is perhaps the most powerful paradigm currently available. Specialized neural networks may consist of further subcategories of semi-autonomous networks, and so on, which contribute to resetting or perturbing each other's global state. A biological neural sub-network is nevertheless likely to be extremely large; as a dynamical system the time required to reach an attractor from some arbitrary global state will probably be astronomical. This may be demonstrated with a simple $K = 5$ random Boolean network with a few hundred cells. Even when an attractor is reached, it may well turn out to be a long cycle or a quasi-infinite chaotic attractor. The notion of memory simply as attractors seems to be inadequate to account for the extremely fast reaction times in biology.

A discrete dynamical system with synchronous updating categorizes its state space reliably along transient trees, far from equilibrium, as well as at the attractors. A network that has evolved or learned a particular global dynamics may be able to reach memory categories in a few steps, possibly just one. Moreover, the complex transient tree topology in the basin of attraction field makes for a much richer substrate for memory than attractors alone, allowing hierarchies of memory subcategories.

Deterministic transient tree topology, as described in this paper, requires synchronous updating in the network, though *asynchronous* updating might conceivably produce analogous phenomena provided that the updating was deterministic, and not random as in Hopfield's model.[8] Deterministic asynchronous updating could be implemented by adding an extra parameter to a random network model; the *length* of each wire coupling as proposed by Harvey et al.,[7] possibly in discrete intervals. A standard speed of signal transmission applied to a network with variable discrete length wires is equivalent to a network with only single length wires, but coupled to network states *before* the previous time step; for example, a cell at time t might have connections to cells at $t-2$, $t-3, \ldots$ as well as to $t-1$.

A difficulty with this model is that a given network state at a particular instant may have multiple successors and would be likely to recur many times outside the attractor. Alternatively, the states in the system's state space would need to consists of multiple time steps, not just one. Network architecture with "higher order in time"[30] or "historical time reference"[31] may be worth pursuing but is beyond the scope of this paper.

There is evidence that the firing of nearby biological neurons is strictly related in time. Phase locking of spike discharges between neighboring cells has been observed, extending up to 7mm across the cortex.[4,21,27] Synchronous firing may be mediated by interneurons, which lack axons,[20] or mechanisms relying on close physical proximity between neurons (their dendrites, cell bodies, and axons). Gap junction effects (physical connections between neurons made by large macro molecules), and ephaptic interactions (the local electrical field) serve to synchronize local neuronal activity.[18]

A BIOLOGICAL MODEL

A random Boolean network may serve as a model of a patch of semi-autonomous biological neurons whose activity is synchronized. A cell's wiring scheme models that subset of neurons connected to a given neuron. Applying the CA rule to a cell's pseudo-neighborhood models the nonlinear computation that a neuron is said to apply to these inputs to decide whether or not to fire at the next time step. This is far more complex than a threshold function.[20] The biological computation may be a function of the topology of the dendritic tree, the microcircuitry of synaptic placements, and intrinsic membrane properties. Networks *within* cells based on the cytoskeleton of microtubules and associated protein polymers may be involved, suggested by Hameroff et al.[5,6] as the neuron's "internal nervous system." There appears to be no shortage of biological mechanisms that could perform the role of a CA's rule table.

The cells in a random Boolean network are arranged in an orderly array for convenience only, but their location may be arbitrary. One presumes that the actual location of neurons in the brain has been optimized through evolution to achieve high density by minimizing the average length of connections that occupy space and consume resources.

The network has been redrawn as a brain-like model in Figure 13; a semi-autonomous population of 27 idealized neurons distributed in three dimensions. Each neuron (Figure 12) receives a postsynaptic excitatory (1) or inhibitory (0) signal from up to 5 neurons in the population (possibly including itself) via its 5 dendrites ($K = 5$), and computes a response signal to its axon according to its particular 32-bit rule table applied to its pseudo-neighborhood. The updating of axonic response is synchronous, and the process is iterated in discrete time steps.

The network's basin of attraction field is implicit in its wiring rule/scheme. In a patch of biological neurons it would be implicit in the wetware. Recognition

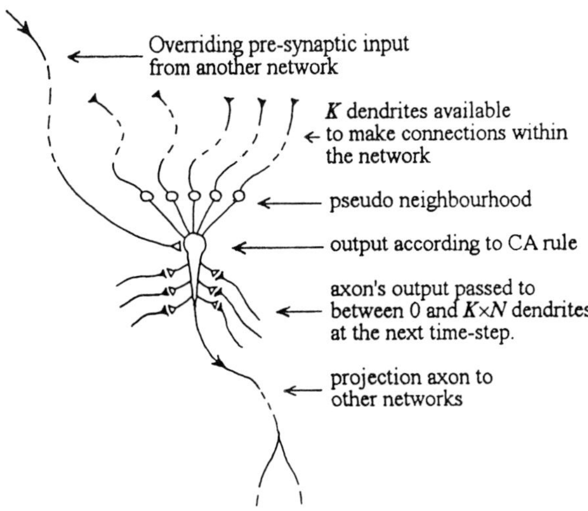

FIGURE 12 A cell of a random Boolean network represented as an idealized neuron with an overriding presynaptic contact from outside the system, and a projection axon to other networks.

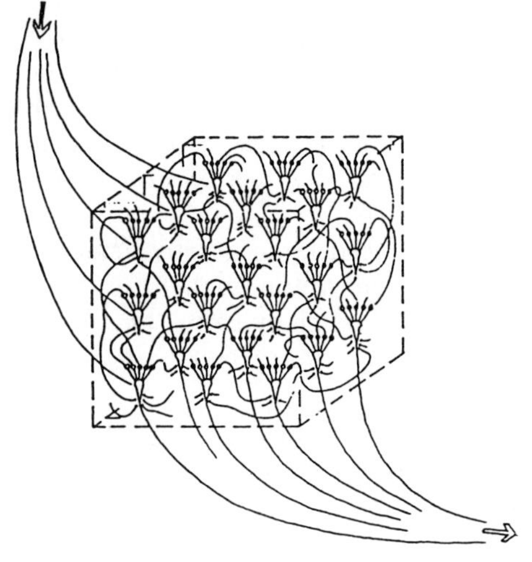

FIGURE 13 A random Boolean network ($N = 27, K = 5$) represented as a semi-autonomous population of idealized neurons. Neurons may have input and output links to other networks.

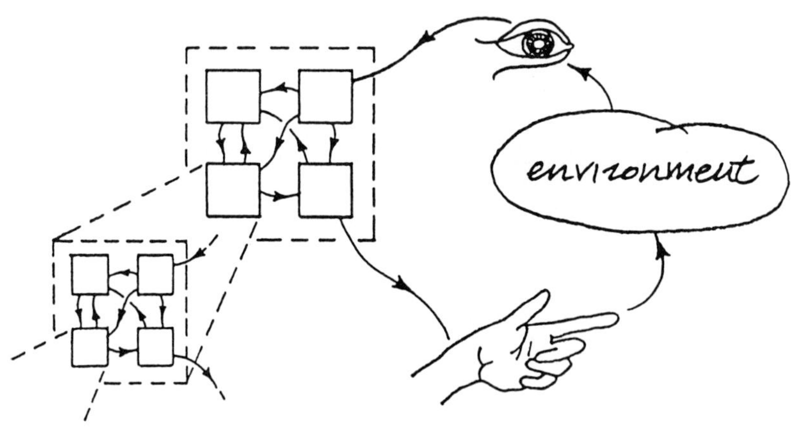

FIGURE 14 A network of weakly coupled semi-autonomous random Boolean networks linked to the environment as a biological model. Each network may consist of a nested hierarchy of networks.

(categorization) of a global state is automatic and direct according to the particular deterministic trajectory (within the transient tree of the basin of attraction) that will inevitably follow. Learning new behavior implies amending the basin of attraction field by adjusting the wiring/rule scheme, analogous to some physical change to the wetware's neural architecture and synaptic function. How the appropriate change might occur in biological networks to achieve a desired novel behavior is unknown. Hameroff et al.[5,6] suggest that cytoskeletal functions may provide retrograde signalling (analogous to backpropagation in artificial neural networks) which may reconfigure intraneuronal architecture.

The system is semi-autonomous because it must be capable of receiving input from outside to reset its global state, and also to communicate its current internal state to the outside via feed-forward channels. Thus each neuron in the model has overriding postsynaptic inputs from outside the population, and a projection axon to targets outside the population.

The model may be elaborated by weakly interconnecting a number of random Boolean networks (or perhaps three-state networks), so that the output of a particular network constituted the overriding input of another. Communication between networks may be *asynchonous*, and at a slower frequency than a particular network's internal synchronous clock. Such an assembly of networks will have implicit in its particular pattern of connections at any instant, a vastly more complex but intangible web of interacting basin of attraction fields—*the ghost in the machine*?

LEARNING ALGORITHMS

Whether or not such a model is biologically plausible, it may be useful in its own right as an artificial neural network where learning (and its side effects) is visualized as alterations to the detailed structure of the basin of attraction field. In networks too large to allow basins, or even fragments of basins, to be computed, the principles still apply. Separate basins in the basin of attraction field, and each node onto which dynamical flow converges, categorize state space. All the network's states other than *Garden of Eden* states are content-addressable memories. Any external input will automatically initiate a dynamical flow along a unique chain of states. Each successive state categorizes states in its transient subtree, far from equilibrium, forming a complex hierarchy of categorization culminating at the attractor. The set of attractors and their branching trees constitute the network's collective memory.

This section sets out learning algorithms that enable a random Boolean network to learn new transitions from experience (and also to forget). Suppose we want to make the state $P1$ the pre-image of state A. Any mismatches between cell values of the actual successor state $B1$ (of the aspiring pre-image, $P1$) and state A can be corrected in one step by either of two methods, adjusting the network's wiring or rule scheme. The two methods have very different consequences.

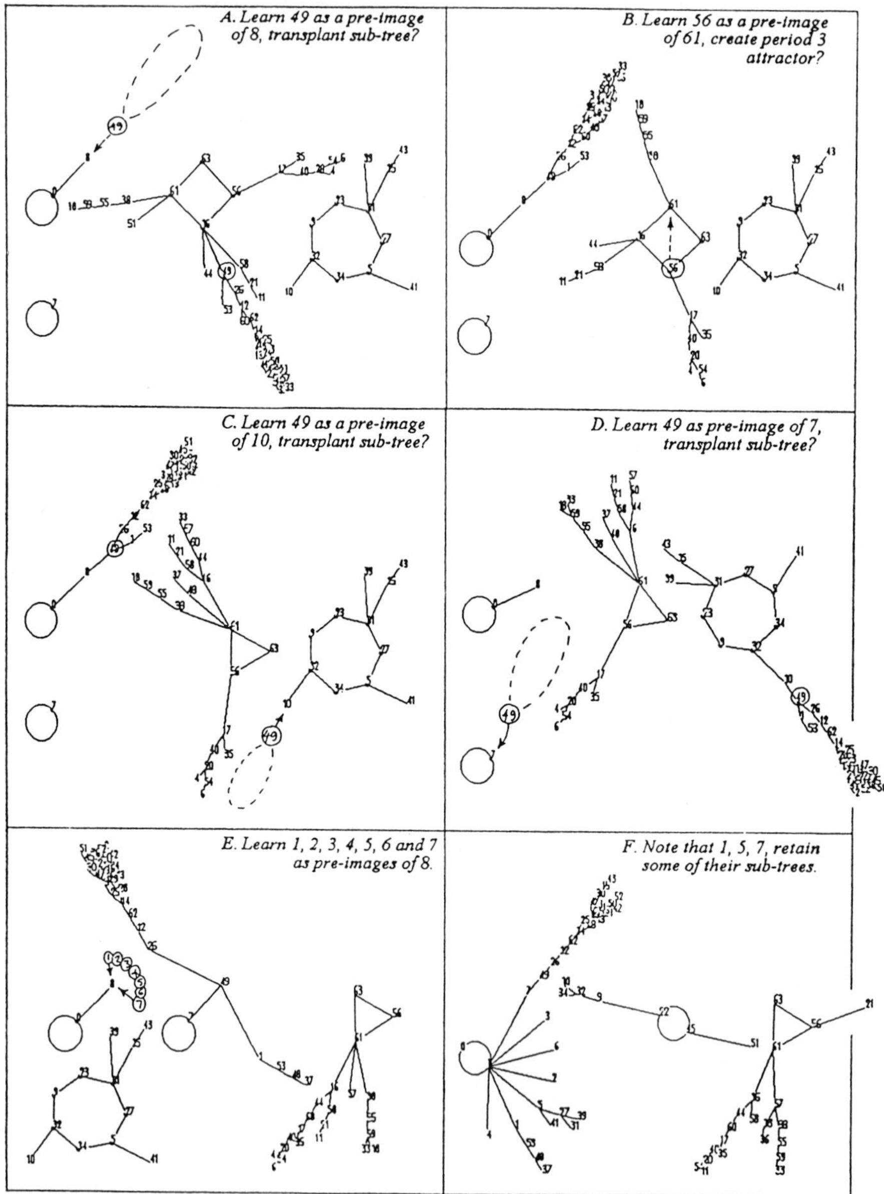

FIGURE 15 A sequence of learning steps, A to F, in a six-cell network. Note the stability of basin structure as well as the side effects of learning at each step. Bit patterns are represented by decimal numbers.

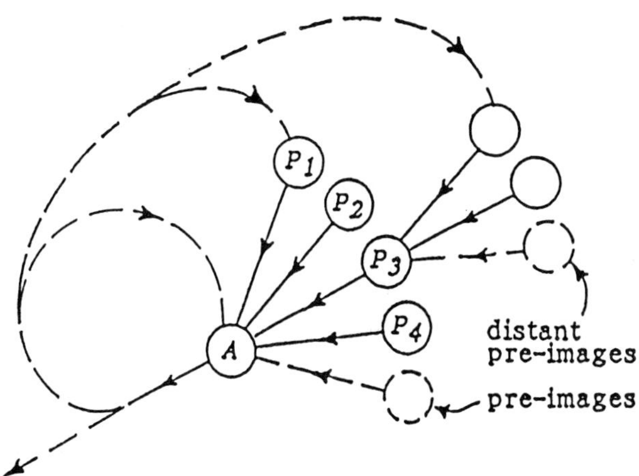

FIGURE 16 States $P1, P2, P3, \ldots$, etc., may be learned as pre-images of the state A. Distant pre-images of A may also be learned, for instance, as pre-images of $P3$. Learning A as a pre-image of itself creates a point attractor. Learning A as a distant pre-image of itself creates a cyclic attractor. If A is learned as the pre-image of some other state in the basin of attraction field, the states flowing into A, its transient subtree, may be fully or partially transplanted along with A.

Before learning starts, a wiring/rule scheme must already be in place. If the relevant transitions in the basin of attraction field are already close to the desired behavior, the side effects of learning will be minimized. The wiring/rule scheme may be selected from an *atlas* of basin of attraction fields as in *The Global Dynamics of Cellular Automata.*[31] It may be pre-evolved from a population of wiring/rule schemes using a genetic algorithm or, in the worst case, assumed at random.

LEARNING BY REWIRING

Consider the state $P1, (P1_1, P1_2, \ldots, P1_N)$, that the network is to *learn* as a pre-image of the given state $A, (A_1, A_2, \ldots, A_N)$. When $P1$ (the aspiring pre-image) is evolved forward by one time step according to its current K-neighbor wiring/rule scheme, its actual successor state is $B1, (B1_1, B1_2, \ldots, B1_N)$. If the states $P1$ and A were selected at random, then $B1$ will probably have mismatches with A in about $N/2$ locations.

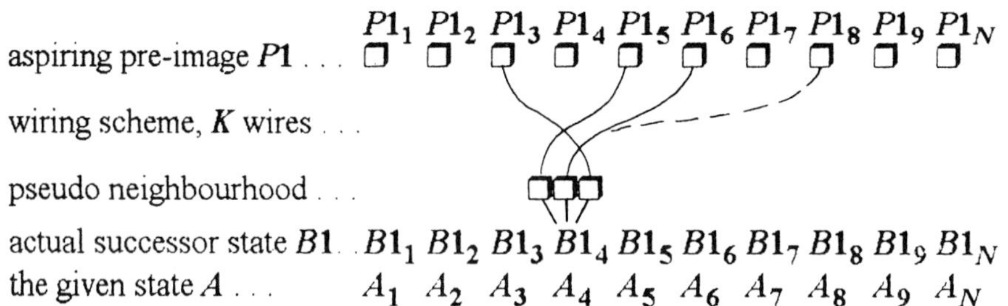

FIGURE 17 Correct a mismatch between $B1_4$ and A_4 by moving one wire.

Suppose that $B1_4 \neq A_4$. One wire, or more if necessary, of the cell's wiring couplings is moved to a new position. I will limit this analysis to single wire moves only. Any move resulting in a pseudo-neighborhood with opposite output (according to that cell's look-up table) will correct the mismatch. This is a stochastic method as there are likely to be many alternative successful options.

Assuming that $P1$ has a roughly equal proportion 0's and 1's, there will be $\approx N/2$ alternative positions where a wire move will pick up a changed value. Given K wires, there will be $\approx K \times N/2$ alternative rewiring options that will change the pseudo-neighborhood. Assuming the rule at cell 4 has $\lambda \approx 0.5$ (the proportion of 1's in the rule table), $\approx 1/2$ of the changed pseudo-neighborhoods will change the value of $B1_4$ so that $B1_4 = A_4$. The number of valid rewiring options to correct each mismatch is therefore $\approx K \times N/4$. The choice for rewiring may initially be selected at random from among the valid options to correct each mismatch, making $P1$ the pre-image of A.

Suppose that another aspiring pre-image, $P2$ (with an actual successor $B2$), is to be learned as a pre-image of A by rewiring, but without forgetting $P1$. If $B2_4 \neq A_4$, there will be $\approx K \times N/4$ valid rewiring options to correct the mismatch. However, $\approx 1/4$ of the options would change the value of $B1_4$, forgetting $P1(\approx 1/2$ the wire moves would change the pseudo-neighborhood of $B1_4$, $\approx 1/2$ of which will have outputs different to $B1_4$).

If several pre-images are to be learned in succession by single wire rewiring, without forgetting any previously learned pre-images, the space of valid rewiring options for a particular cell will be reduced by $\approx 1/4$ for each mismatch correction. The shrinking of the valid rewiring space is sensitive to the initial rewiring choices and to the order of learning. Some rewiring choices will be fitter than others, affecting the capacity of the network to learn that particular set of pre-images.

If the pre-images are close to each other in terms of Hamming distance, there will be fewer mismatches to correct, thus learning capacity increases. Close pre-images are likely be learned by default; the network is able to generalize from examples.

What is the probability of *forgetting* pre-images on the same fan, or transitions elsewhere in the basin of attraction field, as a result of learning by rewiring?

Consider an arbitrary transition $X \to Y$ elsewhere in the basin. The probability of *forgetting* the transition

$$\text{by moving one wire is given by} \quad F_1 + 1 = 1/4$$

$$\text{by moving two wires} \quad F_2 = F_1 + F_1(1 - F_1), \text{ or}$$

$$\text{by moving } n \text{ wires} \quad F_n \sum_{t=0}^{n-1} \left(1 - \frac{1}{4}\right)^i \times \frac{1}{4}.$$

The capacity of the network to learn more states as pre-images of a given state will thus depend on a number of factors: the original wiring/rule scheme, the similarity of the new pre-images to each other, the order of learning, and the choice of rewiring options. However, the network may have additional capacity to learn *distant* pre-images, further upstream in the transient tree as in Figure 16. Note that if the network learns the given state itself as its own pre-image, this will result in a point attractor. If the state is learned as a distant pre-image, this will result in a cyclic attractor with a period equal to the distance.

LEARNING BY MUTATING THE RULE SCHEME

To correct a mismatch between a cell in A and the corresponding cell in the successor of an *aspiring* pre-image, $P1$, by mutating the rule scheme, one specific bit in that cell's rule table is flipped (changed from $0 \to 1$ or $1 \to 0$). There is only one option that is certain to succeed. Adding another aspiring pre-image, by the same method, cannot cause $P1$ to be forgotten. This is because any mismatch between a particular cell in the successor state $B2$ (of the aspiring preimge, $P2$) and A cannot relate to the same rule table entry that was "looked up" to determine $P1$'s successor. Otherwise there would have been no mismatch. Any change to correct the mismatch must be to a different rule table entry; $P1$'s successor cannot be affected. It turns out that there is no limit to the number of pre-images of a given state that can be learned by this method, and no risk of forgetting previously learned pre-images of the state, but of course there may be side effects elsewhere in the basin of attraction field.

As an extreme example, all states in state space can be made the pre-images of any arbitrary state A. The two trivial rules with rule tables consisting of only 0's or only 1's are allocated to cells in A according to whether a cell equals 0 or 1. The result will be a basin of attraction field consisting of a single point attractor. All other states in state space will be *Garden of Eden* pre-images of the point attractor.

Consider the state $P1, (P1_1, P1_2, \ldots, P1_N)$ which the network is to *learn* as a pre-image of the given state $A, (A_1, A_2, \ldots, A_N)$. When $P1$ (the aspiring pre-image) is evolved forward by one time step according to its current K neighbor wiring/rule scheme, its actual successor state is $B1, (B1_1, B1_2, \ldots, B1_N)$.

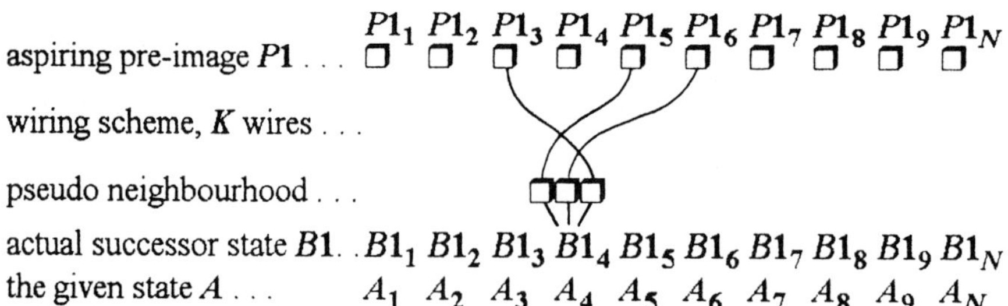

FIGURE 18 Correct a mismatch between $B1_4$ and A_4 by flipping a specific bit in the rule at cell 4.

Suppose that $B1_4 \neq A_4$. To correct the mismatch, the CA rule at cell 4 is mutated by flipping the bit in its rule table corresponding to the pseudo-neighborhood at $B1_4$. There is only one option. As the rule table has 2^K bits, flipping one bit represents a change of $1/2^K$ in the rule table, i.e., for three-neighbor wiring $1/8$, for five-neighbor wiring $1/32$. This is the probability of forgetting another transition elsewhere in the basin of attraction field, outside the immediate pre-image fan where the probability of forgetting $= 0$. For larger K the probability of a rule table bit flip resulting in forgetting some arbitrary transition, $X \rightarrow Y$ becomes smaller at an exponential rate as K increases. It was shown in *The Global Dynamics of Cellular Automata*[31] that a 1-bit mutation (to each cell, i.e., flipping a total of N bits) in a five-neighbor CA resulted generally in a small change in the basin of attraction field.

Consider an arbitrary transition $X \rightarrow Y$ elsewhere in the basin. The probability of *forgetting* the transition

$$\text{by one bit-flip is given by } \; F_1 = \frac{1}{2^K}$$

$$\text{by two bit-flips } \; F_2 = F_1 + F_1(1 - F_1), \text{ or}$$

$$\text{by } n \text{ bit-flips } \; F_n \sum_{i=0}^{n-1} \left(1 - \frac{1}{2^K}\right)^i \times \frac{1}{2^K}$$

If a set of pre-images to be learned are close to each other in terms of Hamming distance, there will be fewer side effects elsewhere in the basin of attraction field. Again, close pre-images are likely be learned by default, and the network is able to generalize from examples.

SCULPTING THE BASIN OF ATTRACTION FIELD

Rewiring has a much greater effect on basin structure than mutating the rule scheme, but in either case the stability of basin structure is noteworthy. By using these methods, point attractors, cyclic attractors, and transient subtrees can be created. Transient subtrees are sometimes transplanted along with the repositioned state (see Figure 15), indicating how learned behavior can be reapplied in a new context. Generalization is present, because bit patterns in the same pre-image fan are likely to be close in Hamming distance to each other, and so may be learned by default from examples. Forgetting involves *pruning* pre-images and transient subtrees, and is achieved by the inverse of the method for learning. Since it is sufficient to create just one mismatch in order to forget, the side effects are minimal as compared with learning.

In Figure 15 I show an arbitrary example (with no particular aim) of visible learning (and side effects) in a six-cell network with regular wiring but randomly allocated rules. At each learning step, a state (or set of states) is made the pre-image of a target state. In Figure 19 I show the basin of attraction field of a 5-cell network that has been taught to segregate 5-bit strings with odd and even parity into two separate basins.

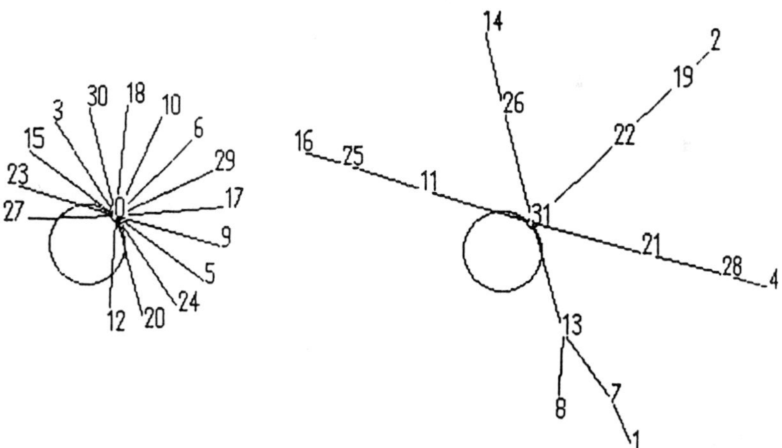

FIGURE 19 The basin of attraction field of a 5-cell parity-sorting network. The first basin has only even parity states the second has only odd parity states. Both basins have point attractors. Bit patterns are represented by decimal numbers.

Combining wiring and rule scheme adjustments may result in a powerful method of cumulative learning (supervised and unsupervised) in random Boolean networks. In future work I anticipate applying the algorithms to simple recognition tasks, to use genetic algorithms to evolve improved start parameters of networks prior to learning, to extend the learning methods to include changes in system size and connectivity, and to investigate assemblies of weakly coupled semi-autonomous networks.

CONCLUSIONS

Whereas CA, embedded as they are in a regular space with homogeneous laws, provide models of processes in physics, including the emergence of self-organization and simple lifelike phenomena, random networks may be appropriate as models for the *non-local* interactions between living subsystems typical of biology. Non-local interactions transcend inanimate physics and are perhaps a key property distinguishing living from non-living matter. Complex non-local networks with emergent properties occur hierarchically at all scales in biology, from networks of genes regulating one another within a cell,[11] to networks regulating entire ecosystems, not to mention human society and technology.

Understanding the dynamics of abstract random networks may provide insights into biological networks in general, where concepts analogous to memory may be ubiquitous; take, for instance, memory in the immune system.

Non-local interactions, shaped by evolution and learning, allow the emergence of unconstrained *categorization* in a system's repertoire of dynamical behavior—its content-addressable memory. The basin of attraction diagrams of random Boolean networks capture the network's memory. The diagrams demonstrate that a complex hierarchy of categorization exists within transient trees, far from equilibrium, providing a vastly richer substrate for memory than attractors alone. In the context of many semi-autonomous weakly coupled networks, the basin field/network relationship may provide a fruitful metaphor for the mind/brain.

ACKNOWLEDGMENTS

I am grateful to Chris Langton, Stuart Kauffman, Steen Rasmussen, Mats Nordahl, Brossl Hasslacher, Crayton Walker, Stuart Hameroff, Joshua Smith, Tim Thomas, Ron Bartlett, Pedro de Oliviera, Inman Harvey, Harry Barrow, and Phil Husbands (especially for his expressions for the probability of forgetting), and many other friends and colleagues at the Santa Fe Institute, the University of Sussex, and elsewhere, for discussions and comments.

REFERENCES

1. Alexander, I., W. Thomas, and P. Bowden. "WISARD, a Radical New Step Forward in Image Recognition." *Sensor Rev.* (1984): 120–124.
2. Ashby, W. R. *Design for a Brain.* Chapman & Hall, 1960.
3. Conway, J. H. "What is Life?" In *Winning Ways for Your Mathematical Plays*, edited by E. Berlekamp, J. H.Conway, and R. Guy, Vol. 2, chap. 25. New York: Academic Press, 1982.
4. Grey, C. M, W. P. König, A. K. Engel, and W. Singer. "Oscillatory Responses in Cat Visual Cortex Exhibit Inter-Columnar Synchronization Which Reflects Global Stimulus Properties." *Nature* **388** (1989): 334–337.
5. Hameroff, S., S. Rasmussen, and B. Mansson. "Molecular Automata in Microtubules: Basic Computational Logic of the Living State." In *Artificial Life*, edited by C. Langton. Santa Fe Institute Studies in the Sciences of Complexity, Proc. Vol. VI, 521–553. Reading, MA: Addison-Wesley, 1989
6. Hameroff, S. R., J. E. Dayhoff, R. Lahoz-Beltra, S. Rasmussen, E. M. Insinna, and D. Koruga. "Nanoneurology and the Cytoskeleton: Quantum Signaling and Protein Conformational Dynamics as Cognitive Substrate." In *Behavioral Neurodynamics*, edited by K. Pribram and H. Szu. Pergamon Press, 1993.
7. Harvey, I., P. Husbands, and D. Cliff. "Issues in Evolutionary Robotics, in From Animals to Animats 2." *Proceedings of the 2nd International Conference on Simulation of Adaptive Behavior*, edited by Mayer, Roitblat, and Wilson, 364–373. MIT Press, 1993.
8. Hopfield, J. J. "Neural Networks and Physical Systems with Emergent Collective Computational Abilities." *PNAS* **79** (1982): 2554–2558.
9. Kauffman, S. A. "Metabolic Stability and Epigenisis in Randomly Constructed Genetic Nets." *J. Theor. Biol.* **22** (1969): 437–467.
10. Kauffman, S. A. "Emergent Properties in Random Complex Systems." *Physica D* **10** (1984): 146–156.
11. Kauffman, S. A. "Requirements of Evolvability in Complex Systems; Orderly Dynamics and Frozen Components." In *Complexity, Entropy, and the Physics of Information*, edited by W. H. Zurek. Santa Fe Institute Studies in the Sciences of Complexity, Proc. Vol. VIII, 151–192. Reading, MA: Addison-Wesley, 1991.
12. Koestler, A. *The Ghost in the Machine.* Pelican Books, 1967.
13. Langton, C. G. "Studying Artificial Life with Cellular Automata." *Physica D* **22** (1986): 120–149.

14. Langton, C. G. "Computation at the Edge of Chaos: Phase Transitions and Emergent Computation." *Physica D* (1990): 12–37
15. Langton, C. G. "Life at the Edge of Chaos." In *Artificial Life II*, edited by C. Langton et al. Santa Fe Institute Studies in the Sciences of Complexity, Proc. Vol. X. Reading, MA: Addison-Wesley, 1991.
16. Li, W. "Phenomenology of Non-Local Cellular Automata." *J. Stat. Phys.* **68** (1992): 829–882.
17. Li, W., and M. G. Nordahl. "Transient Behavior of Cellular Automata Rule 110." Working Paper 92-03-016, Santa Fe Institute , 1992.
18. McCormick, D. A. "Membrane Properties and Neurotransmitter Actions." In *The Synaptic Organization of the Brain*, edited by G. M. Shepherd, 32–66, 3rd ed. Oxford University Press, 1990.
19. von Neuman, J. *Theory of Self Reproducing Automata*, edited and completed by A. W. Burks. University of Illinois Press, 1966.
20. Shepherd, G. M., ed. *The Synaptic Organization of the Brain*, 3rd ed. Oxford University Press, 1990.
21. Singer, W., Response Synchronization of Cortical Neurons; An Epiphenomenon or a Solution to the Binding Problem?" *IBRO News* **19** (1991): 6–7.
22. Vichniac, G. Y. "Boolean Derivatives on Cellular Automata." *Physica D* **45** (1990): 63–74.
23. Walker, C. C., and W. R.Ashby. "On the Temporal Characteristics of Behavior in Certain Complex Systems." *Kybernetik* **3** (1966): 100–108.
24. Walker, C. C. "Behavior of a Class of Complex Systems: The Effect of System Size on Properties of Terminal Cycle." *Cybernetics* (1971): 55–67.
25. Walker, C. C. "Stability of Equilibrial States and Limit Cycles in Sparsely Connected, Structurally Complex Boolean Nets." *Complex Systems* **1** (1987): 1063–1086.
26. Walker, C. C. "Attractor Dominance Patterns in Random-Structure Cellular Automata." *Physica D* (1990): to be published in.
27. Wasserman, G. S. *Isomorphism, Task Dependance, and the Multiple Meaning Theory of Neural Coding.* Biological Signals, 1992.
28. Weisbuch, G., R. De Boer, and A. S. Perelson. "Localized Memories in Idiotypic Networks." *J. Theor. Biol.* **146** (1990): 483–499.
29. Wolfram, S. "Statistical Mechanics of Cellular Automata." *Rev. Mod. Phys.* **55(3)** (1983): 601–664.
30. Wolfram, S. "Universality and Complexity in Cellular Automata." *Physica D* **10** (1984): 1–35.
31. Wuensche, A., and M. J. Lesser. *The Global Dynamics of Cellular Automata; An Atlas of Basin of Attraction Fields of One-Dimensional Cellular Automata* (diskette included). Santa Fe Institute Studies in the Sciences of Complexity, Ref. Vol. I. Reading, MA: Addison-Wesley, 1992.

32. Wuensche, A. "The Ghost in the Machine; Basin of Attraction Fields of Disordered Cellular Automata Networks." Working Paper 92-04-017, Santa Fe Institute, 1992.
33. Wuensche, A. "Basins of Attraction in Disordered Networks." In *Artificial Neural Networks*, edited by I. Alexander and J. Taylor, 2nd ed. Elsevier, 1992.
34. Wuensche, A. "Memory, Far from Equilibrium; Basins of Attraction of Random Boolean Networks." Proceedings, European Conference on Artificial Life '93, Centre for Nonlinear Phenomena and Complex Systems, Université Libre de Bruxelles, 1993.
35. Wuensche, A. "Complexity in One-Dimentional Cellular Automata; Gliders, Basins of Attraction and the Z Parameter." Cognitive Science Research Paper, University of Sussex, 1993.

Carlo C. Maley
545 Technology Square, NE43-803, Massachusetts Institute of Technology, Cambridge, MA, 02139; e-mail: cmaley@ai.mit.edu

The Computational Completeness of Ray's Tierran Assembly Language

INTRODUCTION

One of the strengths of Ray's model, Tierra, is the open-ended nature of evolution in the model. It is therefore important to show that Tierran evolution is not limited by computational incompleteness. Ray implies that the assembly language of his model Tierra is computationally complete, due to similarity with other assembly languages that have been proven to be computationally complete.[1] Although this conclusion may seem reasonable at first, there is one important difference between the Tierran language and most other assembly languages: Tierran lacks an instruction for reading input. I present a formal proof of the computational completeness of the Tierran language by showing how a Tierran program, or organism, can implement a Turing Machine. Thus, on closer examination, the idiosyncrasies of the Tierran language turn out to be theoretically insignificant though perhaps practically crippling. I conclude by suggesting the implementation of an instruction for reading input. This minor expansion to the model has significant implications, including the possibility for self-repairing genomes and behavior that is closely coupled to the organism's environment.

COMPUTATIONAL COMPLETENESS

A computer language, along with the processor that executes it, is considered computationally complete if it is theoretically equivalent to a Universal Turing Machine.[1] The Turing Machine is a mathematical formalism that defines the theoretical limitations of computing. The computer is one instantiation of a machine that is theoretically equivalent to a Turing Machine. In other words, anything a computer can do, a Turing Machine can do, and vice versa. The Turing Machine is useful, not because someone would ever want to program it to do useful things, but because it is a rigourously defined mathematical construct that can be used to assess the power and limitations of computing.

THE SUSPICION

The basic model of computational execution is that a processor executes a program that takes some data as input and produces some data as output. In this way, a program can be seen as a function that maps a domain of input onto a range of output. Most computer languages have some instructions for reading input as well as instructions for writing output. However, the Tierran assembly language combines the two operations in the instruction MOV_IAB (short for "move instruction from address B to address A"). The MOV_IAB instruction copies the contents of memory address B into the contents of memory address A. Thus, reading from B and writing to A happen in a single instruction. It is therefore impossible to examine directly the contents of a memory address.

The inability to directly examine the memory could pose serious problems for designing a program that would read input and produce one of a number of outputs depending on that input. Dr. Ray implies that the Tierran language is computationally complete based on similarities with other assembly languages that have been proven to be computationally complete.[1] However, these other languages all have separate instructions for input and output, usually in the form of READ and WRITE instructions, and the proofs of their computational completeness depend on these separate instructions. The lack of independent input and output instructions in the Tierran language suggests the need for an argument to support the computational

[1]The Universal Turing machine is a specific kind of Turing Machine that can emulate any other kind of Turing Machine. It performs the emulation by working from an encoding of a specific Turing Machine that has been placed on the Universal Turing Machine's memory tape. In this way, a Universal Turing Machine is programmed to act like another form of Turing Machine. It follows from the fact that a Universal Turing Machine is a member of the class of Turing Machines, that if a Tierran can emulate any Turing Machine, it can emulate a Universal Turing Machine.

completeness of the the language that is not based on these somewhat dissimilar models.

PROVING COMPUTATIONAL COMPLETENESS

Perhaps the easiest way to prove that a computational system is computationally complete is to show that it is no more or less powerful, computationally, than a Turing Machine.[2] A constructive proof of computational completeness consists of two parts. First, it must be shown that a Universal Turing Machine can simulate the behavior of the computational system in question. This establishes that a Turing Machine is at least as powerful as the computational system. Second, the computational system must be able to simulate the behavior of Turing Machines, establishing the fact that the computational system is at least as powerful as a Turing Machine. If these two parts can be proven, the computational system must be equivalent to a Turing Machine.[3]

TURING MACHINE EMULATION OF A TIERRAN

Here the computational system in question is a Tierran organism. A Tierran consists of two distinct parts, a virtual processor and a program to be executed by that processor. In order to understand the power and limitations of a Tierran, I will specify the relevant aspects of both the virtual processor and the Tierran assembly language.

THE VIRTUAL PROCESSOR. The virtual processor of a Tierran consists of a number of registers and a stack. A register is a small amount of memory space that can be manipulated quickly. In a typical computer processor, the processor can only manipulate numbers that are held in registers. For example, in order to add two numbers, those numbers must first be loaded into two registers. After the numbers have been added, the sum is automatically placed in a register where it can then be either used again, or stored for future use in the normal memory space. In contrast,

[2] There is no known mathematical formalism that is more computationally powerful than a Turing Machine.

[3] Alternatively, by Theorem 7.9 of Hopcroft and Ullman,[2] we need only prove that we can program a Tierran to simulate a two-counter machine, but this is only slightly easier than simulating a Turing Machine for a Tierran. The significant difference is that instead of writing a symbol to the type, the Tierran would have to keep two counters. A counter can easily be simulated by incrementing or decrementing numbers held in the Tierran's registers.

a stack is a larger form of temporary memory storage with the property that only the number most recently stored on the stack is immediately accessible.[4]

The Tierran virtual processor has two registers, Ax and Bx, for holding addresses of locations in the memory space. It also has two registers, Cx and Dx, for holding integers. In addition, the virtual processor has a stack which can hold up to ten numbers, either integers or addresses. Finally, it has a register called the "instruction pointer" (**IP**) which holds the address of the next assembly language instruction to be executed by the processor. For the purposes of the proof, I will be generous and ignore the finite limitations on stack size and memory space in Tierra.

THE TIERRAN LANGUAGE. The Tierran language consists of 32 instructions, each a single word with no arguments or modifiers. The lack of arguments is achieved by defining the instructions so that they will always operate on the contents of specific registers.[5] The description of each Tierran instruction follows. The instructions in **bold** are the only ones used in the implementation of a Turing Machine.

NOP_0: "No operation 0" does not affect the state of the processor, acting as a template pattern for the purpose of addressing locations in the program code.

NOP_1: "No operation 1" again does not affect the state of the processor, but can form binary patterns along with NOP_0.

OR1: "Binary OR for the first bit" reverses the first (lowest order) bit of the integer in the Cx register.

SHL: "Bitwise shift left" shifts the bits in the Cx register one digit to the left and introduces a 0 to the newly cleared lowest order bit, effectively doubling the number in the register. A combination of OR1's and SHL's can construct any number in the Cx register.

ZERO: "Zero" clears the number in the Cx register by setting it to zero.

IF_CZ: "If Cx is zero" is a conditional branch point instruction, which will only execute the following instruction if there is a zero in the Cx register. Otherwise, it will skip over the instruction that follows IF_CZ.

SUB_AB: "Subtract Bx from Ax" finds the distance from the location stored in Ax to the location stored in Bx, and puts the result in the Cx register.

SUB_AC: "Subtract Cx from Ax" adjusts the address in Ax backwards by the number of units held in the Cx register.

INC_A: "Increment Ax" moves the address in Ax one unit forward.

[4] A stack is similar to a Pez candy or dining hall plate dispenser. You can only see the candy on top, and if you decide that you want the third candy from the top, you must first remove, or "pop," the top two candies before having access to the desired candy. However, the dining hall plate dispenser is a better analogy, because you can also "push" things onto the top of the stack, making the previous plate inaccessible.

[5] In effect, the lack of arguments makes the Tierran language closer to a machine code than an assembly language.

INC_B: "Increment Bx" moves theaddress in Bx one unit forward.

INC_C: "Increment Cx" adds 1 to the number in Cx.

DEC_C: "Decrement Cx" subtracts 1 from the number in Cx (zero is the lower bound).

PUSH_AX: "Push Ax onto the stack" puts the address in Ax on the stack.

PUSH_BX: "Push Bx onto the stack" puts the address in Bx on the stack.

PUSH_CX: "Push Cx onto the stack" puts the number in Cx on the stack.

PUSH_DX: "Push Dx onto the stack" puts the number in Dx on the stack.

POP_AX: "Pop the stack into Ax" removes the number on top of the stack and stores it in Ax.

POP_BX: "Pop the stack into Bx" removes the number on top of the stack and stores it in Bx.

POP_CX: "Pop the stack into Cx" removes the number on top of the stack and stores it in Cx.

POP_DX: "Pop the stack into Dx" removes the number on top of the stack and stores it in Dx.

JMP: "Jump" forces execution to jump to the address specified by the template pattern that follows the JMP instruction by loading that address into the IP register.

JMPB: "Jump backwards" works the same way as JMP except the address must be behind the location of the JMPB instruction.

CALL: "Call a procedure" stores the current location of the IP, as an offset from the starting location of the program, on the stack and then jumps the execution to the location specified by the template pattern following the CALL instruction.

RET: "Return from a procedure" returns execution to the location stored on the top of the stack, which should be the location following the last CALL instruction to be executed.

MOV_CD: "Move the contents of Cx into Dx" copies the contents of Cx and stores the number in Dx.

MOV_AB: "Move the contents of Ax into Bx" copies the contents of Ax and stores the address in Bx.

MOV_IAB: "Move the instruction at location Bx to location Ax" copies the instruction at the location held in Bx, and writes it into the location held in Ax.

ADR: "Find the address" stores the address of the location specified by the template pattern following the ADR instruction, in the Ax register.

ADRB: "Find the address behind" acts the same as ADR except that the specified location must be behind the ADRB instruction.

ADRF: "Find the address in front" acts the same as ADR except that the specified location must be in front of the ADRF instruction.

MAL: "Allocate memory" allocates a continuous block of unused memory the size of the number held in Cx. The address of the start of this block is stored in the Ax register.

DIVIDE: "Divide the cell" introduces a new Tierran to the system, and removes the "mother's" ability to alter the "daughter's" program code. A Tierran may read and execute any instruction found in the memory space but can only write into its own genome as well as the genome of its daughter before "cell division." These restrictions on the Tierran's ability to alter the data in the soup are supposed to be analogous to the protection afforded by a cell wall.

To show that a Turing Machine is at least as powerful as a Tierran, I must show that a Universal Turing Machine can simulate the behavior of a Tierran. This has been done separately by both Dr. Ray and myself, since we have programmed a computer, which is a Universal Turing Machine, to run the model Tierra. Thus, the formal proof is the program code of the Tierra model.[6]

TIERRAN EMULATION OF A TURING MACHINE

The question under examination is whether a Tierran is as powerful as a Universal Turing Machine. To prove this constructively, I must show that a Tierran can be programmed to be a Turing Machine.

DEFINITION OF A TURING MACHINE. A Turing Machine ($\mathbf{M}$) can be denoted by a seven-tuple, $\mathbf{M} = (\mathbf{Q}, \Sigma, \mathbf{T}, \mathbf{f}, \mathbf{q_0}, \mathbf{b}, \mathbf{F})$ where[2]:

$\mathbf{Q}$ is the set of possible *states*;

$\mathbf{T}$ is the finite set of allowable *tape symbols*;

$\mathbf{b}$, a symbol in $\mathbf{T}$, is the *blank* symbol;

$\mathbf{\Sigma}$, a subset of $\mathbf{T}$ not including $\mathbf{b}$, is the set of *input symbols*;

$\mathbf{f}$ is the *next move function*, a mapping from $\mathbf{Q} \times \mathbf{T}$ to $\mathbf{Q} \times \mathbf{T} \times \{L, R\}$ ($\mathbf{f}$ may be undefined for some arguments);

$\mathbf{q_0}$, a state in $\mathbf{Q}$, is the *start state*; and

$\mathbf{F}$, a subset of $\mathbf{Q}$, is the set of *final states.*

The Turing Machine thus has a finite set of states $\mathbf{Q}$. It is conceived as a machine, which at any given time is "in" one of the states of $\mathbf{Q}$ (see Figure 1). The Machine begins processing in the state $\mathbf{q_0}$. After reading a symbol on an infinite one-dimensional tape, it may change state, write a symbol on the tape, and move the machine head that reads from and writes on the tape either left (L) or right (R), according to the function $\mathbf{f}$. For a thorough treatment of the theory of Turing Machines, see Hopcroft and Ullman.[2]

[6] The code for the Tierra model can be copied via anonymous FTP from the node `life.shls.udel.edu`.

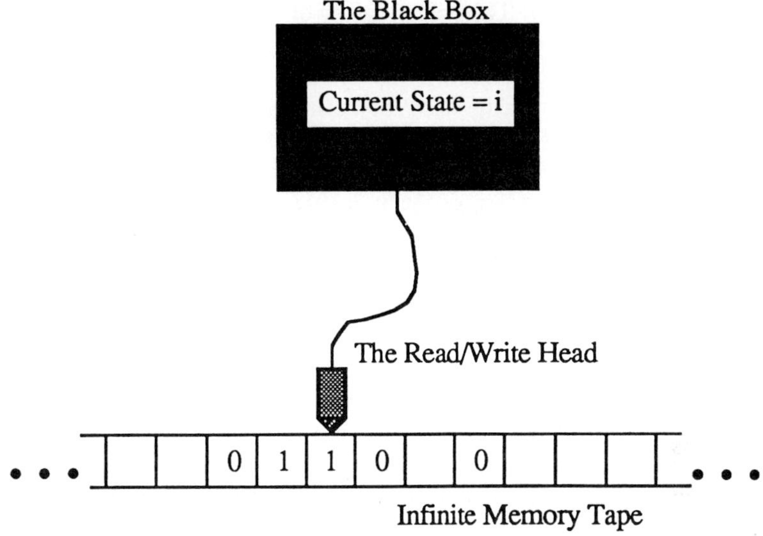

FIGURE 1 An intuitive conceptualization of a Turing Machine. The next move function is contained in the black box. The state of the machine changes depending on the symbol read by the read/write head on the infinite memory tape.

CONSTRUCTION OF A TURING MACHINE. To construct a Turing Machine, I will show how a Tierran can simulate each part of the Turing Machine. A Tierran genotype is a series of adjacent instructions. In this case the Tierran's genotype will contain both the program for simulating the Turing Machine and the tape of input symbols. The two possible input symbols on the tape are ZERO and INC_C. Before reading a tape symbol the Cx register is initialized with a 1, so that after an input symbol is read the contents of Cx will be either 0, 1, or 2, depending on whether a ZERO, blank, or an INC_C was read. So $\Sigma = \{0, 2\}$, effectively a binary code, as in a traditional computer. Let the blank symbol **b** = NOP_0, so that **T** = { ZERO, INC_C, NOP_0}. The read/write head of the Turing Machine can be simulated with the **IP** of the Tierran. After every input symbol there is a CALL to the procedure which implements the **f** function. For example:

```
NOP_0          : an input symbol.
CALL
```

(Template pattern for the **f** function.)

```
ZERO           : the input symbol.
CALL
....
```

Let $\mathbf{Q} = \{\mathbf{q_0}, \mathbf{q_1}, \ldots, \mathbf{q_n}\}$. The current state of the Turing Machine is indicated by the contents of Cx, which holds the integer **i**, if the Turing Machine is in state $\mathbf{q_i}$. Thus, $\mathbf{q_0}$ is simulated by holding a zero in the Cx register. **F** is simply a set of integers. The register Cx can be used for both reading input and for storing the current state by using the Stack and the Dx register to hold one of the two numbers in temporary storage. The following algorithm might be used to swap two numbers, n1 and n2, when n1 is in the Cx register and n2 is in the Dx register:

```
Starting State    : Cx = n1, Dx = n2, Stack = empty.
PUSH_DX           : Cx = n1, Dx = n2, Stack = n2.
MOV_CD            : Cx = n1, Dx = n1, Stack = n2.
POP_CX            : Cx = n2, Dx = n1, Stack = empty.
```

In this way, a Tierran Turing Machine can juggle both the input number and the current state number without losing any information, as long as the stack is not full. Cx is the register that is most easily manipulated, both for loading numbers, via the ZERO, OR1, SHL, DEC_C, and INC_C, and for branching using the IF_CZ instruction. To call one of a variety of procedures, depending on the value of some variable, the Tierran organism must load the number into the Cx register, then keep decrementing it using DEC_C and testing if it is zero, using the IF_CZ.

CALL PROCEDURE **n**:

```
Current State              : Cx = n.
IF_CZ
CALL
(Template pattern for procedure if n = 0.)
DEC_C                      : Cx = n - 1.
IF_CZ
CALL
(Template pattern for procedure if n = 1.)
DEC_C                      : Cx = n - 2.
...
IF_CZ                      : Cx = n - i.
CALL
(Template pattern for procedure if n = i.)
DEC_C                      : Cx = n - (i + 1).
```

and so on.

This is the algorithm that would be used when the Turing Machine needs to determine its current state as well as when the Turing Machine determines whether it read a ZERO, INC_C, or NOP_O as input.

The only remaining aspect of the Turing Machine that needs to be simulated is **f**. The *next move function* is really the heart of a Turing Machine. I now have the algorithmic tools to determine the current state and the current input. These are the two arguments to **f**. It is then trivial to use the two arguments to call a procedure $\mathbf{f}(n, t)$. This procedure must be able to change the current state, write a symbol on the tape, and move the "read/write head" either left or right one input symbol on the tape.[7] I will give an algorithm to deal with each of these challenges.

CHANGING STATE. Changing the current state is easy. Any number can be constructed, within the limits of the register by first clearing the Cx register with a ZERO instruction and then using OR1 and SHL to construct the bits of the number.

WRITING A SYMBOL. Writing a symbol to the tape can be done by calling one of three procedures: "write ZERO," "write INC_C," or "write NOP_0." All three procedures have a similar structure. The current position of the read/write (r/w) head of the simulated Turing Machine is on top of the stack.

WRITING A SYMBOL:

```
(Template pattern for the symbol to be written.)
INC_A               : This is a dummy instruction that will never get
                    executed.  It is necessary to separate the template
                    pattern from the symbol to be written, which may be
                    NOP_0.
(Symbol to be written) : either ZERO, INC_C, or, NOP_0.
(Template pattern identifying the procedure)

⇒                   : The starting point for the execution of the
                    procedure.

ADRB
(Template pattern for symbol to be written)
                    : Loads Ax with address of the above INC_A.
INC_A               : Ax = the address of the "Symbol to be written."
MOV_AB              : Ax = Bx = location of symbol to be written.

POP_AX              : Ax = r/w head position, Bx = location of symbol to
                    be written.
PUSH_AX             : Stack = r/w head position, Ax = r/w head position,
                    Bx = location of symbol to be written.
MOV_IAB             : finally writes the symbol to the tape.
```

MOVE THE READ/WRITE HEAD. The final challenge is to implement the movement of the read/write head of the Turing Machine. The read/write head was the **IP** before the CALL instruction was executed. Since an address is just an integer, it can be manipulated in the Cx register. The algorithm for moving the read/write head one input symbol to the left follows. Again, the current position of the read/write head has been placed on the top of the stack by the CALL to the **f** procedure. Moving one input symbol to the right requires only that the DEC_C instruction be replaced by the INC_C instruction in the following algorithm.

[7]The tape has CALL and pattern instructions between the input symbols, so moving the read/write head one input symbol in either direction requires skipping over those instructions.

MOVE READ/WRITE HEAD LEFT:

```
Current State: Cx = n, Stack = r/w head.
POP_BX     : Bx = r/w head, Cx = n, Stack = empty.
PUSH_CX    : Bx = r/w head, Cx = n, Stack = n.
PUSH_BX    : Bx = r/w head, Cx = n, Stack = n, r/w head.
POP_CX     : Bx = r/w head, Cx = r/w head, Stack = n.
DEC_C      : Bx = r/w head, Cx = r/w head - 1, Stack = n.
...
(DEC_C will have to be repeated m times in order to skip over the
CALL and pattern template instructions that separate the input
symbols.)
           : Bx = r/w head, Cx = r/w head -m, Stack = n.
PUSH_CX    : Bx = r/w head, Cx = r/w head - m,
                       Stack = n, r/w head - m.
POP_BX     : Bx = r/w head - m, Cx = r/w head - m, Stack = n.
POP_CX     : Bx = r/w head - m, Cx = n, Stack = empty.
PUSH_BX    : Cx = n, Stack = r/w head - m.
(Then I prime Cx for reading the next input symbol, and return to
reading the input.)
MOVCD      : Cx = n, Dx = n, Stack = r/w head - m.
ZERO       : Cx = 0, Dx = n, Stack = r/w head - m.
INC_C      : Cx = 1, Dx = n, Stack = r/w head - m.
RET        : Cx = 1, Dx = n, Stack = empty, IP = new r/w head
           position = the old r/w head - m.
```

I now have algorithms for reading input, calculating the current state, and branching to the appropriate procedure for executing $\mathbf{f}(n,t)$. The procedure $\mathbf{f}(n,t)$ itself is composed of three algorithms: calculating a new state, writing a symbol to the tape, and moving the read/write head. Combining all of these algorithms is a trivial if tedious exercise resulting in a fully functional Turing Machine. Thus, a Tierran is theoretically as powerful as a Turing Machine and so is also computationally complete.

LIMITATIONS OF TIERRANS

Although the Tierran language is theoretically computationally complete, the Tierran's behaviors are highly limited by the nature of the instruction set. At present, examining the contents of a memory cell can only be accomplished by executing the instruction in that cell and then examining the change in the Tierran's state. This only works well for instructions that have easily determinable effects on the Tierran's state, like ZERO. This is how I implemented the reading of input symbols in the Tierran Turing Machine. Ray has given Tierrans freedom to read (in the sense that they may find locations using the ADR family of instructions) and copy the instructions of any other Tierran in the Tierran memory space. However, the freedom to examine instructions is of little use when the Tierran can only determine the memory contents indirectly. It would be a rather simple addition to break the MOV_IAB instruction into a READ and a WRITE instruction.

Furthermore, Tierrans are restricted to writing into their own genome and their daughter's genome prior to "cell division." This restriction is analogous to the protection afforded by cell walls and prevents the Tierrans from killing each other off in the same way as their conceptual ancestors, the Core Warriors, eradicate each other.[3,4,5,6] Yet, without a READ instruction, writing into one's own genome is largely a blind and suicidal exercise. With a READ instruction, self-modification could be a practical and perhaps an adaptive exercise for a Tierran.

As a final example of the power of having separate READ and WRITE instructions, there follows the outline of an algorithm for error checking in reproduction. Like the analogous DNA repair mechanisms, such a program would severely diminish the rate of mutations due to copying errors in reproduction.

```
READ parental gene.
WRITE it into the daughter's genome.
READ the daughter's gene that was just written. Compare it to the parental gene.
If they are different, loop to the top and try again.
```

CONCLUSION

I have shown that the Tierran language is computationally complete. This means that Tierrans may theoretically evolve to perform any task a digital computer can perform. However, this is only a theoretical possibility. In practice, a Tierran is severely limited by the nature of the Tierran language, as well as the size of the memory space and the disruption of mutations.[8] By splitting the MOV_IAB instruction into a READ and a WRITE instruction, Tierrans would gain the possibility of examining and reacting more intelligently to their environment.

[8]As the size of an organism grows, so does the probability that the organism will be disrupted by a deleterious mutation before it can complete on iteration of its algorithm.

REFERENCES

1. Aho, A. V., J. E. Hopcroft, and J. D. Ullman. *The Design and Analysis of Computer Algorithms.* Reading, MA: Addison-Wesley, 1974.
2. Dewdney, A. K. "Computer Recreations: In the Game Called Core Wars, Hostile Programs Engage in a Battle of Bits." *Sci. Am.* **250** (1984): 15–19.
3. Dewdney, A. K. "Computer Recreations: A Core War Bestiary of Viruses, Worms and Other Threats to Computer Memories." *Sci. Am.* **252** (1985): 14–19.
4. Dewdney, A. K. "Computer Recreations: A Program Called MICE Nibbles Its Way to Victory at the First Core War Tournament." *Sci. Am.* **256** (1987): 8–11.
5. Dewdney, A.K. "Computer Recreations: Of Worms, Viruses, and Core War." *Sci. Am.* **260** (1989): 90–93.
6. Hopcroft, J. E., and J. D. Ullman. *Introduction to Automata Theory, Languages, and Computation.* Reading, MA: Addison-Wesley, 1979.
7. Ray, T .S. "An Approach to the Synthesis of Life." In *Artificial Life II,* edited by C. G. Langton, C. Taylor, J. Farmer, and S. Rasmussen. Santa Fe Institute Studies in the Sciences of Complexity, Proc. Vol. X, 371–408. Redwood City, CA: Addison-Wesley, 1991.

Nils A. Baas
Department of Mathematical Sciences, NTH, University of Trondheim, N-7034 Trondheim, Norway; e-mail: baas@imf.unit.no

Emergence, Hierarchies, and Hyperstructures

Emergence often occurs in the study of artificial life. We introduce a general framework for studying emergence and give a formal definition.

We distinguish between two major types of emergence: deducible and observational, and interpret, for instance, Gödel's incompleteness theorem in logic as a case of observational emergence.

We show how this naturally leads to higher-order structures and hierarchies—through a kind of evolution. Furthermore we unify these ideas in the general notion of a hyperstructure which bridges the gap between treelike and network structures. This may be a useful notion in the study of self-organization and artificial life. We look for causes of emergence and corresponding mathematical notions, and emphasize the importance of nonlinearity and limits.

INTRODUCTION

The notions of emergence and hierarchy have a long history in science, and recently they have played a major role in the description and discussion of both natural and artificial life forms. They also naturally enter descriptions of evolution and complexity. However, there are many definitions and approaches that may seem quite different from each other, though the ideas behind are often similar. A synthesis and unifying approach seems desirable. Here we will suggest a general framework for studying emergence and hierarchies extending them to a new notion which we call a *hyperstructure*. This will be a unifying concept. *Hyperstructures* may lead to a better understanding of the problems connected with emergence and hierarchies, clarifying problems and leading to new ideas and constructions.

In order to understand a complex structure or system, we must be able to describe it in a reasonable way. We think that *hyperstructures* will be very useful in doing that.

EMERGENCE

"The whole is more than the sum of its parts."

This old philosophical statement has, for more than 2000 years, generated a lot of discussion in philosophy and science—and still does. The reason is that it really pinpoints a very fundamental scientific problem.[4,38] Intuitively emergence is used as a name for creation of new structures and properties. Therefore it has been intimately connected with the theory of evolution. We will not discuss the history and development of emergence—nor present-day approaches, but we refer the reader to the bibliography and the general references. Which are the basic ingredients we need in order to explain what we mean by emergence? We need some primitive objects or entities. For something *new* to be created we need some dynamics or better *interaction* between the entities. But to register that something *new* has come into existence, we need mechanisms to observe the entities. This is intuitively the idea we have in mind and would like to express in a more formal language. But we can only represent events for which we have "meters."[36]

We will start out with a general notion of *structures* as our primitive objects or entities. A structure may, in our sense, be of an abstract or physical nature, e.g., systems, organizations organisms, machines, spaces, fields, symmetries, concepts, etc. Furthermore we assume that we have some kind of observational mechanism (or family of such) in order to evaluate, observe, and describe the structures. This could be an internal mechanism of the system as well as an external. Of course, such mechanisms can be defined in many ways—classically as numerical functions but, as we have seen in quantum mechanics, more sophisticated ones may be needed (operators). In fact they may be very general (functors in a category theory setting).[19,35]

We want to give a general procedure for how to construct a new structure from a family of old ones. So we start out with a family of structures

$$\{S_i\},\ i \in J \text{ (some index set, finite or infinite).}$$

Then we apply our observational mechanisms—*Obs*—to obtain properties of the structures

$$S_i,\ Obs(S_i).$$

Next we subject the S_i's to a family of interactions—*Int*—using the properties registered under the observation (which could be a dynamic process). Hence we get a *new* kind of structure:

$$S = R(S_i,\ Obs(S_i),\ Int)_{i\in J}$$

where R stands for the result of the construction process. We call S a *second-order structure*—as opposed to the S_i's which we consider *first-order structures*—the primitives of the theory. Specifying the observational mechanism is basic and we may also require some stability of the observed properties. The interactions may be caused by the structures themselves or imposed by external factors. Basically we are describing how to form a totality out of a family of structures.

In mathematics we have many examples of such constructions. Of particular interest in connection with emergence is that of *limits* in so-called categories. We will discuss this later. But we do not want to restrict ourselves to this situation. Also our notion of observation plays a crucial role. *Obs* is related to the creation of new categories in the systems (categories here in the philosophical meaning of the word). In some cases the structures and the interactions may not be separable; *Obs* will then act upon them jointly.

Let us prepare the notation for further steps. First-order structures—our primitives—denoted by:

$$S^1_{i_1},\ i_1 \in J_1,$$

second-order structures:

$$S^2 = R(S^1_{i_1},\ Obs^1,\ Int^1),$$

and families:

$$\{S^2_{i_2}\},\ i_2 \in J_2.$$

Second-order structures may now be observed by observational mechanisms Obs^2 (they may be equal, overlap, or disjoint from Obs^1), and may also observe the first-order structures of which they consist. (We may allow first-order structures to be considered also as second-order structures—$S^2 = R(S^1)$, in the same way that in set theory an element of a set x may also be considered as a subset $\{x\}$.)

The collection S^2—which is more than a mere aggregate—is considered as a new unity, whose properties we measure by Obs^2.

In this quite general setting we propose the following:

DEFINITION OF EMERGENCE.

$$P \text{ is an emergent property of } S^2$$

iff

$$P \in Obs^2(S^2), \text{ but } P \notin Obs^2\,(S^1_{i_1}) \text{ for all } i_1.$$

We call S^2 an *emergent structure.* So in this sense we may say that:

"The whole is more than the sum of its parts."

This is what we will call *first-order emergence.*

When we obtain families of second-order structures, new interactions between these may also occur (and be observed). In this sense second-order interactions may emerge (as new laws of the system). Second-order interactions may also be introduced externally using the new emergent properties. It is important to be aware of both these alternatives. It is also possible in this framework for observational mechanisms to emerge.

It may be useful to distinguish between two different notions of emergence—for quite deep reasons, both mathematically and philosophically.

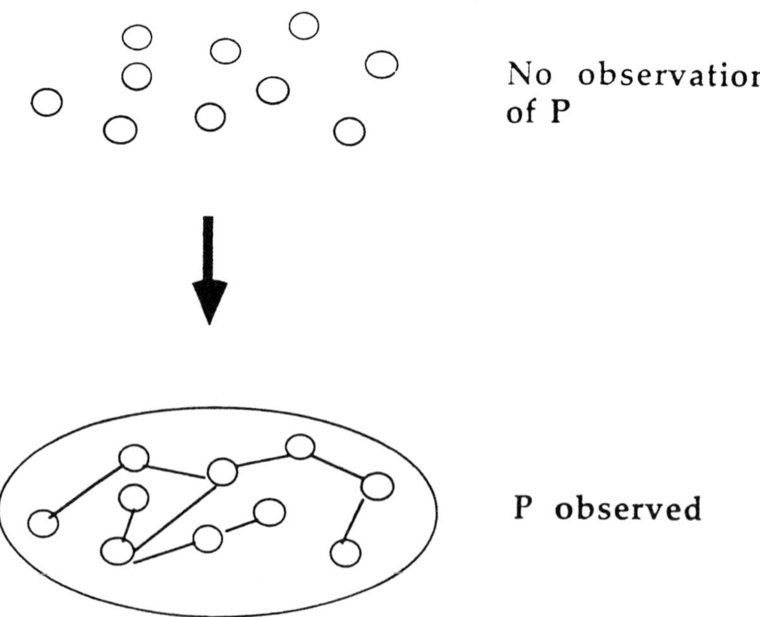

FIGURE 1 Emergence.

A. *Deducible or computable emergence* means that there is a deductional or computational process or theory D such that $P \in Obs^2(S^2)$ can be determined by D from $(S^1_{i_1}, Obs^1, Int^1)$

B. *Observational emergence* if P is an emergent property, but cannot be deduced as in **A**.

Let us indicate a few examples.

DEDUCIBLE EMERGENCE.

1. Coupled or compositional structures (finite or infinite) where the components interact in known ways such that we can deduce or calculate the *new* composite properties. This is the case in many types of engineering constructions.
2. Nonlinear dynamical systems where, for example, simple systems interact to produce *new* and complex behavior. Chaotic dynamical systems may theoretically be of type A, but in practice may be usefully thought of as type B, and hence cases of borderline emergence.
3. Phase—transitions—emergence in the thermodynamic limit. Broken symmetries.[2,3]
4. A simple and explicit example is the following. Given two simple dynamical systems:
$$S^1_1 : \dot{x} = ax,$$
$$S^1_2 : \dot{y} = -bx.$$
$Obs = Obs^1 = Obs^2$ = set of various properties of the system—including existence of periodic solutions. P is the property of having a periodic solution. Therefore $P \notin Obs(S^1_1)$ and $P \notin Obs(S^1_2)$. We then let the systems interact—to form a predator/prey model à la Lotka-Volterra:
$$S^2 : \begin{array}{l} \dot{x} = ax - cxy \\ \dot{y} = -by + dxy \end{array}$$
$$(a,\ b,\ c,\ d > 0).$$
It is a simple well-known fact that $P \in Obs(S^2)$, hence is an emergent property (of type A) caused by the nonlinear couplings or interactions.
5. Depending on our choice of observables we can also create emergence linearly. Let us observe oscillatory solutions. We may consider linear systems with no oscillation, but when coupled linearly we observe oscillatory solutions.
This shows the importance of Observables in our theory, and also how it gives great flexibility.
6. In topology a manifold is a space which is glued together by locally euclidean pieces.
$$M = \cup_{i=1}^{n} V_i / \sim,$$

To study or observe such spaces one uses for instance cohomology theories K^*. We know that $K^*(V_i) = 0$ (trivial), but $K^*(M)$ may be very complex. Let $Obs = Obs^1 = Obs^2 =$ set of properties including measuring the nontrivial complexity of K^*. $P =$ property of nontrivial K^*. Then $P \notin Obs(V_i)$, but

$$P \in Obs(M).$$

In this sense we may say that the nontriviality of $K^*(M)$ has emerged from the glueing construction. Lots of similar examples exist in topology and algebra where for example glueing is replaced by limits (see later section).

7. The Scott model of the λ-calculus.
Let X be a set; $[X \to X] =$ the set of mappings from X to X. Consider the *set* equation:

$$X = [X \to X].$$

Within set theory this is not solvable. In constructing a model of the λ-calculus D. Scott considered spaces D_i with a special topology and continuous mappings $[D_i \to D_i]$. He started with a very simple set D_0, formed:

$$D_1 = [D_0 \to D_0]$$

and, inductively,

$$D_{i+1} = [D_i \to D_i]$$

in such a way that they formed an inverse system

$$D_0 \leftarrow D_1 \leftarrow \ldots \leftarrow D_i \leftarrow \ldots \leftarrow D_\infty$$

where D_∞ is the so-called inverse limit.
For all i

$$D_i \neq [D_i \to D_i],$$

but in the limit

$$D_\infty = [D_\infty, D_\infty].$$

Let us put this into our framework. Set

$$S_i^1 = D_i$$
$$Int^1 = \leftarrow .$$

$Obs(D) = Obs^1(D) = Obs^2(D) =$ set of properties of D—including being a solution of the equation $D = [D \to D]$
$P =$ property of being a solution of $D = [D \to D]$. Then we have:

$$P \notin Obs(D_i), \quad \text{but}$$

$$P \in Obs(D_\infty).$$

So we may say that the property of being a solution to the equation emerged through the inverse limit construction. For details, see Hindley and Seldin[27] and Stoy.[49]

In these examples we assume that the various properties can be decided by well-defined procedures. Therefore *Obs* is really a procedure or algorithm leading to a set of properties.

Some of these examples may look rather abstract and formal, but we think they will help us in better understanding the nature of emergent structures and properties.

OBSERVATIONAL EMERGENCE.

1. A profound example here is Gödel's theorem[26] which states that in some formal systems there are statements which are true, but this cannot be deduced within these systems. Here observation is the truth function. For example, first-order predicate calculus is complete in the sense that every true statement can be deduced from the axioms. If we now add further axioms to cover the theory of arithmetic, Gödel's theorem says that there will be true statements that cannot be deduced. Here we see the additional axioms as "added interactions" among the well-formed expressions.
2. Consider a dynamical system in some space. View the dynamics as an interaction, and associated fractal-like sets like attractors, repellers, Julia sets, etc. as newly created second-order structures. *Obs* decides membership in these sets. Penrose[40] raised the question: Is the Mandelbrot set decidable (i.e., can membership be algorithmically decided; in our terminology, is it deducible)? In order to make the question mathematically meaningful, it had to be put into a new theory of computation over the reals developed by Blum, Shub, and Smale.[15] They showed that most Julia sets are undecidable. In 1991, Shishikura[47] completed the proof of the conjecture that the Mandelbrot set is also undecidable. It is a deep result. (In 1990, D. Sullivan sketched a proof.)
   So in a suitable choice of structure in our framework we may view Julia sets (most of them) and the Mandelbrot set as observationally emergent structures.
3. Another example of the difference in our two notions of emergence is the following. In a formal string system we have a syntactic relation:

$$s = R(s_1.,,,.s_k).$$

Take as our *Obs* a semantic "meaning function" M. The question about semantic compositionality is then:
Does there exist a (computable) function F such that

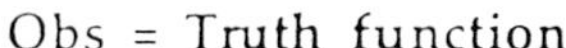

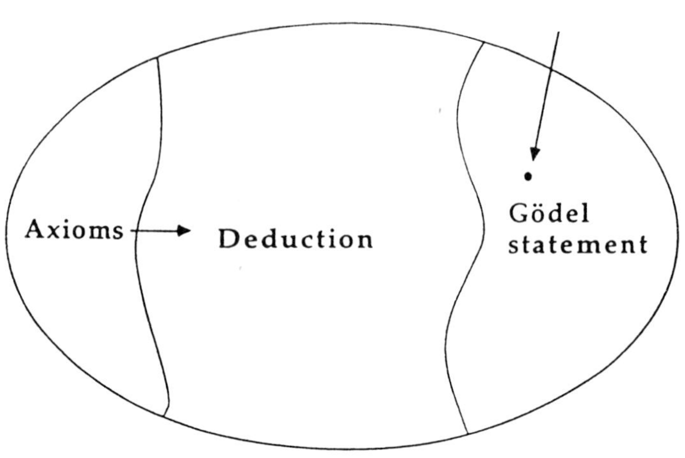

FIGURE 2 Gödel's Theorem.

$$M(s) = F(M(s_1), \ldots, M(s_k))?$$

If yes, $M(s)$ is *deducibly emergent.*
If no, $M(s)$ is *observationally emergent.*

We have here only considered two basic classes of emergence. Clearly both can be further refined and subdivided according to additional complexity conditions. The notion of observational emergence is quite profound as the examples show. Gödel's theorem as observational emergence may at first appear as a rather negative result for constructive purposes. But it is not! We have incorporated the observational mechanisms in such a way that they can be usefully taken into account in further constructions—even if they are not deducible or computable. This is an important aspect which will be used in our *hyperstructure* concept.

Our discussion here shows that even in formal, abstract systems profound emergence may occur—particularly in artificial life models. This seems to be contrary to the claim by Cariani.[17]

We here give a general framework for studying emergence, but we have not discussed the detailed causes of emergence. Let us mention as an observation two basic causes:

1. the interactions are nonlinear—meaning that there is no superposition principle.
2. The number of structures (systems, entities) is often large (number $\to \infty$).

When these two conditions are satisfied, emergent phenomena will often occur. Emergence in our sense may also be viewed as a process of self-organization.

We will end this section with an old provacative question: Does emergence in some cases—as contended by Mill, Lewes, and Morgan—take the rank among the *"laws of nature"*? (Interesting aspects of emergence may be found.[1,4,5,16,21,23,28,30,31,37,38,41,44,46] Some of the material in this paper has also appeared in Baas.[11,12,13,14])

HYPERSTRUCTURES

In our definition of emergence we considered structures at two levels: primitives or first-order structures and second-order structures.

We may say that a structure formed as S^2 is a *complex* structure relative to its primitives. The "observer"—which could be just the environment of a system—

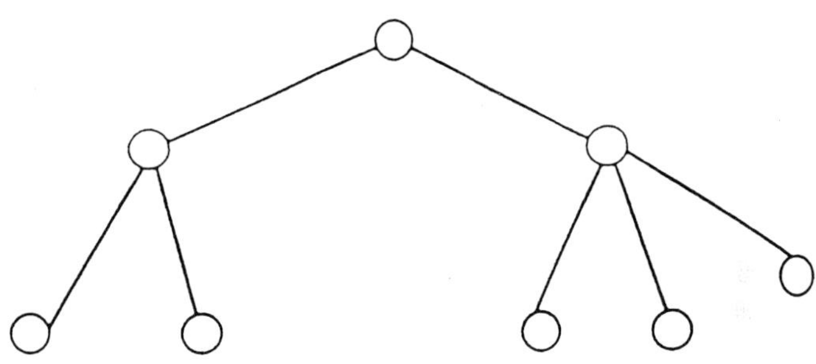

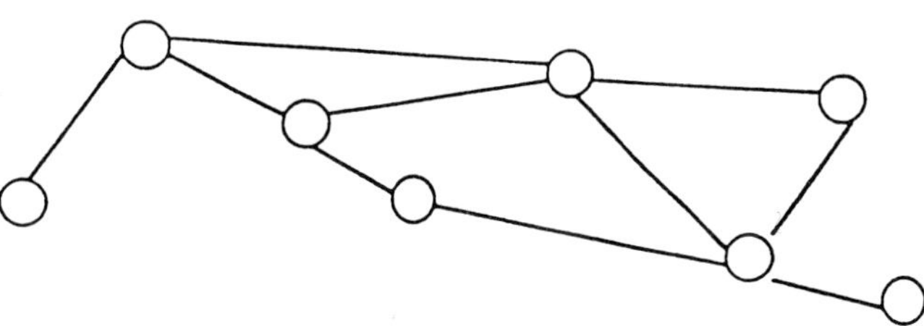

FIGURE 3 Tree structure and network structure.

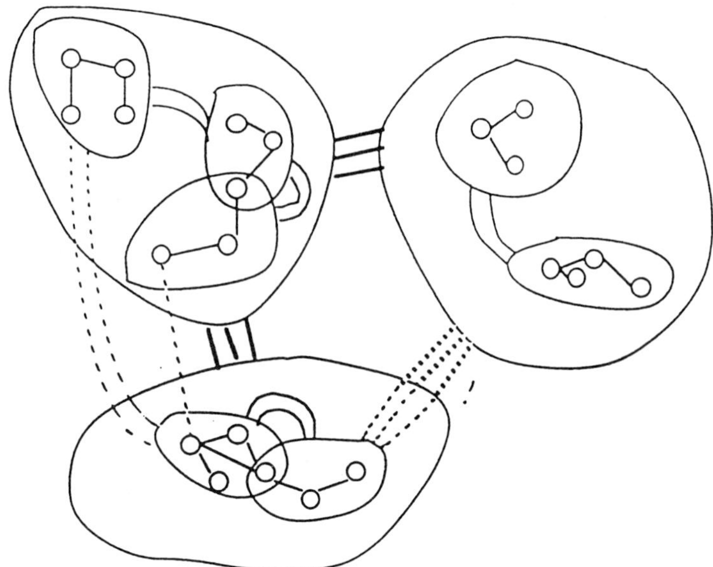

FIGURE 4 Hyperstructure: Example of third-order interaction graph of a hyperstructure allowing cumulative interactions (dashed lines) and even overlapping aggregates.

represents a kind of *selection* mechanism giving rise to a process creating *new levels of structure*. This also gives *adaptation* relative to *Obs* (or environment). So we have a mechanism of *evolution* towards a higher-order structure. Level structures are often thought of as *hierarchies*, but hierarchies with the traditional tree-type structure are insufficient for our purposes—though they will appear as special cases of our construction (see Figure 3). We will return to this later.

Then we proceed to form families of second-order structures. We see them form *new* second-order properties and interactions, giving:

$$S^3 = R(S^2_{i_2}, Obs^2, Int^2).$$

In general Int^2 may be interactions based on both first- and second-order properties.

This process proceeds and we see that at each level of construction *new* properties or *new* behavior may emerge, giving room for *new interactions*, and hence each level is necessary in order to get the last level's properties. The process of *Obs* is essential and shows that this construction is much more sophisticated than a recursive procedure—which may be considered a special case. The new interactions may not be meaningful at lower levels. But it is important to emphasize that the interactions may *emerge* or be introduced *externally*.

Finally, we form the *N-th order structure:*

$$S^N = R(S^{N-1}_{i_{N-1}},\ Obs^{N-1},\ Int^{N-1}),\ i_{N-1} \in J_{N-1}.$$

This is a short way of writing an Nth order structure which in principle depends on all lower levels in a cumulative way which is more general than a recursive structure. The whole process we call: *"the higher-order structure principle"*—the result of which we call a *hyperstructure*—in this case of order N.

The underlying interaction graph in Figure 4 is much more subtle than merely a trees (as for ordinary hierarchies) or just general graphs or networks.

We think that this is a very general, useful, and important process which may be applied to a variety of structures (systems) in order to obtain new—*fundamentally new!*—structures with new emergent properties. Through the level structure we also get orders of emergence and complexity—indicating a classification of such structures. *Nth-order emergent properties are those of* S^N*, which is a cumulative structure.*

We will not provide technically detailed rules of the interactions at the various levels, but just refer to Figure 4 which describes the intuition. It is worth noting that we allow (not require!) overlapping aggregates and that creation of high-level interactions may cause changes in lower-level interactions. Hyperstructures allow for both "downward" and "upward" causation. As a model one may think of interactions of organizations and social systems. If needed, interactions may also be deleted (or forgotten) so that the system is only partially cumulative. In the next section we will discuss in more detail how the classical notion of hierarchy is extended.

It is our conviction that *complexity often takes the form of a hyperstructure.* This is an extension of Simon's thoughts about complexity and hierarchies.[48] The formation principle we have described for hyperstructures may be very useful in understanding, for instance, cognitive and constructional processes. It is a framework for studying problems of *complex design.*[18] In a sense it reflects the principle of evolution where observation and interaction are due to the various environments and the structures themselves. But in our model we allow for introducing new observations and interactions and controlling them, hence being able to influence the evolution and even "speed up" evolution. This flexibility clearly makes the structure more easily adaptable to various changes, and the *Obs* and *Int* mechanisms will allow for creativity (of unpredictable nature), learning, etc.[22]

It is important to notice that the principle can be used to synthesize new structures by varying the *Obs* and *Int* structures. To a large extent we can vary the "laws" or rules at each level, but in real-life evolution they are due to nature. *Obs* and *Int* may even vary continuously—by some kind of sensor device. In other structures, *Obs*, for example, may play no role—like in higher-order logic considered syntactically.

An important way to look at the interactions is as communication or information processing. For every level the information-carrying capacity increases.

In the successive formation of levels in a hyperstructure we may have to impose more conditions depending on the situation—especially conditions on *stability* and *autonomy* seem to be relevant.

Let us give some examples of known structures that may naturally be considered as hyperstructures—but without details.

1. Higher-order organisms viewed as evolutionary products.
2. Hypercycles in the sense of Eigen and Schuster.[20] Here the *Obs* mechanism register the properties leading to intercyclic coupling and hence higher order cycles.
3. Coupled nonlinear dynamical systems may often be viewed as hyperstructures through the appearance of various kinds of attractors.
4. Higher-order logic and set theory. In the syntactic build up no use of *Obs* is made. *Obs* would normally be thought of as part of the semantic structure. In a sense "oracles" may be considered as observers. Furthermore, higher-order (or high-level) languages and metaconstructions may also be viewed as hyperstructures.
5. Higher-order graphs and networks—formalizing the structure illustrated in Figure 4. This also extends the already existing notion of a hypergraph—which in our terminology is a second-order structure.
6. Various types of hyperspaces in mathematics. These are spaces of the form 2^X (or suitable subsets or iterations of this construction). $2^X = \{\text{set of subsets of } X\}$. Fractals are naturally studied in spaces of this type.
7. Cognitive processes where we successively build up knowledge by "observations" and "interactions."
8. General technological constructions designed in a modular or levelwise way and metatype constructions in systems theory.

Hyperstructures should be thought of a language—a framework—to describe complex problems, solutions, and situations. We may also generate a whole series of interesting questions and notions trying to associate hyperstructures with well-known structures, systems, concepts, symbols, signs, etc., like hypernetworks, hyperlanguages, hyperdynamics (e.g., renormalization), hyperalgorithms, hypercomputation (extending Emergent Computation[21]), hyperforms, and hyperpatterns. Also physical and social systems may be organized in this way leading, for example, to hypermaterials and hyperorganizations. In all the cases we will have to specify *Obs* and *Int*.

This is a very general and broad program, and further progress will depend on the study of *specific* examples of these structures from various angles: mathematical, physical, chemical, biological, and computational.

One can also think of organizing populations of machines or robots in such a way that their interactive dynamics lead to completely new properties and behavior; see Baas.[12]

What about the relation to *Life* and *A-Life*? We are convinced that the notion of *hyperstructure is essential in describing both real and artificial life forms*, and hence could be a useful framework in future A-Life work. We think that many A-Life models can fruitfully be put into a hyperstructure framework— i.e., the models of Fontana,[24,20,24] Hogeweg,[29] Rasmussen, et al.,[42] and Ray.[43] Hyperstructures help us understand the evolution and synthesis of complexity which is also basic to understanding the principles of life.

In Baas,[8,9,10] we discussed *cancer* from a hierarchical point of view, related to the concepts in the next section.

Let us conclude this section with a very intriguing question we have been concerned with for many years: Let us assume that higher order organisms is governed by some kind of hyperstructure.

Is *cancer* basically caused by a breakdown or failure of this hyperstructure?

Could we also along these lines introduce artificial cancer in the study of artificial life? We will return to this elsewhere.

SELF-ORGANIZATION

Emergence is closely related to self-organization. It seems reasonable to require that a self-organizing system should exhibit emergent properties. In many models of self-organization and artificial life[24,42,43] new structures arise or emerge. But, when we look at them carefully, these structures are basically what we here call second-order structures. Higher-order structures as we have defined them or as we see them in nature seem more difficult to produce from first principles only. In some cases one may produce higher-order structures by introducing external factors or changing boundary conditions. In other cases, one may ask whether the models considered are too simple, just need further study, or are lacking some principal ingredients. However, some models look promising for producing third order structures from first principles under certain conditions.

In any case it is vital to understand the conditions under which higher-order structures (hyperstructures) occur and, furthermore, when interactions between hyperstructures automatically emerge. Hence the evolutionary process can continue, meaning the structure is *evolvable.* Conditions for *evolvability* are important to study. The structure of the "environment" will also play an important role here.

It would be very interesting to see new models in this area where higher-order hyperstructures naturally emerge. Our point of view is twofold:

1. to understand the *generation* of hyperstructures from first principles, and
2. to understand how to guide (externally) the *design* of the hyperstructure—representing, in a sense, *man-made evolution* (see Baas[12]).

HIERARCHIES

One of the most remarkable things about living systems is their levelwise organization—in the sense that the system can be divided into units, each unit consisting of subunits and each subunit consisting of sub-subunits, etc. We are only able to observe a finite number of levels. This kind of organization is often called hierarchical.

Let us look at a few examples of intuitive hierarchies in various fields:

Biology: Organisms $<$ organs $<$ tissues $<$ cells $<$ macromolecules $<$...
Linguistics: Texts $<$ sentences $<$ words $<$ phonemes $<$...
Technology: Composed technological products.
Administration: Division of labour or chain of command.

In general these systems consist of basic units, these units form higher-order units at another level, and so on.

By a hierarchical system we therefore understand a sequence of sets—the levels—

$$X_1, X_2, \ldots, X_n,$$

and an ordering

$$X_1 < X_2 < \ldots, < X_n,$$

where X_i denotes the units of level i. We think of X_1 as the most global level and X_n the most local.

The following definition is quite useful:

DEFINITION [8] For a given set X we define two types of hierarchical structures.

1. X is a **division hierarchy** if there is associated with it a system of levels X_1, X_2, ..., X_n such that $X_n = X$ and the ordering is being realized by a composition of mappings

$$X_1 \leftarrow X_2 \leftarrow \ldots \leftarrow X_n$$

or, even more generally, $X_i < X_{i+1}$ meaning a mapping $s_i : X_i \rightarrow 2^{X_{i+1}}$ ($=$ set of subsets of X_{i+1}). An example is the biology hierarchy mentioned above.
2. X is a **control hierarchy** if $X = \cup_{i=1}^{n} X_i$ and we have an ordering

$$X_1 < X_2 < \ldots, < X_n$$

being realized as in 1. An example here is the command hierarchy mentioned above. Graphically these systems are represented by a tree in Figure 5.

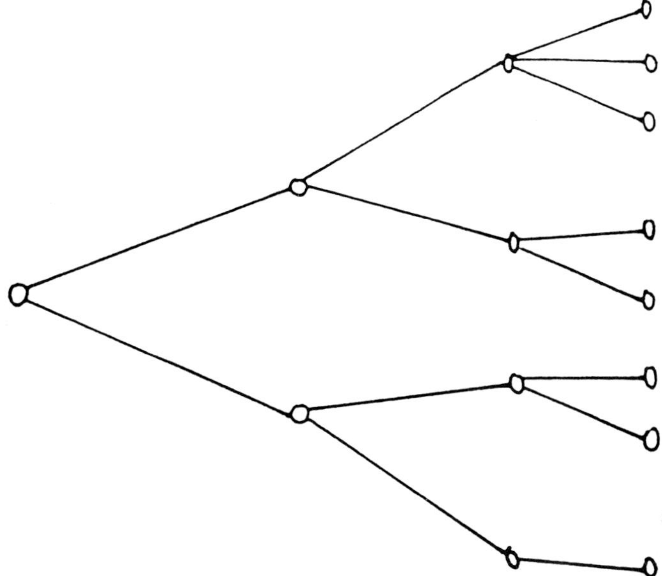

FIGURE 5 Level: $1 < 2 < 3 < 4$.

So hierarchies in this sense are special, but important cases of hyperstructures and occur in most taxonomic situations. The levels are described as mere aggregates without specifying interactions and observation mechanisms. We should also keep in mind that this gives a description of a hierarchy in existence—not of how it was constructed or evolved—as hyperstructures do. Given a hierarchy we would like to describe each level by its properties which we collect in a property—or state—space:

$$X_i \mapsto S_i .$$

These spaces should be thought of as semantic representations of the X_i's.

In many hierarchies it is reasonable to assume a compositionality (or modularity) principle for properties which we will formulate as given by a composition of mappings:

$$S_1 \leftarrow S_2 \leftarrow \ldots \leftarrow S_n .$$

This corresponds to Frege-type compositionality of meaning as used in linguistics.

Therefore it is our proposal that many hierarchical systems should be described by:

1. $X_1 < X_2 < \ldots < X_n$ (syntactic/structural part).
2. $S_1 \leftarrow S_2 \leftarrow \ldots \leftarrow S_n$ (semantic/functional part).

Here the orderings "$<$" distinguish the levels of entities and the mappings "$\rightarrow$" relate the properties or states of the various levels.

The model will then depend on how we describe the levels—which parameters we choose—and how they fit together in the space S_i. We must decide the structure of the S_i's and the mappings.

It is clearly important to have a model or description of how the levels are connected. In Koestler's[32] terminology each level has self-assertive and integrative properties. The mappings describe the balance between these. The processes at each level have to respect these mappings as a kind of level-connecting law.

In previous papers,[7,8,9,10] we have studied a mathematical model of hierarchical systems by assuming that the S_i's are smooth manifolds and the mappings are smooth as well. This enables us to apply powerful mathematics[7,49] and introduce interesting stability notions for hierarchical systems and relate these to biological systems. We suggested a model of cancer as an instability or breakdown of the hierarchical structure. This we now extend to a hyperstructure model.

How does this relate to emergence and hyperstructures? Intuitively we would like to see higher levels and their properties emerge from the lower levels. We have seen that hierarchies are special cases of hyperstructures and we suggest the following synthesis:

A level structure

$$X_1 < X_1 < ... < X_n$$

is always formed as a **hyperstructure**.

The properties registered successively by the *Obs*—mechanisms are collected in the spaces S_i. How these spaces are related will, of course, depend on the degree of emergence—deducible or observational, but this remains to be studied in examples. However, the model we suggest is that in the case where we have a compositionality principle the relations are given as a composition of mappings:

$$S_1 \leftarrow S_2 \leftarrow ... \leftarrow S_n.$$

But in other cases the state—or property spaces—may be built up by a more complicated hyperstructure principle.

Let us end this section on hierarchies with some interesting quotations—which in our opinion also extend to hyperstructures.

A. Koestler:[32] "We do not know what forms of life exist, but we can safely assume that whenever there is life, it is hierarchically organized."

H.A. Simon:[48] "Among possible complex forms, hierarchies are the ones that have the time to evolve.

"My central theme is that complexity frequently takes the form of hierarchy..., and hierarchy, I shall argue, is one of the central structural schemes the architect of complexity uses."

LIMITS AND NONLINEARITY

The formation principle for hyperstructures is basically that as soon as we have a family (collection) of structures, objects, or systems interacting (nonlinearly), new phenomena may be observed and a new higher-order structure—a next-level object—has been formed. Often in the most interesting cases the number of structures is large or infinite. So in studying them by finite means we have to use limits.

But limits have often been used in creating new structures and properties. Let us mention some examples which are so familiar that we may easily overlook the use of limits:

1. Real number as limits of rationals.
2. Derivatives and integrals are defined by limits.
3. Phase transitions by thermodynamic limit (number of particles $\rightarrow \infty$).
4. Limit sets in dynamical systems.
5. Fractals are often defined as limits, as are other exotic geometrical objects.

The ultimate framework for studying limits is **category theory**[35,45]—a mathematical discipline which is very abstract, but has had a great influence in algebraic topology, algebraic geometry and algebra.

So the question is:

Is it possible to study emergence by limits in categories?

Indeed it is and this has been done by Ehresmann and Vanbremeersch.[19] We will explain this idea.

A category is a mathematical structure consisting of

1. objects: $X, Y, \ldots$
2. morphisms between objects, represented by arrows: $X \rightarrow Y$,

satisfying certain conditions.[19,35,45] A system of objects and morphisms form a diagram (or pattern) as in Figure 6.

Under certain conditions one can then construct a new object in the category by glueing the whole diagram together. This object is called the (direct) limit—($\varinjlim$)—of the diagram, and in some respects reflect the behavior of the whole diagram. It is in a sense a universal construction.

The *emergent properties* are then defined as the **new** properties appearing in the limit object.[16] This process can be iterated to take limits of limits again to form higher-order objects.

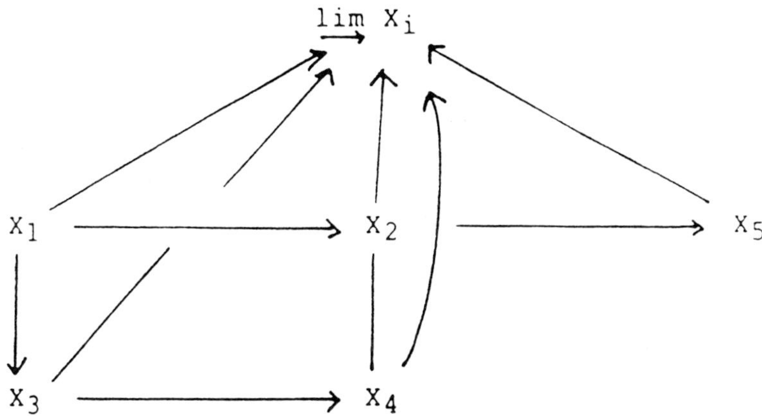

FIGURE 6 Diagram with direct limit.

In our hyperstructure terminology this corresponds to letting arrows stand for interactions and $R = \varinjlim$ (the construction process). But there is no observable mechanism involved in the build-up process—as we have. We think this is a crucial point and it can easily be added to the category theoretical framework, which is well suited for further mathematical treatment. For example if $X = \varinjlim X_i$, then P is emergent in our sense iff

$$P \in Obs(X), \text{ but } \mathrm{P} \notin \mathrm{Obs}(\mathrm{X_i}), \text{ for all i.}$$

Limits are therefore good models of complex, emergent objects.

In the categorical sense, limits also cover finite constructions by taking the index set to be finite.

The interactions are represented by the arrows and this is where the nonlinearities may come in. So very often limits and nonlinearities act together in order to create emergent objects and properties.

RECURSION AND HYPERSTRUCTURES

In many (mathematical) systems and models objects are constructed by iterative procedures or, as we often say, by recursion (some use the term inductive procedure). We start with some primitive objects, use certain rules to form new objects, then apply the rules again to the new objects, etc. In this way we construct a universe of recursively defined objects.

This is how we define the well-formed objects (strings, expressions) of formal systems as, for example, in logic. Also formal languages and string systems of various kinds are generated in this way, and dynamical systems define the development recursively. Furthermore, in languages where the string approach is replaced by category theory, i.e., general type theories,[28] the corresponding languages are defined in terms of similar recursions schemes. Many models in A-Life research are based on this kind of procedure—i.e., see Fontana,[24] Hogeweg,[29] and Rasmussen.[42]

But many processes in nature appear not to be strictly recursive—for example, biological processes (evolution) and cognitive processes. What seems to be missing is a combination of evolution, adaptation, and recursion.

Hyperstructures combine these processes. So in many of the systems mentioned and models considered, our proposal is that instead of using only recursion, replace it (or extend it) with suitably chosen hyperstructures. We then get formal systems and languages with well-formed expressions (formulae) produced as or by hyperstructures.

This would work as follows.

We start out with the primitive objects as our primitive structures, then as Int we use the old recursive iteration process. At the same time we observe and register the system by a suitably chosen Obs mechanism.

Obs may be made (by nature or by us) such that certain structures in the process give a special signal for selection. This will then represent adaptation for specific purposes.

Then these structures—now second order—are being subjected to the same or **new** iterative rules and Obs, etc., according to the hyperstructure scheme chosen.

The introduction of Obs and Int is a way to enrich the structures. This will give new structures with new—emergent or higher-order—properties, and should be very relevant for the study and generation of artificial life structures.

CONCLUSION

Our aim here has been to introduce hyperstructures as a general unifying concept for studying emergence, hierarchics, and higher-order structures. We also think they will be useful in specifying the formation or construction of complex objects in general. Synthesizing complexity—like in Artificial Life models—can be done according to a hyperstructure scheme.

The main purpose of this paper is to give an introduction to our ideas on emergence and hyperstructures. Many of the ideas are at an initial stage and need further elaboration. Our goal is to develop a comprehensive theory and we will return to this elsewhere.

"Life is a property of form, not matter, a result of the organization of matter rather than something that inheres the matter itself," writes C. Langton.[34]

Hyperstructures represent in our opinion such an organizational scheme. It seems interesting to speculate how to form various kinds of matter (at the atomic, molecular level and upwards) into a hyperstructured material through some kind of action. Such "hypermaterials" could possibly represent a new phase of matter bridging the gap between living and nonliving matter.

ACKNOWLEDGMENTS

I would like to thank the Norwegian Research Council (NF/NTNF) for financial support in connection with this work, and Dr. Steen Rasmussen for his enthusiastic support and many discussions.

REFERENCES

1. Ablowitz, R. "The Theory of Emergence." *Philos. Sci.* **6** (1939): 1–16.
2. Anderson, P. W. "More Is Different." *Science* **177** (1972): 393–396.
3. Anderson, P. W. "Is Complexity Physics? Is It Science? What is It?" *Phys. Today* **July** (1991): 9–11.
4. Angyal, A. "The Structure of Wholes" *Philos. Sci.* **6** (1939): 25–37.
5. Assad, A. M., and N. Packard. "Emergent Colonization in an Artificial Ecology." Preprint, University of Illinois, January 1992.
6. Baas, N. A. "On the Models of Thom in Biology and Morphogenesis." *Math. Biosci.* **17** (1973): 178–187.
7. Baas, N. A. "Structural Stability of Composed Mappings." Notes: The Institute for Advanced Studies, Princeton, 1974.
8. Baas, N. A. "Hierarchical Systems: Foundations of a Mathematical Theory and Applications, Parts I and II." Report, University of Trondheim, Norway, 1975.
9. Baas, N. A. "A General Theory of Cancer." Report, University of Trondheim, Norway, 1977.
10. Baas, N. A. "Hierarchical Systems and Models in Biology." In *Proceedings of Symposium in Biomathematics*, 41–46. Stockholm: Swedish National Science Research Coundil, May 1982.
11. Baas, N. A. "Hyperstructures—A Framework for Emergence, Hierarchies, Evolution and Complexity." In *Proceedings of Conference Emergence dans les Modeles de la Cognition,* edited by A. Grumbach et al., 67–93. Paris: Telecom, 1992.

12. Baas, N. A. "Hyperstructures as Tools in Nanotechnology and Nanobiology." In *Towards a Nanobiology: Coherent and Emergent Phenomena in Biomolecular Systems*, edited by S. Rasmussen, S. R. Hameroff, J. Tuzinki, and P. A. Hansson. Cambride, MA: MIT Press, in press.
13. Baas, N. A. "Emergence, Hierarchies, Hyperstructures, and SelfOrganization." In *Proceedings of ISSS Conference.* International Society for the Systems Sciences, Louisville, Kentucky, July 1993.
14. Baas, N. A. "Emergence and Higher-order Structures." In *Proc. of Intl. Conf. On Systems Research, Informatics and Cybernetics*, edited by edited by G. E. Lasker, in press. Baden-Baden, Sidney, AU: International Institute for Advanced Studies in Systems Research and Cybernetics, University of Windor, Windsor, Ontario, Canada, August 1993.
15. Blum, L., M. Shub, and S. Smale "On a Theory of Computation and Complexity over the Real Numbers: NP Completeness, Recursive Functions and Universal Machines." *Bull. Amer. Math. Soc.* **21(1)** (1989): 1–46.
16. Cariani, P. "Emergence and Artificial Life." In *Artificial Life II*, edited by C. Langton, C. Taylor, J. Farmer and S. Rasmussen. Santa Fe Institute Studies in the Sciences of Complexity, Proc. Vol. X, 775–797. Redwood City, CA: Addison Wesley, 1991.
17. Churchland, P. M. *A Neurocomputational Perspective.* Cambridge, MA: MIT Press, 1989.
18. Dawkins, R. *The Blind Watchmaker.* Essex, UK: Longman, 1986.
19. Ehresmann, A.C. and J. P. Vanbremeersch. "Hierarchical Evolutive Systems: A Mathematical Model for Complex Systems." *Bull. Math. Bio.* **49(1)** (1987): 13–50.
20. Eigen, M. and P. Schuster. *The Hypercycle: A Principle of Natural Self-Organization.* Berlin: Springer-Verlag, 1979.
21. Emmeche, C. "Modeling Life: A Note on the Semiotics of Emergence and Computation in Artificial and Natural Living Systems." In *Biosemiotics: The Semiotic Web*, edited by T. A. Sebeok and J. Uniker-Sebeok. Berlin: Mouton de Gruyter, 1992.
22. Farmer, J. D., and N. H. Packard. "Evolution, Games and Learning: Models for Adaptation in Machines and Nature." *Physica D* **22** (1986): VII, XII.
23. Fernandez, J., A. Moreno, and A. Etxeberria. "Life as Emergence: The Roots of a New Paradigm in Theoretical Biology." *World Futures* **32** (1991): 133–149.
24. Fontana, W. "Algorithmic Chemistry." In *Artificial Life II,* edited by C. G. Langton, et al., 159–209. Santa Fe Institute Studies in the Sciences of Complexity, Proc. Vol X. Redwood City, CA: Addison-Wesley, 1991.
25. Forrest, S., ed. "Emergent Computation." *Physica D* **42** (1990): 1–452.
26. Gödel, K. "On Formally Undecidable Propositions of Principia Mathematica and Related Systems I." In *Collected Works*, 145–195. Oxford, UK: Oxford University Press, 1986.

27. Hindley, J. R. and J. P. Seldin *Introduction to Combinators and λ-Calculus.* In London Math. Soc. Student Texts., Vol. 1, edited by E. B Davies. London, UK: Cambridge University Press, 1986.
28. Hofstadter, D. R. *Gödel, Escher, Bach: An Eternal Golden Braid .* Harmondsworth, Middlesex, UK: Penguin Press, 1979.
29. Hogeweg, P. "Mirror Beyond Mirror, Puddles of Life." In *Artificial Life*, edited by C. G. Langton. Santa Fe Institue Studies in the Sciences of Complexity, Proc. Vol. VI, 297–316. Redwood City, CA: Addison-Wesley, 1988.
30. Kampis, G. *Self-Modifying Systems in Biology and Cognitive Science.* Oxford, UK: Pergamon Press, 1991.
31. Klee, R. L. "Microdeterminism and Concepts of Emergence." *Phil. Sci.* **51** (1984): 44–63.
32. Koestler, A. *Janus* London, UK: Hutchinson, 1979.
33. Lambek, J. and P. J. Scott. *Introduction to Higher Order Categorical Logic.* Cambridge, UK: Cambridge University Press, 1986.
34. Langton, C. G. "Artificial Life." In *Artificial Life*, edited by C. G Langton. SFI Studies in the Sciences of Complexity, Proc. Vol. VI, 1–48. Redwood City, CA: Addison-Wesley, 1989.
35. MacLane, S. *Categories for the Working Mathematician.* Berlin: Springer-Verlag, 1971.
36. Minch, E. "Representation of Hierarchical Structure in Evolving Networks." Thesis, State University of New York at Binghamton, 1988.
37. Morgan, Lloyd C. *Emergent Evolution.* London: Williams & Norgate, 1923.
38. Nagel, E. *The Structure of Science: Problems in the Logic of Scientific Explanation.* New York: Columbia University Press, 1961.
39. Nakai, I. "Topologial Stability Theorem for Composite Mappings." *Ann. Inst. Fourier (Grenoble)* **39(2)** (1989): 459–500.
40. Penrose, R. *The Emperor's New Mind.* Oxford, UK: Oxford University Press, 1989.
41. Pepper, S. C. "Emergence." *Philos.* **23** (1926): 241–245.
42. Rasmussen, S., C. Knudsen, and R. Feldberg "Dynamics of Programmable Matter." In *Artificial Life II*, edited by C. G. Langton et al. Santa Fe Institute Studies in the Sciences of Complexity, Proc. Vol. X, 211–254. Redwood City, CA: Addison-Wesley, 1991.
43. Ray, T. S. "An Approach to the Synthesis of Life." In *Artificial Life II*, edited by C. G. Langton et al. Santa Fe Institute Studies in the Sciences of Complexity, Proc. Vol. X, 371–408. Redwood City, CA: Addison-Wesley, 1991.
44. Requardt, M. "Gödel, Turing, Chaitin and the Question of Emergence as a Meta-Principle of Modern Physics: Some Arguments Against Reductionism." *World Futures* **32** (1991): 185–196.
45. Rydeheard, D. E. and R. M. Burstall. *Computational Category Theory.* London: Prentice Hall, 1988.
46. Salthe, S. N. *Evolving Hierarchical Systems.* New York: Columbia University Press, 1985.

47. Shishikura, M. "The Hausdorff Dimension of the Boundary of the Mandelbrot Set and Julia Sets." Preprint. Institute for Mathematical Sciences, SUNY Stony Brook, NY. 1991.
48. Simon, H. A. "The Architecture of Complexity." *Proc. Am. Phil. Soc.* **106** (1962): 467–482.
49. Stoy, J. E. *Denotational Semantics: The Scott-Strachey Approach to Programming Language Theory.* Cambridge, MA: MIT Press, 1977.

Stevan Harnad
Laboratoire Cognition at Mounement, URA CNRS 116, I.B.H.O.P., Université d'Aix Marseille II, 13388 Marseille cedex 13, France

Artificial Life: Synthetic vs. Virtual

Artificial life can take two forms: synthetic and virtual. In principle, the materials and properties of synthetic living systems could differ radically from those of natural living systems yet still resemble them enough to be alive if they were grounded in the relevant causal interactions with the real world. Virtual (purely computational) "living" systems, in contrast, are just ungrounded symbol systems that are systematically interpretable as if they were alive; in reality they are no more alive than a virtual furnace is hot. Virtual systems are better viewed as "symbolic oracles" that can be used (interpreted) to predict and explain real systems, but not to instantiate them. The vitalistic overinterpretation of virtual life is related to the animistic overinterpretation of virtual minds and is probably based on an implicit (and possibly erroneous) intuition that living things have actual or potential mental lives.

THE ANIMISM IN VITALISM

There is a close connection between the superstitious and even supernaturalist intuitions people have had about both life and mind. I think the belief in an immaterial or otherwise privileged "vital force" was, consciously or unconsciously, always parasitic on the mind/body problem, which is that conceptual difficulty we all have in integrating or equating the mental world of thoughts and feelings with the physical world of objects and events, a difficulty that has even made some of us believe in an immaterial soul or some other nonmaterial animistic principle. The nature of the parasitism was this: Whether we realized it or not, we were always imagining living things as having mental lives, either actually or potentially. So our attempts to account for life in purely physical terms indirectly inherited our difficulty in accounting for mind in purely physical terms.

It is for this reason that I believe all positive analogies between the biology of life and the biology of mind are bound to fail: We are invited by some[1,3,2] to learn a lesson from how wrong-headed it was to believe that ordinary physics alone could not explain life—which has now turned out to be largely a matter of protein chemistry—and we are enjoined to apply that lesson to current attempts to explain the mind—whether in terms of physics, biochemistry, or computation—and to dismiss the conceptual dissatisfaction we continue to feel with such explanations, on the grounds that we have been wrong in much the same way before.

But the facts of the matter are actually quite the opposite, I think. Our real mistake has been in inadvertently conflating the problem of life and the problem of mind, for in bracketing mind completely when we consider life, as we do now, we realize there is nothing left that is special about life relative to other physical phenomena—at least nothing more special than whatever is special about biochemistry relative to other branches of chemistry. But in the case of mind itself, what are we to bracket?

THE MIND/BODY PROBLEM: AN EXTRA FACT

The disanalogy and the inappropriateness of extrapolating from life to mind can be made even clearer if we face the mind/body problem squarely: If we were given a true, complete description of the kind of physical system that has a mind, we could still continue to wonder forever (a) why the very same description would not be equally true if the system did not have a mind, but just looked and behaved exactly *as if* it had one and (b) how we could possibly know that every (or any) system that fits the description really has a mind (except by *being* the system in question). The fact of the matter is, we all know there really are mental states, yet that fact is the very one that eludes such a true, complete description and, hence, must be taken entirely on faith (a state of affairs that rightly arouses some skepticism in many of us[15,26,27]).

Now let me point out the disanalogy between the special empirical and conceptual situation just described for the case of mind modeling and the parallel case of life modeling, which instead involves a true, complete description of the kind of physical system that is alive; then, for good measure, this disanalogy will be extended to the third parallel case, matter modeling, which involves this time a true complete description of any physical system at all (e.g., the world of elementary particles or even the universe as a whole). If we were given a true, complete description of the kind of physical system that is alive, could we still ask (a) why the very same description would not be equally true if the system were not alive, but just looked and behaved exactly *as if* it were alive? And could we go on to ask (b) how we could possibly know that every (or any) system that fits that description was really alive? I suggest that we could not raise these questions (apart from general worries about the truth or completeness of the description itself which, it is important to note, is *not* what is at issue here) because in the case of life there just *is* no further fact—like the existence of mental states—that exists independently of the true complete physical description.

Consider the analogous case of physics: Could we ask whether, say, the planets *really* move according to relativistic mechanics, or merely look and behave exactly *as if* they did? I suggest that this kind of question is merely about normal scientific underdetermination—the uncertainty there will always be about whether any empirical theory we have is indeed true and complete. But if we assume it as given (for the sake of argument) that a theory in physics is true and complete, then there is no longer any room left for any conceptual distinction between the "real" and "as if" case, because there is no further fact that distinguishes them. The facts—all objective, physical ones in the case of both physics and biology—have been completely exhausted.

Not so in the case of mind, where a subjective fact, and a fact about subjectivity, still remains, and remains unaccounted for.[26,27] Although the force of the premise that our description is true and complete logically excludes the possibility that any system that fits that description will fail to have a mind, the *conceptual* possibility is not only there, but amounts to a complete mystery as to *why* the description should be true and complete at all (even if it is); for there is certainly nothing in the description itself—which is all objective and physical—that either entails or even shows why it is probable that mental states should exist at all, let alone be accounted for by the description.

I suggest that it is this elusive extra (mental) fact that made people wrongly believe that special vital forces, rather than just physics, would be needed to account for life. But, modulo the problem of explaining mind (setting aside, for the time being, that mind, too, is a province of biology), the problem of explaining life (or matter) has no such extra fact to worry about. So there is no justification for being any more skeptical about present or future theories of life (or matter) than is called for by ordinary considerations of underdetermination, whose resolution is rightly relativized and relegated to the arena of rival theories fighting it out to see which will account the most generally (and perhaps the most economically) for the most data.

SEARLE'S CHINESE ROOM ARGUMENT AGAINST COMPUTATIONALISM

How much worry does this extra fact warrant on its own turf, then, in empirical attempts to model the mind? None, I would be inclined to say (being a methodological epiphenomenalist who believes that only a fool argues about the unknowable[7,8,15]), but there does happen to be one prominent exception to the viability of the methodological strategy of bracketing the mind/body problem even in mind modeling, and that exception also turns out to infect a form of life modeling; so in some respects we are right back where we started, with an animistic consideration motivating a vitalistic one! Fortunately, this exception turns out to be a very circumscribed special case, easily resolved in principle. In practice, however, resolving it has turned out to be especially tricky, with most advocates of the impugned approach (computationalism) continuing to embrace it, despite counterevidence and counterarguments, for essentially hermeneutic (i.e., interpretational) reasons[5,6,13,14]: It seems that the conceptual appeal of systems that are systematically interpretable *as if* they were alive or had minds is so strong that it overrides not only our natural skepticism (at least insofar as mind modeling is concerned) but even the dictates of both common sense and ordinary empirical evidence—or so, at least, I will try to show.

The special case I am referring to is Searle's[30] infamous Chinese Room Argument against computationalism (or the computational theory of mind). I am one of the minority (possibly as few as two) who think Searle's Argument is absolutely right. Computationalism[5] holds that mental states are just computational states—not just any computational states, of course, only certain special ones, in particular, those that are sufficient to pass the Turing Test (TT), which calls for the candidate computational system to be able to correspond with us as a pen pal for a lifetime, indistinguishable from a real pen pal. The critical property of computationalism that makes it susceptible to empirical refutation by Searle's thought experiment is the very property that made it attractive to mind modellers: its implementation independence. According to computationalism, all the physical details of the implementation of the right computational states are irrelevant, just as long as they are implementations of the right computational states; for then each and every implementation will have the right mental states. It is this property of implementation independence—a property that has seemed to some theorists[29] to be the kind of dissociation from the physical that might even represent a solution to the mind/body problem—that Searle exploits in his Chinese Room Argument. He points out that he, too, could become an implementation of the TT-passing computer program—after all, the only thing a computer program is really doing is manipulating symbols on the basis of their shapes—by memorizing and executing all the symbol manipulation rules himself. He could do this even for the (hypothetical) computer program that could pass the TT in Chinese, yet he obviously would not be understanding Chinese under these conditions; hence, by transitivity of implementation-independent properties (or their absence), neither would the (hypothetical) computer implementation be understanding Chinese (or anything). So much for the computational theory of mind.

It is important to understand what Searle's argument does and does not show. It does *not* show that the TT-passing computer cannot possibly be understanding Chinese. Nothing could show that, because of the other mind problem[9,15]: The only way to know that for sure would be to *be* the computer. The computer might be understanding Chinese, but if so, then that could only be because of some details of its particular physical implementation (that its parts were made of silicon, maybe?), which would flatly contradict the implementation-independence premise of computationalism (and leave us wondering about what is special about silicon). Searle's argument also does *not* show that Searle himself could not possibly be understanding Chinese under those conditions. But—unless we are prepared to believe either in the possibility that (1) memorizing and manipulating a bunch of meaningless symbols could induce multiple personality disorder (a condition ordinarily caused only by early child abuse), giving rise to a second, Chinese-understanding mind in Searle, an understanding of which he was not consciously aware,[6,14,19] or, even more far-fetched, that (2) memorizing and manipulating a bunch of meaningless symbols could render them consciously understandable to Searle—the emergence of either form of understanding under such conditions is, by ordinary inductive standards, about as likely as the emergence of clairvoyance (which is likewise not impossible).

So I take it that, short of sci-fi special pleading, Searle's argument that there is nobody home in there is valid for the very circumscribed special case of any system that is purported to have a mind purely by virtue of being an implementation-independent implementation of a TT-passing computer program. The reason his argument is valid is also clear: Computation is just the manipulation of physically implemented symbols on the basis of their shapes (which are arbitrary in relation to what they can be interpreted to mean); the symbols and symbol manipulations will indeed bear the weight of a systematic interpretation, but that interpretation is not *intrinsic* to the system (any more than the interpretation of the symbols in a book is intrinsic to the book). It is projected onto the system by an interpreter with a mind (as in the case of the real Chinese pen pal of the TT-passing computer); the symbols do not mean anything *to* the system, because a symbol-manipulating system is not the kind of thing that anything means anything to.

There remains, however, the remarkable property of computation that makes it so valuable, namely, that the right computational system will indeed bear the weight of a systematic semantic interpretation (perhaps even the TT). Such a property is not to be sneezed at, and Searle does not sneeze at it. He calls it "Weak Artificial Intelligence" (to contrast it with "Strong Artificial Intelligence," which is the form of computationalism his argument has, I suggest, refuted). The practitioners of Weak AI would be studying the mind—perhaps even arriving at a true, complete description of it—using computer models; they could simply never claim that their computer models actually had minds.

SYMBOLIC ORACLES

But how can a symbol system mirror most, perhaps all, of the properties of mind without actually having a mind? Wouldn't the correspondence be too much to be just coincidental? By way of explication, I suggest that we think of computer simulations as symbolic oracles. Consider a computer simulation of the solar system; in principle, we could encode in it all of the properties of all of the planets and the sun, all of the relevant physical laws (discretely approximated, but as closely as we like) such that the solar system simulation could correctly predict and generate all the positions and interactions of the planets far into the future, simulating them in virtual or even real time. Depending on how thoroughly we had encoded the relevant astrophysical laws and boundary conditions, there would be a one-to-one correspondence between the properties of the simulation that was interpretable as a solar system—let us call it the "virtual" solar system—and the properties of the real solar system. Yet none of us would, I hope, want to say that there was any, say, motion, or mass, or gravity in the virtual solar system. The simulation is simply a symbol system that is systematically interpretable *as if* it had motion, mass, or gravity; in other words, it is a symbolic oracle.

This is also the correct way to think of a (hypothetical) TT-passing computer program. It would really just be a symbol system that was systematically interpretable as if it were a pen pal corresponding with someone. This virtual pen pal may be able to predict correctly all the words and thoughts of the real pen pal it was simulating, oracularly, till doomsday, but in doing so it is no more thinking or understanding than the virtual solar system is moving. The erroneous idea that there is any fundamental difference between these two cases (the mental model and the planetary model) is, I suggest, based purely on the incidental fact that thinking is unobservable, whereas moving is observable; but gravity is not observable either, and quarks and superstrings even less so; yet none of those would be present in a virtual universe either. For the virtual universe, like the virtual mind, would really just be a bunch of meaningless symbols—squiggles and squoggles—that were syntactically manipulated in such a way as to be systematically interpretable *as if* they thought or moved, respectively. This is certainly stirring testimony to the power of computation to describe and predict physical phenomena and to the validity of the Church/Turing thesis,[4,20] but it is no more than that. It certainly is not evidence that thinking is just a form of computation.

So computation is a powerful, indeed oracular, tool for modeling, predicting, and explaining planetary motion, life, and mind, but it is not a powerful enough tool for actually *implementing* planetary motion, life, or mind, because planetary motion, life and mind are not mere implementation-independent computational phenomena. I will shortly return to the question of what might be powerful enough, but first let me try to tie these considerations closer to the immediate concerns of this conference on artificial life.

VIRTUAL LIFE: "AS IF" OR REAL?

Chris Langton once proposed to me an analogue of the TT or, rather, a generalization of it from artificial mind to artificial life: Suppose, he suggested, that we could encode all of the initial conditions of the biosphere around the time life evolved, and, in addition, we could encode the right evolutionary mechanisms—genetic algorithms, game of life, what have you—so that the system actually evolved the early forms of life, exactly as it had occurred in the biosphere. Could it not, in principle, go on to evolve invertebrates, vertebrates, mammals, primates, man, and then eventually even Chris and me, having the very conversation we were having at the time in a pub in Flanders—indeed, could it not go on and outstrip us, cycling at a faster pace through the same experiences and ideas Chris and I would eventually go on to arrive at in real time? And if it could do all that, and if we accept it as a premise (as I do) that there would be not one property of the real biosphere, or of real organisms, or of Chris or me, that would not also be present in the virtual world in which all this virtual life, and eventually these virtual minds, including our own, had "evolved," how could I doubt that the virtual life was real? Indeed, how could I even distinguish them?

Well, the answer is quite simple, as long as we do not let the power of hermeneutics loosen our grip on one crucial distinction: the distinction between objects and symbolic descriptions of them. There may be a one-to-one correspondence between object and symbolic description, a description that is as fine grained as you like, perhaps even as fine grained as can be. But the correspondence is between properties that, in the case of the real object, are what they are intrinsically, without the mediation of any interpretation, whereas in the case of the symbol description, the only "objects" are the physical symbol tokens and their syntactically constrained interactions; the rest is just our interpretation of the symbols and interactions *as if* they were the properties of the objects they describe. If the description is complete and correct, it will always bear the full weight of that interpretation, but that still does not make the corresponding properties in the object and the description *identical* properties; they are merely computationally equivalent under the mediation of the interpretation. This should be only slightly harder to see in the case of a dynamic simulation of the universe than it is in the case of a book full of static sentences about the universe.

In the case of real heat and virtual heat or real motion and virtual motion the distinction between identity and formal equivalence is clear. A computer simulation of a fire is not really hot and nothing is really moving in a computer simulation of planetary motion. In the case of thinking, I have already argued that we have no justification for claiming an exception merely because thinking is unobservable (and besides, Searle's "periscope" shows us that we would not find thinking in there even if we *became* the simulation and observed for ourselves what is unobservable to everyone else). What about life? But we have already seen that life—once we bracket the mind/body problem—is not different from any other physical phenomenon, so virtual life is no more alive than virtual planetary motion moves or virtual gravity attracts. Chris Langton's virtual biosphere, in other words, would be just another symbolic oracle: yet another bunch of systematically interpretable squiggles and squoggles: Shall we call this "Weak Artificial Life"? (Sober,[31] as I learned after writing this paper, had already made this suggestion at the Artificial Life II meeting in 1990.)

SENSORIMOTOR TRANSDUCTION, ROBOTICS, THE "TOTAL TURING TEST," AND SYMBOL GROUNDING

I did promise to return to the question of what more might be needed to *implement* life if computation, for all its power and universality, is not strong enough. Again, we return to the Chinese Room Argument for a hint, but this time we are interested in what kind of system Searle's argument *cannot* successfully show to be mindless: We do not have to look far. Even an optical transducer is immune to Searle, for if someone claimed to have a system that could really see (as opposed to being merely interpretable *as if* it could see), Searle's "periscope" would already fail.[11] For, if Searle tried to implement the putative "seeing" system without seeing (as he had implemented the putative "understanding" system without understanding), he would have only two choices. One would be to implement only the *output* of the transducer (if we can assume that its output would be symbols) and whatever symbol manipulations were to be done on that output, but then it would not be surprising if Searle reported that he could not see, for he would not be implementing the whole system, just a part of it. All bets are off when only parts of systems can be implemented. Searle's other alternative would be to actually look at the system's scene or screen in implementing it, but then, alas, he *would* be seeing. Either way, the Chinese Room strategy does not work. Why? Because mere optical transduction is an instance of the many things there are under the sun—including touch, motion, heat, growth, metabolism, photosynthesis, and countless other "analog" functions—that are *not* just implementations of implementation-independent symbol manipulations. And only the latter are vulnerable to Searle's argument.

This immediately suggests a more exacting variant of the Turing Test—what I've called the Total Turing Test (TTT)—which, unlike the TT, is not only immune to Searle's argument but reduces the level of underdetermination of mind modeling to the normal level of underdetermination of scientific theory.[11,15] The TT clearly has too many degrees of freedom, for we all know perfectly well that there is a lot more that people can do than be pen pals. The TT draws on our linguistic capacity, but what about all our robotic capacities, our capacity to discriminate, identify, and manipulate the objects, events, and states of affairs in the world we live in? Every one of us can do that; so can animals.[10,15] Why should we have thought that a system deserved to be assumed to have a mind if all it could generate was our pen pal capacity? Not to mention that there is good reason to believe that our linguistic capacities are *grounded* in our robotic capacities. We do not just trade pen pal symbols with one another; we can each also identify and describe the objects we see, hear, and touch, and there is a systematic *coherence* between how we interact with them robotically and what we say about them linguistically.

My own diagnosis is that the problem with purely computational models is that they are *ungrounded*. There may be symbols in there that are systematically interpretable as meaning "cat," "mat," and "the cat is on the mat," but in reality they are just meaningless squiggles and squoggles apart from the interpretation we project onto them. In an earlier conference in New Mexico,[12] I suggested that the symbols in a symbol system are ungrounded in much the same way the symbols in a Chinese-Chinese dictionary are ungrounded for a non-speaker of Chinese: He could cycle through it endlessly without arriving at meanings unless he already had grounded meanings to begin with (as provided by a Chinese-English dictionary, for an English speaker). Indeed, *translation* is precisely what we are doing when we interpret symbol systems, whether static or dynamic ones, and that is fine, as long as we are using them only as oracles, to help us predict and explain things. For that, their systematic interpretability is quite sufficient. But if we actually want them to *implement* the things they predict and explain, they must do a good deal more. At the very least, the meanings of the symbols must somehow be grounded in a way that is independent of our projected interpretations and in no way mediated by them.

The TTT accordingly requires the candidate system, which is now a robot rather than just a computer, to be able to interact robotically with (i.e., to discriminate, identify, manipulate, and describe) the objects, events, and states of affairs that its symbols are systematically interpretable as denoting, and it must be able to do it so its symbolic performance coheres systematically with its robotic performance. In other words, not only must it be capable, as any pen pal would be, of talking about cats, mats, and cats on mats indistinguishably from the way we do, but it must also be capable of discriminating, identifying, and manipulating cats, mats, and cats on mats exactly as any of us can; and its symbolic performance must square fully with its robotic performance, just as ours does.[16,18]

The TT was a tall order; the TTT is an even taller one. But note that whatever candidate successfully fills the order *cannot* be just a symbol system. Transduction and other forms of analog performance and processing will be essential components in its functional capacity, and subtracting them will amount to reducing what is left to those mindless squiggles and squoggles we have kept coming up against repeatedly in this discussion. Nor is this added performance capacity an arbitrary demand. The TTT is just normal empiricism. Why should we settle for a candidate that has less than our full performance capacities? The proper time to scale the model down to capture our handicaps and deficits is only *after* we are sure we've captured our total positive capacity (with the accompanying hope that a mind will piggyback on it)—otherwise we would be like automotive (reverse) engineers (i.e., theoretical engineers who do not yet know how cars work but have real cars to study) who were prepared to settle for a functional model that has only the performance capacities of a car without moving parts, or a car without gas: The degrees of freedom of such "handicapped" modeling would be too great; one could conceivably go on theory-building forever without ever converging on real automobile performance capacity that way.

A similar methodological problem unfortunately also affects the TTT modeling of lower organisms: If we knew enough about them ecologically and psychologically to be able to say with any confidence what their respective TTT capacities were, and whether we had captured them TTT-indistinguishably, lower organisms would be the ideal place to start; but unfortunately we do not know enough, either ecologically or psychologically (although attempts to approximate the TTT capacities of lower organisms will probably still have to precede or at least proceed apace with our attempts to capture human TTT capacity).

THE TTT VERSUS THE TTTT

Empiricists might want to counter that the proper degrees of freedom for mind modeling are neither those of the TT nor the TTT but the TTTT (Total-Total Turing Test), in which the candidate must be empirically indistinguishable from us not only in all of its macrobehavioral capacities, but also in all of its microstructural (including neural) properties.[1,3,2] I happen to think that the TTTT would be supererogatory, even overly constraining, and that the TTT already narrows the degrees of freedom sufficiently for the branch of reverse engineering that mind modeling belongs to. There is still a kind of implementation independence here, too, but not a computationalist kind: There is no reason to believe that biology has exhausted all of the possibilities of optical transduction, for example. All optical transducers must transduce light, to be sure, but apart from that, there is room for a lot more possibilities along the continuum on which the human retina and the *Limulus ommatidia* represent only two points.

The world of objects and the physics of transducing energy from them provide the requisite constraints for mind modeling, and every solution that manages to generate our TTT capacity within those constraints has (by my lights) equal claim to our faith that it has mental states—I do not really see the one that is TTTT-indistinguishable from us as significantly outshining the rest. My reasons for believing this are simple: We are blind to Turing-indistinguishable differences (that is why there is an other—minds problem and a mind/body problem). By precisely the same token, the Blind Watchmaker is likewise blind to such differences. There cannot have been independent selection pressure for having a mind, since selection pressure can operate directly only on TTT capacity.

Yet a case *might* be made for the TTTT if the capacity to survive, reproduce, and propagate one's genes is an essential part of our TTT capacity, for that narrows down the range of eligible transducers still further, and these are differences that evolution is *not* blind to. In this case, the TTTT might pick out microstructural features that are too subtle to be reflected in individual behavioral capacity, and the TTT look-alikes lacking them might indeed lack a mind.

My own hunch is nevertheless that the TTT is strong enough on its own (although neuroscience could conceivably give us some clues as to how to pass it), and I am prepared to extend human rights to any synthetic candidate that passes it, because the TTT provides the requisite constraints for grounding symbols, and that is strong enough grounds for me. I doubt, however, that TTT capacity can be second-guessed *a priori*, even with the help of symbolic oracles. Perhaps this is the point where we should stop pretending that mind modelers can bracket life and that life modelers can bracket mind. So far, only living creatures seem to have minds. Perhaps the constraints on the creation of synthetic life will be relevant to the constraints on creating synthetic minds.

These last considerations amount only to fanciful speculation, however; the only lesson Artificial Life might take from this paper is that it is a mistake to be too taken with symbol systems that display the formal properties of living things. It is not that only natural life is possible; perhaps there can be synthetic life, too, made of radically different materials and operating on radically different functional principles. The only thing that is ruled out is "virtual" or purely computational life, because life (like mind and matter) is not just a matter of interpretation.

CODA: ANALOG "COMPUTATION"

Let me close with a remark on analog computation. Searle's Chinese Room Argument applies only to computation in the sense of finite, discrete symbol manipulation. Searle cannot implement a transducer because the transduction of physical energy is not just the manipulation of finite discrete symbols. Searle would even be unable to implement a parallel, distributed system like a neural net (though he could implement a discrete serial simulation of one[17]). Indeed, every form of analog computation is immune to Searle's argument. Is analog computation also immune to the symbol grounding problem? I am inclined to say it is, if only because "symbols" usually mean discrete, physically tokened symbols. Some people speak of "continuous symbols," and even suggest that the (real) solar system is an (analog) computer if we choose to use it that way. [21,22,23,24,25] Here we are getting into a more general (and, I think, vaguer) sense of "computing" that I do not think says much about either life or mind one way or the other. For, if the solar system is computing, then everything is computing, and hence *whatever* the brain—and any synthetic counterpart of it—is actually doing, it too is computing. When "implementation independence" becomes so general as to mean obeying the same differential equations, I think the notion that a system is just the implementation of a computation becomes rather uninformative. So take my argument against the overinterpretation of virtual life to be applicable only to finite, discrete symbol systems that are interpretable as if they were living, rather than to systems that obey the same differential equations as living things, which I would regard as instances of synthetic rather than virtual life.

REFERENCES

1. Churchland, P. M. *Matter and Consciousness: A Contemporary Introduction to the Philosophy of Mind.* Cambridge, MA: MIT Press,1984.
2. Churchland, P. M. *The Nature of Mind and the Structure of Science: A Neurocomputational Perspective.* Cambridge, MA: MIT Press, 1989.
3. Churchland, P. S. *Neurophilosophy: Toward a Unified Science of the Mind-Brain.* Cambridge, MA: MIT Press, 1986.
4. Davis, M. *Computability and Unsolvability.* Manchester, NY: McGraw-Hill, 1958.
5. Dietrich, E. "Computationalism." *Soc. Epistemology* **4** (1990): 135–154.
6. Dyer, M. G. "Intentionality and Computationalism: Minds, Machines, Searle and Harnad." *Exper. & Theor. Art. Int.* **2(4)** (1990): 303–319.
7. Harnad, S. "Neoconstructivism: A Unifying Theme for the Cognitive Sciences." In *Language, Mind & Brain,* edited by T. Simon and R. Scholes, 1–11, Hillsdale, NJ: Erlbaum, 1982.
8. Harnad, S. "Consciousness: An Afterthought." *Cog. & Brain Theory* **5** (1982): 29–47.
9. Harnad, S. "Verifying Machine's Minds (Review of J. T. Culbertson, *Consciousness: Natural and Artificial.* NY: Libra, 1982.)." *Contemp. Psych.* **29** (1984): 389–391.

10. Harnad, S. "The Induction and Representation of Categories." In *Categorical Perception: The Groundwork of Cognition,* edited by S. Harnad, 535–565. New York: Cambridge University Press, 1987.
11. Harnad, S. "Minds, Machines and Searle." *Theor. & Exper. Art. Int.* **1** (1990): 5–25.
12. Harnad, S. "The Symbol Grounding Problem." *Physica D* **42** (1990): 335–346.
13. Harnad, S. "Against Computational Hermeneutics (Invited commentary on Eric Dietrich's Computationalism)." *Soc. Epistemology* **4** (1990): 167–172.
14. Harnad, S. "Lost in the Hermeneutic Hall of Mirrors (Invited commentary on Michael Dyer *Minds, Machines, Searle and Harnad.*)." *Exper. & Theor. Art. Int.* **2** (1990): 321–327.
15. Harnad, S. "Other Bodies, Other Minds: A Machine Incarnation of an Old Philosophical Problem." *Minds & Machines* **1** (1991): 43–54.
16. Harnad, S. "Connecting Object to Symbol in Modeling Cognition." In *Connectionism in Context,* edited by A. Clark and R. Lutz, [need page numbers]. Berlin: Springer-Verlag, 1992.
17. Harnad, S. "Grounding Symbols in the Analog World with Neural Nets." Special Issue on Connectionism and Computationalism. *Think* (1993): in press.
18. Harnad, S., S. J. Hanson, and J. Lubin, "Categorical Perception and the Evolution of Supervised Learning in Neural Nets." In *Working Papers of the AAAI Spring Symposium on Machine Learning of Natural Language and Ontology*, edited by D. W. Powers and L. Reeker, 65–74. AAAI,1991. Presented at Symposium on Symbol Grounding: Problems and Practice, Stanford University, March 1991; also reprinted as Document D91-09, Deutsches Forschungszentrum fur Kuenstliche Intelligenz GmbH Kaiserslautern, FRG.
19. Hayes, P., S. Harnad, D. Perlis, and N. Block "Virtual Symposium on the Virtual Mind." *Minds & Machines* (1992): 217–238.
20. Kleene, S. C. *Formalized Recursive Functionals and Formalized Realizability.* Providence, RI: American Mathematical Society, 1969.
21. MacLennan, B. J. "Technology Independent Design of Neurocomputers: The Universal Field Computer" In *Proceedings, IEEE First International Conference on Neural Networks*, edited by M. Caudill and C. Butler, 39–49. New York: Institute of Electrical and Electronic Engineers, Vol. 2, 1987.
22. MacLennan, B. J. "Logic for the New AI." In *Aspects of Artificial Intelligence* edited by J. H. Fetzer, 163-192. Dordrecht: Kluwer, 1988.
23. MacLennan, B. J. "Continuous Symbol Systems: The Logic of Connectionism." In *Neural Networks for Knowledge Representation and Inference*, edited by Daniel S. Levine and Manuel Aparicio IV. Hillsdale, NJ: Lawrence-Erlbaum, in press.
24. MacLennan, B. J. "Characteristics of Connectionist Knowledge Representation." *Infor. Sci.* (in press).

25. MacLennan, B. J. "Grounding Analog Computers." Special Issue on Machine Learning, *Think* (in press).
26. Nagel, T. "What is It Like to Be a Bat?" *Phil. Rev.* **83** (1974): 435–451.
27. Nagel, T. *The View From Nowhere.* New York: Oxford University Press, 1986.
28. Newell, A. "Physical Symbol Systems." *Cog. Sci.* **4** (1986): 135–83.
29. Pylyshyn, Z. W. *Computation and Cognition.* Cambridge, MA: Bradford Books, 1984.
30. Searle, J. R. "Minds, Brains and Programs." *Beh. & Brain Sci.* **3** (1980): 417–424.
31. Sober "Learning from Functionalism: Prospects for Strong AL." In *Artificial Life II* edited by C. G. Langton et al. Santa Fe Institute Studies in the Science Of Complexity, Proc Vol. X, 749–766. Redwood City, CA: Addison-Wesley, 1992.
32. Turing, A. M. "Computing Machinery and Intelligence." In *Minds and Machines* edited by A. Anderson. Engelwood Cliffs NJ: Prentice Hall, 1964.

Claus Emmeche
CONNECT, The Niels Bohr Institute, Blegdamsvej 17, DK-2100 Copenhagen Denmark;
e-mail: emmeche@connect.nbi.dk

Is Life as a Multiverse Phenomenon?

Artificial Life research has revealed one thing about living phenomena that may have deep implications for theoretical biology: Life is not as coherent a phenomenon as usually conceived of. The paper explores this "deconstructive move" of Artificial Life and raises some fundamental questions about the modeling relations used in typical Alife simulations with implications for the concept of emergence and computation. It is argued that distinct concepts of computation that may not be put on the same footing are in play in the language game of A- and B-life research.

INTRODUCTION: IS ARTIFICIAL LIFE POSSIBLE?

When posing the question "is artificial life possible?," our immediate answer is that on the one hand: of course it is—people make it, and indeed very interesting and even breathtaking structures have already been constructed, such as "animats," self-reproducing patterns and the other things we have seen already. In this sense we are forced to take artificial life as a fact (at least as a fact about a new branch of research), nearly in the same way that the philosopher Kant took the theoretical physics of his days, Newtonian physics, as a matter of fact, and then asked: What are the conditions of possibility for this kind of theoretical science? On the other hand, the situation differs from Kant's. Artificial Life does not confront us with an analogy of theoretical mechanics within the field of biology. We face a curious situation: It is not obvious to the majority of biologists that Artificial Life is possible at all, at least in the purely computational sense of "software life." Probably, most biologists would never call these artificial constructs "living." Why not? Because the intuitive notions of life and living systems within biology implies, among other things, that *living beings are a result of a long, ongoing evolutionary process* that have created *autonomous* organisms, single-celled and multicelled, that are highly organized, open (nonequilibrium), material thermodynamic systems based on *metabolism* and some kind of genetic information supported by *macromolecules*, that only metaphorically resemble a computer program. It is not that biologists do not think there are some general principles in their field, some kind of "logic of life" (cf. F. Jacob[20]) or universal forms of the processes that characterizes living phenomena. The genetic code, and the structure of heredity, is an example. But for biologists, life is not a question of pure form, or formal processes; one might characterize the biological way of looking at things as life seen as a *formal-and-material phenomenon.*

So the question about the possibility of artificial life immediately raises three other questions: (1) what counts as "artificial life?"; (2) what is "real life" after all (and what kind of biology do we invent)?; and (3) after what criteria can we claim not only to model, modify and experiment with existing life, but also to realize, i.e., "synthesize" life, or different modes of living behavior? The modeling situation should be considered more closely when simulating formal models of natural systems and artificial universes. Let us first shortly state our basic opinion and simple conjectures on these questions before we proceed to give the full arguments: (1) Artificial Life is not a unitary field of research; that is, Alife can be very different categories of phenomena. (2) "Real life" (i.e., biological life) is not a unitary phenomenon either; in fact, Alife research may help to deconstruct this presupposition of the life sciences and theoretical biology. (3) There are three problems regarding the criteria for "aliveness": (a) Criteria for different sorts of living phenomena may be explicated in order to evaluate the theoretical relevance of Alife models; these criteria, however, may never reflect the richness of various forms of life that we see in the real world, described by some as our poor lonely "single" example: Earthly Life. (b) The criteria of self-reproduction (as a criteria for the presence of life) is not fulfilled (and is not likely to be fulfilled) in nontrivial instances of current Alife. More problematic, however, is; (c) Criteria may not be useful at all for evaluating the strong claims of possible construction of life in other media.

Let us consider these claims one by one.

1. ARTIFICIAL LIFE IS MANY THINGS

That Artificial Life is not a unitary field of research but includes very different phenomena is fairly obvious. There are at least these different ideas of what constitute an artificial living system (for a more methodologically oriented classification, see Taylor[34]).

1.1 TRIVIAL VERSIONS OF "ARTIFICIAL" "LIFE"

1.1.1. Alife as different (mathematical, conceptual, physical) *models* of living systems. This is a rather trivial (or "weak") version of the Alife research program: Everybody seems to agree upon the possibility of modeling living phenomena; these models can be computational or not. However, a closer look reveals disagreement with respect to the adequacy of the formal, computational approach: it is a model, not the real thing. If we chose computational (or, if you like, informational) models, is that just for representational convenience, or is it because we think that the thing modeled is also in some sense computational, or informational? This nontrivial question has not yet been clarified.

1.1.2. Artificially *modified* living organisms (e.g., by gene splicing or cell fusion). Domestic animals and food crops are life modified by humans and, thus, are "artificial" systems produced for the purpose of man.

1.2 NONTRIVIAL VERSIONS OF ALIFE

(This is Alife as "realized life" *de novo*, or by nonbiological means.)

1.2.1. Alife as *computer life* (i.e., computational systems with emergent lifelike phenomena in a range of complexity that in the eye of the beholder seems to approach Nature's own organisms). (This is the "strong version" of Alife.) Many levels of life and even chemistry (cf. "algorithmic chemistry") are included in this concept.

1.2.2. Alife as *animats* (and robots). This is a branch of neocybernetics.

a. A subgroup is "*evolutionary robotics*," a vision of man-made creation of autonomous systems, that can live a life of their own by exploiting material and energy resources to maintain, rebuild, and "reproduce" their own kind, and eventually becoming semiautonomous agents that still depend on human society as a source of components for rebuilding and reproduction of their own kind.

b. "*Weak version robotic Alife*." In this group we find robots and animats built for technical purposes which are seen to behave in a "lifelike" manner.

This, however, is recognized to be a by-product of our interpretation, the behavior of these systems represents a category quite distinct from the behavior of the carbon-based biological cells and organisms.

1.2.3. *Chemical Alife* (i.e., attempts to make real material systems with lifelike characteristics, eventually as *in vitro* models of prebiotic processes; primitive metabolic systems; Eigen hypercyclic systems; replicating micelles; etc.). The goal is to turn *in vitro* experiments into life *in vivo de novo.*

Now, one problem with all these individually interesting disciplines is that their integration into a coherent framework of investigation is still very weak. The skeptic will say that they, in principle, have so little in common that integration might never be achieved (compare Belew[3]). The common ideas of self-organization, emergence, and related concepts are vague and ambiguous. In the long run, it may be a problem for a research program not to study a coherent set of phenomena, but something that may be fundamentally different categories of systems that do not have much in common, except some broad methodological common principles. For the present, it is hardly known to which extent there is general agreement about a set of principles that may constitute "the hard core" of the Alife research program. So Artificial Life seems to be quite different categories of phenomena. What about real life?

2. REAL LIFE AS A MULTIPLE PHENOMENON

It is sometimes said that we all have some basic intuitions of what biological life is because, unlike stars and galaxies, we are ourselves living beings and know life "from within." But can one be so certain about the givenness of this intuition? Surely, we have a "view from within"; but that is not a concept of life (it is rather a qualitative aspect of living consciousness). Have humans beings always been thinking of life as a unitary phenomenon? Obviously not.

"Primitive" (which frankly reads non-occidental) animistic thinking often describes the whole world and its parts as living, but this does not discriminate life in our modern sense. In later and more systemmatic natural philosophies we find concepts, such as the Psyche in Aristotle, which might be compared with a modern generic concept of life as a set of distinct characteristics (or formal properties) unique to living beings as opposed to nonliving matter. (The *psyche* in Aristotle was in fact a differentiated level-specific concept, distinguishing the vegetative, the animate, and the specific human psyche—a valuable idea, even in modern accounts of psychogenesis.[12])

In the Middle Ages, there was still no science of biology in the modern sense but a lot of categorizing activity, and Nature was understood as distinct domains or Kingdoms. There were no unifying concepts of life. According to Foucault, life as a general category with a socially valid meaning did not exist, except perhaps, as one character in the universal distribution of beings or "the Great Chain of Being." Up to the end of the eighteenth century, only living beings existed, forming "several classes, in the series of all things in the world."[1]

The generic concept of life was first developed by some natural historians in the late sixteenth century (such as J. Blumenbach, F. Vicq d'Azyr, A.-L. de Jussieu, and L. J. M. Daubenton) who promoted the idea that *organization* was such an important distinct feature that separated the living from the inorganic nature, and that this difference was more fundamental than the difference between the animal and plant kingdoms. In the following century, the degree of organization became an important key to the study of a natural (as opposed to arbitrary) classification of the order of living Nature. Lamarck temporalized the static view of a chain of being and offered the revolutionary vision that the more complex could have originated from the less complex. In 1802 he coined the term *biology*, by which he wanted to denote the study of all pertaining to "living bodies, their organization, their developmental processes, and their structural complexity" (G. Treviranus and K. F. Burdach independently invented the same term in 1800). Both Treviranus and Lamarck implied that they had identified a new field of research rather than give a new name to an old. In Lamarck it was a part of his struggle against the "imperialism" of the mathematical physicists whose ideas intruded every field and rejected the validity of the methods of natural history at that time. Biology became widely known as one of the "higher sciences" through the works of August Comte in the 1830s, and biology in the modern sense began to take form in the last part of this century, especially after Darwin and the new physiology.

During this whole period, one can find incredible variation as to what was seen to constitute the "basic facts" of living phenomena. We will not deal with this complicated story. Instead, we shall put forward the idea that Artificial Life can be seen as a *deconstruction* of our present rational conceptions of life as a unitary phenomenon, constituted by a single universal set of "generic" properties.[2] Alife research reveals that our concept of life is not a single one, as we would like to think, and that no simple set of fundamental criteria can decide the status of our models and constructs when these are already embedded in specific preconceptions of what constitutes the aliveness of natural and artificial creatures. This "deconstructive move" is a consequence of the inquiry of fundamental Alife research. It may help one to realize at least the following different "conceptual models" of life as a biological phenomenon exists: the organism as a living animal; the cell as the simplest living

[1]See Foucault,[15] p. 160.

[2]The term deconstruction is borrowed from literary criticism[30] and originates in the philosophy of Jacques Derrida, who hoped for "deconstructing" western metaphysics since Plato by so to speak undermine its foundation from within by its own concepts, oppositions and ideas. Deconstruction is both destructive and constructive but, at the same time neither, because Derrida questions the idea that critical reading or construction of alternatives can proceed in a language that is not from the beginning infected by metaphysical ideas. Here, we do not claim to use the term in the original Derridaean sense.

thing; life as an abstract phenomenon; and life in the cybernetic sense as a machine process that can be made by natural selection or by an engineer. Other ideas of life exist as well, but the focus here is restricted to the more or less rational ones.

We normally think of life in the biological sense as having something to do with "good old-fashioned biological organisms" (GOFBO), either single cells or made up of cells, that have a metabolism, constituted by specific macromolecules, etc. But this is already "too much"; we do not need all these details if we are interested in simply drawing a boundary between the living and nonliving world. First and foremost, we have the old prototypic concept of *life* as a cluster of characteristics, more or less motivated by our knowledge of biology. There are many versions of this prototypical GOFBO concept (e.g., Mayr,[28] Farmer & Belin[13]). Life has irritability and metabolism; it can self-reproduce, feed, and develop; and it is vulnerable to illness and death. A given instance may not have all of these properties, but it will have many.

We can then see what is implied by these properties such as self-reproduction or metabolism. To do so we have to consult the disciplines of biology and acquire knowledge of cells, membranes, enzyme systems, genetically coded information, and other vital structures and processes. If anything can be said to be alive, it will have to be organized on the biological level as a cell or as a system of cells (even viruses require a cell as a necessary part of their cycle of existence and, thus, should be considered as a pathological subcomponent of the cell). Thus the next step is a more specific conception of life described at a lower level (the cell and its macromolecular organization). We call this concept of life MOMACE: modern macromolecular-based cells. Molecular biologists seldom care to define life: they know it when they have it, and they know that its complexity is immense when compared with ordinary organic chemistry.

Logically, the two concepts are not equivalent: there is no simple connection between the prototypic old concept of life and the modern concept that explicitly or implicitly involves the existence of cells. Already we can see that the biology of our time has not one, but two conceptions of life. In fact, Sagan[33] drew attention to five different definitions of life, based on physiology, metabolism, biochemistry, genetics, and thermodynamics. Now, if we want to construct life artificially by abstracting its logical form (to use von Neumann's[35] and Langton's[24] words), and subsequently realize this form in other media, it is most likely that we get two different things depending on our point of departure.

From the first concept, we get neocybernetic life: animats, robots, and other lifelike devices,[29] i.e., ROLI: robotic life. That is, machines that are living in the sense of seeming to be autonomous in their movements, with sensors and effectors and an internal structure that coordinates input information and output behavior. (This is, of course, not autonomy in the strict sense of a cell as a "component system"[21] or an "autopoietic system."[27])

What do we get from the definition of life taken from molecular biology? This is not quite clear. Biologically, this definition is more fundamental. Attempts to "realize its content artificially" would lead to a material copy of a cell. Maybe we first have to create "chemical artificial life" before we can make a complete artificial cell. But if an artificial material cell—if we really want it to be *living* (in the sense of being an autonomous self-reproducing metabolizing unit)—has to be made up of the same kind of biochemical compounds as a natural cell (the same types of compounds such as DNA, proteins, carbohydrates, etc.), then it will probably be just another instance of the same kind. The point in making it will in some sense disappear, provided that it will have the same level of complexity. It will be merely a replica. Of course, there is a point in trying to create a primitive self-maintaining version of a metabolizing cell-system,[2,8] if possible.

One might say that we should make a formal version of the cell, eventually realized in a kind of cellular automata model[23] in two or three dimensions, realizing all the formal properties of the physical cell in the cellular automata space (whatever we mean by formal properties in this context). There has been attempts to formalize the property of self-reproduction (though only with success for trivial instances of self-reproduction, cf. Kampis[21]), and this approach could be extended to other properties of life. This is precisely the idea of "strong Alife": any lifelike phenomenon can be realized in other media, because life is a question of form, not constituent materials; it is an abstract phenomenon, a form, or coherent process-structure; and it is an informational structure emerging from lower-level local interactions. But what is revealed here is that this notion functions as a separate intuitive conceptual model of life that we might call ABLI: abstract life.

So life seems to be a group of phenomena if we list these conceptual models of life, some of which are directly connected with the ideas of the Alife research program:

1. Good Old-Fashioned Biological Organisms (GOFBO), i.e., life as a list of properties known partly from the common sense of daily life, partly from the life sciences. Most often this is life conceived of as animals—thus GOFBA: Good Old-Fashioned Biological Animals.
2. Modern Macromolecular-based Cells (MOMACE) as characterized by molecular biology: You know it when you have it in your test tube.
3. Abstract Life (ABLI), i.e., life as a space-time pattern that "realizes" some formal properties of biosystems either within a biochemical medium, or in a symbolic formal space.
4. Robotic Life (ROLI), i.e., neocybernetic life, animats, nanorobotic life, and so on.

One could even add (5) CYBERlife—the idea of creating lifelike structures in a virtual reality to which we can relate through hypermedia—and (6) other nonscientific definitions and conceptions of life.

From the point of view of traditional biology (as the study of general principles of life), this is rather surprising. Artificial Life may help us to see that the idea of universality of the fundamental principles of life may be a presupposition, a metaphysical prejudice with a questionable basis. Traditional biology has been haunted by a lot of dualisms and metaphysical contradictions[31,36] pertaining to the methods of investigation as well as the subject matter: the dualisms between structure and process, form and function, part and whole, inheritance and environment, contingency and necessity, holism and reductionism, vitalism and mechanism, energy and information, concept and metaphor. The construction of Artificial Life may help to dissolve these dualisms, or combine or reinvent them in more fruitful ways, thus giving rise to new ideas about the nature of living beings. In this perspective Artificial Life can be seen as a new way of "reading" the science of biology. We may call it a deconstructive reading: Alife actuates a deconstruction of the Good Old-Fashioned Biological Life. The very opposition between living and dead nature, the organismic and the inorganic domain—that has been constitutive for the whole science of biology since its definition in the beginning of the nineteenth century—may dissolve or be reconstructed in a new framework, inspired by insights gained from Artificial Life as well as other disciplines that may reveal new common principles for living as well as nonliving organization.

3. ARE THERE CRITERIA FOR "VALID" ALIFE CONSTRUCTIONS?

But even if the "universal" conception of life has to be deconstructed as a metaphysical presupposition, and even if "real life" is not a unitary phenomenon, aren't we stuck with some criteria for the usefulness and validity of Alife systems compared with natural systems? Or, do we embrace the methodological credo of philosophers as Paul Feuerabend that "anything goes"? How do we prevent the field from degenerating into mere technicalities or becoming a business of developing new computer tools and toys and games? If Alife extends the field of biology to "life-as-it-could-be" by the constructive creation of any imaginable artificial universe, then how do we know "life-as-it-could-*never*-be?"

Here, it becomes apparent that the evaluation of the artificially constructed "possible life" has to take recourse to one of the mentioned ideas of what constitute the generic properties of life, eventually expressed as a definition of life. And then, if we define life from the point of view of molecular biology/MOMACE, where life is seen as a cell-bound biochemical phenomenon with material as well as formal properties, it is clear that the computational approach can never hope to realize living processes, but only to model life. The "strong version" of Alife will fail.

Strong version of Alife seems more or less to assume that GOFBO as well as MOMACE are out-dated and really without any theoretical foundation: GOFBO is very close to a commonsense "folk biological" conception of life, and MOMACE is not theoretically satisfying either, with no universal validity—it is a so-called "carbon chauvinistic" conception of life. In fact, we do not know if the presence of CHNOPS (not to speak of proteins or nucleic acids) is a universal requirement for the emergence of "lifelike" phenomena other places in the universe. And the *raison dêtre* of "strong Alife" is precisely to overcome the shortcomings of present theoretical biology: That it is based on one single example.

So if Alife research wants to invent or construct "life-as-it-could-be," there is this hurdle: In principle one can simulate any process on a computer (there may be specific limits to the simulation of complex systems, but let us assume general appropriations for the sake of the argument); then how do we know if a new artificial phenomenon constitutes life as it could be in a possible material world? How do we put realistic physical constraints on evolution of behavior and learning in models? And how do we know if the virtual reality that we have constructed constitutes instances of possible "life" or a complicated but purely nonliving physical universe? We have seen many interesting "self-organizing" phenomena in complex nonlinear but purely physical systems that, in some respects, behave as life would. But nobody claim these phenomena to constitute genuine life.

As our ultimate standards for what constitute interesting lifelike phenomena in a virtual universe (possible life), we only have our prescientific ideas and traditional biological intuitions of what makes up real life. An objection would be that we do have a solution for that problem, namely, a list of agreed-upon criteria of the presence of living processes. However, this seems to be a list of problems, rather than a key to the solution. Let us consider an example, the list from Farmer and Belin,[13] which comprises the following set of properties:

Life (as defined by Farmer & Belin[13]):

1. A pattern in spacetime (rather than a specific material object).
2. Self-reproduction, in itself or in a related organism.
3. Information storage of a self-representation.
4. Metabolism that converts matter/energy.
5. Functional interactions with the environment.
6. Interdependence of parts within the organism (preserve the identity of the organism; ability to die).
7. Stability under perturbations of the environment.
8. The ability to evolve.
9. Autonomy. [added feature, C.E.]

Apart from the fact that such lists will always be somewhat arbitrary and express ones chosen frame of description, it is not complete. An important property is added here: life as an autonomous phenomenon. This criterion reflects the evolutionary fact that life is not a predesigned but a naturally evolved phenomenon, and the ecological fact that life is usually not dependent upon us for its existence, so an artificially created organism should be able to go on living a life of its own within a natural environment. Even domesticated animals, which cannot be sure to be able to do that, nevertheless have a reminiscent inherited degree of autonomy and a potential for colonizing other habitats. The authors want the list to be specific for life and so do not include growth. Growth is not a specific property: "there are many inanimate structures such as mountains, crystals, clouds, rust, or garbage dumps that have the ability to grow. Many mature organisms do not grow."[3] But exactly the same sort of obstacles applies to the other properties. But there are other problems as well.

The first criterion can be interpreted to capture the constant turnover of constituent material in any living organism. A large part of the material of our bodies is replaced within a few years, so the "stuff" changes over time but its "pattern" is the same. The turnover is described in detail by biochemistry as metabolism, which is the fourth criterion. Another interpretation is that the first criterion expresses the idea of the medium independence of life, i.e., life as a formal (informational, functional, or form) property.[24] This idea seems to be at odd with an immediate intuition nourished by molecular biology[9] that life is a process found in a set of concrete material systems—living cells—and that it should be studied in close connection with the study of the material components that make up these systems. Furthermore, this intuition tells that the physical properties of these systems impose constraints on the set of possible forms of life that can be realized by this set. For example, properties of the double-lipid layer of the cell membrane and the various membrane-bound proteins determine what kind of material substances between the organism and the external environment are allowed to be exchanged, thus constraining the set of possible inputs to the general metabolism of the cell. Another example is gene regulation: In order to study mechanisms of gene regulation, one has to look at the way the "information" is being "processed" at the level of macromolecular properties of ribosomes, DNA, mRNA, etc. Here one cannot separate the "form" of the process from the molecular "workings" of the system.[9] That does not mean that life could not exist in other media composed of other materials (Farmer and Belin notes that it is "easy to conceive of other forms of life, in different media, with a variety of different reproductive and developmental mechanisms"[4]), but these materials will impose other constraints on the actual properties of this "other life." What is nonexistent in computational Alife is the study of what kind of physical/chemical constraints will harness the lives of the various computer organisms. As long as the latter are only programs, deficit of constraints other than the abstract computational ones, they do not have the reality character of natural organisms.

The second criterion of life, self-reproduction (SR), is important, but it invalidates as a candidate any example of Artificial Life—whether it is ABLI's computer organisms (Langton's self-reproducing loop, for example[23,24]) or the various devices of ROLI. This does not detract from the value of these examples; on the contrary, it negatively shows the power of real nontrivial SR! We shall not give detailed arguments,[5,21] but only state that SR (in the full-fledged sense of what an amoeba cell does when it grows and divides itself into two autonomous living beings) is not achieved yet by artificial means. One must distinguish *replication* of information (replication of DNA in a cell or the replication of a spoken message) from

[3] See Farmer & Belin,[13] p. 818.
[4] See Farmer & Belin.[13] p. 817.

self-reproduction. Biochemical replication is a necessary but not sufficient condition for self-reproduction. Furthermore, the formal machine SR is equivalent to copying and propagation of an informational structure, but, of course, it does not involve the reproduction of the physical machine that supports the process; in short: it is a formal model of (a specific theory) of the process, not the phenomenon itself.

Natural self-reproduction is complete in the sense that the information needed for guiding the process is fully contained in and integrated with the cell or organism being reproduced. In a computational (machine) model of SR (whether it is based on "universal construction" as in von Neumann,[35] or on more elegant designs such as the loop structure of Langton[23,24]), the "reproduced" entities—visualized on a screen as specific configurations of automata states duplicating themselves—do not really contain as intended all the information needed for determining the process of reproduction. From a purely formal view this might be the case, but the physical machine (that realizes the process and which is not reproduced) supports the embedding universe of the reproducing automata and acts as a co-determiner of the process, but is not itself determined by it. One might try to overcome this difficulty by simulating the physical machine in a higher level model of a self-reproducing system (as suggested by von Neumann) but, in such a model, there is still an additional external machine whose determination does not depend on the process of reproduction. The information responsible for "self-reproduction" is not completely localized within the configuration, and the external additional specification (by the embedding "universe" and supporting machine) is equally important for the process.[21] In an autonomous living system, we cannot make the distinction between the entity being reproduced and an ultimate machine whose properties do not depend on the process of reproduction and which is not reproduced itself. DNA is not an external "knowledge" or "description" of the cell, but forms an integral part of the very system; the informational determination is a tacit activity that expresses this information causally (as emphasized by Kampis and Csányi[22]). It is the intrinsic and causal property of the cell that explains why real SR is complete, while modeled SR involves external sign relations between observer and the system modeled.

Robotic SR has not, as far as we know, been achieved. One can probably some day construct machines that drive through a supermarket of components, and collect and assemble these components to make new machines of the same sort. But this sort of ROLI-SR is very different from SR in the biological sense, which is coupled to a process of macromolecular component production in a network of processes which maintain the cell/organism in the very same process as new subunits are produced (cf. Kampis's component systems and Maturana and Varela's autopoietic systems). In contrast, ROLI-SR is based on "allopoietic" production of the components to the SR-supermarket of the machines. Again, this does not mean that the problem of ROLI-SR is not interesting. On the contrary, the idea of colonizing the Moon with a society of machines that have the right complexity to maintain and reproduce themselves should be investigated, and the reasons why it eventually may be in principle (and not just in practice) impossible to realize should be made explicit (probably we have here the whole "frame problem" of Artificial Intelligence blown up to its ultimate dimension).

The other criteria make other troubles[11] not to be discussed here. A fundamental problem common for all criteria when used in the context of computational "strong Alife" is that they are really not criteria for life in the usual sense (of GOFBO or MOMACE), but that they already represent another concept of life, namely ABLI, and thus their relevance as a kind of "conceptual anchor cable" to the physical world of known plants and animals is dubious. "But hesitate," the Alife objector might say, "we didn't want them to be criteria for carbon-based life in the normal sense! We wanted to create new forms of life; lifeforms in other media." But that doesn't help. They are not useful at all for evaluating the strong claim of possible construction of life in a formal domain, because these criteria derive from another world than the world of formal properties, and they do not seem to make sense in the latter domain. One should not forget that the strong version of Alife—that human beings can create life out of nonliving artifacts—is really a radical claim. Alife is not yet life, and may never be. One is tempted to say that what is being studied in Artificial Life for the present, at least in the computational part of the research program, is quite another object: It is not even life as an abstract phenomenon, but the life of abstract concepts ascribed to a specific interpretation of formal computational structures. However, one should remember that contemporary biology does not have to be the only source of systematic knowledge of life, and that neither the concept of life nor the concept of computation are that clear.

WHAT IS BIOLOGICAL COMPUTATION?

The idea of changing a system's state as a kind of computational updating process is so obvious to computer scientists that they often tend to see other systems as well changing state in this way. The brain computes its next states; every single cell of the body takes as input various messenger molecules from other cells and computes which other molecules will be synthesized, etc. If information is intrinsic to life, it is important to ask if this information (e.g., the genetic information in the cell's DNA, or the supposed propagating patterns of conformational switches in the cytoskeleton[16]) is involved in computationlike processes. This question remains unresolved. Another paper[11] deals with the origin of computational powers in natural systems, the "grounding" of semantical information and Langton's cellular automata approach to the origin of computation. What these discussions reveal is that the very concept of computation is not necessarily so well defined as normally conceived of, and we shall shortly sketch why.

In a sense, we simply face different kinds of computations realized in various systems: One kind has the normal intentional structure known from human and machine-mediated human computation (we can define a computer as an interpreted, formal, automatic system[17]); another kind is not conceptual but is found to take place in living cells; and a third one, under special circumstances, may occur in physical nature as a self-organizing phenomenon which can be modeled in the abstract space of cellular automata. One could add an additional kind which may not even deserve the name of computation but which characterizes a highly advanced form of mathematical reasoning that cannot—due to the limitations inherent in formal systems—be done merely by computational or algorithmic approaches, e.g., reasoning about the truth value (true/false) of some propositions in number theory which, by Gödel's incompleteness theorem, can be shown to be undecidable in a given formal system.[19] Instead of kinds of computation, we prefer to talk of four different concepts of computation (coc 1–4), that can be listed as:

coc 1. The formal, or algorithmic, concept of computation, which has its theoretical footing in the notion of a Universal Turing Machine.[7]

coc 2. An informal, intuitive, or "mathematical" concept of computation (and in general: reasoning about numbers) that is not bounded by the known limitations of formal systems. It points simply to the fact that mathematics cannot be reduced to automatic manipulation of symbol tokens; there is more to numbers (and, hence, to computing) than the properties that can be accounted for by formal theories of computation.[5]

coc 3. A biological concept of computation. This seems to be a quasi-theoretical concept that can be understood in many ways: for example, as problem solving by learning and adaptability[6]; as molecular processing of information in cells[1,16]; or as computation by neural networks.

coc 4. A physical concept of computation, that might be nonrepresentationalistic. The entities that cooperate in computational enterprises are patterns that can transmit, store, and modify information,[14,25,26] but these patterns seemingly do not have to "stand for" anything, as long as no functional constraints are imposed from a higher level.

How these different concepts of computation relate to each other is not clear, and have not, as far as we know, been analyzed yet. From a formal point of view (cf. coc 1), or from an anthropocentric, conceptual perspective (that might embrace coc 2), one is not forced to accept the notion of nonrepresentational computation (coc 4). From an biological perspective (coc 3), there is a point in restricting language-

[5]A rude indication of the difference between coc 1 and coc 2 is the number π. You can give an algorithm for generating the decimal expansion of π, to get an approximal value expressed as a rational number with a decimal fraction which may be arbitrarily long, within the limits of the physical possible. This number, however, is not π—in that sense, no human being have ever "seen" π. The algorithm is, so to speak, just the "translation" of the mathematical concept of an irrational and transcendental number to a computationally convenient expression of it. Strictly speaking, one cannot compute π but only some of its approximations.

and concept-dependent processes (such as formal and mathematical computation) to the level of human society which is outside the realm of biology, and to preserve the term "information" for simplier kinds of sign transfer. One may easily be lead to a "restrictivist" approach to computation and dismiss nondoctrinary notions (coc 3 and coc 4) because they are not specifiable within a formal setting. The fact that important research has been devoted to understanding the principal physical limits on computation[4] is not in conflict with such a formalistic stance. The study of these physical limits (on formal computation) do not imply "a physical concept of computation" in the sense of coc 4, but is concerned with, for instance, how fast and energetically "cheaply" one can realize formal computations in physical devices. In spite of the restrictivistic temptation, it is too preliminary to ascribe explanatory monopoly to just one theory of computation, especially as we are seeing a lot of interesting approaches in computer science and Artificial Life resulting in new ideas; e.g., Rasmussen, Knudsen and Feldberg.[32] As these authors observe, "a useful computation theory for natural systems has yet to be formulated" (p. 220, ibid.).

We need a general theory of realized computation in natural systems, and neither Artificial Life simulations, nor theoretical computer science, nor physics or biology provide such a theory yet. Philosophical reflections may help to clarify epistemological and model-theoretic issues and to frame the functional properties of biological information within a greater evolutionary frame.

REFERENCES

1. Albrecht-Buehler, Guenter. "Is Cytoplasm Intelligent, Too?" *Cell & Muscle Motil.* **6** (1985): 1–21.
2. Bagley, R. J., J. D. Farmer, S. A. Kauffman, N. H. Packard, A. S. Perelson, and I. M. Stadnyk. "Modeling Adaptive Biological Systems." *BioSystems* **23** (1989): 113–138.
3. Belew, R. K. "Artificial Life: A Constructive Lower Bound for Artificial Intelligence." *IEEE Expert* **6(1)** (1991): 8–14, 53–59.
4. Bennett, Charles H., and Rolf Landauer. "The Fundamental Physical Limits of Computation." *Sci. Am.* **253(1)** (1985): 48–56.
5. Cariani, Peter. *On the Design of Devices with Emergent Semantic Functions.* Ph.D. Dissertation, Department of Systems Science, State University of New York at Binghamton. Ann Arbor, MI: University Microfilms, 1989.
6. Conrad, Michael. "Molecular Computing." In *Advances in Computers*, edited by Marchall C. Yovits, Vol. 31, 235–324. London: Academic Press, 1990.
7. Davis, Martin. "What is A Computation?" In *Mathematics Today: Twelve Informal Essays*, edited by Lynn Arthur Steen, 241–267. New York: Springer-Verlag, 1978.

8. Dyson, F. *Origins of Life.* Cambridge: Cambridge University Press, 1985.
9. Emmeche, C. "The Problem of Medium-Independence in Artificial Life." In *Complexity, Chaos, and Biological Evolution*, edited by Erik Mosekilde and Lis Mosekilde. NATO ASI Series B, Vol. 270, 247–257. New York: Plenum, 1991.
10. Emmeche, C. "Modeling Life: A Note on the Semiotics of Emergence and Computation in Artificial and Natural Living Systems." In *Biosemiotics: The Semiotic Web 1991*, edited by Thomas A. Sebeok and Jean Umiker-Sebeok, 77–99. Berlin: Mouton de Gruyter, 1992.
11. Emmeche, C. "Life as A Multiverse Phenomenon: The Biosemiotics of Computation." Unpublished.
12. Engelsted, Niels "What is the Psyche and How did It Get into the World?" In *Essays in General Psychology*, edited by N. Engelsted, L. Hem, and J. Mammen, 13–48. Aarhus: Aarhus University Press, 1989.
13. Farmer, J. Doyne, and Aletta d'A. Belin. "Artificial Life: The Coming Evolution." In *Artificial Life II*, edited by Christopher G. Langton, Charles Taylor, J. Doyne Farmer, and Steen Rasmussen. Santa Fe Institute Studies in the Sciences of Complexity, Proc. Vol. X, 815–838. Reading, MA: Addison-Wesley, 1992.
14. Forrest, Stephanie, ed. "Emergent Computation." *Physica D* (special issue) **42** (1990).
15. Foucault, M. *The Order of Things.* London: Tavistock, 1970.
16. Hameroff, S., S. Rasmussen, and B. Münsson. "Molecular Automata in Microtubules: Basic Computational Logic of the Living State?" In *Artificial Life*, edited by C. G. Langton. Santa Fe Institute Studies in the Sciences of Complexity, Proc. Vol. VI, 521–553. Redwood City, CA: Addison-Wesley, 1989.
17. Haugeland, J. *Artificial Intelligence: The Very Idea.* Cambridge, MA: MIT Press, 1985.
18. Hoffmeyer, J., and C. Emmeche. "Code-Duality and the Semiotics of Nature." In *On Semiotic Modeling*, edited by Myrdene Anderson and Floyd Merrell, 117–166. Berlin: Mouton de Gruyter, 1991.
19. Hofstadter, Douglas R., and Escher Gödel. *Bach: An Eternal Golden Braid.* London: The Harvester Press, 1979.
20. Jacob, François. *The Logic of Life.* New York: Vintage Books, 1973.
21. Kampis, Georg. *Self-Modifying Systems in Biology and Cognitive Science.* New York: Pergamon Press, 1991.
22. Kampis, G., and V. Csányi. "Life, Self-Reproduction and Information: Beyond The Machine Metaphor." *J. Theor. Biol.* **148** (1991): 17–32.
23. Langton, Christopher G. "Studying Artificial Life With Cellular Automata." *Physica D* **22** (1986): 120–149.
24. Langton, Christopher G. "Artificial Life." In *Artificial Life*, edited by C. G. Langton. Santa Fe Institute Studies in the Sciences of Complexity, Proc. Vol. VI, 1–47. Redwood City, CA: Addison-Wesley, 1992.

25. Langton, Christopher G. "Computation at the Edge of Chaos: Phase Transitions and Emergent Computation." *Physica D* **42** (1990): 12–37.
26. Langton, Christopher G. "Life at the Edge of Chaos." In *Artificial Life II*, edited by Christopher G. Langton, Charles Taylor, J. Doyne Farmer, and Steen Rasmussen. Santa Fe Institute Studies in the Sciences of Complexity, Proc. Vol. X, 41–91. Reading, MA: Addison-Wesley, 1992.
27. Maturana, Humberto R., and Francisco J. Varela. *Autopoiesis and Cognition.* Boston Studies in the Philosophy of Science, Vol. 42. Dordrecht: D. Reidel, 1980.
28. Mayr, Ernst. *The Growth of Biological Thought.* Cambridge, MA: The Belknap Press of Harvard University Press, 1982.
29. Meyer, Jean-Arcady, and Stewart W. Wilson, eds. *From Animals to Animats.* Cambridge, MA: The MIT Press, 1991.
30. Norris, Christopher. *Deconstruction: Theory and Practice.* London: Routledge, 1991.
31. Oyama, Susan. *The Ontogeny of Information. Developmental Systems and Evolution.* Cambridge: Cambridge University Press, 1985.
32. Rasmussen, Steen, Carsten Knudsen, and Rasmus Feldberg. "Dynamics of Programmable Matter." In *Artificial Life II*, edited by Christopher G. Langton, Charles Taylor, J. Doyne Farmer, and Steen Rasmussen. Santa Fe Institute Studies in the Sciences of Complexity, Proc. Vol. X, 211–254. Reading, MA: Addison-Wesley, 1992.
33. Sagan, Carl. "Life." In *The Encyclopaedia Britannica*, 15th edition, Macropaedia, Vol. 10. London: William Benton, 1973.
34. Taylor, Charles E. "'Fleshing Out' Artificial Life II." In *Artificial Life II*, edited by Christopher G. Langton, Charles Taylor, J. Doyne Farmer and Steen Rasmussen. Santa Fe Institute Studies in the Sciences of Complexity, Proc. Vol. X, 25–38. Redwood City, CA: Addison-Wesley, 1992.
35. von Neumann, John. *Theory of Self-Reproducing Automata*, edited and completed by A. W. Burks. Urbana: University of Illinois Press, 1966.
36. Woodger, J. H. *Biological Principles: A Critical Study.* London: Kegan Paul, 1929.

Brian L. Keeley
Experimental Philosophy Laboratory, Department of Philosophy (0302), University of California at San Diego, 9500 Gilman Drive, La Jolla, CA 92093-0302; e-mail: bkeeley@ucsd.edu

Against the Global Replacement: On the Application of the Philosophy of Artificial Intelligence to Artificial Life

This paper considers itself a complement to the recent wealth of literature suggesting a strong philosophical relationship between artificial life (AL) and artificial intelligence (AI). I seek to point out areas where this analogy seems to break down or where it would lead us to draw hasty conclusions about the philosophical situation of AL. First, I sketch a thought experiment (based on the work of Tom Ray) that purports to suggest how such experiments should be evaluated. In doing so, I suggest that treating AL experiments as if they were just AI experiments applied to a new domain may lead us to see problems (like Searle's "Chinese room") that just aren't there. In the second half of the paper, I take a look at the reasons behind suggesting there is a philosophical relationship between the two fields. I characterize the strong thesis for a translation of AI concepts, metaphors, and arguments into AL as the "global replacement strategy." Such a strategy is only fruitful in as much as there is a strong analogy between AI and AL. I conclude the paper with a discussion of two areas where such a strong analogy seems to break down. These areas relate to liminative materialism and the lack of a "subjective" element in biology. I conclude that the burden of proof lies with one who wishes to import a concept from another discipline into AL, even if that other discipline is AI.

1. INTRODUCTION

In many ways, artificial life (AL) has long been the poor, younger sibling of artificial intelligence (AI). The two fields share many superficial similarities: Where AI can be seen as the synthetic, engineering side of the more analytic theoretical psychology, AL can be seen as the synthetic, engineering side of the more analytic theoretical biology. Both fields make extensive use of the modern digital computer, currently only as *models* but also, practitioners in both fields hope, potentially as *instances* or *examples* of the phenomena they study. The philosophical literature of AL is littered with concepts, metaphors, and arguments taken from AI. Variously, there is mention of AL Turing tests, AL dualism, AL functionalism, AL Chinese rooms, etc., all of which are concepts familiar from decades of discussion in AI.

Some, like Eliot Sober,[21] have even gone as far as to point to a strong analogy between AI and AL, an analogy that seems to vindicate such wholesale philosophical looting of traditional positions in AI. But it is the nature of analogies—even strong analogies—that there are differences between the two related entities. AL *is not* AI. On the basis of these differences, I argue that artificial life would be best served by originating new philosophical positions and metaphors of its own, without haphazardly borrowing such constructions from artificial intelligence. The spirit of this paper is to act as a complement to the growing pool of literature which either documents or implies similarities between AL and AI. Instead, I will try to highlight the dissimilarities between the two endeavors. In particular, I wish to point out areas where these differences are actually advantageous for AL, and where looking at AL through "AI-colored glasses" will lead one to see problems that may not be there.

The paper begins with a thought experiment, which is meant to capture an idealized picture of one of the goals of AL: to create life in a computer. Based loosely on the work of Tom Ray,[18] it is intended to explore the relationship between natural systems and AL programs that purport to exhibit biological phenomena. I hope to determine *the basis* on which we should make the decision of whether a given AL experiment is a genuine example of genuine *artificial* life. In doing so, I suggest the bases for this judgement are different from those traditionally involved in determining whether a system is an example of artificial intelligence. I conclude that treating AL as if it were just AI applied to different natural phenomena leads one to grapple with "Chinese room" objections to AL. However, I argue that the evaluation of AL experiments are sufficiently different to allow them to escape such considerations.

I then turn to the more abstract issue of the proposed analogy between AI and AL. What are the arguments in its favor? More importantly, given such an analogy what license does it give when deciding which concepts and metaphors from AI should be taken up in AL? I argue on the side of caution when "translating" the philosophy of AI into a philosophy of AL, pointing out that doing this properly requires a familiarity with *both* what is analogous and disanalogous between the fields. With this in mind, I end the paper with a discussion of two strong disanalogies between AI and AL: the lack of a viable eliminative materialist position within AL and the lack of anything analogous to the "problem of consciousness" in AL.

2.1 BLOB WORLD VS. BLIP WORLD: AN ARTIFICIAL LIFE METAPHOR

Let us now turn to that old chestnut of philosophical methodology, the thought experiment. In the following, I will consider an idealized example of an AL experiment in order to examine where the epistemological priorities lie, and whether they lie in places where a strong relationship to AI would suggest.

Imagine, if you will, a medium that exhibits some phenomena of interest to biology (Figure 1(a)). Unfortunately, the scale of these phenomena is quite microscopic—it is invisible to the naked eye—requiring the use some kind of "visualizer" that can magnify and regularize the behavior so that it can be seen on a CRT screen. On that screen we see an image consisting of slowly moving circles and some darker masses, all embedded within a heterogeneous medium. As we watch, some of the circles envelope the dark masses, while other circles occasionally split into two more-or-less identical circles. Let us call this medium and its phenomena "blob world."

But now imagine another medium, which also exhibits some interesting behavior (Figure 1(b)). It too is very small and otherwise invisible to the naked eye, so another kind of "visualizer" is required to make the phenomena visible on a CRT screen. What we see on this screen is a column of letters: "0080aaa," "0045aab," "0061acc," etc., next to that are some horizontal bars that are hectically pulsing out and back across the screen. As we watch, new letter combinations come into existence, while others disappear. Though the appropriateness of doing so is not yet apparent, let us call this second medium and its phenomena "blip world."

It should be no surprise when a microbiologist comes around and tells us that blob world is a group of microscopic single-celled organisms feeding and multiplying in a petri dish. And, as she has been recently reading up on the doings in AL, she also tells us that the blip-world output looks a lot like the real-time output of Tom Ray's *Tierra* simulator.[18] (Blip world is not identical to Tierra in all its details—Blip world is simplified for ease of presentation—but they are meant to be identical in their philosophical status. In this sense, Tierra is one of a variety of possible blip worlds. Given the success Ray has had with Tierra, it seems reasonable to expect to see similar research programs in the future.)

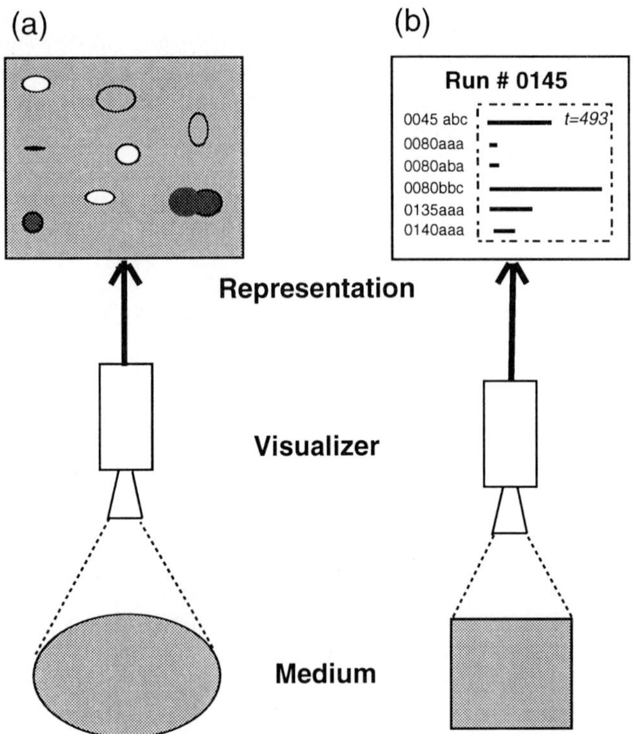

FIGURE 1 (a) First look at Blob world. (b) Blip world.

In blip world, each alpha-numeric string identifies an artificial "organism" which, in turn, is a bit of machine-level computer code. The bar next to the identifier represents the proportion of memory occupied by the token instances of that type of code. Each "organism" is essentially a piece of self-replicating code which contains the instructions required for replicating itself in the medium of RAM. As such, the code is both genotype and phenotype; it is both the instructions for replicating and what is replicated. If allowed, just one of these bits of code would soon replicate itself to the point that it filled up the entire memory with little copies of itself. However, this is prevented by two mechanisms. First, the code is not allowed to replicate itself perfectly. Every now and then it messes up and writes a "0" instead of a "1" or vice versa. In this way, *mutations* of the initial seed code enter the population. Just as with natural organisms, most of these mutations are fatal, in that they do not lead to code capable of self-replication, but some do turn out to be viable in this sense. Second, in order to keep the successfully replicating code from overrunning the system, a proportion (determined by age) is culled each generation. We can imagine that blip world exhibits the same interesting behavior as

Tierra, including "parasites" that locate themselves next to "hosts" and trick these hosts into copying the parasite's code instead of their own, and "hyperparasites" that play a similar trick on the parasites. We also see extended periods of stasis in the pool of different types of code interspersed with spurts of tumultuous change as new types compete with and replace the old.

Blip worlds like Tierra are AL simulations. We are called upon to evaluate the claim that what is going on in these worlds is similar enough to what is going on in *real* biological systems, such as the petri dish, that the predicate "alive" or "biological" ought to be applied to each with equal force. In essence, the claim is that *blip world contains life*, just as biologists agree that blob world does. The only relevant difference, so goes the claim, is that blip world exhibits manmade or *artificial* life, whereas *natural* life is going on in blob world. The only relevant difference between the two situations is one of origins. To evaluate this claim, the two kinds of systems need to be scrutinized in order to determine any relevant dissimilarities or asymmetries between the two situations. It should be kept in mind that the task here is not to determine *whether the claim is true*, but to say *in virtue of what* it is or is not. This latter task is the philosophical one that we must confront. Only after we have determined the basis on which the decision of "life" or "not life" is to be made, can we turn to specific details of a specific system (like Tierra) and attempt to make the decision.

The first difference in these two scenarios is in what is displayed on the screens. With blob world, we see a picture of the petri dish, whereas with blip world we see some kind of data chart. It is like the difference between seeing William S. Burroughs through the lense of a video camera and reading his biography. Clearly, one feels, the two scenarios must be markedly different. In blob world, "real" eating and reproducing is going on. We actually can see it on the screen. But in blip world, we see some kind of symbol manipulation and are treated to the results of these computations on its output screen. At best, only simulated—*as if*—eating and reproducing is going on.

However, this conclusion is hasty. The behavior of the two scenarios are indeed visualized differently. But this is due primarily to the different temporal scales of the two situations. Let us call the representation given in blob world a *window* representation (WR), an as-accurate-as-possible representation of the "look" of the system. It provides the viewer with a "window" on the medium. It is what we imagine we would see if we were miniaturized, or if the petri dish and all of its inhabitants were magically enlarged to the size of a swimming pool. Let us call the representation of blip world a *dynamic time-course* representation (DTCR): a representation of the long-term, gross dynamics of the system represented in aggregate, statistical form.

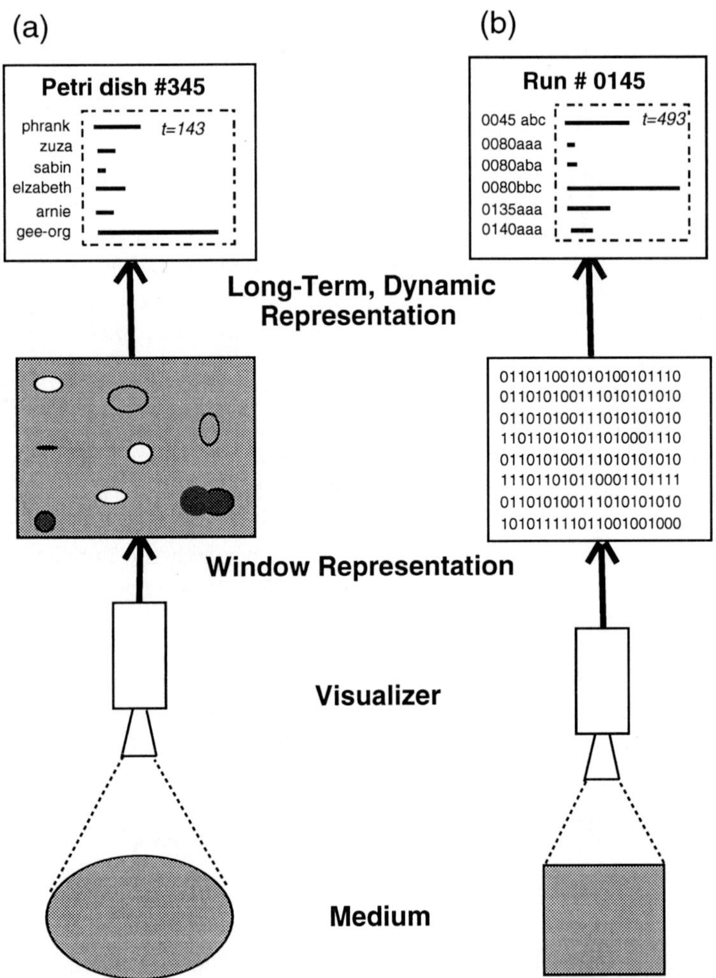

FIGURE 2 (a) A more complete look at blob world and (b) blip world.

If these representations were somehow uniquely and exclusively tied to the systems at hand, this indeed would be an important difference between them. But that one of these representations is commonly and preferentially used with each respective system is just an artifact of what we find most informative about each system and which representation is easiest to generate. For instance, a DTCR of blob world could be generated by identifying the individual cells and keeping track of their movement and reproduction (Figure 2(a)). In other words, if we were patient enough, we could keep track of lineages (perhaps by chemical tagging) and the percentage of the blob world that each lineage occupied. Admittedly it would be difficult to generate such data (especially in real time), but there is no reason *in principle* why it could not be done.

A critical point is that the dynamics displayed in the DTCRs of blip world and blob world show similarities. On the basis of this kind of similarity alone, we are led to believe that AL is possible. Artificial life's biggest claim to fame is that computer models of biological systems are often remarkably good at capturing the gross, high-level dynamics of biological systems. The literature is packed with computer models that capture population dynamics, the evolution of cooperative behavior, speciation, learning, etc. Often, what these models capture are such examples of the "look and feel" of biological systems, but some systems, in particular blip worlds, capture more.

Just as a DTCR of blob world can be produced, it is fairly trivial to produce a WR of blip world (Figure 2(b)). It would be a bit map of memory: a plane of 1's and 0's blinking on and off at a very high rate. These numbers represent the patterns of high and low voltages present in the memory of the computer. Where the WR of the petri dish is made up of "blobs," the WR of this alleged "electronic petri dish" would be made up of "blips." A blip world WR would be pretty meaningless to most viewers, and this is why this representation is rarely used to display the behavior of blip-world systems like Tierra.

But the patterns *are there to be seen*, if one could train oneself to see them. Properly trained, one would see certain strings of bits are more numerous than others, as the more successful codes (and their "children") copied themselves. If one watched closely enough, new types would be seen arising in the population, as mutation and selection occurs. The existence of these patterns is another crucial similarity between blip world and blob world and, as discussed below, this similarity is lacking or unimportant for similarly constructed AI models.

The asymmetry in the representational forms is then an accident of the combined effects of the dynamics of the systems involved, the limits of our perceptual capabilities, and our familiarity with types of representations. It is more informative to see the time-course data of blip world, and it is relatively easy to generate. With life in Petri dishes, such aggregate data is hard to produce. Also familiarity with blob world WRs makes it easier to see the behavior in which we are interested using that kind of representation.

2.2 WHAT OUGHT TO BE MADE FROM THIS METAPHOR?

In many ways then, the situations with blip world and blob world are analogous. There are indeed differences between them: (a) blip world seems to behave at a much higher speed and (b) behavior in blip world is much more easily quantified than that of blob world. We might also add (c) the form of energy used by both systems is different as well—blip-world organisms use electricity where blob-world organisms use sugars and sunlight. However, one feels that these differences are not relevant to the question of whether blip world is truly biological. We can imagine genuine living systems whose metabolisms and life cycles occurred at a much higher rate than that of life on earth. We can imagine having developed the technical know-how to produce measuring devices capable of producing from petri dishes the kinds of DTCRs we can generate so easily for blip world. Similarly, the details of how living systems convert energy into useful behavior also seems to be accidental and not an essential property of life. That blip world is different on these counts only illustrate that, if it is truly biological, it will be a *different* biology from life-as-we-know-it.

Some might say that so far I have overlooked an important difference between the two systems. It might be argued that blip world is "merely a simulation," that all that is going on in blip-world systems is mere symbol manipulation. The crux of this complaint can be traced to John Searle's now classic 1980 paper, "Minds, Brains, and Programs,"[19] in which he claims to refute what he calls "strong artificial intelligence." Strong artificial intelligence (AI) is the claim that an appropriately programmed computer can be an instance of a truly conscious, intelligent system.

Searle carries out this refutation through the use of his "Chinese room" thought experiment, which purports to show that mere rule following and symbol manipulation is not sufficient for true understanding, meaning, or intentionality. A conclusion Searle draws from his arguments is that the best that an AI program could ever be is a *model* or *simulation* of meaningful behavior, but never an *instance* of it. For the purposes of this paper, I am going to accept what I feel is a major conclusion of Searle's argument: a system cannot be said to exhibit a property such as "intelligence" (or, in the case of AL, "life") by virtue of its computational properties alone. As Searle might put it, computational properties are not the proper kind of causal properties to instantiate real intelligence (or life).

In his paper in these proceedings, Stevan Harnad has suggested just such an application of Searle's argument to the endeavor of artificial life.[13] Specifically, his claim is that, unless it is *grounded* (hooked up to the world with sensors and effectors), the best that an AL computer program could ever be is a *simulation* of life, never an *instance* of it. It would seem that such a criticism is supposed to apply to blip-world programs like Tierra. (However, it is difficult to be sure, as he never mentions any specific AL research by name.) Blip world is not hooked up to the world outside the computer in any way significantly different from the way in which traditional AI models are. Apparently, we are invited to draw similar conclusions about the reality of such AL models as Searle and Harnad draw about such AI models.

However, to draw this quick conclusion is to fall into the trap of looking at AL models as if they are simply AI models applied to a different domain. The two situations certainly look alike: a computer program crunching away on a program and throwing data up on a screen that bears a striking resemblance to what some natural phenomena would throw up on a screen and, if that resemblance is close enough, concluding that the computer is instantiating that natural phenomena as well. And, if Searle has refuted this argument in AI, surely he has done so in AL, as well.

Wrong. To see why, consider how the position known as "functionalism" is used in AI. Particularly in its original Turing test form, functionalism embodies the claim that, to some degree of abstraction, what a system is made of does not matter in the determination of whether it is "intelligent," "conscious," "intentional," etc. All that matters is whether it *behaves* in the correct way. The Turing test[23] sets out a strict procedure for determining what is legitimate behavioral evidence for making this judgement: answers to questions input to the system via a teletype. Modern versions of functionalism substitute other behaviors in place of those of the Turing test, such as Harnad's suggestion that the system be able to sense the world and execute robotic behaviors, but the multiple realizability thesis is maintained by throwing out of court any evidence based directly upon *how* the system produces its behavior. (The "multiple realizability thesis" is the claim that, in some sense, what a system is made out of is irrelevant to whether it is an instance of some phenomenon.) In traditional AI functionalism, we are called upon to determine whether a system's gross behavioral output meets some criteria, and then an attribution ("intelligent," "conscious") is *projected back* onto the specific physical system that generated the behavior.

However, this is not how the claim of "life" is decided in the case of blip worlds. I argue that whether blip world contains living things is *not* determined on the basis of what is displayed in either WR or DTCR, as would be the case if blip world were an AI system. *Blip world is evaluated as living or not on the basis of what behavior it exhibits in the medium, the RAM.* If the behavior of the medium is sufficiently like that of the Petri dish, then we call it biological, living, or whatever. Given that both of the media are not directly observable, and given the confidence that the production of the representations does not introduce any artifacts, then the evaluation, in practice, will be carried out by comparing the behavior of these representations. But it should always be clear that the both the WR and the DTCR only can be considered "lifelike" in virtue of the lifelike behavior of the medium that gives rise to them. In AL, the physical medium is judged to be lifelike or not and then that attribution is projected *forward* (not *backward*, as in AI) to the representations that system generates.

If there is an analogous position in the philosophy of AI for this position, it would be what Andy Clark[4] has dubbed *microfunctionalism.* This position, developed as a defense of the more biologically plausible approach to AI known as *connectionism*, reels in the more free-ranging liberalism of traditional functionalism, arguing that it *does* matter what a system is made of and how it produces its surprising behavior. According to microfunctionalism, evidence about the mechanisms utilized by a system to generate its behavior can be legitimately used to determine whether a system should be attributed with a property, such as "intelligence" or "consciousness."

In a similar way, microfunctionalism for AL argues that it is crucial to look inside the computer—at the behavior of the medium—to see what behavior there is and how it is produced. If, upon looking into the system, variables and sets of rules (in other words, symbols and procedures for manipulating those symbols) are found, then it can be concluded (with Searle and Harnad) that the system is merely a simulation of life. If instead a system that exhibits physical properties relevantly similar to those of "real" living systems is discovered, then we may conclude that there is indeed life in blip world.

Indeed it may turn out to be the case that we will decide that what is going on in the RAM of a specific blip-world system like Tierra is just not similar enough to natural life to warrant the claim of artificial *life*. For instance, the Tierra organisms lack both development (they lack anything that resembles morphogenesis) and metabolism, and biologists may decide that these features are indeed crucial to its characterization as a true biological system. And perhaps, as Michael Dyer (in conversation) has suggested, the physics of the internal world of a computer is just too simple and regular, compared to that of the Terrestrial world, in which life as we know it has developed, to support the complex entities typically associated with life. However, such a decision must be made primarily on the basis of continuing work within theoretical biology.

In any case, I have proposed an answer to the philosophical question I set out when I introduced the blip-world vs. blob-world thought experiment: When deciding whether a particular blip-world program is truly biological, in virtue of what is that decision made? I have argued that it is in virtue of blip world's *physical* properties (not its *computational* properties) that it exhibits relevantly biological behavior. While it is true that the medium in which this behavior is found is a "computer," we should never forget that our computer is not some kind of Platonic "purely computational system"; it is a very down-to-earth physical system, a machine. Not everything that a computer can do is "computational" in nature. My NeXT computer workstation can not only simulate a paperweight, it can actually instantiate one as well. It can not only simulate the heat output of a NeXT workstation, as a NeXT workstation it also produces real (not simulated) heat.

The claim here is that what is going on inside a computer running a blip-world program is *not* a computational simulation of life. It is an automated physical procedure for seeding the computer's memory with appropriate physical patterns of high and low voltages, and for appropriately visualizing the resulting dynamics. The fact that all this is going on in a medium that is typically used to perform operations that are interpretable systematically in computational terms is irrelevant.

This claim is highly counterintuitive. I am suggesting that if blip world is judged to be alive, it will be on the basis of its physical, not its computational, properties. The blip world I have described exhibits the property of self-replication in the same way that my workstation exhibits the property of producing heat. Real, physical self-replication is going on inside the computer's RAM, as certain patterns of high and low voltages manipulate neighboring locations until they exhibit an identical pattern of high and low voltages. This is not simulated or *as if* self-replication; this is *instantiated* self-replication!

(Note that I are not claiming that the simulation is not occurring on a computer [this is obvious], I am only claiming that such a simulation is not making use of the well-known computational properties of the computer. Similarly, the use of my NeXT computer to determine the heat output of an identical make of computer is to use a computer as non-computational simulation. Perhaps the term "model" [as in the scale models used by architects, or the models built by children] is a better term for this situation.)

I hope that I have shown that the application of traditional AI philosophical analysis to *prima facie* similar situations in AL can be misleading, saddling AL with problems and concerns (like the Chinese room) that it can do well without. However, the application of AI thinking to AL is an appealing one, and presumably has its utility, as in the case of microfunctionalism. To what extent, and in which situations, is such a comparison fruitful? This question is the topic of the second half of this paper.

3. ANALOGIES AND STRATEGIES

In "Learning from Functionalism—The Prospects for Strong Artificial Life," Elliot Sober[21] explores the following analogy: "Artificial intelligence is to psychology as artificial life is to biology." With this analogy (which I call the *Sober analogy*) he sketches a variety of positions and concerns from the traditional philosophy of AI as they would appear in the philosophy of AL. He discusses "strong" and "weak" AL, biological dualism and identity theory, biological multiple realizability, etc. Sober eventually argues for a functionalist approach to biology and AL that parallels the prominent philosophical position of the same name found in psychology and AI.[1]

Sober is not alone in seeing parallels between AI and AL. In his seminal essay introducing the first AL proceedings, Chris Langton[16] follows a similar path (see, in particular, Figure 11, p. 40). There he notes a similarity between connectionist AI (where relatively complicated "intelligent" behavior is generated using a relatively simple structural substrate as in connectionism) and AL modeling (where relatively complicated "living" behavior is generated using a relatively simple substrate as in cellular automata). Similarly, in that same volume, Pattee[17] notes, "It is clear from this workshop [Artificial Life I] that artificial life studies have closer roots in artificial intelligence and computational modeling than in biology itself."

While this evidence indicates a connection between AL and AI, Sober is arguing for a close relationship between the philosophical situations of each field.

[1]Traditionally, the philosophy of AI has been seen as a specialization of the more general set of concerns of the philosophy of psychology. Similarly, one would expect that a "philosophy of AL" would be a specialization of the philosophy of biology. In lieu of the cumbersome phrasing "philosophies of psychology and AI" and "philosophies of biology and AL," I will refer to only the "philosophy of AI" and the "philosophy of AL" for the sake of brevity.

While Sober argues for an AL version of functionalism, others discuss an AL "Turing test,"[1] an AL "Chinese room,"[13,15] and an AL hardware-software distinction.[8] Given that AL is generally free of philosophical discussion (some would say *refreshingly* free), these examples suggest that Sober is not alone in pointing out a deep philosophical connection between AI and AL.

The Sober analogy is an appealing one, and there is no doubt a lot of truth in it. Where AI is the synthetic, engineering counterpart of the more analytic science of theoretical psychology, AL is the synthetic, engineering counterpart of the more analytic science of theoretical biology. Both AI and AL make extensive use of the digital computer and computer models of their respective phenomena. Both AI and AL argue that there is no reason why we can't build artificial examples of what have been phenomena of purely natural origin.

However, analogies are not incredibly useful by themselves. They just suggest that there are similarities (and differences) between two things. A methodology based on analogy would be more useful. One wants to turn a simple logical relationship into a methodology. For a new endeavor like AL, such a methodology might include setting out the set of important philosophical metaphors, positions, and distinctions to be used in that endeavor. I feel that in the above examples—the application of traditional AI distinctions to AL—imply just such a methodology. In its most extreme form, this implicit strategy (which I call the *Global Replacement Strategy*, or GRS for short) involves taking thirty years of avid discussion in the philosophy of AI and translating it into what then will be the "philosophy of AL." This strategy, apparently supported by the Sober analogy, gives AL a way of generating a complete and well-worked-out philosophical landscape, merely by taking the canon of the philosophy of AI and (stealing a concept from word processing) *globally replacing* all occurrences of "intelligence" with "life."

This extreme application of the Sober analogy is not without its merits. It allows the still-embryonic AL to take advantage of the large philosophical armory that AI has struggled to develop over the better part of three decades. AL can dispense with doing any of this hard work for itself. In a mere five years since its inception, so goes the GRS argument, Sober has given AL a rich and varied philosophical tapestry of positions, arguments, and metaphors to rival that of any other, more established endeavor.

However, no matter how appealing it might seem, GRS is not the best course for the AL community to take. There is good reason to believe that there is much to be gained by originating a novel philosophy of AL, with little derivation from traditional philosophies of psychology and AI. As illustrated in the blip-world vs. blob-world example, thinking of AL in traditional AI terms can lead one one astray. This example illustrates the dangers of the GRS, but a more general account of its hazards is needed.

4.1 DISANALOGIES BETWEEN LIFE AND MIND

Like all such analogies, the Sober analogy does not claim that the central phenomena of psychology ("mind") and biology ("life") are identical, but it does suggest that the way that these phenomena are (or should be) handled in their respective domains are significantly parallel. GRS is calculated to use that parallelism, turning it into a constructive strategy for defining the proper philosophical problem space of AL. But while attractive, there is a problem with this picture. The existence of any important disanalogies between the domains of psychology and biology would point to large areas of concerns that would resist the simple translation of one field into the other. In the remainder of this paper, I will consider what I believe are the two most important differences between the phenomena of life and mind: the lack of a strong eliminative materialist position in biology and the lack of a biological concern with the subjective.

4.2 FOLK BIOLOGY AND ELIMINATIVE MATERIALISM

When looking at the arguments of those who wish to allege a strong analogy, it is often more instructive to note what the author *fails* to mention, rather than what he actually does. Among the positions traditionally available to the philosopher of AI (and, *mutatis mutandi*, to the would-be philosopher of AL), Sober mentions dualism, identity theory, and functionalism, among others. But one position he fails to mention is *eliminative materialism* (EM). Originally argued by Paul Feyerabend[9,10,11] and currently championed by Stephen Stich[22] and Paul M. Churchland,[2,3] EM is primarily a thesis about proper scientific explanation. In particular, it seeks to reject the notion that scientific explanation must be carried out in terms of our folk scientific conception of ourselves. A "folk theory" is just another name for our commonsense notions about a particular domain. For example, Aristotlean physics might be considered an explication of ancient Greek folk physics, a physics in which rocks fall because they desire to return to the place of their origin, and where heavier objects fall faster than lighter ones. Folk *psychology* would consist of the myriad rules of behavior humans use in their everyday relations with one another. (See Churchland[2] for a sketch of these rules.) Central to this folk theory is the liberal attributions of "beliefs," "desires," "moods," etc. to the entities that make up the domain of psychology: other people, pets, fictional characters, etc. In folk theories the issue is *not* whether they are useful abstractions or whether they are important to our day-to-day dealings with the world. (They are essential. Just reflect on the central role that folk psychological attribution plays in our justice system.) The issue is whether these common sense theories have any special status within science. In the case of contemporary scientific physics, it is accepted that folk physics has no special status. If physicists can explain the motion of bodies without anthropomorphizing them, then physics should do so.

The status of folk *psychology* is a different can of worms. As mentioned above, Paul Churchland has argued that not only *can* folk psychology be banished from a mature scientific psychology, but the time has come to actually do so. In making his case against folk psychology, he mounts a three-pronged attack. First, he reminds us that we should not only assess a theory on its successes but also on its failings. And there is a large inventory of presumably psychological phenomena that folk psychology simply fails to address adequately, including the nature and dynamics of mental illness, creative imagination, sleep, perceptual illusions, and learning, just to name a few. Second, he argues that the history of folk psychology does not give one reason to hope for the future of the endeavor. Churchland writes that "the story [of folk psychology] is one of retreat, infertility, and decadence." It is a paradigm case of a degenerating research programme. Finally, Churchland outlines reasons for believing that folk psychology cannot be integrated easily with the rest of scientific explanations. Particularly, it seems to be very much at odds with the one field that it would presumably have the closest associations: neuroscience. On the weight of all three of these deficits, Churchland argues that the days of folk psychology in scientific psychology are numbered.

However, Churchland's EM in psychology is not without its objectors. Indeed, it is probably safe to say that it is still a minority view amongst philosophers of psychology. Some, like Dan Dennett[6,7] and Terence Horgan and James Woodward,[14] have argued that folk notions such as "belief" and "desire" should or must play a role in our scientific psychological explanations. For years, Dennett has argued the importance of an "intentional system" to psychological explanations. An intentional system is one which is "reliably and voluminously predicted" via the attribution of "beliefs," "desires," and other common sense notions to that system. And Dennett argues cogently that humans and other animals are just such systems. This being the case, a scientific psychology must employ concepts from folk psychology.

Horgan and Woodward take a slightly different approach. They argue that the case against folk psychology is overstated, that folk psychology is actually quite a good scientific explanation of psychology, regardless of its perported failings. In any case, they also argue that EM places too stringent restrictions on how folk psychology should be integrated with our other scientific beliefs. That neuroscience cannot capture the basic notions of folk psychology in its theory is no reason to reject it in favor of neuroscience.

But for all this heated debate over the importance of folk theory to psychology, we do not find anything even vaguely similar to biological theory. It is not clear whether such a debate is possible. The primary problem is determining whether a folk theory of biology exists in the first place. And, if a "folk biology" can be rounded up for the purpose, will its fate be more like that of folk psychology or folk physics?

One might look for a folk biology in the lore of the "common person," that general framework of common sense and rules-of-thumb that has served our species so well through the ages. Aside from common sense *psychological* knowledge about natural phenomena (e.g., "Always avoid contact with female bears when they are with their cubs, as mama bears are prone to protective violence when they *believe* their young are threatened"; "My dog is standing next to the door because he *wants* to go out"; etc.), there seems to be little of what might be called specifically *biological* knowledge.

There is a good deal of folk knowledge of *breeding*, such as the old maxim that "like breeds like." The dangers of inbreeding and the knowledge that like animals will only mate with like animals have apparently been well known to breeders for centuries. Our first candidate for a folk biology, then, would be some version of the science of breeding. Indeed, part of the inspiration for Charles Darwin's *Origin of Species*[5] was the great diversity of pigeons that breeders had raised (even without knowledge of Mendelian genetics).

It is appropriate that Darwin's ground-breaking work should be mentioned, as its title names what arguably may be the central notion of any possible folk biology: the concept of a "species." The notion that the biological world is made up of distinct kinds of creatures is probably the first principle of common sense biology. The *Old Testament*, Native American mythology, and many other creation stories share the common feature that distinct kinds of creatures were created separately. Perhaps the biggest job of a scientific biology, from Aristotle onwards, has been the Herculean task of simply cataloging all the kinds of creatures found in our incredibly diverse ecosystem. The notion of distinct species is so central to our notion of what biology should be that this was what Darwin felt he had to explain with his theory of natural selection.

Along with this notion of diversity in the biological realm, perhaps another central notion to folk biology would be that the biological world constitutes a fundamentally different set of things, i.e., that there is something distinct and special about biological entities that separates them from the rest of the furniture of the universe. This notion of an essential difference between living and nonliving things is perhaps best captured in concept of the "vital spirit," the substance that is the essence of the living. Possession of this spirit makes a truly living cell different from a nonliving collection of the same chemicals. Though the popularity of the belief in some kind of nonmaterial animating "spirit" has suffered in this century, the spirit of the issue survives in the demands that society places on biologists and medical doctors to come up with reliable criteria of "life" and "death."

We are beginning to see that at least it is *possible* that something answers to the name "folk biology." It would have an ontology (that the world consists of the "biological" and the "nonbiological," and that the biological world is made up of distinct kinds or "species"). It would also have rules for the behavior between the elements of this ontology (like the laws of breeding). Folk biology might not seem to have the richness typically attributed to folk psychology (most of the breeding rules would seem to delineate all the things with which a given species *cannot* breed), but that might be because I simply have not adequately characterized it here. But one can imagine that a likely story might be put together. For the sake of argument, let us assume that such a likely story could be generated.

Even if the existence of folk biology is granted, it must be noted that, unlike the situation in psychology, there does not seem to be anybody interested in arguing for folk biology as the necessary or appropriate language of biological explanation.

Where there is vociferous debate in the philosophy of psychology, there is only silence in the philosophy of biology.

If the proceeding discussion has any cogency, it indicates that the current state of biology on the issue of eliminative materialism and the role of folk theory is different from that of psychology. This, in turn, indicates an area of disanalogy within the Sober analogy. However, this is not the most striking difference between the study of the mind and the study of life.

4.3 LACK OF THE SUBJECTIVE IN BIOLOGY

Here we come to the most striking difference between psychology and biology, a difference that probably underlies many of the other differences I have already sketched above. Psychological explanation has to explain *more* than just the behavior of psychological systems. One of the things that makes psychology such a difficult endeavor is that in addition to the straightforward *behavioral*, third-person phenomena which stand in need of explanation, in the case of humans at least, there seem to be additional *experiential*, first-person phenomena. Part of the burden of psychology is to explain (or explain away) phenomena related to the *prima facie* claim that psychological systems exhibit attention, intentionality, consciousness, self-consciousness, a "point of view," or the property of being "something-it-is-likely-to-be" that entity, qualia, or any other of the constellation of concepts relating to the subjective nature of the psychological. Indeed, it seems plausible that this element of the psychological is what makes it so resistent to mechanistic or reductionistic explanation. It is the difficulty of even conceiving of a *conscious mechanism* that hampers the would-be psychological mechanist. Whatever consciousness is, apparently no collection of third-person facts about it would ever be complete; after science has done its best, there will still remain first-person facts inaccessible to the traditional scientific method.

It is not our place here to assess or take sides in the role or nature of consciousness in psychology. It needs only to be noted that, like the debate over eliminative materialism and folk psychology, there is no analogous concern in biology. Perhaps we should be thankful, for this is one less obstacle for theoretical biology to overcome or for AL to worry about. Biological phenomena, unlike their psychological counterparts, seem to be exclusively of the behavioral, third-person variety. There is no worry that, after describing all the physical parameters of the system, there still will be "something else." Now, determining what the correct parameters actually are and understanding exactly how biological systems produce the relevant behavior is a tough enough job on its own, but at least the phenomena in question are *there*—waiting to be measured, probed, and replicated. (I should note that Stevan Harnad makes many of these same points, more eloquently than I, in his contribution to these proceedings.[13])

To summarize the discussion so far, Sober proposed that the relationship between AL and biology was analogous to that between AI and psychology. This seems to be a prominent point of view within the AL community. In fact, there seems to be support for the even stronger claim that the philosophy of AL should be the philosophy of AI translated into biological terms, a strategy I call the "global replacement strategy." However, in the last several pages, we have seen that a variety of issues and debates endemic to the philosophy of AI—those relating to eliminative materialism and the subjective nature of mind—seem to have no counterpart in biology. These issues cannot be discarded as being minor side issues within the philosophy of AI. Quite to the contrary, if the amount of ink spilled over them is any indication, they are among the most central philosophical issues of that endeavor. But, if these important issues cannot be translated into the philosophy of AL, what does this indicate about the general usefulness of GRS? It indicates that whatever the alleged validity and usefulness of translating concepts, problems, and metaphors from AI into AL as a constructive strategy, the GRS is clearly too extreme. For all the similarities between AI and AI, as indicated by the Sober analogy, the phenomena of intelligence and life are sufficiently different to preclude any kind of straightforward relationship between the two sciences.

5. CONCLUSION

In this paper, I have tried, through a variety of means, to suggest that a relationship between artificial intelligence and artificial life is not as useful as it might seem at first. Until this point in time, the philosophical discussion within AL has been littered with references to positions, metaphors, and arguments made popular within the history of AI. However, with the notable exception of Sober's 1991 paper, we have seen little discussion specifically of the methodology of importing concepts from AI into AL. By and large, the justification for this procedure simply has been accepted on the basis of the close intellectual ties between the two fields and their respective practitioners. The spirit of this paper is not that of a refutation of this methodology, but as a caution against its unreflective overuse.

We should not be surprised if concepts from AI are useful to AL. In this paper, I mention that Clark's concept of "microfunctionalism" is just such a useful construction. This particular example is not surprising in that Clark uses microfunctionalism in conjunction with his arguments for "connectionist" AI, an approach to AI that is arguably more *biological* than traditional approaches. It is the biological motivation of this position that leads me to suggest its usefulness to AL. Contrary to the GRS, it is not *necessarily* useful to AL because it is a concept from AI. One needs to make an argument for its usefullness beyond that provided by the Sober analogy. Such a burden of proof is laid upon anyone wishing to use any concept from AI in AL. The Sober analogy merely indicates a relationship between the two disciplines, and one should expect a *sharing* of ideas between them, not an eclipse of one by the other.

ACKNOWLEDGMENTS

I would like to thank Aaron Sloman, Marcus Peschl, Derek Smith, Tom Ray, Ron Chrisley, and Inman Harvey for enlightening discussion on the topics of this paper. Georg Schwarz and Sandra Mitchell both read drafts and, in the process of disagreeing with most of what I had to say, offered valuable criticism. I also would like to thank the members of UCSD's Experimental Philosophy Lab, for listening to me present this material in many forms. Earlier versions of this paper were presented at the 1992 Comparative Approaches to Cognitive Science Summer School (Aix-en-Provence), the University of Birmingham School of Computing Science, and at Artificial Life III, itself.

REFERENCES

1. Bedau, M. A., and N. H. Packard. "Measurement of Evolutionary Activity, Teleology, and Life." In *Artificial Life II* , edited by C. Langton, C. Taylor, J. D. Farmer, and S. Rasmussen. Santa Fe Institute Studies in the Sciences of Complexity, Proc. Vol. X, 431–461. Redwood City, CA: Addison-Wesley, 1991.
2. Churchland, P. M. *Scientific Realism and the Plasticity of Mind.* Cambridge Studies in Philosophy. Cambridge, MA: MIT Press, 1979.
3. Churchland, P. M. *A Neurocomputational Perspective: The Nature of Mind and Structure of Science.* Cambridge, MA: MIT Press, 1989.
4. Clark, A. *Microcognition: Philosophy, Cognitive Science, and Parallel Distributed Processing.* Cambridge, MA: MIT Press, 1989: 34–36.
5. Darwin, C. *On the Origin of Species*, 1st ed. 1859.
6. Dennett, D. C. *Brainstorms: Philosophical Essays on Mind and Psychology.* Cambridge, MA: MIT Press, 1978.
7. Dennett, D. C. *The Intentional Stance.* Cambridge, MA: MIT Press, 1987.
8. Farmer, D. F., and A. d'A. Belin. "Artificial Life: The Coming Evolution." In *Artificial Life*, edited by C. G. Langton. Santa Fe Institute Studies in the Sciences of Complexity, Proc. Vol. VI, 815–840. Redwood City, CA: Addison-Wesley, 1989.
9. Feyerabend, P. "Explanation, Reduction and Empiricism." In *Scientific Explanation, Space and Time*, edited by H. Feigl and G. Maxwell. Minnesota Studies in the Philosophy of Science, Vol. 3, 28–97, 1962.
10. Feyerabend, P. "Materialism and the Mind-Body Problem." *Rev. Metaphys.* **17** (1963): 49–66.
11. Feyerabend, P. "Mental Events and the Brain." *J. Phil.* **60** (1963): 295–296.

12. Harnad, S. "The Symbol Grounding Problem." *Physica D* **42** (1990): 335–346.
13. Harnad, S. "Artificial Life: Synthetic vs. Virtual." This volume.
14. Horgan, T., and J. Woodward. "Folk Psychology is Here to Stay." *Phil. Rev.* **XCIV** (April 1985): 197–226.
15. Laing, R. "Artificial Organisms: History, Problems, Directions." In *Artificial Life*, edited by C. G. Langton. Santa Fe Institute Studies in the Sciences of Complexity, Proc. Vol. VI, 49–61. Redwood City, CA: Addison-Wesley, 1989.
16. Langton, C. G. "Artificial Life." In *Artificial Life*, edited by C. G. Langton. Santa Fe Institute Studies in the Sciences of Complexity, Proc. Vol. VI, 1–47. Redwood City, CA: Addison-Wesley, 1989.
17. Pattee, H. H. "Simulations, Realizations, and Theories of Life." In *Artificial Life*, edited by C. G. Langton. Santa Fe Institute Studies in the Sciences of Complexity, Proc. Vol. VI, 63–77. Redwood City, CA: Addison-Wesley, 1989.
18. Ray, T. "An Approach to the Synthesis of Life." In *Artificial Life II*, edited by C. Langton, C. Taylor, J. D. Farmer, and S. Rasmussen. Santa Fe Institute Studies in the Sciences of Complexity, Proc. Vol. X, 371–408. Redwood City, CA: Addison-Wesley, 1991.
19. Searle, J. R. "Minds, Brains, and Programs." *Behav. & Brain Sci.* **3** (1980): 417–424.
20. Searle, J. R. "Consciousness and Cognition." *Behav. & Brain Sci.* **13** (1990).
21. Sober, E. "Learning from Functionalism: Prospects for Strong Artificial Life." In *Artificial Life II*, edited by C. Langton, C. Taylor, J.D. Farmer, and S. Rasmussen. Santa Fe Institute Studies in the Sciences of Complexity, Proc. Vol. X, 749–765. Redwood City, CA: Addison-Wesley, 1991.
22. Stich, S. "From Folk Psychology to Cognitive Science." In *The Case Against Belief.* Cambridge, MA: MIT Press, 1983.
23. Turing, A. M. "Computing Machinery and Intelligence." In *Minds and Machines*, edited by A. Anderson. Engelwood Cliffs, NJ: Prentice Hall, 1964.

Index

A

B

C

G

H

N

O

P

Q

R

S

T